Basic Principles
and Calculations
in
Chemical Engineering

PRENTICE-HALL INTERNATIONAL SERIES
IN THE PHYSICAL AND CHEMICAL ENGINEERING SCIENCES

NEAL R. AMUNDSON, EDITOR, *University of Minnesota*

ADVISORY EDITORS

ANDREAS ACRIVOS, *Stanford University*
JOHN DAHLER, *University of Minnesota*
THOMAS J. HANRATTY, *University of Illinois*
JOHN M. PRAUSNITZ, *University of California*
L. E. SCRIVEN, *University of Minnesota*

Basic Principles
and Calculations
in
Chemical Engineering
Third Edition

David M. Himmelblau

Professor of Chemical Engineering
University of Texas

PRENTICE-HALL, INC., Englewood Cilffs, New Jersey

Library of Congress Cataloging in Publication Data

HIMMELBLAU, DAVID MAUTNER,
    Basic principles and calculations in chemical
engineering.
    (Prentice-Hall international series in the physical
and chemical engineering sciences)
    Includes bibliographies.

    1. Chemical engineering—Tables, calculations, etc.
I. Title.
TP151.H5 1974   660.2   73-1696
ISBN 0-13-066472-3

© 1974, 1967, 1962 by Prentice-Hall, Inc.
Englewood Cliffs, New Jersey

10 9 8 7 6

Printed in the United States of America

PRENTICE-HALL, INC.
PRENTICE-HALL INTERNATIONAL, UNITED KINGDOM AND EIRE
PRENTICE-HALL OF CANADA, LTD. CANADA

To Betty

Contents

Preface

This text is intended to serve as an introduction to the principles and techniques used in the field of chemical, petroleum, and environmental engineering. You will find that the chapters have been organized essentially into a review of fundamental terms, an explanation of how to make material and energy balances, and a review of certain aspects of applied physical chemistry. Each chapter has an information flow diagram that shows how the topics discussed in the chapter relate to the objective of being able to solve successfully problems involving material and energy balances.

Chemical engineers have always been proud of their flexibility and broad background. In line with this philosophy, the goal of this book is to help students achieve the ability to solve a variety of practical problems involving material and energy balances. More than that, it guides the reader into forming generalized patterns of attack in problem-solving which can be used successfully in connection with unfamiliar types of problems. The text is designed to acquaint the student with a sufficient number of fundamental concepts so that he can (1) continue with his training, and (2) start finding solutions to new types of problems on his own. It offers practice in finding out what the problem is, defining it, collecting data, analyzing and breaking down information, assembling the basic ideas into patterns, and, in effect, doing everything but testing the solution.

Some attention has been focused recently on the principles of educational psychology in the preparation of teaching aids (such as textbooks and programmed texts) and in self-paced instruction. One basic principle is that a student should know component concepts and principles before attacking a more complex principle. Because the essential initial concepts are missed when a textbook starts with the general case and subsequently simplifies, in this text the topics are presented in order of easy assimilation rather than in a strictly logical order. The organization is such that easy material is alternated with difficult material in order to give a "breather" after passing over each hump. For example, discussion of unsteady-state (lumped) balances has been deferred until the final chapter because experience has shown that most students lack the mathematical and engineering maturity to absorb these problems simultaneously with the steady-state balances.

A second principle of educational psychology is to reinforce the learning experience by providing detailed guided practice following each new principle. We all have found from experience that there is a vast difference between having a student understand a principle and establishing his ability to apply it. By the use of numerous detailed examples following each brief section of text, it is hoped that straightforward, orderly methods of procedure can be instilled along with some insight into the principles involved. Furthermore, the wide variety of problems at the end of each chapter, about one fourth of which are accompanied by answers, offer practice in the application of the principles explained in the chapter.

Emphasis has been placed on a few fundamental principles, stated both in words, and mathematical symbols, rather than on technology or memorization of formulas. The text does not describe the chemical process industries per se, although certain aspects of these are introduced by the use of problems. Also, the topics of vapor-liquid equilibria, chemical equilibria, and kinetics have been omitted because of a lack of space and because an appreciation of these topics requires more background information and maturity on the part of the student than has been assumed for the topics that are discussed. On the other hand, the topic of real gases has been included because far too much attention is devoted in scientific courses to the ideal gases, leaving a very misleading impression. Those who wish to skip this particular section can do so without interfering with any of the subsequent material. More topics have been included in the text than can be covered in one semester so that the instructor has some choice as to pace of instruction and topics to include. Many references are given at the end of each chapter in the hope that those who need supplementary information will pursue their interest further. With the assistance of a syllabus, the text fits in very neatly with self-paced instruction.

Two problems perplex an author in preparing a new text. One concerns the extent to which metric units should be used. In view of the state of the metrication program in the U.S., I have compromised by incorporating SI units in about ten to twenty percent of the examples and problems, enough to acquaint a student with their usage, but not enough to give him a feeling for their magnitude in terms of his everyday experiences. Because graduating engineers will apparently still be using the traditional units for years in instruments and in design, it is still too early to convert an under-graduate text to the SI system of units even though some research journals are doing so.

A second perplexing problem is to what extent and in what manner should problems involving the preparation of or use of computer codes be introduced into the text. If computer techniques are to be integrated into the classroom successfully, it is wise to start early in the game, but on the other hand the preparation (as opposed to use) of computer programs assumes that the student will have had some instruction in programming prior to, or concurrently with, the use of this text. The selection of appropriate problems and the illustration of good computer habits, pointing out instances in which computer solutions are not appropriate as well as instances when they are, is important. Again, a compromise has been reached. At the end of each chapter you will find a few problems calling for the preparation of a complete computer code. Other problems that are more appropriately solved with the aid of a computer program, particularly a "canned" library code, than with a slide rule or desk computer, are designated by an asterisk (*) after the problem number. I have found that the use of canned computer programs reduces the tedium of many of the trial and error solution techniques and regenerates interest in the learning process.

I would indeed be ungrateful if I did not express many thanks to the hundreds of students who have participated in the preparation of both the original and the revised editions of this book. Various editions have been used for eleven years as a text for the introductory course in chemical engineering at the University of Texas (given at the sophomore level), and many modifications have been introduced as a result of the experience of the students who have used it. Drs. D. R. Paul and J. Stice have been kind enough to contribute a number of new problems, and far too many instructors using the text have contributed their corrections and suggestions for me to list them all by name. However, I do wish to express my appreciation for their kind assistance.

D.M.H.

Austin, Texas

Basic Principles
and Calculations
in
Chemical Engineering

Introduction to Engineering Calculations

1

The allocation of resources in the United States and elsewhere is beginning a period of transition. Air and water, for example, at one time were the economist's favorite examples of "free goods." However, people are accepting the fact that air and water indeed are natural resources of ever-increasing value and are not simply infinite sources and sinks for man's uses in production and consumption. For you to learn how to appreciate and treat the problems that will arise in our modern technology, and especially in the technology of the future, it is necessary to learn certain basic principles and practice their application. This text describes the principles of making material and energy balances and illustrates their application in a wide variety of ways.

We begin in this chapter with a review of certain background information. You have already encountered most of the concepts to be presented in basic chemistry and physics courses. Why, then, the need for a review? First, from experience we have found it necessary to restate these familiar basic concepts in a somewhat more general and clearer fashion; second, you will need practice to develop your ability to analyze and work engineering problems. To read and understand the principles discussed in this chapter is relatively easy; to apply them to different unfamiliar situations is not. An engineer becomes competent in his profession by mastering the techniques developed by his predecessors—then, perhaps, he can pioneer new ones.

This chapter begins with a discussion of units, dimensions, and conversion factors, and then goes on to review some terms you should already be acquainted with, such as

(a) Mole and mole fraction.
(b) Density and specific gravity.
(c) Measures of concentration.
(d) Temperature.
(e) Pressure.

It then provides some clues as to "how to solve problems" which should be of material aid in all the remaining portions of your work. Finally, the principles of stoichiometry are reviewed, and the technique of handling incomplete reactions is illustrated. Figure 1.0 shows the relation of the topics to be discussed to

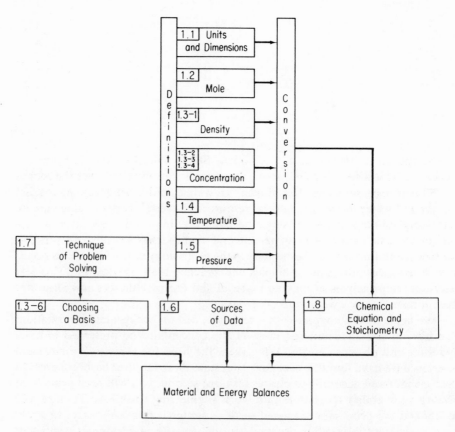

Fig. 1.0. Heirarchy of topics to be studied in this chapter (section numbers are in the upper left-hand corner of the boxes).

each other and to the ultimate goal of being able to solve problems involving both material and energy balances. At the end of the chapter you will find a check list of levels of skill you should possess after completing the chapter.

1.1 Units and dimensions

At some time in every student's life comes the exasperating sensation of frustration in problem solving. Somehow, the answers or the calculations do not come out as expected. Often this outcome arises because of inexperience in the handling of units. The use of units or dimensions along with the numbers in your calculations requires more attention than you probably have been giving to your computations in the past, but it will help avoid such annoying experiences. The proper use of dimensions in problem solving is not only sound from a logical viewpoint—it will also be helpful in guiding you along an appropriate path of analysis from what is at hand through what has to be done to the final solution.

Dimensions are our basic concepts of measurement such as *length, time, mass, temperature,* etc; *units* are the means of expressing the dimensions, as *feet* or *centimeters* for length, or *hours* or *seconds* for time. Units are associated with some quantities you may have previously considered to be dimensionless. A good example is *molecular weight,* which is really the mass of one substance per mole of that substance. This method of attaching units to all numbers which are not fundamentally dimensionless has the following very practical benefits:

(a) It diminishes the possibility of inadvertent inversion of any portion of the calculation.
(b) It reduces the calculation in many cases to simple ratios, which can be easily manipulated on the slide rule.
(c) It reduces the intermediate calculations and eliminates considerable time in problem solving.
(d) It enables you to approach the problem logically rather than by remembering a formula and plugging numbers into the formula.
(e) It demonstrates the physical meaning of the numbers you use.

The rule for handling units is essentially quite simple: treat the units as you would algebraic symbols. For example, you cannot add, subtract, multiply, or divide different units into each other and thus cancel them out—this may be done only for like units. You may be able to add pounds to pounds and calories to calories—subtract them, multiply them, or divide them—but you cannot divide 10 pounds by 5 calories and get 2 any more than you can change 2 apples into 2 bananas. The units contain a significant amount of information content that cannot be ignored.

EXAMPLE 1.1 Dimensions and units

Add the following:

(a)   1 foot + 3 seconds

(b)   1 horsepower + 300 watts

Solution:

The operation indicated by

$$1 \text{ ft} + 3 \text{ sec}$$

has no meaning since the dimensions of the two terms are not the same. One foot has the dimensions of length, whereas 3 sec has the dimensions of time. In the case of

$$1 \text{ hp} + 300 \text{ watts}$$

the dimensions are the same (energy per unit time) but the units are different. You must transform the two quantities into like units, such as horsepower, watts, or something else, before the addition can be carried out. Since 1 hp = 746 watts,

$$746 \text{ watts} + 300 \text{ watts} = 1046 \text{ watts}$$

EXAMPLE 1.2 Conversion of units

If a plane travels at twice the speed of sound (assume that the speed of sound is 1100 ft/sec), how fast is it going in miles per hour?

Solution:

$$\frac{2}{} \left| \frac{1100 \text{ ft}}{\text{sec}} \right| \frac{1 \text{ mi}}{5280 \text{ ft}} \left| \frac{60 \text{ sec}}{1 \text{ min}} \right| \frac{60 \text{ min}}{1 \text{ hr}} = 1500 \frac{\text{mi}}{\text{hr}}$$

or

$$\frac{2}{} \left| \frac{1100 \text{ ft}}{\text{sec}} \right| \frac{60 \dfrac{\text{mi}}{\text{hr}}}{88 \dfrac{\text{ft}}{\text{sec}}} = 1500 \frac{\text{mi}}{\text{hr}}$$

You will note in Example 1.2 the use of what is called the *dimensional equation*. It contains both units and numbers. The initial speed, 2200 ft/sec, is multiplied by a number of ratios (termed *conversion factors*) of equivalent values of combinations of time, distance, etc., to arrive at the final desired answer. The ratios used are simple well-known values and thus the conversion itself should present no great problem. Of course, it is possible to look up conversion ratios, which will enable the length of the calculation to be reduced; for instance, in Example 1.2 we could have used the conversion factor of 60 mi/hr equals 88 ft/sec. However, it usually takes less time to use values you know than to look up shortcut conversion factors in a handbook. Common conversion ratios are listed in Appendix A.

The dimensional equation has vertical lines set up to separate each ratio, and these lines retain the same meaning as an $\times$ or multiplication sign placed between each ratio. The dimensional equation will be retained in this form throughout most of this text to enable you to keep clearly in mind the significance of units in problem solving. It is recommended that units always be written down next to the associated numerical value (unless the calculation is very simple) until you become quite familiar with the use of units and dimensions and can carry them in your head.

At any point in the dimensional equation you can determine the consolidated net units and see what conversions are still required. This may be carried out formally, as shown below by drawing slanted lines below the dimensional equation and writing the consolidated units on these lines, or it may be done by eye, mentally canceling and accumulating the units.

$$\frac{2 \times 1100 \text{ ft}}{\text{sec}} \left| \frac{1 \text{ mi}}{5280 \text{ ft}} \right| \frac{60 \text{ sec}}{1 \text{ min}} \left| \frac{60 \text{ min}}{1 \text{ hr}} \right.$$

$$\frac{\text{ft}}{\text{sec}} \qquad \frac{\text{mi}}{\text{sec}} \qquad \frac{\text{mi}}{\text{min}}$$

EXAMPLE 1.3 Use of units

Change 400 in.³/day to cm³/min.

Solution:

$$\frac{400 \text{ in.}^3}{\text{day}} \left| \left(\frac{2.54 \text{ cm}}{1 \text{ in.}} \right)^3 \right| \frac{1 \text{ day}}{24 \text{ hr}} \left| \frac{1 \text{ hr}}{60 \text{ min}} \right. = 4.56 \frac{\text{cm}^3}{\text{min}}$$

In this example note that not only are the numbers raised to a power, but the units also are raised to the same power.

> There shall be one measure of wine throughout our kingdom, and one measure of ale, and one measure of grain . . . and one breadth of cloth. . . . And of weights it shall be as of measures.

So reads the standard measures clause of the *Magna Carta* (June 1215). The standards mentioned were not substantially revised until the nineteenth century. When the American Colonies separated from England, they retained, among other things, the weights and measures then in use. It is probable that at that time these were the most firmly established and widely used weights and measures in the world.

No such uniformity of weights and measures existed on the European continent. Weights and measures differed not only from country to country but even from town to town and from one trade to another. This lack of uniformity led the National Assembly of France during the French Revolution to enact a decree (May 8, 1790) that called upon the French Academy of Sciences to act in concert with the Royal Society of London to "deduce an invariable standard for all of the measures and all weights." Having already an adequate system of weights and measures, the English did not participate in the French under-taking. The result of the French endeavor has evolved into what is known as the metric system.

The metric system became preferred by scientists of the nineteenth century, partly because it was intended to be an international system of measurement, partly because the units of measurement were theoretically supposed to be independently reproducible, and partly because of the simplicity of its decimal nature. These scientists proceeded to derive new units for newly observed phy-sical phenomena, basing the new units on elementary laws of physics and relating them to the units of mass and length of the metric system. The Americans and the British adapted their system of measurements to the requirements of new technology as time went on for both business and commerce as well as scien-tific uses despite the fact that the other countries, one after another, turned toward the metric system.

Problems in the specification of the units for electricity and magnetism led to numerous international conferences to rectify inconsistencies and culminated in 1960 in the eleventh General Conference on Weights and Measures adopting the SI (Système International) system of units. As of this date (1973) the United States is the last large country not employing or engaged in transforming to some form of the SI units.

Table 1.1 shows the most common systems of units used by engineers in the last few decades. Note that the SI, the cgs, the fps (English absolute), and the British engineering systems all have three basically defined units and that the fourth unit is derived from these three defined units. Only the American engi-neering system has four basically defined units. Consequently, in the American engineering system you have to use a conversion factor, g_c, a constant whose numerical value is not unity, to make the units come out properly. We can use Newton's law to see what the situation is with regard to conversion of units:

$$F = Cma \qquad (1.1)$$

where[1] F = force
$\quad m$ = mass
$\quad a$ = acceleration
$\quad C$ = a constant whose numerical value and units depend on those selected for F, m, and a

In the cgs system the unit of force is defined as the dyne; hence if C is selected to be $C = 1$ dyne/(g)(cm)/sec^2, then when 1 g is accelerated at 1 cm/sec^2

$$F = \frac{1 \text{ dyne}}{\frac{\text{(g)(cm)}}{\text{sec}^2}} \left| \frac{1 \text{ g}}{} \right| \frac{1 \text{ cm}}{\text{sec}^2} = 1 \text{ dyne}$$

[1]A list of the nomenclature is included at the end of the book.

TABLE 1.1 COMMON SYSTEMS OF UNITS

	Length	Time	Mass	Force	Energy	Temperature	Remarks
Absolute (Dynamic) Systems							
Cgs	cm	sec	gram	dyne*	erg, joule, or calorie	°K, °C	Formerly common scientific
Fps (foot-lb-sec or English absolute)	ft	sec	lb	poundal*	ft poundal	°R, °F	
SI (also MKS)	meter	sec	kilogram	newton (N)*	joule (J)	K, °C	Internationally adopted units for ordinary and scientific use
Gravitational Systems							
British engineering	ft	sec	slug*	pound weight	Btu (ft)(lb) $(ft)(lb_f)$	°R, °F	
American engineering	ft	sec	pound mass (lb_m)	pound force (lb_f)	Btu or (hp)(hr)	°R, °F	Used by chemical and petroleum engineers in the U.S.

*Unit derived from basic units; all energy units are derived.

Similarly, in the SI system in which the unit of force is defined to be the newton (N), if $C = 1$ N/(kg)(m)/sec², then when 1 kg is accelerated at 1 m/sec²,

$$F = \frac{1 \text{ N}}{\dfrac{(\text{kg})(\text{m})}{\sec^2}} \left| \frac{1 \text{ kg}}{} \right| \frac{1 \text{ m}}{\sec^2} = 1 \text{ N}$$

However, in the American engineering system we ask that the numerical value of the force and the mass be essentially the same at the earth's surface. Hence, if a mass of 1 lb_m is accelerated at g ft/sec², where g is the acceleration of gravity (about 32.2 ft/sec² depending on the location of the mass), we can make the force be 1 lb_f by choosing the proper numerical value and units for C:

$$F = C \frac{1 \text{ lb}_m}{} \left| \frac{g \text{ ft}}{\sec^2} \right. = 1 \text{ lb}_f \tag{1.2}$$

Observe that for Eq. (1.2) to hold, the units of C have to be

$$C \multimap \frac{\text{lb}_f}{\text{lb}_m \left(\dfrac{\text{ft}}{\sec^2} \right)}$$

A numerical value of 1/32.174 has been chosen for the constant because g equals the numerical value of the average acceleration of gravity at sea level at 45° latitude when the latter is expressed in ft/sec². The acceleration of gravity, you may recall, varies a few tenths of 1 percent from place to place on the surface of the earth and changes considerably as you rise up from the surface as in a rocket. With this selection of units and with the number 32.174 employed in the denominator of the conversion factor, the inverse of C is given the special symbol g_c:

$$g_c = 32.174 \frac{(\text{ft})(\text{lb}_m)}{(\sec^2)(\text{lb}_f)} \tag{1.3}$$

Division by g_c achieves exactly the same result as multiplication by C. You can see, therefore, that in the American engineering system we have the convenience that the numerical value of a pound mass is also that of a pound force if the numerical value of the ratio g/g_c is equal to 1, as it is approximately in most cases. Similarly, a 1-lb mass also usually weighs 1 lb (weight is the force required to support the mass at rest). However, you should be aware that these two quantities g and g_c are not the same. Note also that the pound (mass) and the pound (force) are not same units in the American engineering system even though we speak of *pounds* to express force, weight, or mass. Furthermore, if a satellite with a mass of 1 lb (mass) "weighing" 1 lb (force) on the earth's surface is shot up to a height of 50 mi, it will no longer "weigh" 1 lb (force), although its mass will still be 1 lb (mass).

In ordinary language most people, including scientists and engineers, omit the designation of "force" or "mass" associated with the pound or kilogram but pick up the meaning from the context of the statement. No one gets confused by the fact that a man is six feet tall but has only two feet. Similarly, one should interpret the statement that a bottle "weighs" 5 kg as meaning the bottle has a mass of 5 kg and is attracted to the earth's surface with a force equal to a weight of 5 kg. Additional details concerning units and dimensions can be found in the articles by Silberberg and McKetta,[2] and by Whitney,[3] in various books,[4,5,6] and in the references at the end of this chapter. For systems used in the U. S. S. R., consult *Meas. Tech.* (USSR) (*English Trans.*), no. 10 (April), 1964.

EXAMPLE 1.4   Use of g_c

One hundred pounds of water are flowing through a pipe at the rate of 10.0 ft/sec. What is the kinetic energy of this water in $(ft)(lb_f)$?

Solution:

$$\text{kinetic energy} = K = \tfrac{1}{2}mv^2$$

Assume the 100 lb of water means the mass of the water. (This would be numerically identical to the weight of the water if $g = g_c$.)

$$K = \frac{1}{2}\left|\frac{100\ lb_m}{}\right|\left(\frac{10.0\ ft}{sec}\right)^2\left|\frac{}{32.174\ \dfrac{(ft)(lb_m)}{(sec^2)(lb_f)}}\right| = 155\ (ft)(lb_f)$$

EXAMPLE 1.5   Use of g_c

What is the potential energy in $(ft)(lb_f)$ of a 100-lb drum hanging 10 ft above the surface of the earth with reference to the surface of the earth?

Solution:

$$\text{potential energy} = P = mgh$$

Assume the 100 lb means 100 lb mass; g = acceleration of gravity = 32.2 ft/sec².

$$P = \frac{100\ lb_m}{}\left|\frac{32.2\ ft}{sec^2}\right|\frac{10\ ft}{}\left|\frac{}{32.174\ \dfrac{(ft)(lb_m)}{(sec^2)(lb_f)}}\right| = 1000\ (ft)(lb_f)$$

Notice that in the ratio of g/g_c, or 32.2 ft/sec² divided by 32.174 (ft/sec²)(lb_m/lb_f), the numerical values are almost equal. A good many people would solve the problem by saying 100 lb × 10 ft = 1000 (ft)(lb) without realizing that in effect they are canceling out the numbers in the g/g_c ratio.

---

[2] I. H. Silberberg and J. J. McKetta, Jr., *Petroleum Refiner*, v. 32, 179–183 (April 1953), 147–150 (May 1953).

[3] H. Whitney, "The Mathematics of Physical Quantities," *American Math. Monthly*, v. 75, 115, 227 (1968).

[4] D. C. Ipsen, *Units, Dimensions, and Dimensionless Numbers*, McGraw-Hill, New York, 1960.

[5] S. J. Kline, *Similitude and Approximation Theory*, McGraw-Hill, New York, 1965.

[6] E. F. O'Day, *Physical Quantities and Units* (*A Self-Instructional Programmed Manual*), Prentice-Hall, Englewood Cliffs, N. J., 1967.

EXAMPLE 1.6   Weight

What is the difference in the weight in Newtons of a 100-kg rocket at height of 10 km above the surface of the earth, where $g = 9.76$ m/sec^2, as opposed to its weight on the surface of the earth, where $g = 9.80$ m/sec^2?

Solution:

The weight in Newtons can be computed in each case from Eq. (1.1) with $a = g$ if we ignore the tiny effect of centripetal acceleration resulting from the rotation of the earth (less than 0.3 percent):

$$\text{weight difference} = \frac{100 \text{ kg}}{} \left| \frac{(9.80 - 9.76) \text{ m}}{\text{sec}^2} \right| \frac{1 \text{ N}}{\frac{(\text{kg})(\text{m})}{\text{sec}^2}} = 4.00 \text{ N}$$

Note that the concept of weight is not particularly useful in treating the dynamics of long-range ballistic missiles or of earth satellites because the earth is both round and rotating.

You should develop some facility in converting units from the SI and cgs systems into the American engineering system and vice versa, since these are the three sets of units in this text. Certainly you are familiar with the common conversions in the American engineering system from elementary and high school work. Similarly, you have used the SI and cgs system in physics, chemistry, and mathematics. If you have forgotten, Table 1.2 lists a short selection of essential conversion factors from Appendix A. Memorize them. Common abbreviations also appear in this table. Unit abbreviations are written in lowercase letters, except for the first letter when the name of the unit is derived from a proper name. The distinction between uppercase and lowercase letters should be followed, even if the symbol appears in applications where the other lettering is in uppercase style. Unit abbreviations have the same form for both singular and plural, and they are not followed by a period (except in the case of inches).

Other useful conversion factors will be discussed in subsequent sections of this book.

One of the best features of the SI system is that units and their multiples and submultiples are related by standard factors designated by the prefixes indicated in Table 1.3. Prefixes are not preferred for use in denominators (except for kg). The strict use of these prefixes leads to some amusing combinations of noneuphonious sounds, such as nanonewton, nembujoule, and so forth. Also, some confusion is certain to arise because the prefix M can be confused with m as well as with M $\approx$ 1000 derived from the Roman numerical. When a compound unit is formed by multiplication of two or more other units, its symbol consists of the symbols for the separate units joined by a centered dot (for example, N·m for newton meter). The dot may be omitted in the case of familiar units such as watthour (symbol Wh) if no confusion will result, or if the symbols are separated by exponents as in N·m^{2}kg^{-2}. Hyphens should not be used in symbols

TABLE 1.2 BASIC CONVERSION FACTORS

Dimension	American Engineering	Conversion: American Engineering to cgs	SI and cgs	Conversion: American Engineering to SI
Length	12 in. = 1 ft 3 ft = 1 yd 5280 ft = 1 mi	2.54 cm = 1 in.	10 mm* = 1 cm* 100 cm* = 1 m	3.28 ft = 1 m
Volume	1 ft^3 = 7.48 gal	$(2.54 \text{ cm})^3 = (1 \text{ in.})^3$	1000 cm^3* = 1 l*	35.31 ft^3 = 1.00 m^3
Density	1 ft^3 H$_2$O = 62.4 lb$_m$	—	1 cm^3* H$_2$O = 1 g 1 m^3 H$_2$O = 1000 kg	—
Mass	1 ton$_m$ = 2000 lb$_m$	1 lb = 454 g	1000 g = 1 kg	1 lb = 0.454 kg
Time	1 min = 60 sec 1 hr = 60 min	—	1 min* = 60 sec 1 hr* = 60 min*	—

*An acceptable but not preferred unit in the SI system.

TABLE 1.3 SI PREFIXES

Factor	Prefix	Symbol	Factor	Prefix	Symbol
10^{12}	Tera	T	10^{-1}	Deci*	d
10^9	Giga	G	10^{-2}	Centi*	c
10^6	Mega	M	10^{-3}	Milli	m
10^3	Kilo	k	10^{-6}	Micro	μ
10^2	Hecto*	h	10^{-9}	Nano	n
10^1	Deka*	da	10^{-12}	Pico	p
			10^{-15}	Femto	f
			10^{-18}	Atto	a

*Avoid except for areas and volumes.

for compound units. Positive and negative exponents may be used with the symbols for units. If a compound unit is formed by division of one unit by another, its symbol consists of the symbols for the separate units either separated by a solidus or multiplied by using negative powers (for example, m/s or $m \cdot s^{-1}$ for meters per second). We shall not use the center dot for multiplication in this text but shall use parentheses around the unit instead, a procedure less subject to misinterpretation in handwritten material.

EXAMPLE 1.7 Application of dimensions

A simplified equation for heat transfer from a pipe to air is

$$h = 0.026G^{0.6}/D^{0.4}$$

where h = the heat transfer coefficient in Btu/(hr)(ft^2)(°F)

G = mass rate of flow in lb$_m$/(hr)(ft^2)

D = outside diameter of the pipe in ft

If h is to be expressed in cal/(min)(cm^2)(°C), what should the new constant in the equation be in place of 0.026?

Solution:

To convert to cal/(min)(cm^2)(°C) we set up the dimensional equation as follows:

$$h = \frac{0.026G^{0.6}\text{Btu}}{D^{0.4}(\text{hr})(\text{ft}^2)(°\text{F})} \left| \frac{252\ \text{cal}}{1\ \text{Btu}} \right| \frac{1\ \text{hr}}{60\ \text{min}} \left| \left(\frac{1\ \text{ft}}{12\ \text{in.}}\right)^2 \right.$$

$$\left| \left(\frac{1\ \text{in.}}{2.54\ \text{cm}}\right)^2 \right| \frac{1.8°\text{F}}{1°\text{C}} = 2.11 \times 10^{-4}\ \frac{G^{0.6}\ \text{cal}}{D^{0.4}(\text{min})(\text{cm}^2)(°\text{C})}$$

(*Note:* See Sec. 1.4 for a detailed discussion of temperature conversion factors.) If G and D are to be used in units of

$$G': \frac{\text{g}}{(\text{min})(\text{cm}^2)}$$

$$D': \text{cm}$$

then an additional conversion is required:

$$h' = \cfrac{2.11 \times 10^{-4} \left[\dfrac{G'(g)}{(min)(cm^2)} \middle| \dfrac{1\,lb_m}{454\,g} \middle| \dfrac{60\,min}{1\,hr} \middle| \left(\dfrac{2.54\,cm}{1\,in.}\right)^2 \middle| \left(\dfrac{12\,in.}{1\,ft}\right)^2 \right]^{0.6}}{\dfrac{1}{\left[D'(cm) \middle| \left(\dfrac{1\,in.}{2.54\,cm}\right) \middle| \left(\dfrac{1\,ft}{12\,in.}\right) \right]^{0.4}}} = 1.48 \times 10^{-2} \dfrac{(G')^{0.6}\,cal}{(D')^{0.4}(min)(cm^2)(°C)}$$

Dimensional considerations can also be used to help identify the dimensions of terms or quantities in terms in an equation. Equations must be dimensionally consistent; i.e., each term in an equation must have the same net dimensions and units as every other term to which it is added or subtracted. The use of dimensional consistency can be illustrated by an equation which represents gas behavior and is known as van der Waals' equation, an equation which will be discussed in more detail in Chap. 3:

$$\left(p + \frac{a}{V^2}\right)(V - b) = RT$$

Inspection of the equation shows that the constant a must have the dimensions of [(pressure)(volume)^2] in order for the expression in the first set of parentheses to be consistent throughout. If the units of pressure are atm and those of volume are cm^3, then a will have the units specifically of [(atm)(cm)^6]. Similarly, b must have the same units as V, or in this particular case the units of cm^3.

1.2 The mole unit

What is a mole? The best answer is that a *mole* is a certain number of molecules atoms, electrons, or other specified types of particles.[7] In particular, the 1969 International Committee on Weights and Measures approved the mole (sometimes abbreviated as mol) in the SI system as being "the amount of a substance that contains as many elementary entities as there are atoms in 0.012 kg of carbon 12." Thus in the SI and cgs systems the mole contains a different number of molecules than it does in the American engineering system. In the SI and cgs systems a mole has about 6.02×10^{23} molecules; we shall call this a *gram mole* (abbreviated *g mole*). In the American engineering system a *pound mole* (abbreviated *lb mole*) has $6.02 \times 10^{23} \times 454$ molecules. Thus the pound mole and the gram mole represent two different quantities. Here is another way to look at

[7]For a discussion of the requirements of the mole concept, refer to the series of articles in *J. Chem. Educ.*, v. 38, 549–556 (1961).

the mole unit:

$$\text{the g mole} = \frac{\text{mass in g}}{\text{molecular weight}} \qquad (1.4)$$

$$\text{the lb mole} = \frac{\text{mass in lb}}{\text{molecular weight}} \qquad (1.5)$$

or

$$\text{mass in g} = (\text{mol. wt})(\text{g mole}) \qquad (1.6)$$

$$\text{mass in lb} = (\text{mol. wt})(\text{lb mole}) \qquad (1.7)$$

The values of the molecular weights (relative molar masses) are built up from the tables of atomic weights based on an arbitrary scale of the relative masses of the elements. The terms atomic "weight" and molecular "weight" are universally used by chemists and engineers instead of the more accurate terms atomic "mass" or molecular "mass." Since weighing was the original method for determining the comparative atomic masses, as long as they were calculated in a common gravitational field, the relative values obtained for the atomic "weights" were identical with those of the atomic "masses." There have always been and are today some questions in issue about the standard reference element and its atomic weight. Many of the old standards such as H (atomic hydrogen) = 1, O (atomic oxygen) = 100, O = 1 and O = 16 have been discarded, and now $^{12}C = 12$ exactly is used as the reference point on the chemist's scale, i.e., 12.000 g of carbon 12 contain 1 g mole or 6.023×10^{23} atoms (or 0.012 kg in the SI system).

On this scale of atomic weights, hydrogen is 1.008, carbon is 12.01, etc. (In most of our calculations we shall round these off to 1 and 12, respectively.) In your calculations you may attach any unit of mass you desire to these atomic weights, as, for example, a gram atom, a pound atom, a ton atom, etc. Thus a gram atom of oxygen is 16.00 g, a pound atom of oxygen is 16.00 lb, and a kilogram atom of hydrogen is 1.008 kg.

A compound is composed of more than one atom, and the molecular weight of the compound is nothing more than the sum of the weights of the atoms of which it is composed. Thus H_2O consists of 2 hydrogen atoms and 1 oxygen atom, and the molecular weight of water is $(2)(1.008) + 16.000 = 18.02$. These weights are all relative to the ^{12}C atom as 12.0000, and again, any mass unit can be attached to the molecular weight of water, such as 18.02 g/g mole, 18.02 lb/lb mole, etc.

You can compute average molecular weights for mixtures of constant composition even though they are not chemically bonded if their compositions are known accurately. Thus later on in Example 1.12 we shall show how to calculate the average molecular weight of air. Of course, for a material such as fuel oil or coal whose composition may not be exactly known, you cannot determine an exact molecular weight, although you might estimate an approximate average molecular weight good enough for engineering calculations.

EXAMPLE 1.8   Molecular weights

If a bucket holds 2.00 lb of NaOH (mol. wt = 40.0), how many

(a) Pound moles of NaOH does it contain?
(b) Gram moles of NaOH does it contain?

Solution:

Basis: 2.00 lb NaOH

(a) $\dfrac{2.00 \text{ lb NaOH} \mid 1 \text{ lb mole NaOH}}{\mid 40.0 \text{ lb NaOH}} = 0.050 \text{ lb mole NaOH}$

(b_1) $\dfrac{2.00 \text{ lb NaOH} \mid 1 \text{ lb mole NaOH} \mid 454 \text{ g mole}}{\mid 40.0 \text{ lb NaOH} \mid 1 \text{ lb mole}} = 22.7 \text{ g mole}$

or

(b_2) $\dfrac{2.00 \text{ lb NaOH} \mid 454 \text{ g} \mid 1 \text{ g mole NaOH}}{\mid 1 \text{ lb} \mid 40.0 \text{ g NaOH}} = 22.7 \text{ g mole}$

EXAMPLE 1.9   Molecular weights

How many lb of NaOH are in 7.50 g mole of NaOH?

Solution:

Basis: 7.50 g mole NaOH

$\dfrac{7.50 \text{ g moles NaOH} \mid 1 \text{ lb mole} \mid 40 \text{ lb NaOH}}{\mid 454 \text{ g moles} \mid 1 \text{ lb mole NaOH}} = 0.66 \text{ lb NaOH}$

1.3 Conventions in methods of analysis and measurement

There are certain definitions and conventions which we should mention at this time since they will be constantly used throughout this book. If you memorize them now, you will immediately have a clearer perspective and save considerable trouble later on.

1.3.1 Density. Density is the ratio of mass per unit volume, as, for example, gram/cm³ or lb/ft³. It has both a numerical value and units. To determine the density of a substance, you must find both its volume and its mass or weight. If the substance is a solid, a common method to determine its volume is to displace a measured quantity of inert liquid. For example, a known weight of a material can be placed into a container of liquid of known weight and volume, and the final weight and volume of the combination measured. The density (or specific gravity) of a liquid is commonly measured with a hydrometer (a known weight and volume is dropped into the liquid and the depth to which it penetrates into the liquid is noted) or a Westphal balance (the weight of a known slug is compared in the unknown liquid with that in water). Gas densities are quite difficult to measure; one device used is the Edwards balance, which compares the weight of a bulb filled with air to the same bulb when filled with the unknown gas.

In most of your work using liquids and solids, density will not change very much with temperature or pressure, but for precise measurements for common substances you can always look up in handbook the variation of density with temperature. The change in density with temperature is illustrated in Fig. 1.1

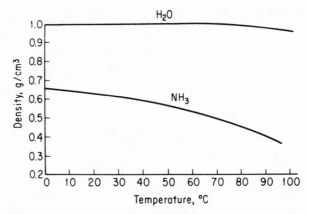

Fig. 1.1. Densities of liquid H_2O and NH_3 as a function of temperature.

for liquid water and liquid ammonia. Figure 1.2 illustrates how density also varies with composition. In the winter you may put antifreeze in your car radiator. The service station attendant checks the concentration of antifreeze by measuring the specific gravity and, in effect, the density of the radiator solution after it is mixed thoroughly. He has a little thermometer in his hydrometer kit in order to be able to read the density at the proper temperature.

1.3.2 Specific Gravity. Specific gravity is commonly thought of as a dimensionless ratio. Actually, it should be considered as the ratio of two densities—that of the substance of interest, A, to that of a reference substance. In symbols:

$$\text{sp gr} = \text{specific gravity} = \frac{(\text{lb/ft}^3)_A}{(\text{lb/ft}^3)_{\text{ref}}} = \frac{(\text{g/cm}^3)_A}{(\text{g/cm}^3)_{\text{ref}}} = \frac{(\text{kg/m}^3)_A}{(\text{kg/m}^3)_{\text{ref}}} \qquad (1.8)$$

The reference substance for liquids and solids is normally water. Thus the specific gravity is the ratio of the density of the substance in question to the density of water. The specific gravity of gases frequently is referred to air, but may be referred to other gases, as will be discussed in more detail in Chap. 3.

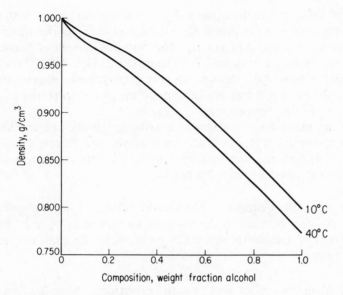

Fig. 1.2. Density of a mixture of ethyl alcohol and water as a function of composition.

Liquid density can be considered to be nearly independent of pressure for most common calculations, but, as just mentioned, it varies somewhat with temperature; therefore, to be very precise when referring to specific gravity, state the temperature at which each density is chosen. Thus

$$\text{sp gr} = 0.73\frac{20°}{4°}$$

can be interpreted as follows: the specific gravity at 20°C referred to that of water at 4°C is 0.73. Since the density of water at 4°C is very close to 1.0000 in the cgs system, the numerical values of the specific gravity and density in this system are essentially equal. Since densities in the American engineering system are expressed in lb/ft³ and the density of water is about 62.4 lb/ft³, it can be seen that the specific gravity and density values are not numerically equal in the American engineering system.

In the petroleum industry the specific gravity of petroleum products is usually reported in terms of a hydrometer scale called °API. The equation for the API scale is

$$°\text{API} = \frac{141.5}{\text{sp gr}\frac{60°}{60°}} - 131.5 \tag{1.9}$$

or

$$\text{sp gr}\frac{60°}{60°} = \frac{141.5}{°\text{API} + 131.5} \tag{1.10}$$

The volume and therefore the density of petroleum products vary with temperature, and the petroleum industry has established 60°F as the standard temperature for volume and API gravity. The National Bureau of Standards has published the National Standard Petroleum Oil Tables, NBS Circular C410 (1936), which relates API, density, specific gravity, and temperature for all petroleum oils. Thus you may convert any of the above properties at any temperature to any other temperature.

There are many other systems of measuring density and specific gravity which are somewhat specialized as, for example, the Baume (°Be) and the Twaddell (°Tw) systems. Relationships among the various systems of density may be found in standard reference books.

1.3-3 Specific Volume. The specific volume of any compound is the inverse of the density, that is, the volume per unit mass or unit amount of material. Units of specific volume might be ft^3/lb_m, ft^3/lb mole, cm^3/g, bbl/lb_m, or similar ratios.

1.3-4 Mole Fraction and Weight Fraction. Mole fraction is simply the moles of a particular substance divided by the total number of moles present. This definition holds for gases, liquids, and solids. Similarly, the weight fraction is nothing more than the weight of the substance divided by the total weight of all substances present. Mathematically these ideas can be expressed as

$$\text{mole fraction} = \frac{\text{moles of } A}{\text{total moles}} \tag{1.11}$$

$$\text{weight fraction} = \frac{\text{weight of } A}{\text{total weight}} \tag{1.12}$$

Mole percent and weight percent are the respective fractions times 100.

1.3-5 Analyses. The analyses of gases such as air, combustion products, and the like are usually on a dry basis—i.e., water vapor is excluded from the analysis. Such an analysis is called an *Orsat* analysis. In practically all cases the analysis of the gas is on a volume basis, which for the ideal gas is the same as a mole basis. Consider a typical gas analysis, that of air, which is approximately

$$\begin{array}{r} 21\% \text{ oxygen} \\ \underline{79\% \text{ nitrogen}} \\ 100\% \text{ total} \end{array}$$

This means that any sample of air will contain 21 percent oxygen by volume and also 21 mole percent oxygen. Percent, you should remember, is nothing more than the fraction times 100; consequently the mole fraction of oxygen is 0.21.

Analyses of liquids and solids are usually given by weight percent, but occasionally by mole percent. In this text analyses of liquids and solids will always be assumed to be weight percent unless otherwise stated.

EXAMPLE 1.10   Mole fraction and weight fraction

An industrial-strength drain cleaner contains 5.00 lb of water and 5.00 lb of NaOH. What is the weight fraction and mole fraction of each component in the bottle of cleaner?

Solution:

Basis: 10.0 lb of total solution

component	lb	weight fraction	mol. wt	lb moles	mole fraction
H_2O	5.00	$\dfrac{5.00}{10.0} = 0.500$	18.0	0.278	$\dfrac{0.278}{0.403} = 0.699$
NaOH	5.00	$\dfrac{5.00}{10.0} = 0.500$	40.0	0.125	$\dfrac{0.125}{0.403} = 0.311$
Total	10.00	1.000		0.403	1.00

The lb moles are calculated as follows:

$$\frac{5.00 \text{ lb } H_2O}{} \left| \frac{1 \text{ lb mole } H_2O}{18.0 \text{ lb } H_2O} = 0.278 \text{ lb mole } H_2O \right.$$

$$\frac{5.00 \text{ lb NaOH}}{} \left| \frac{1 \text{ lb mole NaOH}}{40.0 \text{ lb NaOH}} = 0.125 \text{ lb mole NaOH} \right.$$

Adding these quantities together gives the total lb moles.

EXAMPLE 1.11   Density and specific gravity

If dibromopentane has a specific gravity of 1.57, what is its density in lb_m/ft^3, in g/cm^3, and in kg/m^3?

Solution:
Our reference substance is water.

(a)
$$\frac{1.57 \dfrac{g \text{ DBP}}{cm^2}}{1.00 \dfrac{g \text{ } H_2O}{cm^3}} \left| 1.00 \frac{g \text{ } H_2O}{cm^3} = 1.57 \frac{g \text{ DBP}}{cm^3} \right.$$

(b)
$$\frac{1.57 \dfrac{lb_m \text{DBP}}{ft^3}}{1.00 \dfrac{lb_m H_2O}{ft^3}} \left| 62.4 \frac{lb_m H_2O}{ft^3} = 97.9 \frac{lb_m \text{DBP}}{ft^3} \right.$$

(c)
$$\frac{1.57 \text{ g DBP}}{cm^3} \left| \left(\frac{100 \text{ cm}}{1 \text{ m}}\right)^3 \right| \frac{1 \text{ kg}}{1000 \text{ g}} = 1.57 \times 10^3 \frac{kg \text{ DBP}}{m^3}$$

Note how the units of specific gravity as used here clarify the calculation.

EXAMPLE 1.12   Compute the average molecular weight of air

Basis: 100 lb mole of air

component	moles = %	mol. wt	lb	weight %
O_2	21.0	32	672	23.17
N_2*	79.0	28.2	2228	76.83
Total	100		2900	100.00

*Includes Ar, CO_2, Kr, Ne, Xe, and is called atmospheric nitrogen. The molecular weight is 28.2. Table 1.4 lists the detailed composition of air.

The average molecular weight is 2900 lb/100 lb mole = 29.00.

TABLE 1.4 COMPOSITION OF CLEAN, DRY AIR NEAR SEA LEVEL

Component		Percent by volume = mole percent
Nitrogen		78.084
Oxygen		20.9476
Argon		0.934
Carbon dioxide		0.0314
Neon		0.001818
Helium		0.000524
Methane		0.0002
Krypton		0.000114
Nitrous oxide		0.00005
Hydrogen		0.00005
Xenon		0.0000087
Ozone	Summer:	0–0.000007
	Winter:	0–0.000002
Ammonia		0–trace
Carbon monoxide		0–trace
Iodine		0–0.000001
Nitrogen dioxide		0–0.000002
Sulfur dioxide		0–0.0001

1.3-6 Basis. Have you noted in the previous examples that the word *basis* has appeared at the top of the computations? This concept of basis is vitally important both to your understanding of how to solve a problem and also to your solving it in the most expeditious manner. The basis is the reference chosen by you for the calculations you plan to make in any particular problem, and a proper choice of basis frequently makes the problem much easier to solve. The basis may be a period of time—for example, hours, or a given weight of material—such as 5 lb of CO_2 or some other convenient quantity. In selecting a sound basis (which in many problems is predetermined for you but in some problems is not so clear), you should ask yourself the following questions:

(a) What do I have to start with?

(b) What do I want to find out?

(c) What is the most convenient basis to use?

These questions and their answers will suggest suitable bases. Sometimes, when a number of bases seem appropriate, you may find it is best to use a unit basis of 1 or 100 of something, as, for example, pounds, hours, moles, cubic feet, etc. Many times for liquids and solids when a weight analysis is used, a convenient basis is 1 or 100 lb or kg; similarly, 1 or 100 moles is often a good choice if a mole analysis is used for a gas. The reason for these choices is that the fraction or percent automatically equals the number of pounds, kg, or moles, respectively, and one step in the calculations is saved.

EXAMPLE 1.13   Choosing a basis

Aromatic hydrocarbons form 15 to 30 percent of the components of leaded fuels and as much as 40 percent of nonleaded gasoline. The carbon to hydrogen ratio helps to characterize the fuel components. If a fuel is 80 percent C and 20 percent H by weight, what is the C/H ratio in moles?

Solution:
If a basis of 100 lb or kg of oil is selected, then the percent = pounds or kg.

Basis: 100 lb of oil (or 100 kg of oil)

component	$kg = \%$ or $lb = \%$	mol. wt	kg moles or lb moles
C	80	12.0	6.67
H	20	1.008	19.84
	100		

Consequently the C/H ratio in moles is

$$C/H = 6.67/19.84 = 0.33$$

EXAMPLE 1.14   Choosing a basis

Most processes for producing high energy content gas or gasoline from coal include some type of gasification step to make hydrogen or synthesis gas. Pressure gasification is preferred because of its greater yield of methane and higher rate of gasification.

Given that a 50.0 lb test run of gas averages 10.0 percent H_2, 40.0 percent CH_4, 30.0 percent CO, and 20.0 percent CO_2, what is the average molecular weight of the gas?

Solution:
The obvious basis is 50.0 lb of gas ("What I have to start with") but a little reflection will show that such a basis is of no use. You cannot multiply *mole percent*

of this gas times pounds and expect the answer to mean anything. Thus the next step is to choose a "convenient basis" which is 100 lb or kg moles of gas, and proceed as follows:

Basis: 100 lb moles of gas

composition	$\% = lb\ moles$	mol. wt	lb
CO_2	20.0	44.0	880
CO	30.0	28.0	840
CH_4	40.0	16.04	642
H_2	10.0	2.02	20
	100.0		2382

$$\text{average molecular weight} = \frac{2382\ lb}{100\ lb\ moles} = 23.8\ lb/lb\ mole$$

It is important that your basis be indicated at the beginning of the problem so that you will keep clearly in mind the real nature of your calculations and so that anyone checking your problem will be able to understand on what basis they are performed. If you change bases in the middle of the problem, a new basis should be indicated at that time. Many of the problems which we shall encounter will be solved on one basis and then at the end will be shifted to another basis to give the desired answer. The significance of this type of manipulation will become considerably clearer as you accumulate more experience. The ability to choose the basis that requires the fewest steps in solution can only come with practice. You can quickly accumulate the necessary experience if, as you look at each problem illustrated in this text, you determine first in your own mind what the basis should be and then compare your choice with the selected basis. By this procedure you will quickly obtain the knack of choosing a sound basis.

1.3-7 Concentrations. Concentration means the quantity of some solute per fixed amount of solvent or solution in a mixture of two or more components, as, for example,

(a) Weight per unit volume (lb_m of solute/ft^3, g of solute/l, lb_m of solute/bbl, kg of solute/m^3).

(b) Moles per unit volume (lb mole of solute/ft^3, g mole of solute/l, g mole of solute/cm^3).

(c) Parts per million—a method of expressing the concentration of extremely dilute solutions. Ppm is equivalent to a weight fraction for solids and liquids because the total amount of material is of a much higher order of magnitude than the amount of solute; it is essentially a mole fraction for gases.

(d) Other methods of expressing concentration with which you should be familiar are molarity (mole/l), normality (equivalents/l), and molality (mole/1000 g of solvent).

A typical example of the use of these concentration measures is the set of guidelines by which the Environmental Protection Agency defined the extreme levels at which the five most common air pollutants could harm individuals over periods of time.

(a) Sulfur dioxide: 2620 μg/m^3, equal to 1 ppm averaged over a 24-hr period.

(b) Particulate matter: 1000 μg/m^3, or a "coefficient of haze" equal to 8, whichever is stricter, averaged over 24 hr. (Total absence of haze, if it could occur in the earth's atmosphere, would be equal to a zero haze coefficient.)

(c) Carbon monoxide: 57.5 mg/m^3 (50 ppm) when averaged over an 8-hr period; 86.3 mg/m^3 (75 ppm) over a 4-hr average; 144 mg/m^3 (125 ppm) when averaged over 1 hr.

(d) Photochemical oxidants: 800 μg/m^3 (0.4 ppm) when averaged over 4 hr; 1200 μg/m^3 (0.6 ppm) when averaged over 2 hr; 1400 μg/m^3 (0.7 ppm) when averaged over 1 hr.

(e) Nitrogen dioxide: 3750 μg/m^3 (2 ppm) averaged over 1 hr, 938 μg/m^3 (0.5 ppm), 24-hr average.

It is important to remember that in an ideal solution, such as in gases or in a simple mixture of hydrocarbon liquids or compounds of like chemical nature, the volumes of the components may be added without great error to get the total volume of the mixture. For the so-called nonideal mixtures this rule does not hold, and the total volume of the mixture is bigger or smaller than the sum of the volumes of the pure components.

1.4 Temperature

Our concept of temperature probably originated with our physical sense of hot or cold. Attempts to be more specific and quantitative led to the idea of a temperature scale and the thermometer—a device to measure how hot or cold something is. We are all familiar with the thermometer used in laboratories which holds mercury sealed inside a glass tube or the alcohol thermometer used to measure outdoor temperatures.

Although we do not have the space to discuss in detail the many methods of measuring temperature, we can point out some of the other more common techniques with which you are probably already familiar:

(a) The voltage produced by a junction of two dissimilar conductors changes with temperature and is used as a measure of temperature (the *thermocouple*).

(b) The property of changing electrical resistance with temperature gives us a device known as the *thermistor*.

(c) Two thin strips of metal bonded together at one end expand at different rates with change of temperature. These strips assist in the control of the flow of water in the radiator of an automobile and in the operation of air conditioners and heating systems.

(d) High temperatures can be measured by devices called *pyrometers* which note the radiant energy leaving a hot body.

Figure 1.3 illustrates the appropriate ranges for various temperature-measuring devices.

As you also know, temperature is a measure of the thermal energy of the random motion of the molecules of a substance at thermal equilibrium. Temperature is normally measured in degrees Fahrenheit or Celsius (Centigrade). The common scientific scale is the *Celsius* scale,[8] where 0° is the ice point of water and 100° is the normal boiling point of water. In the early 1700s Fahrenheit, a glassblower by trade, was able to build mercury thermometers that gave temperature measurements in reasonable agreement with each other. The *Fahrenheit* scale is the one commonly used in everyday life in the United States. Its reference points are of more mysterious origin, but it is reported that the fixed starting point, or 0° on Fahrenheit's scale, was the temperature of an ice-salt mixture, and 96°, the temperature of the blood of a healthy man, was selected as the upper point because it is easily divisible by 2, 3, 4, 6, and 8, whereas 100° is not. In any case, as now standardized, 32°F represents the ice point and 212°F represents the normal boiling point of water. In the SI system, temperature is measured in kelvins, a unit named after the famous Lord Kelvin. Note that in the SI system the degree symbol is suppressed; i.e., the boiling point of water is 373 K.

The Fahrenheit and Celsius scales are *relative* scales; that is, their zero points were arbitrarily fixed by their inventors. Quite often it is necessary to use *absolute* temperatures instead of relative temperatures. Absolute temperature scales have their zero point at the lowest possible temperature which man believes can exist. As you may know, this lowest temperature is related both to the ideal gas laws and to the laws of thermodynamics. The absolute scale, which is based on degree units the size of those in the Celsius (Centigrade) scale, is called the *Kelvin* scale; the absolute scale which corresponds to the Fahrenheit degree scale is called the *Rankine* scale in honor of a Scottish engineer. The relations between relative temperature and absolute temperature are illustrated in Fig. 1.4. We shall round off absolute zero on the Rankine scale of −459.58°F to −460°F; similarly, −273.15°C will be rounded off to −273°C.

You should recognize that the unit degree, i.e., the unit temperature difference on the Kelvin-Celsius scale, is not the same size as that on the Rankine-Fahrenheit scale. If we let Δ°F represent the unit temperature difference in the Fahrenheit scale, Δ°R be the unit temperature difference in the Rankine scale,

---

[8]As originally devised by Celsius in 1742, the freezing point was designated as 100° Officially, °C now stands for *degrees Celsius*.

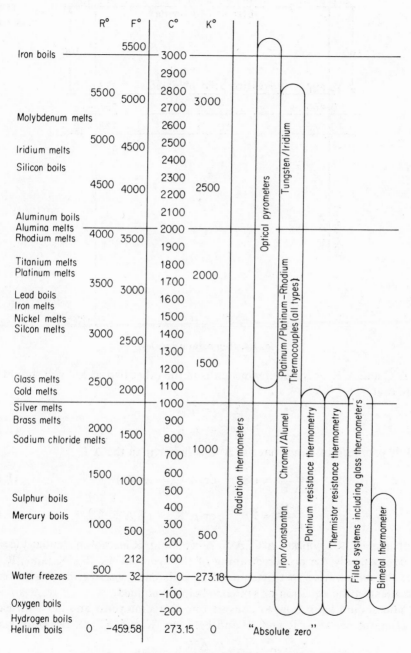

Fig. 1.3. Temperature measuring instruments span the range from near absolute zero to beyond 3000°K. The chart indicates the preferred methods of thermal instrumentation for various temperature regions.

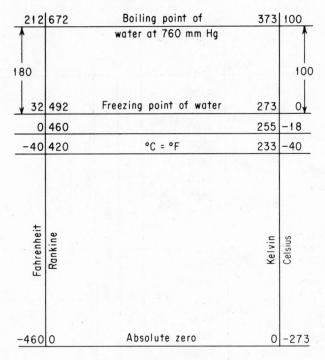

Fig. 1.4. Temperature scales.

and $\Delta°C$ and $\Delta°K$ be the analogous units in the other two scales, you should be aware that

$$\Delta°F = \Delta°R \tag{1.13}$$

$$\Delta°C = \Delta°K \tag{1.14}$$

Also, if you keep in mind that the $\Delta°C$ is larger than the $\Delta°F$,

$$\frac{\Delta°C}{\Delta°F} = 1.8 \quad \text{or} \quad \Delta°C = 1.8 \, \Delta°F \tag{1.15}$$

$$\frac{\Delta°K}{\Delta°R} = 1.8 \quad \text{or} \quad \Delta°K = 1.8 \, \Delta°R \tag{1.16}$$

Unfortunately, the symbols $\Delta°C$, $\Delta°F$, $\Delta°K$, and $\Delta°R$ are not in standard usage, and consequently the proper meaning of the symbols $°C$, $°F$, $°K$, and $°R$, as either the temperature or the temperature difference, *must be interpreted* from the context of the equation or sentence being examined.

You should learn how to convert one temperature to another with ease. The relations between $°R$ and $°F$ and between $°K$ and $°C$ are, respectively,

$$T_{°R} = T_{°F}\left(\frac{1 \, \Delta°R}{1 \, \Delta°F}\right) + 460 \tag{1.17}$$

$$T_{°K} = T_{°C}\left(\frac{1 \, \Delta°K}{1 \, \Delta°C}\right) + 273 \tag{1.18}$$

Because the relative temperature scales do not have a common zero at the same temperature, as can be seen from Fig. 1.4, the relation between °F and °C is

$$T_{°F} - 32 = T_{°C}\left(\frac{1.8 \; \Delta°F}{1 \; \Delta°C}\right) \tag{1.19}$$

After you have used Eqs. (1.17)–(1.19) a bit, they will become so familiar that temperature conversion will become an automatic reflex. During your "learning" period," in case you forget them, just think of the appropriate scales side by side as in Fig. 1.4, and put down the values for the freezing and boiling points of water.

From Fig. 1.4 you will notice that −40° is a common temperature on both the Centigrade and Fahrenheit scales. Therefore the following relationships hold:

$$T_{°F} = (T_{°C} + 40)\left(\frac{1.8 \; \Delta°F}{1 \; \Delta°C}\right) - 40 \tag{1.20}$$

$$T_{°C} = (T_{°F} + 40)\left(\frac{1 \; \Delta°C}{1.8 \; \Delta°F}\right) - 40 \tag{1.21}$$

In Fig. 1.4 all the values of the temperatures have been rounded off, but more significant figures can be used. 0°C and its equivalents are known as *standard conditions of temperature.*

EXAMPLE 1.15 **Temperature conversion**

Convert 100°C to (a) °K, (b) °F, (c) °R.

Solution:

(a)
$$(100 + 273)°C\frac{1 \; \Delta°K}{1 \; \Delta°C} = 373°K$$

or with suppression of the Δ symbol

$$(100 + 273)°C\frac{1°K}{1°C} = 373°K$$

(b)
$$(100°C)\frac{1.8°F}{1°C} + 32 = 212°F$$

(c)
$$(212 + 460)°F\frac{1°R}{1°F} = 672°R$$

$$\text{or} \quad (373°K)\frac{1.8°R}{1°K} = 672°R$$

EXAMPLE 1.16 **Temperature conversion**

The thermal conductivity of aluminum at 32°F is 117 Btu/(hr)(ft²)(°F/ft). Find the equivalent value at 0°C in terms of Btu/(hr)(ft²)(°K/ft).

Solution:
Since 32°F is identical to 0°C, the value is already at the proper temperature. The "°F" in the denominator of the thermal conductivity actually stands for Δ°F

so that the equivalent value is

$$\frac{117 \text{ (Btu)(ft)}}{\text{(hr)(ft}^2)(\Delta°\text{F)}} \bigg| \frac{1.8 \Delta°\text{F}}{1 \Delta°\text{C}} \bigg| \frac{1 \Delta°\text{C}}{1 \Delta°\text{K}} = 211 \text{ (Btu)/(hr)(ft}^2)(°\text{K/ft})$$

or with the Δ symbol suppressed

$$\frac{117 \text{ (Btu)(ft)}}{\text{(hr)(ft}^2)(°\text{F)}} \bigg| \frac{1.8°\text{F}}{1°\text{C}} \bigg| \frac{1°\text{C}}{1°\text{K}} = 211 \text{ (Btu)/(hr)(ft}^2)(°\text{K/ft})$$

1.5 Pressure

Pressures, like temperatures, can also be expressed by either absolute or relative scales. Pressure is defined as "force per unit area." Figure 1.5 shows a column of mercury held in place by a sealing plate. Suppose that the column of mercury

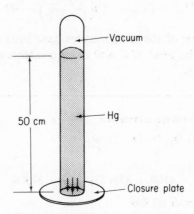

Fig. 1.5. Pressure.

has an area of 1 cm³ and is 50 cm high. From Table D.1 we can find that the sp gr at 20°C and hence the density, essentially, of the Hg is 13.55 g/cm³. Thus the force exerted on the 1 cm² section of that plate by the column of mercury is

$$F = \frac{13.55 \text{ g}}{\text{cm}^3} \bigg| 50 \text{ cm} \bigg| 1 \text{ cm}^2 \bigg| \frac{980 \text{ cm}}{\text{(sec)}^2} \bigg| \frac{1 \text{ kg}}{1000 \text{ g}} \bigg| \frac{1 \text{ m}}{100 \text{ cm}} \bigg| \frac{1 \text{ N}}{1 \frac{\text{(kg)(m)}}{\text{sec}^2}}$$

$$= 6.64 \text{ N}$$

The pressure on the section of the plate covered by the mercury is

$$p = \frac{6.64 \text{ N}}{1 \text{ cm}^2} \bigg| \left(\frac{100 \text{ cm}}{1 \text{ m}}\right)^2 = 6.64 \times 10^4 \frac{\text{N}}{\text{m}^2}$$

If we had started with units in the American engineering system, the pressure could be similarly computed as

$$p = \frac{846 \text{ lb}_m}{1 \text{ ft}^3} \left| \frac{50 \text{ cm}}{} \right| \frac{1 \text{ in.}}{2.54 \text{ cm}} \left| \frac{1 \text{ ft}}{12 \text{ in.}} \right| \frac{32.2 \text{ ft}}{\text{sec}^2} \left| \frac{}{\frac{32.147 (\text{ft})(\text{lb}_m)}{(\text{sec})^2 (\text{lb}_f)}} \right.$$

$$= 1387 \frac{\text{lb}_f}{\text{ft}^2}$$

Whether relative or absolute pressure is measured in a pressure-measuring device depends on the nature of the instrument used to make the measurements. For example, an open-end manometer (Fig. 1.6) would measure a relative pressure, since the reference for the open end is the pressure of the atmosphere at the open end of the manometer. On the other hand, closing off the end of the manometer (Fig. 1.7) and creating a vacuum in the end results in a measurement against a complete vacuum, or against "no pressure." This measurement is called *absolute pressure*. Since absolute pressure is based on a complete vacuum, a fixed reference point which is unchanged regardless of location or temperature or weather or other factors, absolute pressure then establishes a precise, invariable value which can be readily identified. Thus the zero point for an absolute pres-

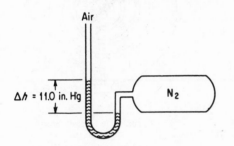

Fig. 1.6. Open-end manometer showing a pressure above atmospheric pressure.

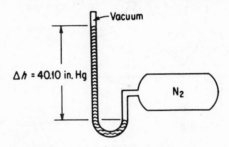

Fig. 1.7. Absolute pressure manometer.

sure scale corresponds to a perfect vacuum, whereas the zero point for a relative pressure scale usually corresponds to the pressure of the air which surrounds us at all times, and as you know, varies slightly.

If the mercury reading is set up as in Fig. 1.8, the device is called a *barometer* and the reading of atmospheric pressure is termed *barometric pressure*.

An understanding of the principle upon which a manometer operates will aid you in recognizing the nature of the pressure measurement taken from it. As shown in Fig. 1.6 for an open-end, U-tube manometer, if the pressure measured is greater than atmospheric, the liquid is forced downward in the leg to which the pressure source is connected and upward in the open leg. Eventually a point of hydrostatic balance is reached in which the manometer fluid stabilizes, and the difference in height of the fluid in the open leg relative to that in the leg attached to the pressure source is exactly equal to

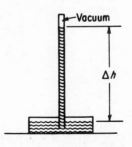

Fig. 1.8. A barometer.

the difference between the atmospheric pressure and the applied pressure in the other leg. If vacuum were applied instead of pressure to the same leg of the manometer, the fluid column would rise on the vacuum side. Again, the difference in pressure between the pressure source in the tank and atmospheric pressure is balanced by the difference in the height of the two columns of fluid. Water and mercury are commonly used indicating fluids for manometers; the readings thus can be expressed in "inches of water" or "inches of mercury." (In ordinary engineering calculations we ignore the vapor pressure of mercury and minor changes in the density of mercury due to temperature changes in making pressure measurements.)

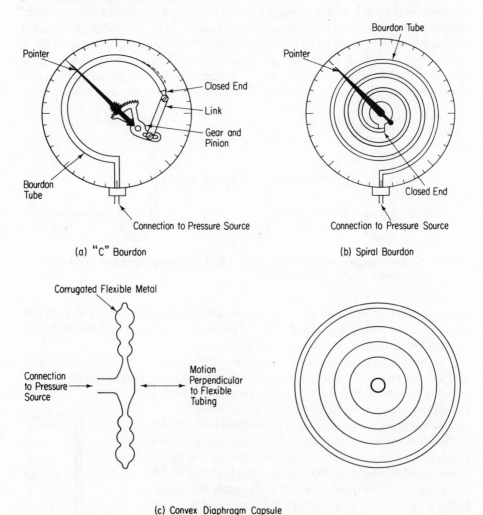

(a) "C" Bourdon

(b) Spiral Bourdon

(c) Convex Diaphragm Capsule

Fig. 1.9. Bourdon and diaphragm pressure-measuring devices.

Another type of common measuring device is the visual *Bourdon gauge* (Fig. 1.9), which normally (but not always) reads zero pressure when open to the atmosphere. The pressure-sensing device in the Bourdon gauge is a thin metal tube with an elliptical cross section closed at one end which has been bent into an arc. As the pressure increases at the open end of the tube, it tries to straighten out, and the movement of the tube is converted into a dial movement by gears and levers. Figure 1.9 also illustrates a diaphragm capsule gauge. Figure 1.10 indicates the pressure ranges for the various pressure-measuring devices.

Pressure scales may be temporarily somewhat more confusing than temperature scales since the reference point or zero point for the relative pressure scales is not constant, whereas in the temperature scales the boiling point or the freezing point of water is always a fixed value. However, you will become accustomed to this feature of the pressure scale with practice.

The relationship between relative and absolute pressure is illustrated in Figs. 1.11 and 1.12 and is also given by the following expression:

$$\text{gauge pressure} + \text{barometer pressure} = \text{absolute pressure} \qquad (1.22)$$

Equation (1.22) can be used only with consistent units. Note that you must add the atmospheric pressure, i.e., the barometric pressure, to the gauge, or relative

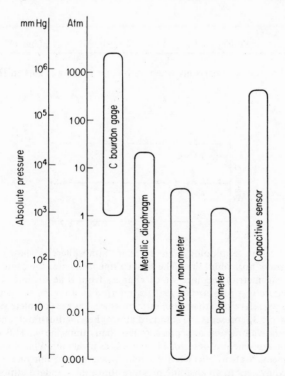

Fig. 1.10. Ranges of application for pressure-measuring devices.

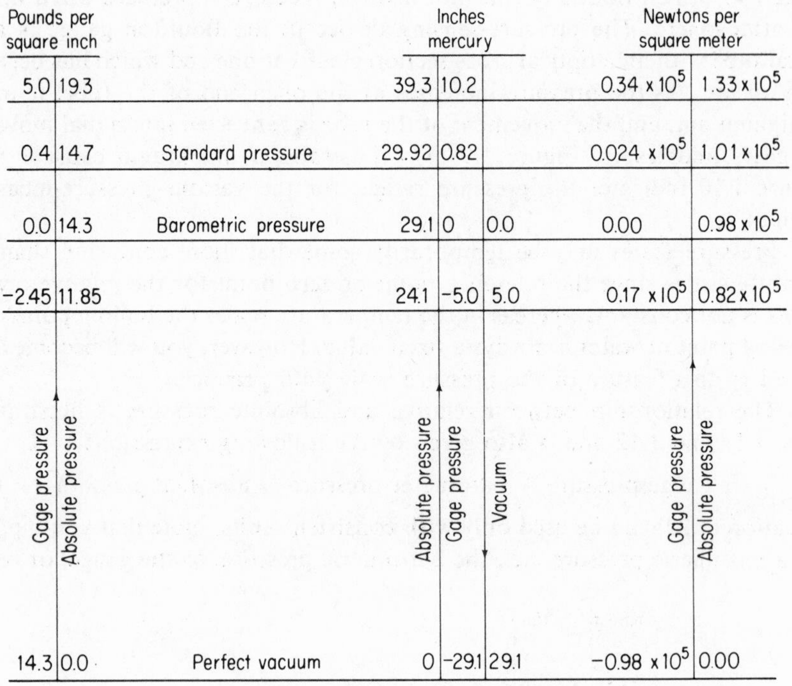

Pounds per square inch			Inches mercury			Newtons per square meter	
5.0	19.3		39.3	10.2		0.34 x 10⁵	1.33 x 10⁵
0.4	14.7	Standard pressure	29.92	0.82		0.024 x 10⁵	1.01 x 10⁵
0.0	14.3	Barometric pressure	29.1	0	0.0	0.00	0.98 x 10⁵
−2.45	11.85		24.1	−5.0	5.0	0.17 x 10⁵	0.82 x 10⁵
Gage pressure	Absolute pressure		Absolute pressure	Gage pressure	Vacuum	Gage pressure	Absolute pressure
14.3	0.0	Perfect vacuum	0	−29.1	29.1	−0.98 x 10⁵	0.00

Fig. 1.11. Pressure comparisons when barometer reading is 29.1 in Hg.

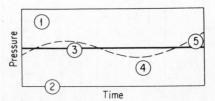

Fig. 1.12. Pressure terminology. The standard atmosphere is shown by the heavy horizontal line. The broken line illustrates the atmospheric (barometric) pressure which changes from time to time. 1 in the figure is a pressure of 19.3 psi referred to a complete vacuum or 5 psi referred to the barometric pressure. 2 is the complete vacuum, while 3 represents the standard atmosphere. 4 illustrates a negative relative pressure or a pressure less than atmospheric. This type of measurement is described in the text as a vacuum type of measurement. 5 also indicates a vacuum measurement, but one which is equivalent to an absolute pressure above the standard atmosphere.

pressure (or manometer reading if open on one end), in order to get the absolute pressure.

Another term applied in measuring pressure which is illustrated in Figs. 1.11 and 1.12 is *vacuum*. In effect, when you measure pressure as "inches of mercury vacuum," you reverse the usual direction of measurement and measure from the barometric pressure down to the vacuum, in which case a perfect vacuum would be the highest pressure that you could achieve. This procedure would be the same as evacuating the air at the top of a mercury barometer and watching the mercury rise up in the barometer as the air is removed. The vacuum system of measurement of pressure is commonly used in apparatus which operate at pressures less than atmospheric—such as a vacuum evaporator or vacuum filter. A pressure which is only slightly below barometric pressure may sometimes be expressed as a "draft" (which is identical to the vacuum system) in inches of water, as, for example, in the air supply to a furnace or a water cooling tower.

As to the units of pressure, we have shown in Fig. 1.11 two common systems: pounds per square inch (psi) and inches of mercury (in. Hg). Pounds per square inch absolute is normally abbreviated "psia," and "psig" stands for "pounds per square inch gauge." Figure 1.11 compares the relative and absolute pressure scales in terms of these two systems of measurement when the barometric pressure is 29.1 in. Hg (14.3 psia) with the SI system of units. Other systems of expressing pressure exist; in fact you will discover there are as many different units of pressure as there are means of measuring pressure. Some of the other most frequently used systems are

> Millimeters of mercury   (mm Hg)
>
> Feet of water   (ft H_2O)
>
> Atmospheres   (atm)
>
> Bars   (bar)
>
> Newtons per square meter (N/m^2)

To sum up our discussion of pressure and its measurement, you should now be acquainted with

(a) Atmospheric pressure—the pressure of the air and the atmosphere surrounding us which changes from day to day.

(b) Barometric pressure—the same as atmospheric pressure, called "barometric pressure" because a barometer is used to measure atmospheric pressure.

(c) Absolute pressure—a measure of pressure referred to a complete vacuum, or zero pressure.

(d) Gauge pressure—pressure expressed as a quantity measured from (above) atmospheric pressure (or some other reference pressure).

(e) Vacuum—a method of expressing pressure as a quantity below atmospheric pressure (or some other reference pressure).

You definitely must not confuse the standard atmosphere with atmospheric pressure. The *standard atmosphere* is defined as the pressure (in a standard gravitational field) equivalent to 14.696 psi or 760 mm Hg at 0°C or other equivalent value, whereas atmospheric pressure is a variable and must be obtained from a barometer each time you need it. The standard atmosphere may not actually exist in any part of the world except perhaps at sea level on certain days, but it is extremely useful in converting from one system of pressure measurement to another (as well as being useful in several other ways to be considered later).

Expressed in various units, the *standard atmosphere* is equal to

1.000	atmospheres (atm)
33.91	feet of water (ft H_2O)
14.7	(14.696, more exactly) pounds per square inch absolute (psia)
29.92	(29.921, more exactly) inches of mercury (in. Hg)
760.0	millimeters of mercury (mm Hg)
1.013×10^5	newtons per square meter (N/m^2)

To convert from one set of pressure units to another, it is convenient to use the relationships among the standard pressures as shown in the examples below.

If pressures are measured by means of the height of a column of liquid[9] other than mercury or water (for which the standard pressure is known), it is easy to convert from one liquid to another by means of the following expression:

$$p = \rho g h \qquad (1.23)$$

where ρ = density of the liquid
g = acceleration of gravity

By equating the same pressure yielded by two liquids with different densities, you will get the ratio between the heights of two columns of liquid. If one of these liquids is water, it is then easy to convert to any of the more commonly used liquids as follows:

$$\rho g h = \rho_{H_2O} g h_{H_2O} \qquad or \qquad \frac{h}{h_{H_2O}} = \frac{\rho_{H_2O}}{\rho}$$

EXAMPLE 1.17 Pressure conversion

Convert 35 psia to in. Hg.

Solution:
It is desirable to use the ratio of 14.7 psia to 29.92 in. Hg, an identity, to carry out this conversion.

[9]Sometimes these liquid columns are referred to as "heads" of liquid.

Basis: 35 psia

$$\frac{35 \text{ psia} \mid 29.92 \text{ in. Hg}}{\mid 14.7 \text{ psia}} = 71.25 \text{ in. Hg}$$

an identity

EXAMPLE 1.18   Pressure conversion

The density of the atmosphere decreases with increasing altitude. When the pressure is 340 mm Hg, how many inches of water is it? How many kilonewtons per square meter?

Solution:

Basis: 340 mm Hg

$$\frac{340 \text{ mm Hg} \mid 33.91 \text{ ft H}_2\text{O} \mid 12 \text{ in.}}{\mid 760 \text{ mm Hg} \mid 1 \text{ ft}} = 182 \text{ in. H}_2\text{O}$$

$$\frac{340 \text{ mm Hg} \mid 1.013 \times 10^5 \text{ N/m}^2 \mid 1 \text{ kN}}{\mid 760.0 \text{ mm Hg} \mid 1000 \text{ N}} = 45.4 \text{ kN/m}^2$$

EXAMPLE 1.19   Pressure conversion

The pressure gauge on a tank of CO_2 used to fill soda water bottles reads 51.0 psi. At the same time the barometer reads 28.0 in. Hg. What is the absolute pressure in the tank in psia? See Fig. E1.19.

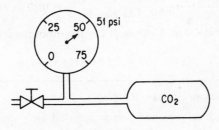

Fig. E1.19.

Solution:
The pressure gauge is reading psig, not psia.

From Eq. (1.22) the absolute pressure is the sum of the gauge pressure and the atmospheric (barometric) pressure expressed in the same units.

Basis: Barometric pressure = 28.0 in. Hg

$$\text{atmospheric pressure} = \frac{28.0 \text{ in. Hg} \mid 14.7 \text{ psia}}{\mid 29.92 \text{ in. Hg}} = 13.78 \text{ psia}$$

(*Note:* Atmospheric pressure does not equal 1 standard atm.) The absolute pressure in the tank is

$$51.0 + 13.78 = 64.78 \text{ psia}$$

EXAMPLE 1.20 Pressure conversion

Air is flowing through a duct under a draft of 4.0 in. H_2O. The barometer indicates that the atmospheric pressure is 730 mm Hg. What is the absolute pressure of the gas in inches of Hg? See Fig. E1.20.

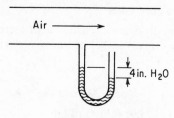

Air ⟶

4 in. H_2O

Fig. E1.20.

Solution:

Again we have to employ consistent units, and it appears in this case that the most convenient units are those of inches of mercury.

Basis: 730 mm Hg of pressure

$$\text{atmospheric pressure} = \frac{730 \text{ mm Hg}}{} \left| \frac{29.92 \text{ in. Hg}}{760 \text{ mm Hg}} \right. = 28.9 \text{ in. Hg}$$

Basis: 4.0 in. H_2O draft (under atmospheric)

$$\frac{4 \text{ in. } H_2O}{} \left| \frac{1 \text{ ft}}{12 \text{ in.}} \right| \frac{29.92 \text{ in. Hg}}{33.91 \text{ ft } H_2O} = 0.29 \text{ in. Hg}$$

Since the reading is 4.0 in. H_2O draft (under atmospheric), the absolute reading in uniform units is

$$28.9 - 0.29 = 28.6 \text{ in. Hg absolute}$$

EXAMPLE 1.21 Vacuum Pressure Reading

Small animals such as mice can live at reduced air pressures down to 3.0 psia (although not comfortably). In a test a mercury manometer attached to a tank as shown in Fig. E1.21 reads 25.4 in. Hg and the barometer reads 14.79 psia. Will the mice survive?

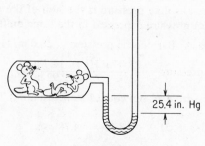

25.4 in. Hg

Fig E1.21.

Solution:

Basis: 25.4 in. Hg below atmospheric

We shall assume that $g/g_c \cong 1.00$ and ignore any temperature corrections to convert the mercury to 0°C. Then, since the vacuum reading on the tank is 25.4 in. Hg below atmospheric, the absolute pressure in the tank is

$$14.79 \text{ psia} - \frac{25.4 \text{ in. Hg}}{} \left| \frac{14.7 \text{ psia}}{29.92 \text{ in. Hg}} \right. = 14.79 - 12.49 = 2.3 \text{ psia}$$

The mice probably will not survive.

1.6   Physical and chemical properties of compounds and mixtures

Publications of industrial associations provide a tremendous storehouse of data. For example, the American Petroleum Institute publishes both the *Manual on Disposal of Refinery Wastes*[10] and the *Technical Data Book-Petroleum Refining*[11]. Tables 1.5 1.6 and 1.7 are typical of the tables that can be found in the former on p. 2-4, and p. 13-11.

If you know the chemical formula for a pure compound, you can look up many of its physical properties in standard reference books, such as

  (a) Perry's *Chemical Engineers' Handbook*[12]
  (b) *Handbook of Physics and Chemistry*[13]
  (c) Lange's *Handbook*[14]
  (d) *The Properties of Gases and Liquids.*[15]

Also, specialized texts in your library frequently list properties of compounds. For example, the book *Fuel Flue Gases*[16] lists considerable data on gas mixtures which are primarily of concern in the gas utility field. Appendix D in this text tabulates the chemical formula, molecular weight, melting point, boiling point, etc., for most, but not all, of the compounds involved in the problems you will find at the end of the chapters.

Many of the materials we talk about and use every day are not pure com-

[10]American Petroleum Institute, Div. of Refining, 1271 Ave. of the Americas, New York, N.Y., 1969.

[11]*ibid.*, 2nd ed., 1970.

[12]J. H. Perry, ed, *Chemical Engineers' Handbook*, 4th ed., McGraw-Hill, New York, 1963.

[13]*Handbook of Chemistry and Physics*, Chemical Rubber Publishing Co., Cleveland, Ohio, annually.

[14]N. A. Lange, *Handbook of Chemistry*, McGraw-Hill, New York, annually.

[15]R. C. Reid and T. K. Sherwood, *The Properties of Gases and Liquids*, 2nd ed., McGraw-Hill, New York, 1966.

[16]*Fuel Flue Gases*, American Gas Association, New York, 1941.

pounds, and it is considerably more trouble to obtain information about the properties of these materials. Fortunately, for materials such as coal, coke, petroleum products, and natural gas—which are the main sources of energy in this country—tables and formulas are available in reference books and handbooks which list some of the specific gross properties of these mixtures. Some typical examples of natural gas analyses and ultimate analyses of petroleum and petroleum products as shown in Tables 1.5, 1.6, 1.7, 1.8, and 1.9 point out the variety of materials that can be found.

TABLE 1.5 DISSOLVED SOLIDS IN FRESH WATERS OF THE UNITED STATES

Component	Concentration (mg/l)		
	In 5% of the waters is less than	In 50% of the waters is less than	In 95% of the waters is less than
Total dissolved solids	72.0	169.0	400.0
Bicarbonate, HCO_3	40.0	90.0	180.0
Sulfate, SO_4	11.0	32.0	90.0
Chloride, Cl	3.0	9.0	170.0
Nitrate, NO_3	0.2	0.9	4.2
Calcium, Ca	15.0	28.0	52.0
Magnesium, Mg	3.5	7.0	14.0
Sodium, Na, and Potassium, K	6.0	10.0	85.0
Iron, Fe	0.1	0.3	0.7

SOURCE: *Manual on Disposal of Refinery Wastes*, Amer. Pet. Inst., Div. of Refining, pp. 2–4; New York, 1969.

TABLE 1.6 CHARACTERISTICS OF WATERS WHICH SUPPORT A VARIED
AND PROFUSE FISH FAUNA—A CRITERION OF QUALITY

Characteristic	Value of Characteristic		
	In 5% of the waters is less than	In 50% of the waters is less than	In 95% of the waters is less than
pH value	6.7	7.6	8.3
	Concentration (mg/l)		
Dissolved oxygen, O_2	5.0	6.8	9.8
Carbon dioxide, CO_2			
Free	0.1	1.5	5.0
Fixed	8.0	45.0	95.0
Ammonia, NH_3	0.5	1.5	2.5
	Mho (25°C)		
Specific conductivity	$(50)(10^{-6})$	$(270)(10^{-6})$	$(1100)(10^{-6})$

SOURCE: *Manual on Disposal of Refinery Wastes*, Amer. Pet. Inst., Div. of Refining, pp. 2–4; New York, 1969.

TABLE 1.7 REFINERY BIOLOGICAL TREATMENT UNIT FEED CHARACTERISTICS

	Ranges reported*
Chlorides, mg/l	200–960
COD, mg/l	140–640
BOD_5, mg/l	97–280
Suspended solids, mg/l	80–450
Alkalinity, mg/l as $CaCO_3$	77–210
Temperature, °F	69–100
Ammonia nitrogen, mg/l	56–120
Oil, mg/l	23–130
Phosphate, mg/l	20–97
Phenolics, mg/l	7.6–61
pH	7.1–9.5
Sulfides, mg/l	1.3–38
Chromium, mg/l	0.3–0.7

SOURCE: *Manual on Disposal of Refinery Wastes*, Amer. Pet. Inst., Div. of Refining, pp. 2–4. New York, 1969.

*Values are the averages of the minima and maxima reported by 12 refineries treating total effluent. Individual plants have reported data well outside many of these ranges.

TABLE 1.8 TYPICAL DRY GAS ANALYSES

Type	Analysis (vol. %—excluding water vapor)									
	CO_2	O_2	N_2	CO	H_2	CH_4	C_2H_6	C_3H_8	C_4H_{10}	$C_5H_{12}^+$
Natural gas	6.5					77.5	16.0			
Natural gas, dry*	0.2		0.6			99.2				
Natural gas, wet*	1.1					87.0	4.1	2.6	2.0	3.4
Natural gas, sour†	(H_2S 6.4)					58.7	16.5	9.9	5.0	3.5
Butane							2.0	3.5	75.4	*n*-butane
									18.1	isobutane
								Illuminants		
Reformed refinery oil	2.3	0.7	4.9	20.8	49.8	12.3	5.5	3.7		
Coal gas, by-product	2.1	0.4	4.4	13.5	51.9	24.3		3.4		
Producer gas	4.5	0.6	50.9	27.0	14.0	3.0				
Blast furnace gas	5.4	0.7	8.3	37.0	47.3	1.3				
Sewage gas	22.0		6.0		2.0	68.0				

SOURCE: *Fuel Flue Gases*, American Gas Association, New York, 1941, p. 20.

*Dry gas contains much less propane (C_3H_8) and higher hydrocarbons than wet gas does.

†*Sour* implies that the gas contains significant amounts of hydrogen sulfide.

TABLE 1.9 ULTIMATE ANALYSIS OF PETROLEUM CRUDE

Type	Sp gr	At °C	Weight %				
			C	H	N	O	S
Pennsylvania	0.862	15	85.5	14.2			
Humbolt, Kan.	0.921		85.6	12.4			0.37
Beaumont, Tex.	0.91		85.7	11.0	2.61		0.70
Mexico	0.97	15	83.0	11.0	1.7		
Baku, U.S.S.R.	0.897		86.5	12.0		1.5	

SOURCE: Data from W. L. Nelson, *Petroleum Refinery Engineering*, 4th ed., McGraw-Hill, New York, 1958.

Since petroleum and petroleum products represent complex mixtures of hydrocarbons and other organic compounds of various types together with various impurities, if the individual components cannot be identified, then the mixture is treated as a uniform compound. Usually, the components in natural gas can be identified, and thus their individual physical properties can be looked up in the reference books mentioned above or in the Appendix in this text. As will be discussed in Chap. 3 under gases, many times the properties of a pure gas when mixed with another gas are the sum of the properties of the pure components. On the other hand, liquid petroleum crude oil and petroleum fractions are such complicated mixtures that their physical properties are hard to estimate from the pure components (even if known) unless the mixture is very simple. As a result of the need for methods of predicting the behavior of petroleum stocks, empirical correlations have been developed in recent years for many of the physical properties we want to use. These correlations are based upon the °API, the Universal Oil Products characterization factor K, the boiling point, and the apparent molecular weight of the petroleum fraction. These parameters in turn are related to five or six relatively simple tests of the properties of oils. Some of the details of these tests, the empirical parameters, and the properties that can be predicted from these parameters will be found in the *API Technical Data Book* and in Appendix K.

1.7 Technique of solving problems

If you can form good habits of problem solving early in your career, you will save considerable time and avoid many frustrations in all aspects of your work, in and out of school. In solving material and energy balance problems, you should

(a) Read the available information through thoroughly and understand what is required for an answer. Sometimes, as in life, the major obstacle is to find out what the problem really is.

(b) Determine what additional data are needed, if any, and obtain this information.

(c) Draw a simplified picture of what is taking place and write down the available data. You may use boxes to indicate processes or equipment, and lines for the flow streams.

(d) Pick a basis on which to start the problem, as discussed in Sec. 1.3-6.

(e) If a chemical equation is involved, write it down and make sure it is balanced.

By this time you should have firmly in mind what the problem is and a reasonably clear idea of what you are going to do about it; however, if you have not seen exactly how to proceed from what is available to what is wanted, then you should

(f) Decide what formulas or principles are governing in this specific case and what types of calculations and intermediate answers you will need to get the final answer. If alternative procedures are available, try to decide which are the most expedient. If an unknown cannot be found directly, give it a letter symbol, and proceed as if you knew it.

(g) Make the necessary calculations in good form, being careful to check the arithmetic and units as you proceed.

(h) Determine whether the answer seems reasonable in view of your experience with these types of calculations.

Problems which are long and involved should be divided into parts and attached systematically piece by piece.

If you can assimilate this procedure and make it a part of yourself—so that you do not have to think about it step by step—you will find that you will be able to materially improve your speed, performance, and accuracy in problem solving.

> The major difference between problems solved in the classroom and in the plant lies in the quality of the data available for the solution. Plant data may be of poor quality, inconclusive, inadequate, or actually conflicting, depending on the accuracy of sampling, the type of analytical procedures employed, the skill of the technicians in the operation of analytical apparatus, and many other factors. The ability of an engineer to use the stoichiometric principles for the calculation of problems of material balance is only partly exercised in the solution of problems, even of great complexity, if solved from adequate and appropriate data. The remainder of the test lies in his ability to recognize poor data, to request and obtain usable data, and, if necessary, to make accurate estimates in lieu of incorrect or insufficient data.[17]

[17]B. E. Lauer, *The Material Balance*, Work Book Edition, 1954, p. 89.

1.8 The chemical equation and stoichiometry

As you already know, the chemical equation provides a variety of qualitative and quantitative information essential for the calculation of the combining weights of materials involved in a chemical process. Take, for example, the combustion of heptane as shown below. What can we learn from this equation?

$$C_7H_{16} + 11\,O_2 \longrightarrow 7\,CO_2 + 8\,H_2O \tag{1.25}$$

First, make sure that the equation is balanced. Then we can see that 1 mole (*not* lb_m or kg) of heptane will react with 11 moles of oxygen to give 7 moles of carbon dioxide plus 8 moles of water. These may be lb moles, g moles, kg moles, or any other type of mole, as shown in Fig. 1.13. One mole of CO_2 is formed from each $\frac{1}{7}$ mole of C_7H_{16}. Also, 1 mole of H_2O is formed with each $\frac{7}{8}$ mole of CO_2. Thus the equation tells us in terms of moles (*not* mass) the ratios among reactants and products.

Stoichiometry (stoi-ki-om-e-tri)[18] deals with the combining weights of elements and compounds. The ratios obtained from the numerical coefficients in the chemical equation are the stoichiometric ratios that permit you to calculate the moles of one substance as related to the moles of another substance in the chemical equation. If the basis selected is to be mass (lb_m, kg) rather than moles, you should use the following method in solving problems involving chemical equations: (a) Calculate the number of moles of the substance equivalent to the basis using the molecular weight; (b) change this quantity into the moles of the desired product or reactant by multiplying by the proper stoichiometric ratio, as determined by the chemical equation; and (c) then change the moles of product or reactant to a weight basis. These steps are indicated in Fig. 1.14 for the above reaction. You can combine these steps in a single dimensional equation, as shown in the examples below, for ease of slide rule calculations; however, the order of operations should be kept the same.

An assumption implicit in the above is that the reaction takes place exactly as written in the equation and proceeds to 100 percent completion. When reactants, products, or degree of completion of the actual reaction differ from the assumptions of the equation, additional data must be made available to indicate the actual status or situation.

EXAMPLE 1.22 Stoichiometry

In the combustion of heptane, CO_2 is produced. Assuming that it is desired to produce 500 lb of dry ice per hour and that 50 percent of the CO_2 can be converted into dry ice, how many pounds of heptane must be burned per hour?

[18]From the Greek *stoicheion*, basic constituent, and *metrein*, to measure.

$$C_7H_{16} \quad + \quad 11O_2 \quad \longrightarrow \quad 7CO_2 \quad + \quad 8H_2O$$

Qualitative information

heptane	reacts with	oxygen	to give	carbon dioxide	and	water

Quantitative information

1 molecule of heptane	reacts with	11 molecules of oxygen	to give	7 molecules of carbon dioxide	and	8 molecules of water
6.023×10^{23} molecules of C_7H_{16}	+	$11(6.023 \times 10^{23})$ molecules of O_2	$\longrightarrow$	$7(6.023 \times 10^{23})$ molecules of CO_2	+	$8(6.023 \times 10^{23})$ molecules of H_2O
1 g mole of C_7H_{16}	+	11 g moles of O_2	$\longrightarrow$	7 g moles of CO_2	+	8 g moles of H_2O
1 kg mole of C_7H_{16}	+	11 kg moles of O_2	$\longrightarrow$	7 kg moles of CO_2	+	8 kg moles of H_2O
1 lb mole of C_7H_{16}	+	11 lb moles of O_2	$\longrightarrow$	7 lb moles of CO_2	+	8 lb moles of H_2O
1 ton mole of C_7H_{16}	+	11 ton moles of O_2	$\longrightarrow$	7 ton moles of CO_2	+	8 ton moles of H_2O
1(100) g of C_7H_{16}	+	11(32) g of O_2	=	7(44) g of CO_2	+	8(18) g of H_2O
100 g		352 g		308 g		144 g

$$452 \text{ g} = 452 \text{ g}$$
$$452 \text{ kg} = 452 \text{ kg}$$
$$452 \text{ ton} = 452 \text{ ton}$$
$$452 \text{ lb} = 452 \text{ lb}$$

Fig. 1.13. The chemical equation.

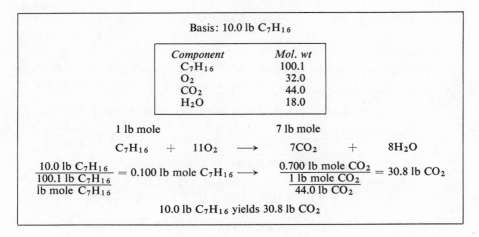

Fig. 1.14. Stoichiometry.

Solution:

Basis: 500 lb dry ice (or 1 hr)

Mol. wt heptane $= 100$

Chemical equation as in Fig. 1.13

$$\frac{500 \text{ lb dry ice}}{} \left| \frac{1 \text{ lb } CO_2}{0.5 \text{ lb dry ice}} \right| \frac{1 \text{ lb mole } CO_2}{44 \text{ lb } CO_2} \left| \frac{1 \text{ lb mole } C_7H_{16}}{7 \text{ lb mole } CO_2} \right.$$

$$\left| \frac{100 \text{ lb } C_7H_{16}}{1 \text{ lb mole } C_7H_{16}} \right. = 325 \text{ lb } C_7H_{16}$$

Since the basis of 500 lb dry ice is identical to 1 hr, 325 lb of C_7H_{16} must be burned per hour. Note that pounds are converted first to moles, then the chemical equation is applied, and finally moles are converted to pounds again for the final answer.

EXAMPLE 1.23 Stoichiometry

Corrosion of pipes in boilers by oxgen can be alleviated through the use of sodium sulfite. Sodium sulfite removes oxygen from boiler feedwater by the following reaction:

$$2 Na_2SO_3 + O_2 \longrightarrow 2 Na_2SO_4$$

How many pounds of sodium sulfite are theoretically required to remove the oxygen from 8,330,000 lb of water (10^6 gal) containing 10.0 parts per million (ppm) of dissolved oxygen and at the same time maintain a 35 percent excess of sodium sulfite? See Fig. E1.23.

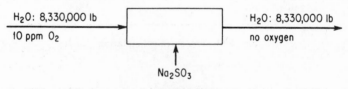

Fig. E1.23.

Solution:

Additional data: mol. wt of Na_2SO_3 is 126.

Chemical equation: $2 Na_2SO_3 + O_2 \longrightarrow 2 Na_2SO_4$

Basis: 8,330,000 lb H_2O with 10 ppm O_2 or 83.3 lb O_2

$$\frac{8,330,000 \text{ lb } H_2O}{} \left| \frac{10 \text{ lb } O_2}{\underbrace{(1,000,000 - 10 \text{ lb } O_2) \text{ lb } H_2O}_{\textbf{effectively same as 1,000,000}}} \right. = 83.3 \text{ lb } O_2$$

$$\frac{8,330,000 \text{ lb } H_2O}{} \left| \frac{10 \text{ lb } O_2}{10^6 \text{ lb } H_2O} \right| \frac{1 \text{ lb mole } O_2}{32 \text{ lb } O_2} \left| \frac{2 \text{ lb mole's } Na_2SO_3}{1 \text{ lb mole } O_2} \right.$$

$$\left| \frac{126 \text{ lb } Na_2SO_3}{1 \text{ lb mole } Na_2SO_3} \right| \frac{1.35}{1} = 885 \text{ lb } Na_2SO_3$$

EXAMPLE 1.24 Stoichiometry

A limestone analyzes

$$
\begin{array}{ll}
CaCO_3 & 92.89 \\
MgCO_3 & 5.41 \\
Insoluble & 1.70
\end{array}
$$

(a) How many pounds of calcium oxide can be made from 5 tons of this limestone?

(b) How many pounds of CO_2 can be recovered per pound of limestone?

(c) How many pounds of limestone are needed to make 1 ton of lime?

Solution:

Read the problem carefully to fix in mind exactly what is required. Lime will include all the impurities present in the limestone which remain after the CO_2 has been driven off. Next draw a picture of what is going on in this process. See Fig. E1.24.

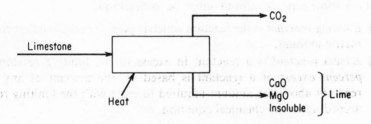

Fig. E1.24.

To complete the preliminary analysis you need the following chemical equations:

$$CaCO_3 \longrightarrow CaO + CO_2$$
$$MgCO_3 \longrightarrow MgO + CO_2$$

Additional data are

	$CaCO_3$	$MgCO_3$	CaO	MgO	CO_2
Mol. wt	100	84.3	56.0	40.3	44

Basis: 100 lb limestone

This basis was selected because lb = %.

comp.	lb = %	lb mole	⟶ lime	lb	$CO_2(lb)$
$CaCO_3$	92.89	0.9289	CaO	52.0	40.9
$MgCO_3$	5.41	0.0641	MgO	2.59	2.82
Insol.	1.70		Insol.	1.70	
Total	100.00	0.9930	Total	56.3	43.7

Note that the total pounds of products equal the 100 lb of entering limestone. Now to calculate the quantities originally asked for:

(a) CaO produced $= \dfrac{52.0 \text{ lb CaO}}{100 \text{ lb stone}} \Big| \dfrac{2000 \text{ lb}}{1 \text{ ton}} \Big| \dfrac{5 \text{ ton}}{} = 5200 \text{ lb CaO}$

(b) CO_2 recovered $= \dfrac{43.7 \text{ lb } CO_2}{100 \text{ lb stone}} = 0.437$ lb, or,

$$\dfrac{0.9930 \text{ lb mole } CaCO_3 + MgCO_3}{100 \text{ lb stone}} \left| \dfrac{1 \text{ lb mole } CO_2}{1 \text{ lb mole } CaCO_3 + MgCO_3} \right.$$

$$\left| \dfrac{44 \text{ lb } CO_2}{1 \text{ lb mole } CO_2} = 0.437 \text{ lb} \right.$$

(c)   limestone required $= \dfrac{100 \text{ lb stone}}{56.3 \text{ lb lime}} \left| \dfrac{2000 \text{ lb}}{1 \text{ ton}} = 3560 \text{ lb stone} \right.$

In industrial reactions you will rarely find exact stoichiometric amounts of materials used. To make a desired reaction take place or to use up a costly reactant, excess reactants are nearly always used. This excess material comes out together with, or perhaps separately from, the product—and sometimes can be used again. Even if stoichiometric quantities of reactants are used, but if the reaction is not complete or there are side reactions, the products will be accompanied by unused reactants as well as side-products or reactants. In these circumstances some new definitions must be understood:

(a) *Limiting reactant* is the reactant which is present in the smallest stoichiometric amount.

(b) *Excess reactant* is a reactant in excess of the limiting reactant. The *percent excess* of a reactant is based on the amount of any excess reactant above the amount required to react with the limiting reactant according to the chemical equation, or

$$\% \text{ excess} = \dfrac{\text{moles in excess}}{\text{moles required to react with limiting reactant}} 100$$

where the moles in excess frequently can be calculated as the total available moles of a reactant less the moles required to react with the limiting reactant. A common term, *excess air*, is used in combustion reactions; it means the amount of air available to react that is in excess of the air theoretically required to *completely* burn the combustible material. The *required* amount of a reactant is established by the limiting reactant and is for all other reactants the corresponding stoichiometric amount. **Even if only part of the limiting reactant actually reacts, the required and excess quantities are based on the entire amount of the limiting reactant.**

Three other terms that are used in connection with chemical reactions have less clear-cut definitions: conversion, selectivity, and yield. No universally agreed upon definitions exist for these terms—in fact, quite the contrary. Rather than cite all the possible usages of these terms, many of which conflict, we shall define them as follows:

(c) *Conversion* is the fraction of the feed or some material in the feed that is converted into products. What the basis in the feed is and into what products the basis is being converted must be clearly specified or endless confusion results. Conversion is somewhat related to the *degree of completion* of a reaction, which is usually the percentage or fraction of the limiting reactant converted into products.

(d) *Selectivity* expresses the amount of a desired product as a fraction or percent of the theoretically possible amount from the feed material converted. Often the quantity defined here as selectivity is called efficiency, conversion efficiency, specificity, yield, ultimate yield, or recycle yield.

(e) *Yield*, for a single reactant and product, is the weight or moles of final product divided by the weight or moles of initial reactant (P lb of product A per R lb of reactant B). If more than one product and more than one reactant are involved, the reactant upon which the yield is to be based must be clearly stated.

The employment of these concepts can best be illustrated by examples.

EXAMPLE 1.25   Limiting reactant and incomplete reaction

Antimony is obtained by heating pulverized stibnite with scrap iron and drawing off the molten antimony from the bottom of the reaction vessel:

$$Sb_2S_3 + 3Fe \longrightarrow 2Sb + 3FeS$$

0.600 kg of stibnite and 0.250 kg of iron turnings are heated together to give 0.200 kg of Sb metal.

Calculate:

(a) The limiting reactant.
(b) The percentage of excess reactant.
(c) The degree of completion (fraction).
(d) The percent conversion.
(e) The selectivity (percent).
(f) The yield.

Solution:
The molecular weights needed to solve the problem and the g moles forming the basis are

	g	*mol. wt*	*g mole*
Sb_2S_3	600	339.7	1.77
Fe	250	55.8	4.48
Sb	200	121.8	1.64
FeS		87.9	

(a) To find the limiting reactant, we examine the chemical reaction equation and note that if 1.77 g mole of Sb_2S_3 reacts, it requires $3(1.77) = 5.31$ g mole of Fe, whereas if 4.48 g mole of Fe react, it requires $(4.48/3) = 1.49$ g mole of Sb_2S_3 to be available. Thus Fe is present in the smallest stoichiometric amount and is the limiting reactant; Sb_2S_3 is the excess reactant.

(b) The percentage of excess reactant is:

$$\% \text{ excess} = \frac{(1.77 - 1.49)}{1.49}(100) = 18.8\% \text{ excess } Sb_2S_3$$

(c) Although Fe is the limiting reactant, not all the limiting reactant reacts. We can compute from the 1.64 g moles of Sb how much Fe actually does react:

$$\frac{1.64 \text{ g mole Sb}}{} \Bigg| \frac{3 \text{ g mole Fe}}{2 \text{ g mole Sb}} = 2.46 \text{ g mole Fe}$$

If by the fractional degree of completion is meant the fraction conversion of Fe to FeS, then

$$\text{fractional degree of completion} = \frac{2.46}{4.48} = 0.55$$

(d) The percent conversion can be arbitrarily based on the Sb_2S_3 if our interest is mainly in the stibnite:

$$\frac{1.64 \text{ g mole Sb}}{} \Bigg| \frac{1 \text{ g mole } Sb_2S_3}{2 \text{ g mole Sb}} = 0.82 \text{ g mole } Sb_2S_3$$

$$\% \text{ conversion of } Sb_2S_3 \text{ to Sb} = \frac{0.82}{1.77}(100) = 46.3\%$$

(e) The selectivity refers to the conversion of Sb_2S_3 (we assume) based on the theoretical amount that can be converted, 1.49 g mole:

$$\text{selectivity} = \frac{0.82}{1.49}(100) = 55\%$$

(f) The yield will be stated as kilograms of Sb formed per kilogram of Sb_2S_3 that was fed to the reaction:

$$\text{yield} = \frac{0.200 \text{ kg Sb}}{0.600 \text{ kg } Sb_2S_3} = \frac{1}{3} \frac{\text{kg Sb}}{\text{kg } Sb_2S_3}$$

EXAMPLE 1.26   Limiting reactant and incomplete reactions

Aluminum sulfate can be made by reacting crushed bauxite ore with sulfuric acid, according to the following equation:

$$Al_2O_3 + 3H_2SO_4 \longrightarrow Al_2(SO_4)_3 + 3H_2O$$

The bauxite ore contains 55.4 weight percent aluminum oxide, the remainder being impurities. The sulfuric acid solution contains 77.7 percent H_2SO_4, the rest being water.

To produce crude aluminum sulfate containing 1798 lb of pure aluminum sulfate, 1080 lb of bauxite ore and 2510 lb of sulfuric acid solution are used.

(a) Identify the excess reactant.

(b) What percentage of the excess reactant was used?

(c) What was the degree of completion of the reaction?

Solution:

The pound moles of substances forming the basis of the problem can be computed as follows:

$$\frac{1798 \text{ lb Al}_2(\text{SO}_4)_3 \mid 1 \text{ lb mole Al}_2(\text{SO}_4)_3}{342.1 \text{ lb Al}_2(\text{SO}_4)_3} = 5.26 \text{ lb mole}$$

$$\frac{1080 \text{ lb bauxite} \mid 0.554 \text{ lb Al}_2\text{O}_3 \mid 1 \text{ lb mole Al}_2\text{O}_3}{1 \text{ lb bauxite} \mid 101.9 \text{ lb Al}_2\text{O}_3} = 5.87 \text{ lb mole}$$

$$\frac{2510 \text{ lb acid} \mid 0.777 \text{ lb H}_2\text{O} \mid 1 \text{ lb mole H}_2\text{SO}_4}{1 \text{ lb acid} \mid 98.1 \text{ lb H}_2\text{SO}_4} = 19.88 \text{ lb mole}$$

(a) Assume that Al_2O_3 is the limiting reactant. Then $5.87 \times 3 = 17.61$ lb mole of H_2SO_4 would be required, which is present. Hence H_2SO_4 is the excess reactant.

(b) The $\text{Al}_2(\text{SO}_4)_3$ actually formed indicates that:

$$\frac{5.26 \text{ lb mole Al}_2(\text{SO}_4)_3 \mid 3 \text{ lb mole H}_2\text{SO}_4}{1 \text{ lb mole Al}_2(\text{SO}_4)_3} = 15.78 \text{ lb mole H}_2\text{SO}_4 \text{ was used}$$

$$\frac{15.78}{19.88}(100) = 79.4\%$$

(c) The fractional degree of completion was

$$\frac{5.26}{5.87} = 0.90$$

EXAMPLE 1.27   The meaning of yield

Your assistant has rushed into your office with a glowing tale of a yield of 108 percent in the manufacture of phthalic anhydride from xylene by the reaction

$$\underset{\text{xylene}}{\text{C}_6\text{H}_4(\text{CH}_3)_2} + 3\text{O}_2 \longrightarrow \underset{\text{phthylic anhydride}}{\text{C}_6\text{H}_4(\text{CO})_2\text{O}} + 3\text{H}_2\text{O}$$

Should you compliment him or deflate his accomplishment?

Solution:

If the claim is that 108 lb of phthalic anhydride (Ph) are produced per 100 lb of xylene (X), the selectivity is not especially good.

Basis: 100 lb xylene

$$\frac{100 \text{ lb X} \mid 1 \text{ lb mole X}}{106 \text{ lb X}} = 0.944 \text{ lb mole X}$$

$$\frac{108 \text{ lb Ph} \mid 1 \text{ lb mole X}}{148 \text{ lb Ph}} = 0.730 \text{ lb mole Ph}$$

$$(100)\frac{0.730 \text{ lb mole X}}{0.944 \text{ lb mole Ph}} = 77.3\%$$

If the claim is that 108 percent selectivity is achieved, the claim must be erroneous.

You should remember that the chemical equation does not indicate the true mechanism of the reaction or how fast or to what extent the reaction will take place. For example, a lump of coal in air will sit unaffected at room temperature, but at higher temperatures it will readily burn. All the chemical equation indicates is the stoichiometric amounts required for the reaction and obtained from the reaction if it proceeds in the manner in which it is written. Also remember to make sure that the chemical equation is balanced before using it.

1.9 Digital computers in solving problems

High-speed digital computers have made a considerable impact on problem solving both in industry and in education. There is no question that a good fraction of today's engineering graduates (who may well be working as engineers in the year 2000) will see the use of computers in their technical work expand dramatically. Consequently, you should know how to use computers in your computations and if possible apply your knowledge to solving some of the more complicated problems in this book. Many routes of approach are used in engineering education to familiarize students with the role of the computer in solving engineering problems. On the one hand, library computer programs ("canned" programs) can be employed to solve specific types of problems, absolving you of most of the programming chores. On the other hand, you can prepare the complete computer code to solve a problem from scratch. Problems marked with an asterisk in this book indicate problems that are especially appropriate for computer solutions by canned programs, and at the end of each chapter are a few problems for which the complete computer program is to be written. Wide differences in computers and the proliferation of programming languages have made the inclusion of detailed computer codes in this text less than particularly meaningful except for one or two solved examples.

You should not attempt to justify the use of a computer as a time-saving device per se when working with a single problem. It is much better to look on the computer and the program as a tool which will enable you, after some experience, to solve complex problems expediently and which may save time and trouble for a simple problem, but may not. Modern, high-speed computers can eliminate repetitive, time-consuming routine calculations that would take years to complete with a slide rule or a desk calculator. Also, computers can essentially eliminate numerical errors once the programs are verified. Typical problems encountered in this text that are most susceptible to computer solution are (a) the solution of simultaneous equations, (b) the solution of one or more nonlinear equations, (c) statistical data fitting, and (d) various types of iterative calculations. Typical industrial problems routinely solved are:

(a) Heat and mass transfer coefficient correlations.
(b) Correlation of chemical structure with physical properties.
(c) Analysis of gaseous hydrocarbon mixtures.

(d) Particle size analysis.

(e) Granular fertilizer formulation.

(f) Evaluation of infrared analysis.

(g) Optimum operating conditions.

(h) Comprehensive economic evaluations.

(i) Chemical-biological coordination and correlation.

(j) Vapor-liquid equilibrium calculations.

(k) Mass spectrometer matrix inversion.

(l) Equipment design.

(m) Preparation of tables for platinum resistance thermometer.

One of the most significant changes in the use of computers has been in the meaning of the word *solution*. It is widely recognized that a mathematical closed-form solution, while unquestionably pleasing aesthetically and relatively universal in application, may not necessarily be interpreted easily. Curves and graphs may constitute a more desirable format for a solution, but a computer and a program essentially consist of a procedure whereby a numerical answer can be obtained to a specific problem. Repetitive solutions must be obtained to prepare tables, graphs, etc., but these, of course, can be prepared automatically with an appropriate computer code.

The use of a guaranteed canned computer program saves the user considerable time and effort relative to programming his own code. Every major computer center has, or should have, an extensive subroutine library, often stored on an on-line library tape, for effecting extensive numerical calculations and/or simulation. Of course, such programs may still be fallible. When an unexpected answer appears, the user of the program must know enough about the problem to be able to account for the discrepancy. Often the trouble lies in the interpretation of what the program is intended to do, or it is caused by improper entry of parameters or data. If you can avoid these difficulties, you can concentrate on stating the problem correctly and estimating the expected answer; the mechanics of obtaining the answer would be left to the computer. Considerably more complex and realistic problems can be solved, furnishing a deeper understanding of the principles involved, if the latter are not obscured by concern with the problem-solving technique.

If you do use the computer, you will be forced to be far more precise than you have been accustomed to being in the past. Because of the nature of the computer languages, i.e., the necessity for very precise grammar and punctuation, it is unusual to solve completely correctly an engineering problem on the first approach. Experience has shown that, even in the most carefully prepared programs, there will be a certain percentage of errors, which may include programming logic, actual coding, data preparation, or the card punching. For this reason, processing of a problem requires prior program verification in which a computer run of a sample problem is executed for which the results are known, can be estimated, or can be calculated by manual methods. To facilitate such a check-out as well as to guard against undetected and possible future data errors, a good program should include self-checking features strategically located in the program. If physically impossible conditions are created, the calculation should be stopped and the accumulated results printed out. You no doubt will require many tries before achieving success. The turn-around time, i.e., the time elapsed between submission of a program to the computer and its return for checking and possible resubmission in the case of error, must be fairly short if problems are to be solved in a reasonable amount of time. The use of remote time-sharing terminals can assist in the speedy resolution of programming errors. You will also be forced to think more logically and in greater detail than you have been accustomed to doing, if you do your own programming, about the problem and how to solve it. The computer is a rigid task master that requires precision in the statement of the problem and the flow of information needed to effect a solution. No ambiguity is permitted.

WHAT YOU SHOULD HAVE LEARNED FROM THIS CHAPTER

1. You should have memorized the common conversion units.
2. You should be able to convert units from the American engineering system to the cgs and SI systems and vice versa with ease, and understand the significance of g_c.
3. You should understand *moles, molecular weights, density, specific gravity, mole* and *weight fraction, temperature,* and *pressure,* and be able to work problems involving these concepts.
4. You should have well in mind the proper approach to problem solving and be able to effectively put into practice the principles discussed in the chapter.
5. You should know how to apply the principles of stoichiometry to problems involving chemical reactions.

SUPPLEMENTARY REFERENCES

General

1. Benson, S. W., *Chemical Calculations*, 2nd ed., John Wiley, New York, 1963.
2. Considine, D. M., and S. D. Ross, eds., *Handbook of Applied Instrumentation*, McGraw-Hill, New York, 1964.
3. Henley, E. J., and H. Bieber, *Chemical Engineering Calculations*, McGraw-Hill, New York, 1959.
4. Hougen, O. A., K. M. Watson, and R. A. Ragatz, *Chemical Process Principles*, Part I, 2nd ed., John Wiley, New York, 1959.
5. Littlejohn, C. E., and F. G. Meenaghan, *An Introduction to Chemical Engineering*, Van Nostrand Reinhold, New York, 1959.

6. Ranz, W. E., *Describing Chemical Engineering Systems*, McGraw-Hill, New York, 1970.

7. Whitwell, J. C., and R. K. Toner, *Conservation of Mass and Energy*, Ginn/Blaisdell, Waltham, Mass., 1969.

8. Williams, E. T., and R. C. Johnson, *Stoichiometry for Chemical Engineers*, McGraw-Hill, New York, 1958.

Units and Dimensions

1. Danloux-Dumesnils, M., *The Metric System*, Oxford University Press, Inc., New York, 1969.

2. Donovan, F., *Prepare Now for a Metric Future*, Weybright and Talley, New York, 1970.

3. Page, C. H., and P. Vigoureux, *The International System of Units (SI)*, NBS Spec. Publ. 330, Jan. 1971. (For sale by the Superintendent of Documents, U.S. Government Printing Office, Washington, D.C. 20402, SD Catalog No. 13: 10: 330, price 50 cents.)

4. Paul, M. A., "The International System of Units," *J. Chem. Doc.*, V. 11, p. 3 (1971).

5. O'Day, E. F., *Physical Quantities and Units*, Prentice-Hall, Englewood Cliffs, N.J., 1967.

6. Klinkenberg, A., "The American Engineering System of Units and Its Dimensional Constant g_c," *Indus. Engr. Chem.*, 61, no. 4, p. 53 (1969).

PROBLEMS[19]

1.1. Electrically driven automobiles have been proposed as one solution to the problem of excessive automobile emissions in large cities. The Ghia Rowan is a sleek Italo-American prototype, an electrically driven system which converts the kinetic energy produced by the car's natural momentum into chemically stored battery energy via the car's braking system. The vehicle is roughly 3 m long, weighs 596 kg, and has a range of about 320 km at a speed of 41–43.5 mi/hr. Powered by two 9-hp electric motors placed under the rear seats and connected to the control box, the Rowan has no differential, transmission system, or clutch assembly.

Rewrite the specifications given so that they are all consistent in (a) the SI system and (b) in the American engineering system.

1.2. Hollywood (AP)—Blonde Jan Eastlund from Europe, a loser in last year's Miss Universe pageant, came up a winner this week with a studio contract and a starring role in her first movie. Jan speaks English with a slight accent, and is baffled by our units of measure. Asked for her measurements, she answered, "My bust is 92, waist 58, and my hips 92." A studio press agent gave an imitation of a

---

[19]An asterisk designates problems appropriate for computer solution. Also refer to the computer problems after Problem 1.104.

man suffering apoplexy. "That can't be right," he cried. "When you say '92', what do you mean—92 what!" "Centimeters, of course," Jan answered. "Is that so bad?" Help out the distraught publicist by providing the correct dimensions in inches.

1.3. A pipeline moving 6142 bbl of oil per hour is moving the equivalent of how many cubic meters per second? (42 gal = 1 bbl.)

1.4. Change the following to the desired units:
   (a) 235 g to pounds.
   (b) 610 l to cubic feet.
   (c) 30 g/l to pounds/cubic feet.
   (d) 14.7 lb/in.^{2} to kg/cm^2

1.5. How many cubic meters are there in a rod of uniform cross section that is 15 in. long by $\frac{1}{4}$ in. in diameter?

1.6. It is proposed to set first-class postage for mail at \$0.10/oz or fraction thereof. For the benefit of a visitor from France, explain the cost in francs (\$1.00 = 5 francs) per gram.

1.7. If gasoline costs \$0.40/U.S. gal, what is the cost in liters? Cubic meters?

1.8. A dime has a mass of approximately 2.50 g. What is its mass in (a) kg, (b) lb, and (c) slugs?

1.9. Convert the following to the desired units:
   (a) 60 mi/hr to m/sec.
   (b) 30 N/m^2 to lb_f/ft^2.
   (c) 16.3 (J)(m) to Btu.
   (d) 4.21 kW to J/sec.

1.10. Find the following:
   (a) The acceleration produced when 3 N act on a mass of 0.016 kg.
   (b) The weight in lb_f of 4.31 slugs.
   (c) The energy accumulated when a force of 5.00 lb_f acts on a mass of 6.00 lb_m for a distance of 4.10 ft.

1.11. Convert the following:
   (a) 60.0 mi/hr to ft/sec.
   (b) 50.0 lb/in.^{2} to kg/m^2.
   (c) 6.20 cm/hr^2 to nm/sec^2.

1.12. If a rocket uses 2 l of liquid oxygen per second as an oxidizer, how many cubic feet per hour of liquid oxygen are used?

1.13. Water is flowing through a pipe of 3.355-in.^{2} cross-sectional area at the rate of 60 gal/min. Find the velocity, ft/sec.

1.14. 400 m/sec is equivalent to how many yards/min?

1.15. A shipment of coffee beans has been received at the dock and the billing given in terms of \$/kg. Provide the conversion factor for the accounting office to convert future billings to \$/lb_m.

1.16. Draw an information flow diagram in which there are four nodes: length, time, mass, temperature. Link the following quantities to the nodes by means of lines (arcs): area, volume, speed, acceleration, force, energy, pressure, density, power, specific heat (heat capacity).

1.17. An elevator which weighs 10,000 lb is pulled up 10 ft between the first and second floors of a building 100 ft high. The greatest velocity the elevator attains is 3 ft/sec. How much kinetic energy does the elevator have in $(ft)(lb_f)$ at this velocity?

1.18. Find the kinetic energy of a ton of water moving at 60 mi/hr expressed as (a) $(ft)(lb_f)$, (b) ergs, (c) joules, (d) (hp)(sec), (e) (watt)(sec), and (f) (liter) (atm).

1.19. Calculate the kinetic and potential energy of a missile moving at 12,000 mi/hr above the earth where the acceleration due to gravity is 30 ft/sec².

1.20. Densities may sometimes be expressed as linear functions of temperature, such as

$$\rho = \rho_o + At$$

where $\rho = lb/ft^3$ at temperature t
       $\rho_o = lb/ft^3$ at temperature t_0
       $t = $ temperature in °F

If the equation is dimensionally consistent, what must the units of A be?

1.21. The equation for pressure drop due to friction for fluids flowing in a pipe is

$$\Delta p = \frac{2fL\rho v^2}{D}$$

where $\Delta p = $ pressure drop
      $v = $ velocity
      $L = $ length of the pipe
      $\rho = $ density of the fluid
      $D = $ diameter of the pipe

Is the equation dimensionally consistent? What are the units of the *friction factor f*? Use units of the ft-lb-sec system for Δp, L, ρ, v, and D; is the equation then consistent?

1.22. The density of a certain liquid is given an equation of the following form:

$$\rho = (A + Bt)e^{CP}$$

where $\rho = $ density in g/cm³
     $t = $ temp. in °C
     $P = $ pressure in atm

(a) The equation is dimensionally consistent. What are the units of A, B, and C?

(b) In the units above,

$$A = 1.096$$
$$B = 0.00086$$
$$C = 0.000953$$

Find A, B, and C if ρ is expressed in lb/ft³, t in °R, and P in $lb_f/in.^2$

1.23. The equation for the flow of water through a nozzle is as follows:

$$q = C\sqrt{\frac{2g}{1 - (d_1/d_2)^4}}\left(A\sqrt{\frac{\Delta p}{\rho}}\right)$$

where $q = $ volume flowing per unit time
     $g = $ local gravitational acceleration

d_1 = smaller nozzle diameter
d_2 = larger nozzle diameter
A = area of nozzle outlet
Δp = pressure drop across nozzle
ρ = density of fluid flowing
C = dimensionless constant

State whether this equation is dimensionally consistent. Show how you arrived at your conclusion.

1.24. Countercurrent gas centrifuges have been used to separate ^{235}U from ^{238}U. The rate of diffusive transport is $K = 2\pi D \rho \bar{r}$.

If K = rate of transport of light component to the center of the centrifuge, g moles/(sec)(cm of height)

D = diffusion coefficient
ρ = molar density, g moles/cm^3
$\bar{r}$ = log mean radius, $(r_2 - r_1)/\ln(r_2/r_1)$, with r in cm

what are the units of D?

1.25. A useful dimensionless number called the *Reynolds number* is

$$\frac{DU\rho}{\mu}$$

where D = the diameter or length
U = some characteristic velocity
ρ = the fluid density
μ = the fluid viscosity

Calculate the Reynolds number for the following cases:

	1	2	3	4
D	2 in.	20 ft	1 ft	2 mm
U	10 ft/sec	10 mi/hr	1 m/sec	3 cm/sec
ρ	62.4 lb/ft^3	1 lb/ft^3	12.5 kg/m^3	25 lb/ft^3
μ	0.3 lb$_m$/(hr)(ft)	0.14×10^{-4} lb$_m$/(sec)(ft)	2×10^{-6} centipoise (cP)	1×10^{-6} centipoise

1.26. The Colburn equation for heat transfer is

$$\left(\frac{h}{CG}\right)\left(\frac{C\mu}{k}\right)^{2/3} = \frac{0.023}{(DG/\mu)^{0.2}}$$

where C = heat capacity, Btu/(lb of fluid)(°F)
μ = viscosity, lb/(hr)(ft)
k = thermal conductivity, Btu/(hr)(ft^2)(°F)/ft
D = pipe diameter, ft
G = mass velocity, lb/(hr)(ft^2) cross section

What are the units of the heat transfer coefficient h?

1.27. Write the formula and calculate the pound molecular weight of (a) silver sulfate, (b) anhydrous barium chloride, (c) propyl benzoate, and (d) thyroxin (thyro-oxyindol).

1.28. Write the formula and calculate the gram molecular weight of (a) nickel bromide, (b) anhydrous chromic sulfate, (c) benzyl alcohol, and (d) cortisone.

1.29. How many pounds of compound are contained in each of the following?
(a) 64 g of BaMnO$_4$.
(b) 200 g moles of ferric ammonium oxalate.

    (c) 40 lb moles of dimethyl amine.

    (d) 11 kg moles of nicotine.

1.30. Convert the following:

    (a) 105 g moles of potassium oxide to g.

    (b) 2 lb moles of aluminum acetate to lb.

    (c) 114 g moles of oxalic acid to lb.

    (d) 3 lb of quinine to g moles.

    (e) 40 g of trichlorobenzene to lb moles.

1.31. (a) How many g moles are represented by 100 g of CO_2?

    (b) Calculate the weight in pounds of 3.5 g moles of nitrogen.

1.32. Comment on the statement "a mole is an amount of substance containing the same number of atoms as 12 g of pure carbon-12."

1.33. How many pounds of compound are contained in each of the following?

    (a) 130 g moles of sodium hydroxide (anhydrous).

    (b) 300 g moles of sodium oleate.

    (c) 165 lb moles of sodium perchromate.

    (d) 62 lb moles of pure nitric acid.

    (e) 36 g moles of cupric tartrate.

    (f) 36 lb moles of cuprous phosphide (Cu_6P_2).

    (g) 72 lb moles of nickel bromide.

    (h) 12 g moles of nitrogen sulfide (N_4S_4).

    (i) 120 lb moles of potassium nitrate.

    (j) 11 g moles of stannic fluoride.

1.34. Convert the following:

    (a) 120 g moles of NaCl to g.

    (b) 120 g of NaCl to g moles.

    (c) 120 lb moles of NaCl to lb.

    (d) 120 lb of NaCl to lb moles.

    (e) 120 lb moles of NaCl to g.

    (f) 120 g moles of NaCl to lb.

    (g) 120 lb of NaCl to g moles.

    (h) 120 g of NaCl to lb moles.

    (i) 120 g moles of NaCl to lb moles.

    (j) 120 g of NaCl to lb.

1.35. The lifting power of a balloon is proportional to the weight lost by replacing air with some other gas. Calculate the lifting power of helium as a percentage of the lifting power of hydrogen.

1.36. The specific gravity of diethanolamine (DEA) at 60°F/60°F is 1.096. On a day when the temperature is 60.0°F, a 1347-gal volume of DEA is carefully metered into a storage tank. This volume corresponds to how many pounds of DEA?

1.37. The density of a certain solution is 8.80 lb/gal at 80°F. How many cubic feet will be occupied by 10,010 lb of this solution at 80°F?

1.38. The density of benzene at 60°F is 0.879 g/ml. What is the specific gravity of benzene at 60°F/60°F?

1.39. If the sp gr of silver oxide is 7.30 with reference to water at 25°C, calculate the sp gr with reference to water at 4°C. To water at 77°F.

1.40. A hydrometer is constructed so that the reference temperature is 77°F and the measurement is made at 77°F. In a handbook the sp gr of 10 percent acetic acid in water is given as 1.0125 $\frac{25°C}{4°C}$. Compute what the value of the reading should be in the hydrometer.

1.41. The normal range of the sp gr of urine is 1.008–1.030 with reference to water at 15°C. Find the range of densities of urine for $\frac{15°C}{25°C}$.

1.42. In a handbook you find that the conversion between °API and density is 0.800 density = 45.28 °API. Is this a misprint?

1.43. A mixture of liquid hydrocarbons contains 10.0 percent *n*-heptane, 40.0 percent *n*-octane, and 50.0 percent *i*-pentane by weight. The specific gravities $(\frac{60°F}{60°F})$ of the pure components are

$$n\text{-heptane} = 0.685$$

$$n\text{-octane} = 0.705$$

$$i\text{-pentane} = 0.622$$

(a) What is the sp gr $(\frac{60°F}{60°F})$ of 93 lb of this mixture?
(b) How many U.S. gallons will be occupied by 130 lb of this mixture?

1.44. A solution of sulfuric acid at 60°F is found to have a sp gr of 1.22. From the tables in Perry's *Chemical Engineer's Handbook*, this is found to be 30 percent by weight H_2SO_4. What is the concentration of H_2SO_4 in the following units: (a) lb mole/gal, (b) lb/ft³, (c) g/l, (d) lb H_2SO_4/lb H_2O, and (e) lb mole H_2O/lb mole total solution?

1.45. Two immiscible liquids are allowed to separate in a vessel. One liquid has a specific gravity of 0.936; the second liquid weighs 9.63 lb/gal. A block which is 9 by 9 by 9 in. and weighs 25.8 lb is dropped into the vessel. Will the block float at the top, stop at the interface where the two liquids are separated, or sink? What fraction of the volume of the block is in one or both of the liquids?

1.46. A hydrometer has a glass stem 12 in. long. The stem is made of glass tubing which weighs 0.042 lb/ft and is 0.400 in. in diameter on the outside. The bulb has a volume of 60 cm³. The entire hydrometer, including the lead shot which is placed in the bulb, weighs 63.0 g. The stem of the hydrometer is graduated in inches from 0 at the top to 12 at the bottom. See Fig. P1.46. Find the calibration equation of the hydrometer, giving the density G in g/cm³ as a function of the reading R from 0 to 12. (The hydrometer is read at the interface of the liquid sample on the stem.) Also give the upper and lower limits of densities in g/cm³ for which this hydrometer may be used.

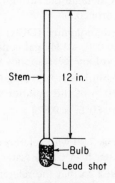

Stem —| 12 in.

Bulb
Lead shot

Fig. P1.46.

1.47. You have 100 lb of gas of the following composition:

$$CH_4 \quad 30\%$$
$$H_2 \quad 10\%$$
$$N_2 \quad 60\%$$

What is the average molecular weight of this gas?

1.48. Calculate the average molecular weight of the following gas mixture (volume percent is given in the column to the right):

$$CO_2 \quad 2.0$$
$$CO \quad 10.0$$
$$O_2 \quad 8.0$$
$$N_2 \quad 75.0$$
$$H_2O \quad \underline{5.0}$$
$$\quad\quad 100.0\%$$

1.49. A mixture of gases is analyzed and found to have the following composition:

$$CO_2 \quad 12.0\%$$
$$CO \quad 6.0$$
$$CH_4 \quad 27.3$$
$$H_2 \quad 9.9$$
$$N_2 \quad 44.8$$

How much will 3 lb moles of this gas mixture weigh?

1.50. The waste gas from a process analyzes 50.0 percent CO_2, 10.0 percent C_2H_4, and 40.0 percent hydrogen, all compositions being volume percent. What is the average molecular weight of the waste gas? What is the composition in weight percent?

1.51. Typical municipal refuse contains the following materials:

SAMPLE MUNICIPAL REFUSE COMPOSITION—U.S. EAST COAST

Physical (wt %)		Rough chemical (wt %)	
Cardboard	7	Moisture	28.0
Newspaper	14	Carbon	25.0
Miscellaneous paper	25	Hydrogen	3.3
Plastic film	2	Oxygen	21.1
Leather, molded	2	Nitrogen	0.5
plastics, rubber		Sulfur	0.1
Garbage	12	Glass, ceramics, etc.	9.3
Grass and dirt	10	Metals	7.2
Textiles	3	Ash, other inerts	5.5
Wood	7	Total	100.0
Glass, ceramics, stones	10		
Metallics	8		
Total	100		

SOURCE: E. R., Kaiser, "Refuse Reduction Processes," in *Proceedings, The Surgeon General's Conference on Solid Waste Management for Metropolitan Washington*, U.S. Public Health Service Publication No. 1729, Government Printing Office, Washington, D.C., July 1967, p. 93.

(a) What is the rough chemical analysis on a water-free basis?

(b) Give a rough estimate of the volume reduction of the municipal waste if it is completely incinerated. Clearly state any assumptions you make.

1.52. The following table shows the annual inputs of phosphorus to Lake Erie.

	Short tons/yr
Source	
Lake Huron	2,240
Land drainage	6,740
Municipal waste	19,090
Industrial waste	2,030
	30,100
Outflow	4,500
Retained	25,600

SOURCE: *23rd Report to Committee on Government Operations*, U.S. Government Printing Office, Washington, D.C., 1970.

(a) Convert the retained phosphorus to concentration in micrograms per liter assuming that Lake Erie contains 1.2×10^{14} gal of water and that average phosphorus retention time is 2.60 yr.

(b) What percentage of the input comes from municipal waste?

(c) What percentage of the input comes from detergents, assuming they represent 70 percent of the municipal waste?

(d) If 10 parts per billion of phosphorus trigger nuisance algal blooms, as has been reported in some documents, would removing 30 percent of the phosphorus in the municipal waste and all the phosphorus in the industrial waste be effective in reducing the eutrophication, i.e., the unwanted algal blooms, in Lake Erie?

(e) Would removing all the phosphate in detergents help?

1.53. It has been suggested that nitric oxide be removed from automobile emissions by adsorption on a cobalt-clay compound. The proposed method requires 1 lb of clay per 7 g of NO. Typical nitric oxide emissions of automobile engines (assuming a speed of 60 mi/hr) amount to approximately 4 g of NO per mile. Assume that simple modifications of the engine can decrease the current NO level of 3000 to 300 ppm, how many pounds of clay would be required to adsorb the remaining 300 ppm? Comment on the practicability of the method.

1.54. Five thousand barrels of 28°API gas oil are blended with 20,000 bbl of 15° API fuel oil. What is the °API (API gravity) of the mixture? What is the density in lb/gal and lb/ft³?

1.55. One barrel each of gasoline (55°API), kerosene (40°API), gas oil (31°API), and isopentane are mixed. What is the composition of the mixture expressed in weight percent and volume percent? What is the API gravity and the density of the mixture in g/cm³ and lb/gal?

1.56. If 100 cm³ of ethyl alcohol is added to 100 cm³ of water at 20°C, the total volume of the mixture is only 193 cm³. In other words, volumes are not additive for this system as they are for hydrocarbon mixtures.

(a) If the proof of alcohol-water mixtures is defined as twice the volume percent alcohol, and volume percent is defined as the volumes of pure alcohol obtainable from 100 volumes of mixture, determine accurately the weight and mole percent alcohol in a 100-proof mixture. The sp gr of the mixture is 0.9303, of pure alcohol 0.7893, and of pure water 0.9982, all at 20°C.

(b) Why is it unscientific to express compositions as *volume percent* in this system? Why is it done commercially?

1.57. Convert 55°C to °F; to °K; to °R.

1.58. Two thermometers are to be checked against a standard thermometer. The standard reads −22°F. What should the other two thermometers read if they are calibrated in °C and °K, respectively?

1.59. Mercury boils at 630°K. What is its boiling temperature expressed in °C? In °F? In °R?

1.60. Convert 90°R to °C; to °F; to °K.

1.61. The inhabitants of Betelgeuse, in the constellation Orion, have contacted us by radio, and in communicating with them we find that the zero of their temperature scale is based on the freezing point of methane (−182.5°C) and that the kindling point of wood (45°F) is 100 on their scale. Derive an equation relating °B as used by the Betelgeusians to °F. What is 122°C in °B?

1.62. Show that

(a) $T_{°F} = 2\{T_{°C} + \frac{1}{10}[160 - T_{°C}]\}$

(b) $T_{°F} = \{[T_{°C} \times 2] - \frac{1}{10}[T_{°C} \times 2]\} + 32$

1.63. Calculate all temperatures from the one value given:

	(a)	(b)	(c)	(d)	(e)	(f)	(g)	(h)
°F	140				1000			
°R			500			1000		
°K		298					1000	
°C				−40				1000

1.64. The emissive power of a black body depends on the fourth power of the temperature and is given by

$$W = AT^4$$

where W = emissive power in Btu/(ft²)(hr)

A = Stefan-Boltzmann constant, 0.171×10^{-8} Btu/(ft²)(hr)(°R)⁴

T = temperature in °R

What is the value of A in the units J/(m²)(sec)(°K)⁴?

1.65.* The temperature scale we use at present is linear, forming an ordered sequence of numbers starting at 0°K,

$$+0°K \cdots 273.16°K \cdots + \infty°K$$

Kelvin proposed the relation

$$T = e^{\psi}$$

where ψ depends on the thermal properties of the system and T is the absolute temperature (°K). It is possible to construct another temperature scale, the ψ scale, or the logarithmic scale. The logarithmic nature of the ψ scale assures that the zero value on the linear absolute scale (T scale) will be reached asymptotically, i.e., $T = 0°K$, when $\psi = -\infty$. Comparing both these scales, the inaccessibility of the absolute zero follows generically, not being imaginable from the linear T scale alone. Plot the ψ vs. T scales, and calculate the value of ψ for $T = 1°K$ and $273°K$.

1.66. According to a letter to the editor of *Chemical and Engineering News*, May 17, 1965, the flashing rate of a firefly and the chirping rate of a cricket are both linearly related to temperature:

$$T(°F) = 0.25x_1 + 39.0 \qquad x_1 = \text{chirps/min}$$
$$T(°F) = 2.32x_2 + 48.1 \qquad x_2 = \text{flashes/min}$$

Estimate the temperature in °R for 132 chirps/min; 11 flashes/min; the average of the two predictions.

1.67. Suppose that a submarine inadvertently sinks to the bottom of the ocean at a depth of 3340 ft. It is proposed to lower a diving bell to the submarine and attempt to enter the conning tower. What must the air pressure be in the diving bell at the level of the submarine in order to prevent water from entering into the bell when the opening valve at the bottom is cracked open slightly? Give your answer in absolute atmospheres. Assume that sea water has a constant density of 63.9 lb_m/ft^3.

1.68. What is the gauge pressure at a depth of 4.50 mi below the surface of the sea if the water temperature averages 60°F? Give your answer in lb (force) per sq. in. The sp gr of sea water at 60°F/60°F is 1.042 and is assumed to be independent of pressure.

1.69. Air in a scuba diver's tank shows a pressure of 44.0 psia. What is the pressure in (a) atm; (b) newtons/(meter)^2; and (c) in. Hg?

1.70. A pressure gauge on a welder's tank gives a reading of 22.4 psig. The barometric pressure is 28.6 in. Hg. Calculate the absolute pressure in the tank in (a) lb_f/ft^2; (b) in. Hg; (c) newtons/(meter)^2; and (d) ft water.

1.71. Air cannot be used for diving at depths of greater than 150 ft because of nitrogen's narcotic effects. Divers cite the "martini law": Every 50 ft of depth is equivalent to drinking one martini. A depth of 1000 ft is equivalent to how many (a) martinis, (b) lb_f/ft^2; (c) newtons/(meter)^2. Assume that the density of sea water is constant at 63.9 lb/ft^3.

1.72. A manometer uses kerosene, sp gr 0.82, as the fluid. A reading of 5 in. on the manometer is equivalent to how many mm Hg?

1.73. The pressure gauge on the steam condenser for a turbine indicates 26.2 in. Hg of vacuum. The barometer reading is 30.4 in. Hg. What is the pressure in the condenser in psia?

1.74. A pressure gauge on a process tower indicates a vacuum of 3.53 in. Hg. The barometer reads 29.31 in. Hg. What is the absolute pressure in the tower in millimeters of Hg?

1.75. The indicating liquid in the manometer shown in Fig. P1.75 is water, and the other liquid is benzene. These two liquids are essentially insoluble in each other. If the manometer reading is $\Delta Z = 14.3$ in. of water, what is the pressure difference in lb/in^2? The temperature is 70°F.

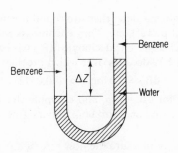

Fig. P1.75.

1.76. Water at 60.0°F is flowing through a pipeline, and the line contains an orifice meter for flow measurement. An orifice meter is a circular plate with a hole in the center that is placed in the pipe. The ends of a U-tube manometer 48.0 in. long are connected across the orifice meter. See Fig. P1.76. If the maximum expected pressure drop across the meter is 3.55 psi, what is the maximum sp gr at 60°F/60°F which the manometer liquid can have when used with this equipment if the liquid is to stay in the manometer?

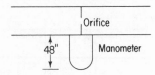

Fig. P1.76.

1.77. Examine Fig. P1.77. The barometric pressure is 720 mm Hg. The density of the oil is 0.80 g/cm³. The Bourdon gauge reads 33.1 psig. What is the pressure (psia) of the gas?

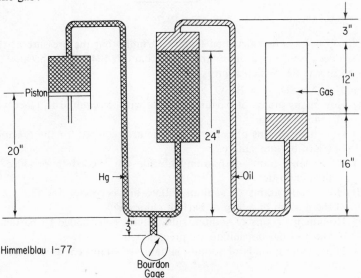

Himmelblau 1-77

Fig. P1.77.

1.78.* Logarithmic scales are not new; they are found for hydrogen ion concentration (pH), decibels (signal power), etc. They encompass wide ranges of a variable on a reasonably compact scale. The deciboyle (dB_o) has been proposed as a pressure unit to encompass the 30 decades over which pressure measurements take place,

$$dB_o = 10 \log_{10} p + 0.0572$$

where p is in atm. Plot dB_o vs. p, and compute the values of deciboyles for a high-vacuum, 10^{-14} atm, and for 60 kilobars (the pressure used in making diamonds).

1.79. (a) An orifice is used to measure the flow rate of a gas through a pipe as shown in Fig. P1.79. The pressure drop across the orifice is measured with a mercury manometer, both legs of which are constructed of $\frac{1}{4}$-in. inner diameter (ID) glass tubing. If the pressure drop across the orifice is equivalent to 4.65 in. Hg, calculate h_2 and h_3 (both in inches) if h_1 is equal to 13.50 in.

     (b) The right glass leg of the manometer in Fig. P1.79 becomes dirty and is replaced with glass tubing which is $\frac{3}{8}$-in. ID. The manometer is filled with the same volume of mercury as in part (a). For the same pressure drop as in part (a), calculate h_2 and h_3 (both in inches).

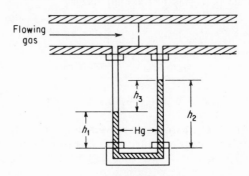

Fig. P1.79.

1.80. John Long says he calculated from a formula that the pressure at the top of Pikes Peak is 9.75 psia. John Green says that it is 504 mm Hg because he looked it up in a table. Which John is right?

1.81. $BaCl_2 + Na_2SO_4 \longrightarrow BaSO_4 + 2NaCl$

     (a) How many grams of barium chloride will be required to react with 5.00 g of sodium sulfate?

     (b) How many grams of barium chloride are required for the precipitation of 5.00 g of barium sulfate?

     (c) How many grams of barium chloride are needed to produce 5.00 g of sodium chloride?

     (d) How many grams of sodium sulfate are necessary for the precipitation of the barium of 5.00 g of barium chloride?

     (e) How many grams of sodium sulfate have been added to barium chloride if 5.00 g of barium sulfate are precipitated?

     (f) How many pounds of sodium sulfate are equivalent to 5.00 lb of sodium chloride?

(g) How many pounds of barium sulfate are precipitated by 5.00 lb of barium chloride?

(h) How many pounds of barium sulfate are precipitated by 5.00 lb of sodium sulfate?

(i) How many pounds of barium sulfate are equivalent to 5.00 lb of sodium chloride?

1.82. $AgNO_3 + NaCl \longrightarrow AgCl + NaNO_3$

(a) How many grams of silver nitrate will be required to react with 5.00 g of sodium chloride?

(b) How many grams of silver nitrate are required for the precipitation of 5.00 g of silver chloride?

(c) How many grams of silver nitrate are equivalent to 5.00 g of sodium nitrate?

(d) How many grams of sodium chloride are necessary for the precipitation of the silver of 5.00 g of silver nitrate?

(e) How many grams of sodium chloride have been added to silver nitrate if 5.00 g of silver chloride are precipitated?

(f) How many pounds of sodium chloride are equivalent to 5.00 lb of sodium nitrate?

(g) How many pounds of silver chloride are precipitated by 5.00 lb of silver nitrate?

(h) How many pounds of silver chloride are precipitated by 5.00 lb of sodium chloride?

(i) How many pounds of silver chloride are equivalent to 5.00 lb of silver nitrate?

1.83. How many grams of chromic sulfide will be formed from 0.718 g of chromic oxide according to the equation

$$2Cr_2O_3 + 3CS_2 \longrightarrow 2Cr_2S_3 + 3CO_2$$

1.84. What weight of Cu_2O would be required to furnish 40 lb of Cu?

1.85. Removal of CO_2 from a manned spacecraft has been accomplished by absorption with lithium hydroxide according to the following reaction:

$$2LiOH(s) + CO_2(g) \longrightarrow Li_2CO_3(s) + H_2O(l)$$

(a) If 2.20 lb of CO_2 are released per day per man, how many pounds of LiOH are required per day per man?

(b) What is the penalty, i.e., the percentage increase, in weight if the cheaper NaOH is substituted for LiOH?

1.86. The Gemini and Apollo spacecrafts carried oxygen as a supercritical fluid. However, for flights of long duration it will be necessary to regenerate oxygen from waste products. Most of the proposed systems are based on the reduction of carbon dioxide with direct or subsequent recovery of the oxygen contained in the carbon dioxide. The processes that appear most promising involve the reduction of carbon dioxide with molecular hydrogen to form water, which is subsequently electrolyzed to produce oxygen. One variation of this process is known as the Sabatier or methanization reaction. The equation for this reaction is

$$CO_2(g) + 4H_2(g) \xrightarrow[200°-260C]{Ni} CH_4(g) + 2H_2O(g)$$

At a temperature of 200°C, essentially all the carbon dioxide is converted. The methane from this reaction can be pyrolyzed, and the resulting hydrogen can be recovered and then recycled:

$$CH_4(g) \longrightarrow C(s) + 2H_2(g)$$

The water from the methanization reaction can be electrolyzed to yield additional hydrogen and the desired oxygen:

$$2H_2O(aq) \xrightarrow{e^-} 2H_2(g) + O_2(g)$$

(a) How many grams of $C(s)$ are formed per gram of $CO_2(g)$?
(b) How many moles of $CH_4(g)$ are formed per gram of $O_2(g)$ recovered?
(c) How many pounds of water as a liquid serve as an intermediate per pound of $CO_2(g)$ converted?

1.87. Some states and cities are placing a ban on laundry detergents containing more than 20 percent phosphates. In reading the detailed provisions of one law, it prohibits the sale of any detergent with a phosphorus pentoxide content of over 20 percent by weight. How much is this when expressed as weight percent sodium triphosphate, the compound found in the detergent?

1.88. Suppose that the following reaction is carried out:

$$2NaOH + H_2SO_4 \longrightarrow Na_2SO_4 + 2H_2O$$

How many pound moles of H_2O will result from the complete reaction of 156 lb of NaOH with the H_2SO_4?

1.89. When sodium nitrate is treated with sulfuric acid, the two most important reactions which occur are

(1) $2NaNO_3 + H_2SO_4 \longrightarrow Na_2SO_4 + 2HNO_3$

(2) $NaNO_3 + H_2SO_4 \longrightarrow NaHSO_4 + HNO_3$

For a given weight of sulfuric acid, which reaction yields the most nitric acid? Which reaction would be used commercially? Explain, with reference to your source of information.

1.90. What is the difference in cost per pound, as a source of HCN, between KCN (98 percent) at 50 cents/lb and a mixture of KCN (65 percent) and NaCN (25 percent) at 60 cents/lb (for the mixture)?

1.91. To neutralize 5 gal of waste iron pickle liquor, 1.25 lb of sodium hydroxide is used. How many pounds of slaked lime, 94 percent $Ca(OH)_2$, are required for 1000 gal of the same liquor?

1.92. Sulfuric acid can be manufactured by the contact process according to the following reactions:

(1) $S + O_2 \longrightarrow SO_2$

(2) $2SO_2 + O_2 \longrightarrow 2SO_3$

(3) $SO_3 + H_2O \longrightarrow H_2SO_4$

You are asked as part of the preliminary design of a sulfuric acid plant with a design capacity of 2000 tons/day of 66° Bé (Baumé) (93.2 percent H_2SO_4 by weight) to calculate the following:

(a) How many tons of pure sulfur are required per day to run this plant?

(b) How many tons of oxygen are required per day?

(c) How many tons of water are required per day for reaction (3)?

1.93. Sulfuric acid production plants add to the already growing problem of atmospheric pollution with sulfur dioxide and sulfur trioxide released from the processing. However, sulfuric acid production is one way of ameliorating the problem of what to do with the sulfur dioxide released in the winning of metals from their ores. For example, sulfuric acid can be produced as a by-product in the manufacture of zinc from sulfide ores. The ore analyzes 65.0 percent ZnS and 35.0 percent inert impurities by weight. The ore is burned in a furnace, and the resulting SO_2 is converted to SO_3 in a catalytic reactor. The SO_3 is then absorbed by water to give sulfuric acid, the final product being 2.0 percent water and 98.0 percent H_2SO_4 by weight. A total of 99.0 percent of the sulfur in the ore is recovered in the acid. The chemical reactions are

$$ZnS + 1.5O_2 \longrightarrow ZnO + SO_2$$

$$SO_2 + 0.5O_2 \longrightarrow SO_3$$

$$SO_3 + H_2O \longrightarrow H_2SO_4$$

(a) Calculate the weight of product sulfuric acid (98.0 percent H_2SO_4 by weight) produced in a zinc plant which processes 186 tons of ore per day.

(b) How much water is required per day?

1.94. In the manufacture of chlorine gas, a solution of sodium chloride in water is electrolyzed in specially designed cells. The overall reaction is

$$2NaCl + 2H_2O \longrightarrow 2NaOH + H_2 + Cl_2$$

At the end of one batch run, the sodium hydroxide in a particular cell was 10.4 percent by weight, and the cell contained 20,000 lb of solution (to the nearest 100 lb). How many pounds of NaCl were required to produce 1850 lb of Cl_2?

1.95. Oxygen can be produced by heating potassium chlorate, which costs 12 cents/lb, or potassium nitrate, which costs 8 cents/lb. Which one produces oxygen at the lower cost?

$$2KClO_3 \longrightarrow 2KCl + 3O_2$$

$$2KNO_3 \longrightarrow 2KNO_2 + O_2$$

1.96. A compound whose molecular weight is 119 analyzes as follows:

C	70.6%
H	4.2
N	11.8
O	13.4

What is the formula?

1.97. A compound whose molecular weight is 103 analyzes as follows:

C	81.5%
H	4.9
N	13.6

What is the formula?

1.98. Seawater contains 65 ppm of bromine in the form of bromides. In the Ethyl-Dow recovery process, 0.27 lb of 98 percent sulfuric acid is added per ton of water, together with the theoretical Cl_2 for oxidation; finally, ethylene (C_2H_4) is united with the bromine to form $C_2H_4Br_2$. Assuming complete recovery and using a basis of 1 lb bromine, find the weights of acid, chlorine, seawater, and dibromide involved.

$$2Br^- + Cl_2 \longrightarrow 2Cl^- + Br_2$$

$$Br_2 + C_2H_4 \longrightarrow C_2H_4Br_2$$

1.99. Ethane, C_2H_6, reacts with pure oxygen to form water and CO_2. If 3 lb moles of C_2H_6 are mixed with 12 lb moles of oxygen and 80 percent of the ethane reacts,
   (a) How many pound moles of C_2H_6 are present in the final mixture?
   (b) How many pound moles of O_2 are present in the final mixture?
   (c) How many pound moles of H_2O are present in the final mixture?
   (d) How many pounds of total final mixture are there?
   (e) Which reactant component, C_2H_6 or O_2, is in excess, and what is the percentage excess over the theoretical requirement for complete combustion?

1.100. A barytes composed of 100 percent $BaSO_4$ is fused with carbon in the form of coke containing 6 percent ash (which is infusible). The composition of the fusion mass is

$BaSO_4$	11.1%
BaS	72.8
C	13.9
Ash	2.2
	100.0%

Reaction:

$$BaSO_4 + 4C \longrightarrow BaS + 4CO$$

Find the excess reactant, the percentage of the excess reactant, and the degree of completion of the reaction.

1.101. One can view the blast furnace from a simple viewpoint as a process in which the principal reaction is

$$Fe_2O_3 + 3C \longrightarrow 2Fe + 3CO$$

but some other undesired side reactions occur, mainly

$$Fe_2O_3 + C \longrightarrow 2FeO + CO$$

After mixing 600.0 lb of carbon (coke) with 1.00 ton of pure iron oxide, Fe_2O_3, the process produces 1200.0 lb of pure iron, 183 lb of FeO, and 85.0 lb of Fe_2O_3. Calculate the following items:
   (a) The percentage of excess carbon furnished, based on the principal reaction.
   (b) The percentage conversion of Fe_2O_3 to Fe.
   (c) The pounds of carbon used up and the pounds of CO produced per ton of Fe_2O_3 charged.

1.102. Aluminum sulfate is used in water treatment and in many chemical processes. It can be made by reacting crushed bauxite with 77.7 weight percent sulfuric acid. The bauxite ore contains 55.4 weight percent aluminum oxide, the remainder being impurities. To produce crude aluminum sulfate containing 2000 lb of pure aluminum sulfate, 1080 lb of bauxite and 2510 lb of sulfuric acid solution (77.7 percent acid) are used.

(a) Identify the excess reactant.

(b) What percentage of the excess reactant was used?

(c) What was the degree of completion of the reaction?

Molecular weights:

$$Al_2O_3 = 101.9$$

$$H_2SO_4 = 98.1$$

$$Al_2(SO_4)_3 = 342.1$$

1.103. Phosgene gas is probably most famous for being the first toxic gas used offensively in World War I, but it is also used extensively in the chemical processing of a wide variety of materials. Phosgene can be made by the catalytic reaction between CO and chlorine gas in the presence of a carbon catalyst. The chemical reaction is

$$CO + Cl_2 \longrightarrow COCl_2$$

Suppose that you have measured the reaction products from a given reactor and found that they contained 3.00 lb moles of chlorine, 10.00 lb moles of phosgene, and 7.00 lb moles CO. Calculate the following:

(a) The percentage of excess reactant used.

(b) The percentage conversion of the limiting reactant.

(c) The lb moles of phosgene formed per lb mole of total reactants fed to the reactor.

1.104. Diborane, B_2H_6, a possible propellant for solid fuel rockets, can be made in a variety of different ways. One of the simplest, but not the cheapest, is to use lithium hydride as follows:

$$6LiH + 2BCl_3 \longrightarrow B_2H_6 + 6LiCl$$

Suppose that 200 lb of LiH are mixed with 1000 lb of BCl_3, and 45.0 lb of B_2H_6 are recovered. Determine the following:

(a) The percentage conversion of LiH to B_2H_6 (based on the BCl_3 that actually reacts).

(b) The yield of B_2H_6 based on the LiH charged.

(c) The pounds of LiCl formed.

PROBLEMS TO PROGRAM ON THE COMPUTER

1.1. Write a program to convert °C into °K, °F, and °R at 1°C intervals. Have the program arranged so that you can do the following:

(a) Read in a starting value of °C.

(b) Read in a stopping value of °C.

(c) Use an IF loop to convert °C to the other temperature scales at 1°C intervals between the starting and stopping values, inclusive.

(d) Print out the four temperatures for each conversion with appropriate column headings.

(e) Use comment lines where appropriate to indicate your procedure in conversion.

1.2. Write a computer program to convert millimeters of Hg into inches of Hg, feet of H_2O, and psia at 2 mm Hg (even-numbered) increments below 760 mm Hg. Have the computer arranged so that you can do the following:

(a) Read in a starting value for the millimeters of Hg.

(b) Stop at 760 mm Hg (the last value to be converted).

(c) Read in the necessary conversion factors such as CPSIA in

$$\text{PSIA} = \text{CPSIA} * \text{HGMM} \quad \text{for psia} = 0.01933 \text{ mm Hg.}$$

(d) Use one-dimensional arrays to store values.

(e) Use a DO loop to compute all values. You must compute the number of iterations required with your program.

(f) Print out the results after all the conversions have been made.

    (1) Print your name at top of first page.

    (2) Print appropriate column headings at the top of each new page.

    (3) Print only 50 lines of conversion results per page.

1.3. Prepare a computer program to give the specific gravity of a fluid as a function of °Bé (Baumé) and °API over the interval 0°–30° in 1° increments. Equation (1.10) gives the relation between sp gr and °API (for liquids less dense than water), while the relationships for °Bé are

liquids more dense than water

$$\text{sp gr} = \frac{145.0}{145.0 - \text{°Bé}}$$

liquids less dense than water

$$\text{sp gr} = \frac{140.0}{130.0 + \text{°Bé}}$$

Read the °API or °Bé into the program as data.

Material Balances **2**

Conservation laws occupy a special place in science and engineering. Common statements of these laws take the form of "mass (energy) is neither created nor destroyed," "the mass (energy) of the universe is constant," "the mass (energy) of any isolated system is constant," or equivalent statements. To refute a conservation law, it would be sufficient to find just one example of a violation. However, Lavoisier and many of the scientists who followed in his path studied chemical changes quantitatively and found invariably that the sum of the weights of the substances entering into a reaction equaled the sum of the weights of the products of the reaction. Thus man's collective experience has been summed up and generalized as the law of the conservation of matter. The only "proof" we have for this law is of a negative type—never in our previous experience has any process been observed that upon thorough examination (including consideration of the precision of the data) has been found to contradict the principle. Of course, we must exclude processes involving nuclear transformations, or else extend our law to include the conservation of both energy and matter. (Insofar as we shall be concerned here, the word *process* will be taken to mean a series of physical operations on or physical or chemical changes in some specified material.)

In this chapter we shall discuss the principle of the conservation of matter and how it can be applied to engineering calculations, making use of the background information discussed in Chap. 1. Figure 2.0 shows the relations between the topics discussed in this chapter and the general objective of making material and energy balances. In approaching the solution of material balance problems, we shall first consider how to analyze them in order to clarify the method and

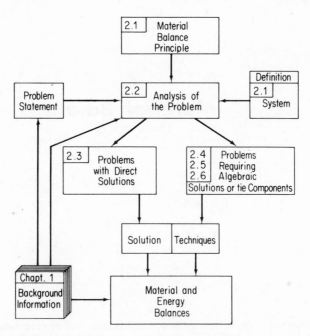

Fig. 2.0. Heirarchy of topics to be studied in this chapter (section numbers are in the upper left-hand corner of the boxes).

the procedure of solution. The aim will be to help you acquire a generalized approach to problem solving so that you may avoid looking upon each new problem, unit operation, or process as entirely new and unrelated to anything you have seen before. As you scrutinize the examples used to illustrate the principles involved in each section, explore the method of analysis, but avoid memorizing each example by rote, because, after all, they are only samples of the myriad problems which exist or could be devised on the subject of material balances. Most of the principles we shall consider are of about the same degree of complexity as the law of compensation devised by some unknown, self-made philosopher who said, "Things are generally made even somewhere or some place. Rain always is followed by a dry spell, and dry weather follows rain. I have found it an invariable rule that when a man has one short leg, the other is always longer!"

In working these problems you will find it necessary to employ some engineering judgment. You think of mathematics as an exact science. For instance, suppose that it takes 1 man 10 days to build a brick wall; then 10 men can finish it in 1 day. Therefore 240 men can finish the wali in 1 hr, 14,400 can do the job in 1 min, and with 864,000 men the wall will be up before a single brick is in place! Your password to success is the famous IBM motto: THINK.

2.1 Material balance

To take into account the flow of material in and out of a system, the generalized law of the conservation of mass is expressed as a material balance. A *material balance* is nothing more than an accounting for mass flows and changes in inventory of mass for a system. Equation (2.1) describes in words the principle of the material balance applicable to processes both with and without chemical reaction:

$$
\begin{Bmatrix} \text{accumulation} \\ \text{within} \\ \text{the} \\ \text{system} \end{Bmatrix} = \begin{Bmatrix} \text{input} \\ \text{through} \\ \text{system} \\ \text{boundaries} \end{Bmatrix} - \begin{Bmatrix} \text{output} \\ \text{through} \\ \text{system} \\ \text{boundaries} \end{Bmatrix} + \begin{Bmatrix} \text{generation} \\ \text{within} \\ \text{the} \\ \text{system} \end{Bmatrix} - \begin{Bmatrix} \text{consumption} \\ \text{within} \\ \text{the} \\ \text{system} \end{Bmatrix}
$$

$$(2.1)$$

In Eq. (2.1) the generation and consumption terms in this text refer to gain or loss by chemical reaction. The accumulation may be positive or negative. You should understand in reading the words in Eq. (2.1) that the equation refers to a time interval of any desired length, including a year, hour, or second or a differential time.

Equation (2.1) reduces to Eq. (2.2) for cases in which there is no generation (or usage) of material within the system,

$$\text{accumulation} = \text{input} - \text{output} \qquad (2.2)$$

and reduces further to Eq. (2.3) when there is, in addition, no accumulation within the system,

$$\text{input} = \text{output} \qquad (2.3)$$

If there is no flow in and out of the system and no generation (or usage), Eq. (2.1) reduces to the basic concept of the conservation of one species of matter within an enclosed isolated system.

Inherent in the formulation of each of the above balances is the concept of a system for which the balance is made. By *system* we mean any arbitrary portion or whole of a process as set out specifically by the engineer for analysis. Figure 2.1 shows a system in which flow and reaction take place; note particularly that

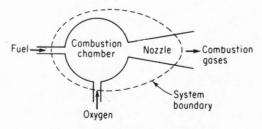

Fig. 2.1. A flow system with combustion.

the system boundary is formally circumscribed about the process itself to call attention to the importance of carefully delineating the system in each problem you work. In nearly all of the problems in this chapter the mass accumulation term will be zero; that is, primarily *steady-state* problems will be considered. (See Chap. 6 for cases in which the mass accumulation is not zero.) For a steady-state process we can say, "What goes in must come out." Illustrations of this principle can be found below.

Material balances can be made for a wide variety of materials, at many scales of size for the system and in various degrees of complication. To obtain a true perspective as to the scope of material balances, examine Figs. 2.2 and 2.3.

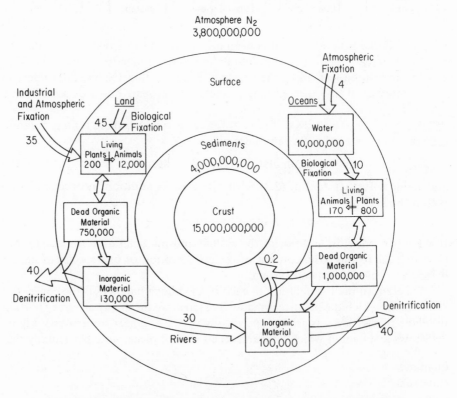

Fig. 2.2. Distribution and annual rates of transfer of nitrogen in the biosphere (in millions of metric tons).

Figure 2.2 delineates the distribution of nitrogen in the biosphere as well as the annual transfer rates, both in millions of metric tons. The two quantities known with high confidence are the amount of nitrogen in the atmosphere and the rate of industrial fixation. The apparent precision of the other figures reflects chiefly an effort to preserve indicated or probable ratios among different inventories. Because of the extensive use of industrially fixed nitrogen, the amount of nitro-

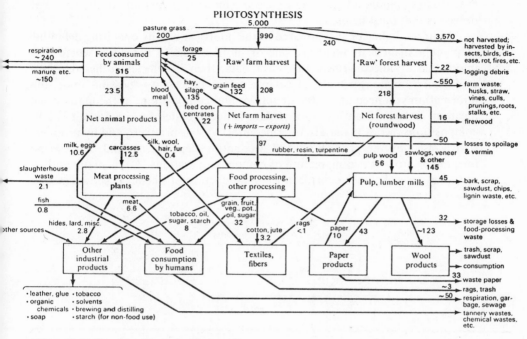

All figures refer to millions of tons of dry organic matter per year.

Fig. 2.3. Production and disposal of the products of photosynthesis. (Taken from A. V. Kneese, R. U. Ayres, and R. C. d'Arge, *Economics and the Environment*, Resources for the Future, Inc., 1755 Massachusetts Ave., Washington, D.C., 1970, p. 32.)

gen available to land plants may significantly exceed the nitrogen returned to the atmosphere by denitrifying bacteria in the soil. Figure 2.3 resulted from an analysis of the organic wastes arising from the processing of food and forest products. Changes in any of the streams can pinpoint sources of improvement or deterioration in environmental quality.

We should also note in passing that balances can be made on many other quantities in addition to mass. Balances on dollars are common (your bank statement, for example) as are balances on the number of entities, as in traffic counts, population balances, and social services.

In the process industries, material balances assist in the planning for process design, in the economic evaluation of proposed and existing processes, in process control, and in process optimization. For example, in the extraction of soybean oil from soybeans, you could calculate the amount of solvent required per ton of soybeans or the time needed to fill up the filter press, and use this information in the design of equipment or in the evaluation of the economics of the process. All sorts of raw materials can be used to produce the same end product and quite a few different types of processing can achieve the same end result so that case studies (simulations) of the processes can assist materially in the financial decisions that must be made.

Material balances also are used in the hourly and daily operating decisions of plant managers. If there are one or more points in a process where it is impossible or uneconomical to collect data, then if sufficient other data are available, by making a material balance on the process it is possible to get the information you need about the quantities and compositions at the inaccessible location. In most plants a mass of data is accumulated on the quantities and compositions of raw materials, intermediates, wastes, products, and by-products that is used by the production and accounting departments and that can be integrated into a revealing picture of company operations. A schematic outline of the material balance control in the manufacture of phenol is shown in Fig. 2.4.

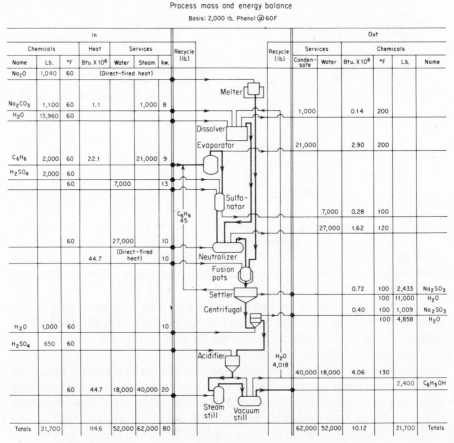

Fig. 2.4. Material (and energy) balances in the manufacture of phenol presented in the form of a ledger sheet. Taken from *Chem. Eng.*, p. 177 (April 1961), by permission.

EXAMPLE 2.1   Water balance for a river basin

Water balances on river basins for a season or for a year can be used to check predicted ground water infiltration, evaporation, or precipitation in the basin. Prepare a water balance, in symbols, for a large river basin, including the physical processes indicated in Fig. E2.1 (all symbols are for 1 yr).

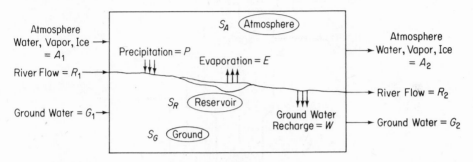

Fig. E2.1.

Solution:

Equation (2.2) applies in as much as there is no reaction in the system. Each term in Eq. (2.2) can be represented by symbols defined in Fig. E2.1. Let the subscript t_2 designate the end of the year and t_1 designate the beginning of the year. If the system is chosen to be the atmosphere plus the river plus the ground, then the accumulation is

$$(S_{At_2} - S_{At_1}) + (S_{Rt_2} - S_{Rt_1}) + (S_{Gt_2} - S_{Gt_1})$$

The inputs are $A_1 + R_1 + G_1$ and the outputs are $A_2 + R_2 + G_2$. Consequently, the material balance is

(a)
$$(S_{At_2} - S_{At_1}) + (S_{Rt_2} - S_{Rt_1}) + (S_{Gt_2} - S_{Gt_1})$$
$$= (A_1 - A_2) + (R_1 - R_2) + (G_1 - G_2)$$

If the system is chosen to be just the river (including the reservoir), then the material balance would be

(b)
$$\underset{\text{accumulation}}{(S_{Rt_2} - S_{Rt_1})} = \underset{\text{input}}{(R_1 + P)} - \underset{\text{output}}{(R_2 + E + W)}$$

and analogous balances could be made for the water in the atmosphere and in the ground.

One important point to always keep in mind is that the material balance is a balance on mass, not on volume or moles. Thus, in a process in which a chemical reaction takes place so that compounds are generated and/or used up, you will have to employ the principles of stoichiometry discussed in Chap. 1 in making a material balance.

Let us analyze first some very simple examples in the use of the material balance.

EXAMPLE 2.2 Material balance

Hydrogenation of coal to give hydrocarbon gases is one method of obtaining gaseous fuels with sufficient energy content for the future. Figure E2.2 shows how a free-fall fluid bed reactor can be set up to give a product gas of high methane content.

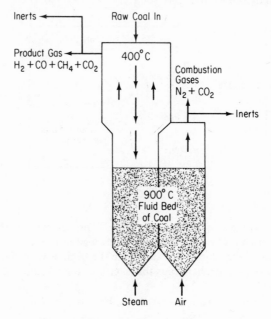

Fig. E2.2.

Suppose, first, that the gasification unit is operated without steam at room temperature (25°C) to check the air flow rates and that cyclones separate the solids from the gases effectively at the top of the unit so that no accumulation (build up) of coal, or air, occurs in the unit. If 1200 kg of coal per hour (assume that the coal is 80 percent C, 10 percent H, and 10 percent inert material) is dropped through the top of the reactor,

(a) How many kilograms of coal leave the reactor per hour?
(b) If 15,000 kg of air per hour is blown into the reactor, how many kilograms of air per hour leave the reactor?
(c) If the reactor operates at the temperatures shown in Fig. E2.2 and with the addition of 2000 kg of steam (H_2O vapor) per hour, how many kilograms per hour of combustion and product gases leave the reactor per hour.

Solution:

Basis: 1 hr

(a) Since 1200 kg/hr of coal enter the reactor, and none remains inside, 1200 kg/hr must leave the reactor.
(b) Similarly, 15,000 kg/hr of air must leave the reactor.

(c) All the material except the inert portion of the coal leaves as a gas. Consequently, we can add up the total mass of material entering the unit, subtract the inert material, and obtain the mass of combustion gases by difference:

$$\frac{1200 \text{ kg coal}}{} \left| \frac{10 \text{ kg inert}}{100 \text{ kg coal}} \right. = 120 \text{ kg inert}$$

entering material	*kg*
Coal	1,200
Air	15,000
Steam	2,000
Total	18,200 − 120 = 18,080 kg/hr of gases

EXAMPLE 2.3 Material balance

(a) If 300 lb of air and 24.0 lb of carbon are placed in a reactor (see Fig. E2.3) at 600°F and after complete combustion no material remains in the reactor, how many pounds of carbon will come out? How many pounds of oxygen? How many pounds total?

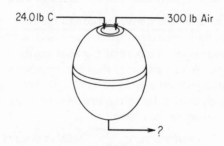

24.0 lb C ———————— 300 lb Air

Fig. E2.3.

(b) How many moles of carbon and oxygen enter? How many leave the reactor?

(c) How many total moles enter the reactor and how many leave the reactor?

Solution:

(a) The total material leaving the reactor would be 324 lb. Since air has 21.0 percent O_2 and 79.0 percent N_2,

Basis: 300 lb air

$$\frac{300 \text{ lb air}}{} \left| \frac{1 \text{ lb mole air}}{29.0 \text{ lb air}} \right| \frac{21.0 \text{ lb mole } O_2}{100 \text{ lb mole air}} = 2.18 \text{ lb mole } O_2$$

$$\frac{24 \text{ lb C}}{} \left| \frac{1 \text{ lb mole C}}{12.0 \text{ lb C}} \right. = 2.00 \text{ lb mole C}$$

The oxygen required to completely burn 24.0 lb of C is 2.00 lb mole; consequently

$$2.18 - 2.00 = 0.18$$

is the number of moles of unused oxygen, and this is equivalent to $(0.18)(32) = 5.76$ lb of O_2 as such leaving the reactor. In addition, 88 lb of CO_2 leave and

$$\frac{2.18 \text{ lb mole } O_2 \mid 79.0 \text{ lb mole } N_2 \mid 28.2 \text{ lb } N_2}{\mid 21.0 \text{ lb mole } O_2 \mid 1 \text{ lb mole } N_2} = 230 \text{ lb } N_2$$

The moles of N_2 are

$$\frac{230 \text{ lb } N_2 \mid 1 \text{ lb mole } N_2}{\mid 28.2 \text{ lb } N_2} = 8.20 \text{ lb mole } N_2$$

Summing up all input and exit streams, we have

in	lb	lb mole	out	lb	lb mole
O_2	70	2.18	O_2	5.76	0.18
N_2	230	8.20	N_2	230	8.20
C	24	2.00	CO_2	88	2.00
Total	324	12.38		324	10.38

There will be no carbon leaving as carbon because all the carbon burns to CO_2; the carbon leaves in a combined form with some of the oxygen.

    (b) The moles of C and O_2 entering the reactor are shown in part (a); no moles of C, as such, leave.

    (c) 12.38 total moles enter the reactor and 10.38 total moles leave.

Note that in the example above, although the total pounds put into a process and the total pounds recovered from a process have been shown to be equal, there is no such equality on the part of the *total* moles in and out, *if* a chemical reaction takes place. What is true is that the number of atoms of an element (such as C, O, or even oxygen expressed as O_2) put into a process must equal the atoms of the same element leaving the process. In Example 2.3 the total moles in and out are shown to be unequal:

	total moles in		total moles out
O_2	2.18	O_2	0.18
N_2	8.20	N_2	8.20
C	2.00	CO_2	2.00
	12.38		10.38

although the atoms of O in (expressed as moles of O_2) equal the atoms of O (similarly expressed as O_2) leaving:

	component moles in			component moles out	
O_2:	O_2		2.18	O_2	0.18
				O_2 in CO_2	2.00
	Total = O_2		2.18	Total = O_2	2.18
N_2:	N_2		8.20	N_2	8.20
C:	C		2.00	C in CO_2	2.00

Keeping in mind the above remarks for processes involving chemical reactions, we can summarize the circumstances under which the input equals the output for *steady-state processes* (no accumulation) as follows:

	type of balance	Equality required for input and output of steady-state process without chemical reaction	with chemical reaction
total	Total mass	Yes	Yes
balances	Total moles	Yes	No*
	Mass of a pure compound	Yes	No*
component	Moles of a pure compound	Yes	No*
balances	Mass of an atomic species	Yes	Yes
	Moles of an atomic species	Yes	Yes

(The asterisk indicates that equality may occur by chance.)

We now turn to consideration of problems in which the accumulation term is zero. The basic task is to turn the problem, expressed in words, into a quantitative form, expressed in mathematical symbols and numbers, and then solve the mathematical statements.

2.2 Program of analysis of material balance problems

Much of the remaining portion of this chapter demonstrates the techniques of setting up and solving problems involving material balances. Later portions of the text consider combined material and energy balances. Since these material balance problems all involve the same principle, although the details of the applications of the principle may be slightly different, we shall first consider a generalized method of analyzing such problems which can be applied to the solution of any type of material balance problem.

We are going to discuss a method of analysis of material balance problems that will enable you to understand, first, how similar these problems are, and second, how to solve them in the most expeditious manner. For some types of problems the method of approach is relatively simple and for others it is more complicated, but the important point is to regard problems in distillation, crystallization, evaporation, combustion, mixing, gas absorption, or drying not as being different from each other but as being related from the viewpoint of how to proceed to solve them.

An orderly method of analyzing problems and presenting their solutions represents training in logical thinking that is of considerably greater value than mere knowledge of how to solve a particular type of problem. Understanding how to approach these problems from a logical viewpoint will help you to develop those fundamentals of thinking that will assist you in your work as an engineer long after you have read this material.

If you want to make a material balance for a system, in general you have to have on hand information dealing with *two* fundamental concepts. One of these is the mass (weight) in all streams of the material entering and leaving the system and present in the system. The other information required is the composition of all the streams entering and leaving the system and the composition of the material in the system. Of course, if a chemical reaction takes place inside the system, the equation for the reaction and/or the extent of the reaction are important pieces of information.

Let us look at some problems in general by means of the "black box" technique. All we do is draw a black box or line around the process and consider what is going into the process and what comes out of the process. This procedure defines the system, i.e., the process or body of matter set out to be analyzed. As previously mentioned, we shall assume that the process is taking place in the steady state, i.e., that there is no accumulation or depletion of material. Even if the process is of the batch type in which there is no flow in and out, let us pretend that the initial material is pushed into the black box and that the final material is removed from the black box. By this hypothesis we can imagine a batch process converted into a fictitious flow process, and talk about the "streams" entering and leaving even though in the real process nothing of the sort happens (except over the entire period of time under consideration). After you do this three or four times, you will see that it is a very convenient technique, although at the beginning it may seem a bit illogical.

Let us start with a box with just three streams entering or leaving (all three cannot enter or leave, of course, in a steady-state process). In the examples in Figs. 2.5 and 2.6, F stands for the feed stream, P stands for product, and W stands for the third stream or "change," but any other symbols would do as well. The procedure of solution depends to some extent on which of the compositions and weights you know. All the problems in Figs. 2.5 and 2.6 have sufficient information shown to effect their solutions, except Fig. 2.5(*d*).

With respect to Fig. 2.5(*a*) we could use a total mass balance to compute the value of W:

$$F = P + W$$
$$100 = 60 + W \quad \text{or} \quad W = 40 \tag{2.4}$$

To determine the composition of W, because no reaction takes place, we can write a mass balance for each of the components: EtOH, H_2O, MeOH [$\omega =$ mass ("weight") fraction].

$$
\begin{array}{ccc}
\textit{balance in} & = & \textit{out} \\
\omega_F F & = & \omega_P P + \omega_W W
\end{array}
$$

$$
\begin{array}{llcr}
\text{EtOH} & (0.50)(100) = (0.80)(60) + \omega_{\text{EtOH},W}(40) & & (2.4a) \\
\text{H}_2\text{O} & (0.40)(100) = (0.05)(60) + \omega_{\text{H}_2\text{O},W}(40) & & (2.4b) \\
\text{MeOH} & (0.10)(100) = (0.15)(60) + \omega_{\text{MeOH},W}(40) & & (2.4c)
\end{array}
$$

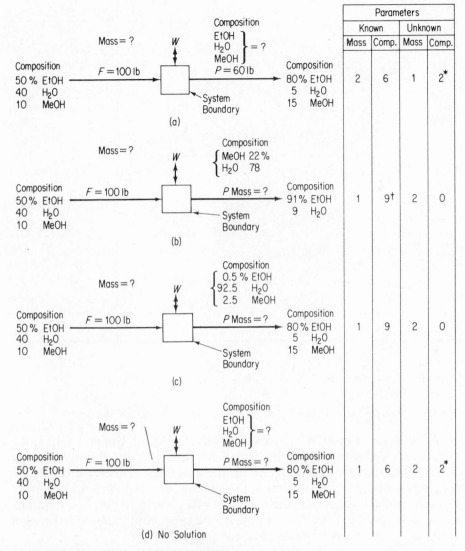

	Parameters			
	Known		Unknown	
	Mass	Comp.	Mass	Comp.
(a)	2	6	1	2*
(b)	1	9†	2	0
(c)	1	9	2	0
(d)	1	6	2	2*

* The third percent can be obtained by difference.

† A stream with only two components has 0% for the third component.

Fig. 2.5. Typical material balance problems (no chemical reaction involved).

Because

$$\omega_{\text{EtOH}} + \omega_{\text{H}_2\text{O}} + \omega_{\text{MeOH}} = 1 \qquad (2.5)$$

by definition, only two of the mass fractions are unknown; the third can be computed by difference. Consequently, only two of the component material balances have to be solved simultaneously, and in this particular instance the equations are not coupled so that they can be solved independently.

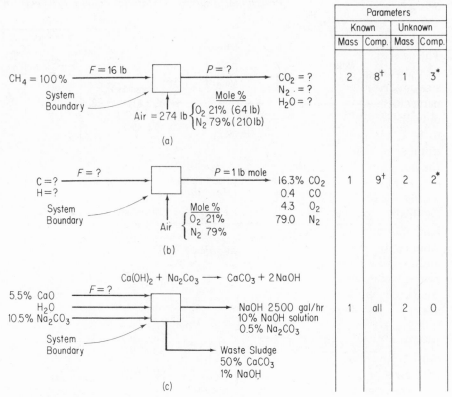

	Parameters			
	Known		Unknown	
	Mass	Comp.	Mass	Comp.

For figure (a):

$CH_4 = 100\%$ $F = 16$ lb $P = ?$

System Boundary

Air $= 274$ lb $\begin{cases} O_2\ 21\%\ (64\ lb) \\ N_2\ 79\%\ (210\ lb) \end{cases}$

Mole %

$CO_2 = ?$
$N_2\ \cdot = ?$
$H_2O = ?$

| 2 | 8† | 1 | 3* |

(a)

For figure (b):

$C = ?$
$H = ?$ $F = ?$ $P = 1$ lb mole

System Boundary

Air $\begin{cases} O_2\ 21\% \\ N_2\ 79\% \end{cases}$ Mole %

16.3% CO_2
0.4   CO
4.3 O_2
79.0 N_2

| 1 | 9† | 2 | 2* |

(b)

For figure (c):

$Ca(OH)_2 + Na_2CO_3 \longrightarrow CaCO_3 + 2NaOH$

5.5% CaO
H_2O $F = ?$
10.5% Na_2CO_3

System Boundary

NaOH 2500 gal/hr
10% NaOH solution
0.5% Na_2CO_3

Waste Sludge
50% $CaCO_3$
1% NaOH

| 1 | all | 2 | 0 |

(c)

*The third percent can be obtained by difference.
†Components not listed are present in 0%.

Fig. 2.6. Typical material balance problems (with chemical reaction).

You should recognize that not all of the four mass balances, Eq. (2.4)–(2.4c), are independent equations. Note how the sum of the three component balances (2.4a), (2.4b), and (2.4c) equals the total mass balance. Consequently, the number of degrees of freedom or the number of independent equations will be equal to the number of components. In many problems you can substitute the total material balance for any one of the component material balances if you plan to solve two or more equations simultaneously.

Since many of the problems you will encounter involving the use of material balances require you to solve for more than one unknown, you should remember that for each unknown you need to have at least one independent material balance or other independent piece of information. Otherwise, the problem is indeterminate. For example, a problem such as illustrated in Fig. 2.5(c) in which the compositions of all the streams are known and the weights of two streams are unknown requires two independent material balances for its solution. What to do when more, or fewer, independent material balance equations are available than unknowns is discussed in Sec. 2.4.

The columns on the right-hand side of Fig. 2.5 tabulate the number of known and unknown parameters, for both the total mass and the compositions. The asterisk designates that one of the three unknown compositions can always be obtained by using the equivalent of Eq. (2.5). Hence in a three-component stream of unknown composition, only two of the three mass fractions (or percents) really need to be evaluated. The dagger indicates that in a stream with only one or two components, the composition of the other components is known to be 0 percent. Note how for each of the first three figures in Fig. 2.5 the number of independent material balances (three in each case) does not exceed the number of unknown quantities.

For problems involving chemical reaction, a total mass balance can be written as well as a mass balance for each atomic species or multiple thereof, i.e., hydrogen expressed as H_2. For example, in Fig. 2.6(a) you can write a total mass (*not* moles) balance and a carbon, a hydrogen, a nitrogen, and an oxygen balance. The carbon balance might be in terms of C, and the hydrogen, nitrogen, and oxygen balances in terms of H_2, N_2, and O_2, respectively. Depending on the problem statement, you would not necessarily have to use each of the balances. Also keep in mind that in using balances on atomic species, both mass and moles are conserved.

Figure 2.6(c) illustrates a case in which the chemical equation is given. However, in solving problems related to Fig. 2.6(a) and (b), chemical equation(s) must be assumed to apply, or else some information given or assumed about the nature of the reaction that takes place.

The strategy to be used in solving material balance problems is very simple. As a general rule, before making any calculations you should

(a) Draw a picture of the process.
(b) Place all the available data in the picture.
(c) See what compositions are known or can be immediately calculated for each stream.
(d) See what masses (weights) are known or can easily be found for each stream. [One mass (weight) can be assumed as a basis.]
(e) Select a suitable basis for the calculations. Every addition or subtraction *must* be made with the material on the *same* basis.
(f) Make sure the system is well defined.

After this has been accomplished, you are ready to make the necessary number of material balances.

As explained previously, you can write

(a) A *total* material balance.
(b) A *component* material balance for each component present.

Not all the balances, however, will be independent.

Problems in which the mass (weight) of one stream and the composition of one stream are unknown can be solved without difficulty by direct addition or direct subtraction. Problems in which all the compositions are known and two or more of the weights are unknown require some slightly more detailed calculations. If a tie component exists which makes it possible to establish the relationship between the unknown weights and the known weights, the problem solution may be simplified. (The tie component will be discussed in detail in Sec. 2.5.) When there is no direct or indirect tie component available, algebra must be used to relate the unknown masses to the known masses.

2.3 Problems with direct solutions

Problems in which one mass (weight) and one composition are unknown can be solved by direct addition or subtraction, as shown in the examples below. There is no need to use formal algebraic techniques. You may find it necessary to make some brief preliminary calculations in order to decide whether or not all the information about the compositions and weights that you would like to have is available. Of course, in a stream containing just one component, the composition is known, because that component is 100 percent of the stream.

In dealing with problems involving combustion, you should become acquainted with a few special terms:

(a) *Flue* or *Stack gas*—all the gases resulting from a combustion process including the water vapor, sometimes known as *wet basis.*

(b) *Orsat analysis* or *dry basis*—all the gases resulting from the combustion process not including the water vapor. (Orsat analysis refers to a type of gas analysis apparatus in which the volumes are measured over water; hence each component is saturated with water vapor. The net result of the analysis is to eliminate water as a component being measured.)

Pictorially, we can express this classification for a given gas as in Fig. 2.7. To convert from one analysis to another, you have to ratio the percentages for the components as shown in Example 2.9.

(c) *Theoretical air* (or *theoretical oxygen*)—the amount of air (or oxygen) required to be brought into the process for *complete* combustion. Sometimes this quantity is called the *required* air (or oxygen).

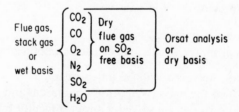

Fig. 2.7. Comparison of gas analyses on different bases.

(d) *Excess air* (or *excess oxygen*)—in line with the definition of excess reactant given in Chap. 1, excess air (or oxygen) would be the amount of air (or oxygen) **in excess of that required for complete combustion** as computed in (c).

Even if only partial combustion takes place, as, for example, C burning to both CO and CO_2, the excess air (or oxygen) is computed as if the process of combustion produced only CO_2. The percent excess air is identical to the percent excess O_2 (a more convenient method of calculation):

$$\% \text{ excess air} = 100\frac{\text{excess air}}{\text{required air}} = 100\frac{\text{excess } O_2/0.21}{\text{required } O_2/0.21} \qquad (2.6)$$

Note that the ratio $1/0.21$ of air to O_2 cancels out in Eq. (2.6). Percent excess air may also be computed as

$$\% \text{ excess air} = 100\frac{O_2 \text{ entering process} - O_2 \text{ required}}{O_2 \text{ required}} \qquad (2.7)$$

or

$$= 100\frac{\text{excess } O_2}{O_2 \text{ entering} - \text{excess } O_2}$$

since

$$O_2 \text{ entering process}$$
$$= O_2 \text{ required for complete combustion} + \text{excess } O_2 \qquad (2.8)$$

The precision of these different relations for calculating the percent excess air may not be the same. If the percent excess air and the chemical equation are given in a problem, you know how much air enters with the fuel, and hence the number of unknowns is reduced by one.

EXAMPLE 2.4   Excess air

A man comes to the door selling a service designed to check "chimney rot." He explains that if the CO_2 content of the gases leaving the chimney rises above 15 percent, it is dangerous to your health, is against the city code, and causes your chimney to rot. On checking the flue gas from the furnace he finds it is 30 percent CO_2. Suppose that you are burning natural gas which is about 100 percent CH_4 and that the air supply is adjusted to provide 130 percent excess air. Do you need his service?

Solution:

Let us calculate the actual percentage of CO_2 in the gases from the furnace assuming that complete combustion takes place. See Fig. E2.4. The 130 percent

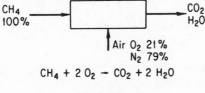

$$CH_4 + 2 O_2 \rightarrow CO_2 + 2 H_2O$$

Fig. E2.4.

excess air means 130 percent of the air required for complete combustion of CH_4. The chemical reaction is

$$CH_4 + 2O_2 \longrightarrow CO_2 + 2H_2O$$

Basis: 1 mole CH_4

By picking a basis, the quantity of the CH_4 stream is fixed. The quantity of air is fixed by the specification of the excess air and is computed as follows: On the basis of 1 mole of CH_4, 2 moles of O_2 are required for complete combustion, or

$$\frac{2O_2}{} \bigg| \frac{1.00 \text{ air}}{0.21 \text{ } O_2} = 9.54 \text{ moles air required}$$

composed of 2 moles O_2 and 7.54 moles N_2. The excess air is $9.54(1.30) = 12.4$ moles air, composed of 2.60 moles O_2 and 9.80 moles N_2.

Summing up our calculations so far we have entering the furnace:

	air	O_2	N_2
Required	9.54	2.00	7.54
Excess	12.4	2.60	9.80
Total	21.94	4.60	17.34

Since 2 moles of the 4.60 moles of entering O_2 combine with the C and H of the CH_4 to form CO_2 and H_2O, respectively, the composition of the gases from the furnace should contain the following:

	moles	%
CH_4	0	0
CO_2	1.00	4.4
H_2O	2.00	8.7
O_2	2.60	11.3
N_2	$\begin{cases} 7.54 \\ 9.80 \end{cases}$	75.6
	22.94	100.0

The salesman's line seems to be rot all the way through.

EXAMPLE 2.5   Excess air

Fuels for motor vehicles other than gasoline are being eyed because they generate lower levels of pollutants than does gasoline. Compressed ethane has been suggested as a source of economic power for vehicles. Suppose that in a test 20 lb of C_2H_4 are burned with 400 lb of air to 44 lb of CO_2 and 12 lb of CO. What was the percent excess air?

Solution:

$$C_2H_4 + 3O_2 \longrightarrow 2CO_2 + 2H_2O$$

Basis: 20 lb C_2H_2

Since the percentage of excess air is based on the complete combustion of C_2H_4 to CO_2 and H_2O, the fact that combustion is not complete was no influence on the definition of "excess air." The required O_2 is

$$\frac{20 \text{ lb } C_2H_4}{} \left| \frac{1 \text{ lb mole } C_2H_4}{28 \text{ lb } C_2H_4} \right| \frac{3 \text{ lb moles } O_2}{1 \text{ lb mole } C_2H_4} = 2.14 \text{ lb moles } O_2$$

The entering O_2 is

$$\frac{400 \text{ lb air}}{} \left| \frac{1 \text{ lb mole air}}{29 \text{ lb air}} \right| \frac{21 \text{ lb moles } O_2}{100 \text{ lb moles air}} = 2.90 \text{ lb moles } O_2$$

The percent excess air is

$$100 \times \frac{\text{excess } O_2}{\text{required } O_2} = 100 \frac{\text{entering } O_2 - \text{required } O_2}{\text{required } O_2}$$

$$\% \text{ excess air} = \frac{2.90 \text{ lb moles } O_2 - 2.14 \text{ lb moles } O_2}{2.14 \text{ lb moles } O_2} \left| 100 \right. = 35.5\%$$

EXAMPLE 2.6 Drying

A wet paper pulp is found to contain 71 percent water. After drying it is found that 60 percent of the original water has been removed. Calculate the following:

(a) The composition of the dried pulp.
(b) The mass of water removed per kilogram of wet pulp.

Solution:

Basis: 1 kg wet pulp

Initially we know only the composition and mass (the chosen basis) of the wet pulp. However, we can easily calculate the composition (100 percent H_2O) and mass of water removed; examine Fig. E2.6.

$$H_2O \text{ removed} = 0.60(0.71) = 0.426 \text{ kg}$$

Thus in essence we now know two masses and two compositions.

(1) By a pulp balance the final amount of pulp in the dried pulp is 0.29 kg and the associated water is

$$\textbf{in} - \textbf{loss} = \textbf{remainder}$$

$$0.71 - 0.426 = 0.284 \text{ kg } H_2O \qquad \textit{water balance}$$

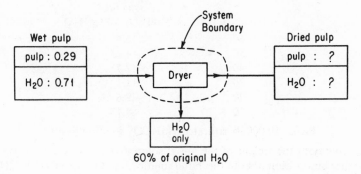

Fig. E2.6.

or, alternatively, since 40 percent of the H_2O was left in pulp,

$$0.40(0.71) = 0.284 \text{ kg } H_2O$$

(2) The composition of the dried pulp is

	kg	%
Pulp (dry)	0.29	50.5
H_2O	0.284	49.5
Total	0.574	100.0

EXAMPLE 2.7 Crystallization

A tank holds 10,000 lb of a saturated solution of $NaHCO_2$ at 60°C. You want to crystallize 500 lb of $NaHCO_3$ from this solution. To what temperature must the solution be cooled?

Solution:
A diagram of the process is shown in Fig. E2.7.

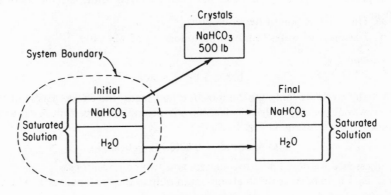

Fig. E2.7.

Additional data are needed on the solubility of $NaHCO_3$ as a function of temperature. From any handbook you can find

	solubility
temp. (°C)	g $NaHCO_3$/100 g H_2O
60	16.4
50	14.45
40	12.7
30	11.1
20	9.6
10	8.15

Basis: 10,000 lb saturated $NaHCO_3$ solution at 60°C

We know both the weight of the initial solution and the weight of the crystals; we know the composition of both the initial solution and the crystals (100%$NaHCO_3$). Thus the weight and composition of the final solution can be calculated by subtraction.

(a) Initial composition:

$$\frac{16.4 \text{ lb NaHCO}_3}{16.4 \text{ lb NaHCO}_3 + 100 \text{ lb H}_2\text{O}} = 0.141 \quad \text{or} \quad 14.1\% \text{NaHCO}_3$$

$$100 - 14.1 = 85.9\% \text{HO}$$

(b) Material balance (component balance on the NaHCO₃):

initial NaHCO₃ − crystals NaHCO₃ = final NaHCO₃
0.141(10,000) −   500   =   910 lb

(c) Final composition and weight:

$$\text{H}_2\text{O} = (0.859)10,000 = 8590 \text{ lb}$$
$$\text{NaHCO}_3 = \qquad\qquad\quad 910 \text{ lb}$$
$$\text{Total} \qquad\qquad\quad 9500 \text{ lb}$$

Since the final solution is still saturated and has

$$\frac{910 \text{ lb NaHCO}_3}{8590 \text{ lb H}_2\text{O}} = \frac{10.6 \text{ g NaHCO}_3}{100 \text{ g H}_2\text{O}}$$

the temperature to which the solution must be cooled is (using a linear interpolation)

$$30°\text{C} - \left[\frac{11.1 - 10.6}{11.1 - 9.6}\right][10°\text{C}] = 27°\text{C}$$

EXAMPLE 2.8 Distillation

A moonshiner is having a bit of difficulty with his still. The operation is shown in Fig. E2.8. He finds he is losing too much alcohol in the bottoms (waste). Calculate the composition of the bottoms for him and the weight of alcohol lost in the bottoms.

Solution:

The information provided gives us the composition and weight of feed and distillate. The composition of the bottoms can be obtained by subtracting the distillate from the feed.

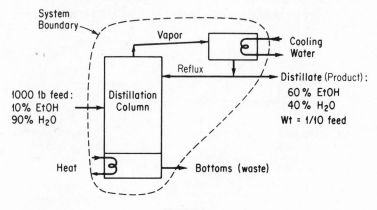

Fig. E2.8.

Basis: 1000 lb feed

		in		out	
				distillate	bottoms (*lb*)
	component	wt %	lb	(*lb*)	(by subtraction)
material balance	{EtOH	10	100	60	40*
	{H₂O	90	900	40	860
	Total	100	1000	100	900

*Alcohol lost.

$$\frac{1000 \text{ lb feed} \mid 1 \text{ lb distillate}}{\mid 10 \text{ lb feed}} = 100 \text{ lb distillate}$$

$$\frac{40 \text{ lb EtOH} \mid 100}{900 \text{ lb bottoms} \mid} = 4.4\% \text{ EtOH}$$

$$\frac{860 \text{ lb H}_2\text{O} \mid 100}{900 \text{ lb bottoms} \mid} = 95.6\% \text{H}_2\text{O}$$

EXAMPLE 2.9 Combustion

The same ethane as used in Example 2.5 is initially mixed with oxygen to obtain a gas containing 80 percent C_2H_6 and 20 percent O_2 that is then burned with 200 percent excess air. Eighty percent of the ethane goes to CO_2, 10 percent goes to CO, and 10 percent remains unburned. Calculate the composition of the exhaust gas on a wet basis.

Solution:

We know the composition of the air and fuel gas; if a weight of fuel gas is chosen as the basis, the weight of air can easily be calculated. However, it is wasted effort to convert to a weight basis for this type of problem. Since the total moles entering and leaving the boiler are not equal, if we look at any one component and employ the stoichiometric principles previously discussed in Chap. 1 together with Eq. (2.1), we can easily obtain the composition of the stack gas. The net generation term in Eq. (2.1) can be evaluated from the stoichiometric equations listed below. The problem can be worked in the simplest fashion by choosing a basis of 100 moles of entering gas. See Fig. E2.9.

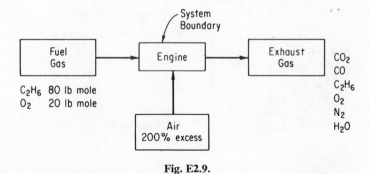

Fig. E2.9.

Basis: 100 lb moles of fuel

$$C_2H_6 + \tfrac{7}{2}O_2 \longrightarrow 2CO_2 + 3H_2O$$
$$C_2H_6 + \tfrac{5}{2}O_2 \longrightarrow 2CO + 3H_2O$$

The total O_2 entering is 3.00 times the required O_2 (100 percent required plus 200 percent excess). Let us calculate the required oxygen:
O_2 (for complete combustion):

$$\frac{80 \text{ lb mole } C_2H_6 \mid 3.5 \text{ lb mole } O_2}{\mid 1 \text{ lb mole } C_2H_6} = 280 \text{ lb mole } O_2$$

Required O_2:

$$280 - 20 = 260 \text{ lb mole } O_2$$

(*Note:* The oxygen used to completely burn the fuel is reduced by the oxygen already present in the fuel to obtain the oxygen required in the entering air.)
Next we calculate the input of O_2 and N_2 to the system:
O_2 entering with air:

$$3(260 \text{ lb mole } O_2) = 780 \text{ lb mole } O_2$$

N_2 entering with air:

$$\frac{780 \text{ mole } O_2 \mid 79 \text{ lb mole } N_2}{\mid 21 \text{ lb mole } O_2} = 2930 \text{ lb mole } N_2$$

Now we apply our stoichiometric relations to find the components generated within the system:

$$\frac{80 \text{ lb mole } C_2H_6 \mid 2 \text{ lb mole } CO_2 \mid 0.8}{\mid 1 \text{ lb mole } C_2H_6 \mid} = 128 \text{ lb mole } CO_2$$

$$\frac{80 \text{ lb mole } C_2H_6 \mid 3 \text{ lb mole } H_2O \mid 0.8}{\mid 1 \text{ lb mole } C_2H_6 \mid} = 192 \text{ lb mole } H_2O$$

$$\frac{80 \text{ lb mole } C_2H_6 \mid 2 \text{ lb mole } CO \mid 0.1}{\mid 1 \text{ lb mole } C_2H_6 \mid} = 16 \text{ lb mole } CO$$

$$\frac{80 \text{ lb mole } C_2H_6 \mid 3 \text{ lb mole } H_2O \mid 0.1}{\mid 1 \text{ lb mole } C_2H_6 \mid} = 24 \text{ lb mole } H_2O$$

To determine the O_2 remaining in the exhaust gas, we have to find how much of the available (800 lb mole) O_2 combines with the C and H.

$$\frac{80 \text{ lb mole } C_2H_6 \mid 3.5 \text{ lb mole } O_2 \mid 0.8}{\mid 1 \text{ lb mole } C_2H_6 \mid} = 224 \text{ lb mole } O_2 \text{ to burn to } CO_2 \text{ and } H_2O$$

$$\frac{80 \text{ lb moles } C_2H_6 \mid 2.5 \text{ lb moles } O_2 \mid 0.1}{\mid 1 \text{ lb mole } C_2H_6 \mid} = 20 \text{ lb mole } O_2 \text{ to burn to } CO \text{ and } H_2O$$

$$244 \text{ lb mole } O_2 \text{ total "used up" by reaction}$$

By an oxygen (O_2) balance we get

$$O_2 \text{ out} = 780 \text{ lb mole} + 20 \text{ lb mole} - 244 \text{ lb mole} = 556 \text{ lb mole } O_2$$

From a water balance we find

$$H_2O \text{ out} = 192 \text{ lb mole} + 24 \text{ lb mole} = 216 \text{ lb mole } H_2O$$

EXAMPLE 2.9 Alternative manner of presentation

Initial Component	lb mole = %	lb mole O₂ Required	Actual lb mole O₂ Used	lb mole formed of			C₂H₆ left	Reactions
				CO₂	H₂O	CO		
C_2H_6	64		$\frac{7}{2}(64) = 224$	128	192			$C_2H_6 + \frac{7}{2}O_2 \longrightarrow 2CO_2 + 3H_2O$
C_2H_6	8 }80	$\frac{7}{2}(80) = 280$	$\frac{5}{2}(8) = 20$	128	24	16		$C_2H_6 + \frac{5}{2}O_2 \longrightarrow 2CO + 3H_2O$
C_2H_6	8						8	$C_2H_6 \longrightarrow C_2H_6$
O_2	20		-20					
	100	260	244	128	216	16	8	

O_2 entering with air: $3(260) = 780$
N_2 entering with air: $780\frac{21}{79} = 2930$
O_2 in exit gas: $780 + 20 - 244 = 556$

air fuel used

Summary of Material Balances

Total mole in ≠ total mole out
Total lb in = total lb out: (each compound has to be multiplied by its molecular weight)

balances on atomic species:
- lb mole C in = lb mole C out: $2(80) = 128 + 16 + 2(8)$
- lb mole H_2 in = lb mole H_2 out: $3(80) = 216 + 3(8)$
- lb mole O_2 in = lb mole O_2 out: $20 + 780 = 128 + \frac{1}{2}(216) + \frac{1}{2}(16) + 556$
- lb mole N_2 in = lb mole N_2 out: $2930 = 2930$

balances on compounds:
- lb mole C_2H_6 in = lb mole C_2H_6 out + lb mole C_2H_6 consumed: $80 = 8 + 72 + 0$
- lb mole CO_2 in = lb mole CO_2 out − lb mole CO_2 generated: $0 = 128 - 128$
- lb mole CO in = lb mole CO out − lb mole CO generated: $0 = 16 - 16$
- lb mole H_2O in = lb mole H_2O out − lb mole H_2O generated: $0 = 216 - 216$
- lb mole N_2 in = lb mole N_2 out + lb mole N_2 consumed: $2930 = 2930 + 0$
- lb mole O_2 in = lb mole O_2 out + lb mole O_2 consumed: $800 = 556 + 244$

The balances on the other compounds—C_2H_6, CO_2, CO, N_2—are too simple to be formally listed here.

Summarizing these calculations, we have

| | | *lb mole* | | *% in* |
component	fuel	air	exhaust gas	exhaust gas
C_2H_6	80	—	8	0.21
O_2	20	780	556	14.42
N_2	—	2930	2930	76.01
CO_2	—	—	128	3.32
CO	—	—	16	0.42
H_2O	—	—	216	5.62
	100	3710	3854	100.00

On a dry basis we would have (the water is omitted)

component	lb mole	%
C_2H_6	8	0.23
O_2	556	15.29
N_2	2930	80.52
CO_2	128	3.52
CO	16	0.44
	3638	100.00

Once you have become familiar with these types of problems, you will find the tabular form of solution illustrated in Example 2.9, on the facing page, a very convenient form to use.

2.4 Material balances using algebraic techniques

As illustrated in Figs. 2.5 and 2.6, problems can be posed or formulated in different ways depending on the type of information available on the process streams and their respective compositions. The problems treated in the previous section were quite easy to solve, once the problem had been converted from words into numbers, because the missing information pertained to a single stream. Only simple addition or subtraction was required to find the unknown quantities. Other types of material balance problems can be solved by writing the balance formally, and assigning letters or symbols to represent the unknown quantities. Each unknown stream, or component, is assigned a letter to replace the unknown value in the total mass balance or the component mass balance, as the case may be. Keep in mind that for each unknown so introduced you will have to write one independent material balance if the set of equations you form is to have a unique solution.

If more than one piece of equipment or more than one junction point is involved in the problem to be solved, you can write material balances for each piece of equipment and a balance around the whole process. However, since the overall balance is nothing more than the sum of the balances about each piece of equipment, not all the balances will be independent. Appendix L discusses how you can determine whether sets of linear equations are independent or not.

Under some circumstances, particularly if you split a big problem into smaller parts to make the calculations easier, you may want to make a material balance about a *mixing point*. As illustrated in Fig. 2.8, a mixing point is nothing more than a junction of three or more streams and can be designated as a system in exactly the same fashion as any other piece of equipment.

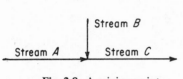

Fig. 2.8. A mixing point.

Some brief comments are now appropriate as to how to solve sets of coupled simultaneous equations. If only two or three linear material balances are written, the unknown variables can be solved for by substitution. If the material balances consist of large sets of linear equations, you will find suggestions for solving them in Appendix L. If the material balance is a nonlinear equation, it can be plotted by hand or by using a computer routine, and the root(s), i.e., the crossings of the horizontal axis, located. See Fig. 2.9. If two material balances have to be solved, the equations can be plotted and their intersection(s) located. With many simultaneous nonlinear equations to be solved, the use of computer routines is essential; for linear equations, computer routines are quick and prevent human error. Making a balance for each component for each defined sys-

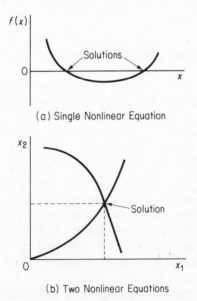

(a) Single Nonlinear Equation

(b) Two Nonlinear Equations

Fig. 2.9. Solution of nonlinear material balances by graphical techniques.

tem, a set of independent equations can be obtained whether linear or nonlinear. Total mass balances may be substituted for one of the component mass balances. Likewise, an overall balance around the entire system may be substituted for one of the subsystem balances. By following these rules, you should encounter no difficulty in generating sets of independent material balances for any process.

Illustrations of the use of algebraic techniques to solve material balance problems can be found below in this section and in the next section (on tie components).

EXAMPLE 2.10 Mixing

Dilute sulfuric acid has to be added to dry charged batteries at service stations in order to activate a battery. You are asked to prepare a batch of new acid as follows. A tank of old weak battery acid (H_2SO_4) solution contains 12.43 percent H_2SO_4 (the remainder is pure water). If 200 kg of 7.77 percent H_2SO_4 are added to the tank, and the final solution is 18.63 percent H_2SO_4, how many kilograms of battery acid have been made? See Fig. E2.10.

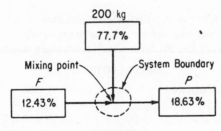

Fig. E2.10.

Solution:
In this problem one mass and three compositions are known; two masses are missing. To assist us, we shall arbitrarily label the masses of the two unknown solutions F and P. With two unknowns it is necessary to set up two independent equations. However, we can write three material balances, any pair of which are independent equations.

Basis: 200 kg of 77.7 percent H_2SO_4 solution

type of balance	in	=	out	
Total	$F + 200$	=	P	(a)
component				
H_2SO_4	$F(0.1243) + 200(0.777) = P(0.1863)$			(b)
H_2O	$F(0.8757) + 200(0.223) = P(0.8137)$			(c)

Use of the total mass balance and one of the others is the easiest way to find P:

$$(P - 200)(0.1243) + 200(0.777) = P(0.1863)$$
$$P = 2110 \text{ kg acid}$$
$$F = 1910 \text{ kg acid}$$

EXAMPLE 2.11 Distillation

A typical distillation column is shown in Fig. E2.11 together with the known information for each stream. Calculate the pounds of distillate per pound of feed and per pound of waste.

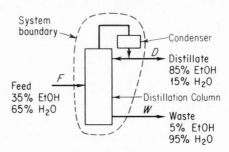

Fig. E2.11.

Solution:

Inspection of Fig. E2.11 shows that all compositions are known, but three weights are unknown. Because only two independent balances can be written, one weight must be chosen as the basis for the problem to effect an algebraic solution.

Basis: 1.00 lb feed

type of balance	in	=	out
total	1.00	=	$D + W$
EtOH	1.00(0.35)	=	$D(0.85) + W(0.05)$
H₂O	1.00(0.65)	=	$D(0.15) + W(0.95)$

Solving for D with $W = (1.00 - D)$,

$$1.00(0.35) = D(0.85) + (1.00 - D)(0.05)$$

$$D = 0.375 \text{ lb/lb feed}$$

Since $W = 1 - 0.375 = 0.625$ lb,

$$\frac{D}{W} = \frac{0.375}{0.625} = \frac{0.60 \text{ lb}}{\text{lb}}$$

EXAMPLE 2.12 Multiple equipment

In your first problem as a new company engineer you have been asked to set up an acetone recovery system to remove acetone from gas that was formerly vented but can no longer be handled that way, and to compute the cost of recovering acetone. Your boss has provided you with the design of the proposed system, as shown in Fig. E2.12. From the information given, list the flow rates (in pounds per hour) of all the streams so that the size of the equipment can be determined, and calculate the composition of the feed to the still.

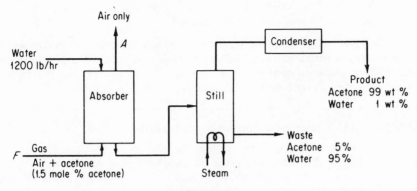

Fig. E2.12.

Solution:

Convert mole percent of the air-acetone feed to weight percent because all the other compositions are given in weight percent:

$$\text{mol. wt acetone} = 58, \qquad \text{mol. wt air} = 29$$

Basis: 100 lb moles gas feed

composition	lb mole	mol. wt	lb	wt %
Acetone	1.5	58	87	2.95
Air	98.5	29	2860	97.05
	100.0		2947	100.00

Now we know the compositions (on a weight basis) of all the streams. There are five streams entering and leaving the entire system (excluding the steam), and the weight of only one is known, leaving four weights as unknowns.

Basis: 1 hr

Balances on complete system (overall balances):

Let

$$
\left.
\begin{aligned}
F &= \text{feed to absorber, lb} \\
A &= \text{air leaving absorber, lb} \\
P &= \text{product, lb} \\
W &= \text{waste, lb}
\end{aligned}
\right\} \text{ the four unknowns}
$$

total $F + 1200 = P + W + A$

acetone $F(0.0295) = P(0.99) + W(0.05)$

water $1200 = P(0.01) + W(0.95)$

air $F(0.9705) = A$

We now have four balances and four unknowns; however, only three of the equations are independent equations, as you can easily discover by adding up the component balances to get the overall balance. What is wrong? It seems as if there is not enough information available to solve this problem.

One additional bit of information would be enough to solve the problem as originally stated. This information might be, for example,

(a) The weight of entering gas per hour product per hour, or waste per hour.
(b) The composition of the feed into the still.

With an extra piece of pertinent information available, a complete solution is possible; without it, only a partial solution can be effected. Observe that making additional balances about the individual pieces of equipment will not resolve the problem, since as many new unknowns are added as independent equations.

EXAMPLE 2.13   Countercurrent stagewise mass transfer

In many commercial processes such as distillation, extraction, absorption of gases in liquids, and the like, the entering and leaving streams represent two different phases that flow in opposite directions to each other, as shown in Fig. E2.13(a). (The figure could just as well be laid on its side.)

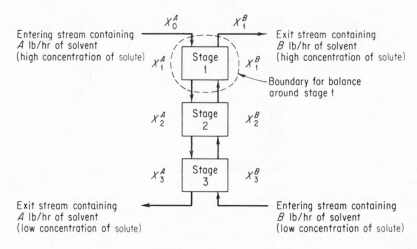

Entering stream containing
A lb/hr of solvent
(high concentration of solute)

Exit stream containing
B lb/hr of solvent
(high concentration of solute)

Boundary for balance
around stage 1

Exit stream containing
A lb/hr of solvent
(low concentration of solute)

Entering stream containing
B lb/hr of solvent
(low concentration of solute)

A, B = lb of stream less solute in the stream
     = lb of solvent

Fig. E2.13(a).

This type of operation is known as *countercurrent* operation. If equilibrium is attained between each stream at each stage in the apparatus, calculations can be carried out to relate the flow rates and concentration of products to the size and other design features of the apparatus. We shall illustrate how a material balance can be made for such type of equipment. The letter X stands for the weight concentration of solute in pounds of solute per pound of stream, solute-free. The streams are assumed immiscible as in a liquid-liquid extraction process.

Around stage 1 the solute material balance is

in   out

$$\frac{A\ \text{lb}}{\text{hr}}\left|\frac{X_0^A\ \text{lb}}{\text{lb}\ A}\right. + \frac{B\ \text{lb}}{\text{hr}}\left|\frac{X_2^B\ \text{lb}}{\text{lb}\ B}\right. = \frac{A\ \text{lb}}{\text{hr}}\left|\frac{X_1^A\ \text{lb}}{\text{lb}\ A}\right. + \frac{B\ \text{lb}}{\text{hr}}\left|\frac{X_1^B\ \text{lb}}{\text{lb}\ B}\right.$$

or

$$A(X_0^A - X_1^A) = B(X_1^B - X_2^B) \tag{a}$$

Around stage 2 the solute material balance is

in   **out**

$$\frac{A \text{ lb}}{\text{hr}} \bigg| \frac{X_1^A \text{ lb}}{\text{lb } A} + \frac{B \text{ lb}}{\text{hr}} \bigg| \frac{X_3^B \text{ lb}}{\text{lb } B} = \frac{A \text{ lb}}{\text{hr}} \bigg| \frac{X_2^A \text{ lb}}{\text{lb } A} + \frac{B \text{ lb}}{\text{hr}} \bigg| \frac{X_2^B \text{ lb}}{\text{lb } B}$$

or

$$A(X_1^A - X_2^A) = B(X_2^B - X_3^B)$$

We could generalize that for any stage number n

$$A(X_{n-1}^A - X_n^A) = B(X_n^B - X_{n+1}^B) \tag{b}$$

Also a solute material balance could be written about the sum of stage 1 and stage 2 as follows:

in   **out**

$$A(X_0^A) + B(X_3^B) = A(X_2^A) + B(X_1^B) \tag{c}$$

or to generalize for an overall solute material balance between the top end and the nth stage,

$$A(X_0^A - X_n^A) = B(X_1^B - X_{n+1}^B)$$

or for the $(n-1)$th stage,

$$A(X_0^A - X_{n-1}^A) = B(X_1^B - X_n^B)$$

Changing signs,

$$A(X_{n-1}^A - X_0^A) = B(X_n^B - X_1^B) \tag{d}$$

If we rearrange Eq. (d) assuming that A, B, X_0^A, and X_1^B are constants (i.e., steady-state operation) and X^A and X^B are the variables as we go from stage to stage, we can write

$$X_{n-1}^A = \frac{B}{A} X_n^B + \left(X_0^A - \frac{B}{A} X_1^B \right) \tag{e}$$

Equations (a)–(e) represent an unusual type of equation, one that gives the relationship between discrete points rather than continuous variables; it is called a *difference* equation. In Eq. (e) the locus of these points will fall upon a line with the slope B/A and an intercept $[X_0^A - (B/A)X_1^B]$, as shown in Fig. E2.13(b).

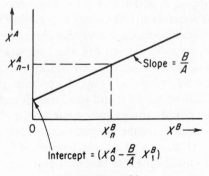

Fig. E2.13(b).

2.5   Problems involving tie components (elements)

A *tie component* (*element*) is material which goes from one stream into another without changing in any respect or having like material added to it or lost from it. If a tie component exists in a problem, in effect you can write a material balance that involves *only two* streams. Furthermore, the material blance has a particularly simple form, that of a ratio. For example, in Fig. 2.5(b) all the MeOH in stream F goes into stream W. Consequently, streams F and W can be related by a MeOH material balance:

$$\omega_{\text{MeOH},F}F = \omega_{\text{MeOH},W}W \qquad \text{or} \qquad \frac{\omega_{\text{MeOH},W}}{\omega_{\text{MeOH},F}} = \frac{F}{W}$$

$$0.10(100) = 0.22W \qquad \text{or} \qquad W = \frac{0.10}{0.22}(100)$$

   To select a tie element, ask yourself the question, What component passes from one stream to another unchanged with constant weight (mass)? The answer is the tie component. Frequently several components pass through a process with continuity so that there are many choices of tie elements. Sometimes the amounts of these components can be added together to give an overall tie component that will result in a smaller percentage error in your calculations than if an individual tie element had been used. It may happen that a minor constituent passes through with continuity, but if the percentage error for the analysis of this component is large, you should not use it as a tie component. Sometimes you cannot find a tie component by direct examination of the problem, but you still may be able to develop a hypothetical tie component or a man-made tie component which will be equally effective as a tie component for a single material. A tie component is useful even if you do not know all the compositions and weights in any given problem because the tie component enables you to put two streams on the same basis. Thus a *partial solution* can be effected even if the entire problem is not resolved.

   In solutions of problems involving tie components it is not always advantageous to select as the basis "what you have." Frequently it turns out that the composition of one stream is given as a percentage composition, and then it becomes convenient to select as the basis 100 lb (or other mass) of the material because then pounds equal percentage and the same numbers can be used to represent both. After you have carried out the computations on the basis of 100 lb (or another basis), then the answer can be transformed to the basis of the amount of material actually given by using the proper conversion factor. For example, if you had 32.5 lb of material and you knew its composition on the basis of percentage, you then could take as your basis 100 lb of material, and at the end of the calculations convert your answer to the basis of 32.5 lb.

   Tie component problems can, of course, be worked by algebraic means, but in combustion problems and in some of the more complicated types of industrial chemical calculations so many unknowns and equations are involved that the use of a tie component greatly simplifies the calculations. Use of a tie component also clarifies the internal workings of a process. Consequently, the examples below illustrate the use of both tie-component and algebraic solutions. Become proficient in both techniques, as you will have to draw upon both of them in your career as an engineer.

EXAMPLE 2.14 Drying

Fish caught by man can be turned into fish meal, and the fish meal can be used as feed to produce meat for man or used directly as food. The direct use of fish meal significantly increases the efficiency of the food chain. However, fish-protein concentrate, primarily for aesthetic reasons, is used as a supplementary protein food. As such, it competes with soy and other oilseed proteins.

In the processing of the fish, after the oil is extracted, the fish cake is dried in rotary drum dryers, finely ground, and packed. The resulting product contains 65 percent protein. In a given batch of fish cake that contains 80 percent water (the remainder is dry cake), 100.0 lb of water is removed, and it is found that the fish cake is then 40 percent water. Calculate the weight of the fish cake originally put into the dryer.

Solution:

From Fig. E2.14 we see that all the compositions and one weight are known and that there is a tie component of BDC.

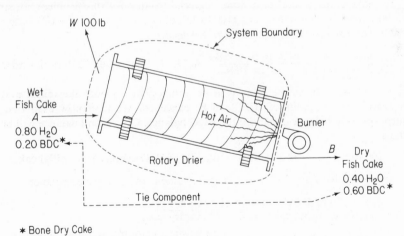

* Bone Dry Cake

Fig. E2.14.

To solve the problem by use of algebra requires the use of two independent material balance equations since there are two unknowns—A and B.

Basis: 100 lb water evaporated

$$\text{in} = \text{out}$$

$$\left. \begin{array}{ll} \text{Total balance} & A = B + W = B + 100 \\ \text{BDC balance} & 0.20A = 0.60B \end{array} \right\} \text{ mass balances}$$

Notice that because the bone dry cake balance involves only two streams, you can set up a direct ratio of A to B, which is the essence of the tie component:

$$B = \frac{0.20A}{0.60} = \frac{1}{3}A$$

Introduction of this ratio into the total balance gives

$$A = \tfrac{1}{3}A + 100$$

$$\tfrac{2}{3}A = 100$$

$$A = 150 \text{ lb initial cake}$$

To solve the same problem by use of the tie component, take as a basis 100 lb of initial material (the 100 lb of H_2O will not serve as a useful basis).

Basis: 100 lb initial fish cake

initial − loss = final

component	% = lb	lb*	lb	%
H_2O	80	100	?	40
BDC	20 ⟵	tie ⟶	20	60
	100		?	100

*All H_2O.

To make either a total or a water material balance for this problem, all streams must be placed on the same basis (100 lb initial cake) by use of the tie component. *Tie final cake to initial cake by a tie component:*

$$\frac{20 \text{ lb BDC}}{100 \text{ lb initial cake}} \bigg| \frac{100 \text{ lb final cake}}{60 \text{ lb BDC}} = 33.3 \text{ lb final cake/100 lb initial cake}$$

Notice that the BDC is the same (the tie component) in both the initial and final stream and that the units of BDC can be cancelled. We have now used one independent material balance equation, the BDC balance. Our second balance will be total balance:

100 lb initial − 33.3 lb final = 66.7 lb of H_2O evap./100 lb initial cake

An alternative way to obtain this same value is to use a water balance

$$\frac{20 \text{ lb BDC}}{100 \text{ lb initial cake}} \bigg| \frac{40 \text{ lb } H_2O \text{ in dry cake}}{60 \text{ lb BDC in dry cake}}$$

$$= 13.3 \text{ lb } H_2O \text{ left/100 lb initial cake}$$

80 lb initial H_2O − 13.3 lb final H_2O

$$= 66.7 \text{ lb } H_2O \text{ evap./100 lb initial cake}$$

Next a shift is made to the basis of 100 lb H_2O evaporated as required by the problem statement:

$$\frac{100 \text{ lb initial cake}}{66.7 \text{ lb } H_2O \text{ evap.}} \bigg| \frac{100 \text{ lb } H_2O \text{ evap.}}{} = 150 \text{ lb initial cake}$$

Another suitable starting basis would be to use 1 lb of bone dry cake as the basis.

Basis: 1 lb BDC

	initial	$-$ loss $=$	final
component	*fraction*	*lb*	*fraction*
H_2O	0.80	100	0.40
BDC	0.20		0.60
	$\overline{1.00}$		$\overline{1.00}$

initial HO $-$ **final H_2O** $=$ **H_2O lost**   water balance

$$\frac{0.80 \text{ lb } H_2O}{0.20 \text{ lb BDC}} - \frac{0.40 \text{ lb } H_2O}{0.60 \text{ lb BDC}} =$$

$$4.0 \quad - \quad 0.67 \quad = 3.33 \text{ lb } H_2O/1 \text{ lb BDC}$$

Again we have to shift bases to get back to the required basis of 100 lb of H_2O evaporated:

$$\frac{1 \text{ lb } \cancel{\text{BDC}}}{3.33 \text{ lb } H_2O} \left| \frac{100 \text{ lb } H_2O \text{ evap.}}{} \right| \frac{1.0 \text{ lb initial cake}}{0.2 \text{ lb } \cancel{\text{BDC}}} = 150 \text{ lb initial cake}$$

You can see that the use of the tie component makes it quite easy to shift from one basis to another. If you now want to compute the pounds of final cake, all that is necessary is to write

$$\begin{array}{r} 150 \text{ lb initial cake} \\ -100 \text{ lb } H_2O \text{ evaporated} \\ \hline 50 \text{ lb final cake} \end{array}$$

Now that we have examined some of the correct methods of handling tie components let us look at another method frequently attempted (not for long!) by some students before they become completely familiar with the use of the tie component. They try to say that

$$0.80 - 0.40 = 0.40 \text{ lb } H_2O \text{ evaporated}$$

$$\frac{100 \text{ lb}}{0.40 \text{ lb}} = 250 \text{ lb cake}$$

On the surface this looks like an easy way to solve the problem—what is wrong with the method? More careful inspection of the calculation shows that two different bases were used in subtracting $(0.80 - 0.40)$. The 0.80 is on the basis of 1.00 lb of initial cake, while the 0.40 is on the basis of 1.00 lb of final cake. You would not say $0.80/A - 0.40/B = 0.40/A$, would you? From our early discussion about units you can see that this subtraction is not a legitimate operation.

EXAMPLE 2.15   Dilution

Trace components can be used to measure flow rates in pipelines, process equipment, and rivers that might otherwise be quite difficult to measure. Suppose that the water analysis in a flowing creek shows 180 ppm Na_2SO_4. If 10 lb of Na_2SO_4 are added to the stream uniformly over a 1-hr period and the analysis downstream where mixing is complete indicates 3300 ppm Na_2SO_4, how many gallons of water are flowing per hour? See Fig. E2.15.

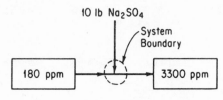

Fig. E2.15.

Tie-component Solution:

Basis: 100 lb initial solution (water $+$ Na_2SO_4 initially present)

After choosing a basis of 100 lb of initial solution, we know three compositions and one weight. A tie component, the water, is present.

composition	initial lb $= \%$		final $\%$
H_2O	99.982	$\xleftrightarrow{\text{tie component}}$	99.670
Na_2SO_4	0.018		0.330
	100.000		100.000

$$\frac{0.33 \text{ lb } Na_2SO_4}{99.67 \text{ lb } H_2O} \left| \frac{99.982 \text{ lb } H_2O}{100 \text{ lb initial solution}} \right.$$

$$= 0.331 \text{ lb } Na_2SO_4/100 \text{ lb initial solution}$$

The tie component of water comprises our first material balance. Now that the exit Na_2SO_4 concentration is on the same basis as the entering Na_2SO_4 concentration, by an Na_2SO_4 balance (our second balance) we find that the Na_2SO_4 added on the same basis is

$$0.331 - 0.0180 = 0.313 \text{ lb}/100 \text{ lb initial solution}$$

Basis: 1 hr

$$\frac{100 \text{ lb initial solution}}{0.313 \text{ lb } Na_2SO_4 \text{ added}} \left| \frac{10 \text{ lb added}}{1 \text{ hr}} \right| \frac{1 \text{ gal}}{8.35 \text{ lb}} = 383 \text{ gal/hr}$$

Algebraic Solution:

Let

$$x = \text{total lb flowing/hr initially}$$

$$y = \text{total lb flowing/hr after addition of } Na_2SO_4$$

total balance $x + 10 = y$ $\Big\}$ mass balances
water balance $0.99982x = 0.99670y$

$$x + 10 = \frac{0.99982x}{0.99670}$$

The water balance involves only two streams and provides a direct ratio between x and y. One of the difficulties in the solution of this problem using a water balance is that the quantity

$$\left[\frac{0.99982}{0.99670} - 1\right] = 0.00313$$

has to be determined quite accurately. It would be more convenient to use an Na_2SO_4 balance,

$$0.00018x + 10 = 0.00330y$$

Solving the total balance and the Na_2SO_4 balance simultaneously,

$$\cdot x = 3210 \text{ lb/hr}$$

$$\frac{3210 \text{ lb}}{\text{hr}} \left| \frac{1 \text{ gal}}{8.35 \text{ lb}} = 385 \text{ gal/hr}\right.$$

EXAMPLE 2.16   Crystallization

The solubility of barium nitrate at 100°C is 34 g/100 g H_2O and at 0°C is 5.0 g /100 g H_2O. If you start with 100 g of $Ba(NO_3)_2$ and make a saturated solution in water at 100°C, how much water is required? If this solution is cooled to 0°C, how much $Ba(NO_3)_2$ is precipitated out of solution?

Tie-component Solution:

Basis: 100 g $Ba(NO_3)_2$

The maximum solubility of $Ba(NO_3)_2$ in H_2O at 100°C is a saturated solution, or 34 g/100 g H_2O. Thus the amount of water required at 100°C is

$$\frac{100 \text{ g } H_2O}{34 \text{ g } Ba(NO_3)_2} \left| \frac{100 \text{ g } Ba(NO_3)_2}{} = 295 \text{ g } H_2O\right.$$

If the 100°C solution is cooled to 0°C, the $Ba(NO_3)_2$ solution will still be saturated, and the problem now can be pictured as in Fig. E2.16. The tie element is the water.

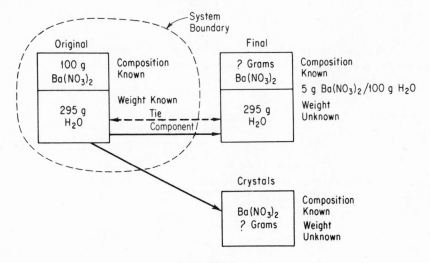

Fig. E2.16.

$$\frac{295 \text{ g } H_2O}{} \left| \frac{5 \text{ g } Ba(NO_3)_2}{100 \text{ g } H_2O} \right. = 14.7 \text{ g } Ba(NO_3)_2 \text{ in final solution}$$

original   —   **final**   =   **crystals**

$100 \text{ g } Ba(NO_3)_2 - 14.7 \text{ g } Ba(NO_3)_2 = 85.3 \text{ g } Ba(NO_3)_2$ precipitated

Algebraic Solution:
Let

$$x = \text{g crystals}$$
$$y = \text{g final solution}$$

H_2O balance $295 = y\left(\dfrac{100}{100 + 5}\right) \dfrac{\text{g } H_2O}{\text{g final solution}}$

Total balance $395 = x + y$

$$y = 295\left(\frac{105}{100}\right) = 310$$

$$x = 395 - y = 85 \text{ g } Ba(NO_3)_2 \text{ crystals}$$

EXAMPLE 2.17   Combustion

The main advantage of catalytic incineration of odorous gases or other obnoxious substances over direct combustion is the lower cost. Catalytic incinerators operate at lower temperatures—500°–900°C compared with 1100°–1500°C for thermal incinerators—and use substantially less fuel. Because of the lower operating temperatures, materials of construction do not need to be as heat resistant, reducing installation and construction costs.

In a test run, a liquid having the composition 88 percent C and 12 percent H_2 is vaporized and burned to a flue gas (fg) of the following composition:

CO_2	13.4
O_2	3.6
N_2	83.0
	100.0%

To compute the volume of the combustion device, determine how many pound moles of dry fg are produced per 100 lb of liquid feed. What was the percentage of excess air used? See Fig. E2.17.

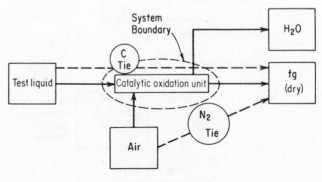

Fig. E2.17.

Tie-Component Solution:

In this problem four streams are present all of whose compositions are known, two entering and two leaving. We have the weights of three streams as unknown, (if we choose one weight as the basis). However, since we do not have to answer any questions about the water stream, two tie elements to relate the test fluid, dry flue gas, and air streams to each other would be sufficient to solve the problem. In examining the data to determine whether a tie element exists, we see that the carbon goes directly from the test fluid to the dry flue gas, and nowhere else, and so carbon will serve as one tie component. The N_2 in the air all shows up in the dry flue gas so N_2 can be used as another tie component. As explained before, there is no reason to convert all the available data to a weight basis, even though the moles of material entering and leaving the oxidation unit are not the same. Let us consider the moles of each atomic specie present. Then we have in 100 lb of test fluid:

$$C: \quad \tfrac{88}{12} = 7.33 \text{ lb mole}$$

$$H_2: \quad \tfrac{12}{2} = 6.00 \text{ lb mole}$$

We can say the pound moles of C entering the process equal the pound moles of C leaving the process since the molecular weight is a constant factor in both inlet and outlet material and can be canceled out. Thus

$$\text{lb C in} = \text{lb C out}$$

$$\text{lb mole C in} = \text{lb mole C out}$$

In step (a) below both of these methods of attack are employed.

In the dry flue gas

$$\text{Basis: 100 lb mole dry fg}$$

$$\frac{13.4 \text{ lb mole C}}{} \left| \frac{12 \text{ lb C}}{1 \text{ lb mole C}} \right. = 161 \text{ lb C}$$

In the test fluid

$$\text{Basis: 100 lb test fluid}$$

$$\frac{88 \text{ lb C}}{} \left| \frac{1 \text{ lb mole C}}{12 \text{ lb C}} \right. = 7.33 \text{ lb mole C}$$

(a$_1$) Using lb C = lb C as the tie component,

$$\frac{100 \text{ lb mole dry fg}}{161 \text{ lb C}} \left| \frac{88 \text{ lb C}}{100 \text{ lb test fluid}} \right. = \frac{54.6 \text{ lb mole dry fg}}{100 \text{ lb test fluid}}$$

(a$_2$) Using lb mole C = lb mole C as the tie component,

$$\frac{100 \text{ lb mole dry fg}}{13.4 \text{ lb mole C}} \left| \frac{7.33 \text{ lb mole C}}{100 \text{ lb test fluid}} \right. = \frac{54.6 \text{ lb mole dry fg}}{100 \text{ lb test fluid}}$$

(b)   The N_2 serves as the tie component to tie the air to the dry flue gas, and since we know the excess O_2 in the dry flue gas, the percentage of excess air can be computed.

$$\text{Basis: 100 lb mole dry fg}$$

$$\frac{83.0 \text{ lb mole } N_2}{} \left| \frac{1.00 \text{ lb mole air}}{0.79 \text{ lb mole } N_2} \right. = \frac{105.0 \text{ lb mole air}}{100 \text{ lb mole dry fg}}$$

$$(105.0)(0.21) = O_2 \text{ entering} = 22.1 \text{ lb mole } O_2$$

Or, saving one step,

$$\frac{83.0 \text{ lb mole } N_2}{} \left| \frac{0.21 \text{ lb mole } O_2 \text{ in air}}{0.79 \text{ lb mole } N_2 \text{ in air}} = \frac{22.1 \text{ lb mole } O_2 \text{ entering}}{100 \text{ lb mole dry fg}} \right.$$

$$\% \text{ excess air} = 100 \frac{\text{excess } O_2}{O_2 \text{ entering } - \text{ excess } O_2} = \frac{(100)3.6}{22.1 - 3.6} = 19.4\%$$

To check this answer, let us work part (b) of the problem using as a basis

Basis: 100 lb test fluid

The required oxygen for complete combustion is

comp.	lb	lb mole	lb mole O_2 required
C	88	7.33	7.33
H_2	12	6	3
			10.33

The excess oxygen is

$$\frac{3.6 \text{ lb mole } O_2}{100 \text{ lb mole fg}} \left| \frac{54.6 \text{ lb mole dry fg}}{100 \text{ lb oil}} = 1.97 \text{ lb mole} \right.$$

$$\% \text{ excess air} = \frac{\text{excess } O_2}{\text{required } O_2} 100 = \frac{1.97(100)}{10.33} = 19.0\%$$

The answers computed on the two different bases agree reasonably well here, but in many combustion problems slight errors in the data will cause large differences in the calculated percentage of excess air. Assuming no mathematical mistakes have been made (it is wise to check), the better solution is the one involving the use of the most precise data. Frequently this turns out to be the method used in the check here—the one with a basis of 100 lb of test fluid.

If the dry flue-gas analysis had shown some CO, as in the following hypothetical analysis,

CO_2	11.9%
CO	1.6
O_2	4.1
N_2	82.4

then, on the basis of 100 moles of dry flue gas, we would calculate the excess air as follows:

O_2 entering with air:

$$\frac{82.4 \text{ lb moles } N_2}{} \left| \frac{0.21 \text{ lb mole } O_2}{0.79 \text{ lb mole } N_2} = 21.9 \text{ lb mole } O_2 \right.$$

Excess O_2:

$$4.1 - \frac{1.6}{2} = 3.3 \text{ lb mole}$$

$$\% \text{ excess air} = 100 \frac{3.3}{21.9 - 3.3} = 17.7\%$$

Note that to get the true excess oxygen, the apparent excess oxygen in dry flue gas, 4.1 lb moles, has to be reduced by the amount of the theoretical oxygen not combining with the CO. According to the reactions

$$C + O_2 \longrightarrow CO_2, \qquad C + \tfrac{1}{2}O_2 \longrightarrow CO$$

for each mole of CO in the dry flue gas, $\frac{1}{2}$ mole of O_2 which should have combined with the carbon to form CO_2 did not do so. This $\frac{1}{2}$ mole carried over into the flue gas and inflated the value of the true excess oxygen expected to be in the flue gas. In this example, 1.6 lb moles of CO are in the flue gas, so that $(1.6/2)$ lb mole of theoretical oxygen are found in the flue gas in addition to the true excess oxygen.

Algebraic Solution:

An algebraic solution of the originally stated problem would proceed as follows:
Let

$$x = \text{lb of test fluid}$$

$$y = \text{lb moles of dry fg}$$

$$z = \text{lb moles of air}$$

$$w = \text{lb moles of } H_2O \text{ associated with the flue gas}$$

One of these unknowns can be replaced by the basis, say $x = 100$ lb of test fluid. Since three unknowns will remain, it will be necessary to use in the solution three independent material balances from among the four component and one total balance that can be written. The balances are

total

$$100 + 29z = 30.25y + 18w \quad \text{where } 30.25 = \text{average mol. wt. of fg}$$

component

$$C \qquad 100(0.88) = \frac{y \text{ lb mole dry fg}}{} \left| \frac{0.134 \text{ lb mole C}}{1 \text{ lb mole dry fg}} \right| \frac{12 \text{ lb C}}{1 \text{ lb mole C}}$$

$$H_2 \qquad 100(0.12) = \frac{w \text{ lb mole } H_2O}{} \left| \frac{1 \text{ lb mole } H_2}{1 \text{ lb mole } H_2O} \right| \frac{2.016 \text{ lb } H_2}{1 \text{ lb mole } H_2}$$

$$N_2 \qquad 0.79z = 0.83y$$

$$O_2 \qquad 0.21z = \frac{w \text{ lb mole } H_2O}{} \left| \frac{0.5 \text{ lb mole } O_2}{1 \text{ lb mole } H_2O} \right.$$

$$+ \frac{y \text{ lb mole dry fg}}{} \left| \frac{0.036 \text{ lb mole } O_2 + 0.134 \text{ mole } O_2}{1 \text{ lb mole dry fg}} \right.$$

Note that the balances for the tie components C and N_2 (as well as H_2) involve only two streams, whereas the O_2 (and total) balance involves three streams. We shall use the C and N_2 balances.

Basis: 100 lb of test fluid

$$y = \frac{88}{(0.134)(12)} = 54.6 \text{ lb mole dry fg}$$

$$z = \frac{0.83}{0.79}y \qquad = 57.5 \text{ lb mole air}$$

$$O_2 \text{ entering} = 57.5(0.21) = 12.1 \text{ lb mole}$$

The remainder of the calculations follow those described above. Can you demonstrate that the dry flue-gas analysis is slightly inaccurate?

EXAMPLE 2.18   Distillation

A continuous still is to be used to separate acetic acid, water, and benzene from each other. On a trial run the calculated data were as shown in Fig. E2.18. Data record-

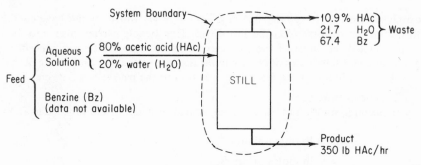

Fig. E2.18.

ing the benzene composition of the feed were not taken because of an instrument defect. Calculate the benzene flow in the feed per hour.

A quick inspection of the flow diagram shows that enough information about weights and compositions would be available to work this problem if only the benzene composition in the feed were known, but it is not. However, we have one added source of information not readily apparent on the surface, and that is that there are *two* tie components from the feed to the waste: water and benzene. A choice now must be made in procedure: the problem can be calculated using the tie components or using algebraic techniques.

Tie-Component Solution:

Take a convenient basis of 100 lb of waste (100 lb of feed would involve more work since the benzene composition is unknown, but 100 lb of aqueous solution— 80 percent HAc, 20 percent H_2O—would be a sound basis). Let $x = $ lb of benzene in the feed/100 lb feed.

Basis: 100 lb waste

composition	*feed* *lb*	*waste* % = *lb*	*product* %
HAc	$(100 - x)0.80$	10.9	100
H_2O	$(100 - x)0.20$	21.7	
Bz	x	67.4	
Total	100	100	

For an initial step let us calculate the quantity of feed per 100 lb of waste. We have the water in the feed (21.7 lb) and the benzene in the feed (67.4 lb); these two materials appear only in the waste and not in the product and can act as tie components. All that is needed is the HAc in the feed. We can use water as the tie component to find this quantity.

$$\frac{21.7 \text{ lb } H_2O}{100 \text{ lb waste}} \Bigg| \frac{0.80 \text{ lb HAc}}{0.20 \text{ lb } H_2O} = 86.8 \text{ lb HAc/100 lb waste}$$

The product now is

$$\textbf{HAc in feed} \; - \; \textbf{HAc in waste} = \textbf{HAc in product} \qquad \textit{HAc balance}$$
$$86.8 \quad - \quad 10.9 \quad = \quad 75.9$$

Take a new basis:

$$\text{Basis: 350 lb HAc product} \equiv 1 \text{ hr}$$

$$\frac{67.4 \text{ lb Bz in feed}}{75.9 \text{ lb HAc in product}} \left| \frac{350 \text{ lb HAc product}}{\text{hr}} \right. = 311 \text{ lb Bz/hr}$$

Algebraic Solution:

To employ algebra in the solution of this problem, we need an independent material balance for each unknown. The unknowns are

$$W = \text{waste, lb}$$

$$F = \text{feed, lb}$$

$$x = \text{lb benzene/100 lb feed}$$

$$\text{Basis: 1 hr} \equiv 350 \text{ lb HAc product}$$

The equations are

Total balance	$F = W + 350$
H_2O balance	$F\left[\dfrac{0.20(100 - x)}{100}\right] = W(0.217)$
HAc balance	$F\left[\dfrac{0.80(100 - x)}{100}\right] = W(0.109) + 350$
Bz balance	$F\left[\dfrac{x}{100}\right] = W(0.674)$

Any three will be independent equations, but you can see the simultaneous solution of three equations involves more work than the use of tie components.

EXAMPLE 2.19 Economics

A chemical company buys coal at a contract price based on a specified maximum amount of moisture and ash. Since you have married the boss's daughter, your first job assignment is as purchasing agent. The salesman for the Higrade Coal Co. offers you a contract for 10 carloads/month of coal with a maximum moisture content of 3.2 percent and a maximum ash content of 5.3 percent at \$4.85/ton (weighed at delivery). You accept this contract price. In the first delivery the moisture content of the coal as reported by your laboratory is 4.5 percent and the ash content is 5.6 percent. The billed price for this coal is \$4.85/ton—as weighed by the railroad in the switchyards outside your plant. The accounting department wants to know if this billing price is correct. Calculate the correct price.

Solution:

What you really pay for is 91.5 percent combustible material with a maximum allowable content of ash and water. Based on this assumption, you can find the cost of the combustible material in the coal actually delivered.

$$\text{Basis: 1 ton coal as delivered}$$

composition	delivered	contract coal
Combustible	0.899	0.915
Moisture plus ash	0.101	0.085
	1.000	1.000

The contract calls for

$$\frac{\$4.85}{\text{ton contract}} \left| \frac{1 \text{ ton contract}}{0.915 \text{ ton comb.}} \right| \frac{0.899 \text{ ton comb.}}{1 \text{ ton del.}} = \$4.76/\text{ton del.}$$

The billed price is wrong.

If the ash content had been below 5.3 percent but the water content above 3.2 percent, presumably an adjustment should have been made in the billing for excess moisture even though the ash was low.

2.6 Recycle, bypass, and purge calculations

Processes involving *feedback* or recycle of part of the product are frequently encountered in the chemical and petroleum industry as well as in nature. Figure 2.10 shows by a block diagram the character of the recycle stream. Figures 2.3 and 2.4 indicate some typical recycle streams in the world at large. As another example, in planning long space missions, all the food and water will have to be provided from stores on board the spacecraft. Table 2.1 lists the daily intake and output per man of solid and water. Since the average internal water consumption amounts to some 5.50 lb/man/day, this leaves an additional requirement of 4.69 lb that must be made up from stores or reclaimed from waste products containing water. Figure 2.11 shows the sources of water and how by the use of several recycle streams the 5.50 lb/day is proposed to be collected.

Many industrial processes employ recycle streams. For example, in some drying operations, the humidity in the air is controlled by recirculating part of the wet air which leaves the dryer. In chemical reactions, the unreacted material may be separated from the product and recycled, as in the synthesis of ammonia. Another example of the use of recycling operations is in fractionating columns where part of the distillate is refluxed through the column to maintain the quantity of liquid within the column.

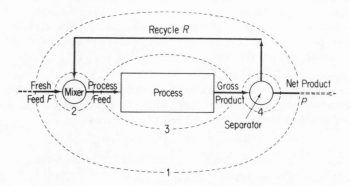

Fig. 2.10. A process with recycle (the numbers designate possible material balances—see text).

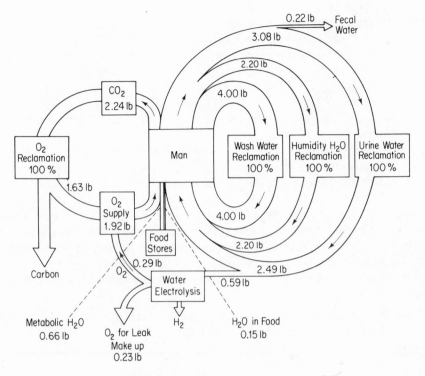

Fig. 2.11. Water and oxygen recycle in a space vehicle.

TABLE 2.1 MAN'S DAILY BALANCE

		Total balance (lb)	Water balance (lb)
Output	Urine (95%H$_2$O)	3.24	3.08
	Feces (75.8%H$_2$O)	0.29	0.22
	Transpired	2.20	2.20
	Carbon dioxide (1.63 lb O$_2$)	2.24	
	Other losses	0.14	
		8.11	5.50
Input	Food	1.5	0.15
	Oxygen	1.92	
	Metabolic water		0.66
	Water	4.69	4.69
		8.11	5.50

Recycle streams initially may be confusing to you, but with a little practice in solving problems involving recycle you should thereafter experience little difficulty. The first point you should grasp concerning recycle calculations is that the process shown in Fig. 2.10 is in the *steady state*—no buildup or depletion of material takes place inside the process or in the recycle stream. The values of F, P, and R are *constant*. The feed to the process itself is made up of two streams, the fresh feed and the recycle material. The gross product leaving the process is separated into two streams, the net product and the material to be recycled. In some cases the recycle stream may have the same composition as the gross product stream, while in other instances the composition may be entirely different depending on how the separation takes place and what happens in the process.

Recycle problems have certain features in common, and the techniques you should use in solving these problems are the familiar ones previously encountered in this chapter in the use of material balances. You can make a material balance on a total basis or for each component. Depending on the information available concerning the amount and composition of each stream, you can determine the amount and composition of the unknowns. If tie components are available, they simplify the calculations. If they are not available, then algebraic methods should be used.

A balance can be written in several ways, four of which are shown by dashed lines in Fig. 2.10:

(a) About the entire process including the recycle stream, as indicated by the dashed lines marked 1 in Fig. 2.10.
(b) About the junction point at which the fresh feed is combined with the recycle stream (marked 2 in Fig. 2.10).
(c) About the process only (marked 3 in Fig. 2.10).
(d) About the junction point at which the gross product is separated into recycle and net product (marked 4 in Fig. 2.10).

Only three of the four balances are independent. However, balance 1 will not include the recycle stream, so that the balance will not be directly useful in calculating a value for the recycle R. Balances 2 and 4 to include R. It would also be possible to write a material balance for the combination of 2 and 3 or 3 and 4 and include the recycle stream.

If a chemical reaction occurs in the process, the stoichiometric equation, the limiting reactant, and the degree of completion all should be considered before beginning your calculations. As discussed in Chap. 1, the term *conversion* as applied to Fig. 2.10 may be either the fraction or percentage of the fresh feed which reacts or the fraction of the feed plus recycle. If 100 lb of substance A in the fresh feed is converted on an overall basis into 40 lb of desired product, 30 lb of waste, 20 lb of a secondary product, and 10 lb of A pass through the process unchanged, the total conversion of A is 90 percent based on the fresh feed. However, the *yield* of primary and secondary products is only 60 lb of products per 100 lb of A. If, in addition, the recycle stream contains 100 lb of A, the total conversion of A on a once-through basis is only 45 percent. You can see that the basis on which the conversion and yield are calculated must always be clearly specified. When the fresh feed consists of more than one material, the yield and conversion must be stated for a single component, usually the limiting reactant, the most expensive reactant, or some similar compound.

You should carefully analyze a problem to ascertain whether there is a simplified method of solution (as illustrated in the following examples), but if such a simplification escapes you, or if there is none, then the standard procedure of setting up algebraic material balances will always be effective, although perhaps long and involved. One psychological stumbling block is the stream of unknown weight or composition which is found to be essential for the solution and for which you have no information. By labeling this stream with a letter— R, X, or whatever—you can proceed to make material balances in the ordinary way and solve for the unknown stream. Of course, for each unknown you set up, you must write an independent equation, and so from the viewpoint of the economic use of your time, it is advisable to minimize the number of unknowns. Some illustrations follow.

EXAMPLE 2.20 Recycle without chemical reaction

A distillation column separates 10,000 lb/hr of a 50 percent benzene-50 percent toluene mixture. The product recovered from the condenser at the top of the column contains 95 percent benzene, and the bottoms from the column contain 96 percent toluene. The vapor stream entering the condenser from the top of the column is 8000 lb/hr. A portion of the product is returned to the column as reflux, and the rest is withdrawn for use elsewhere. Assume the compositions of the streams at the top of the column (V), the product withdrawn (D), and the reflux (R) are identical. Find the ratio of the amount refluxed to the product withdrawn.

Solution:

See Fig. E2.20. All the compositions are known and two weights are unknown. No tie components are evident in this problem; thus an algebraic solution is mandatory.

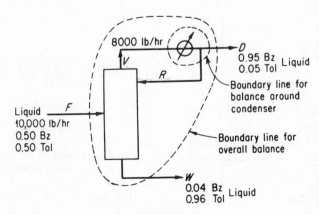

Fig. E2.20.

By making an overall balance (ignoring the reflux stream for the moment), we can find D. With D known, a balance around the condenser will give us R.

Basis; 1 hr

Overall material balances:
Total material:

$$F = D + W$$

$$10,000 = D + W \tag{a}$$

Component (benzene):

$$FX_F = DX_D + WX_W$$

$$10,000(0.50) = D(0.95) + W(0.04) \tag{b}$$

Solving (a) and (b) together,

$$5000 = (0.95)(10,000 - W) + 0.04W$$

$$W = 4950 \text{ lb/hr}$$

$$D = 5050 \text{ lb/hr}$$

Balance around the condenser:
Total material:

$$V = R + D$$

$$8000 = R + 5050$$

$$R = 2950 \text{ lb/hr}$$

$$\frac{R}{D} = \frac{2950}{5050} = 0.584$$

EXAMPLE 2.21   Recycle without chemical reaction

Data are presented in Fig. E2.21(a) for an evaporator. What is the recycle stream in pounds per hour?

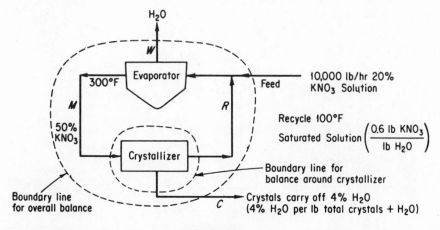

Fig. E2.21(a).

Solution:

In this problem the compositions are all known except for the recycle stream, but this can be easily calculated. On the basis of 1 lb of water, the saturated recycle stream contains $(1.0\,\text{lb H}_2\text{O} + 0.6\,\text{lb KNO}_3) = 1.6\,\text{lb}$ total. The recycle stream composition is

$$\frac{0.6\,\text{lb KNO}_3}{1\,\text{lb H}_2\text{O}}\bigg|\frac{1\,\text{lb H}_2\text{O}}{1.6\,\text{lb solution}} = 0.375\,\text{lb KNO}_3/\text{lb solution}$$

With all compositions known, a search for a tie component shows that one exists for an overall balance—the KNO_3. Temporarily ignoring the interior streams, we can draw a picture as in Fig. E2.21(b). Now we can calculate the amount of crystals (and H_2O evaporated, if wanted).

Basis: 1 hr $\equiv$ 10,000 lb feed

$$\frac{10,000\,\text{lb }F}{1\,\text{lb }F}\bigg|\frac{0.20\,\text{lb KNO}_3}{1\,\text{lb }F}\bigg|\frac{1\,\text{lb crystals}}{0.96\,\text{lb KNO}_3} = 2080\,\text{lb/hr crystals}$$

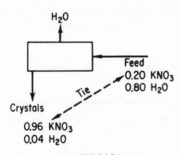

Fig. E2.21(b).

To determine the recycle stream R we need, in the absence of an additional tie component, either

(a) A balance around the evaporator—or
(b) A balance around the crystallizer.

The latter is easier since only three rather than four (the recycle added to the feed is the fourth stream) streams are involved.

Total balance on crystallizer:

$$M = C + R$$
$$M = 2080 + R$$

Component (KNO_3) balance on crystallizer:

$$Mx_M = Cx_C + Rx_R$$
$$0.5M = 0.96C + R(0.375)$$

Solving these two equations,

$$0.5(2080 + R) = 0.375R + 2000$$
$$R = 7680\,\text{lb/hr}$$

EXAMPLE 2.22 Recycle with a chemical reaction

In a proposed process for the preparation of methyl iodide, 2000 lb/day of hydro-iodic acid are added to an excess of methanol:

$$HI + CH_3OH \longrightarrow CH_3I + H_2O$$

If the product contains 81.6 percent CH_3I along with the unreacted methanol, and the waste contains 82.6 percent hydroiodic acid and 17.4 percent H_2O, calculate, assuming that the reaction is 40 percent complete in the process vessel:

(a) The weight of methanol added per day.
(b) The amount of HI recycled.

Solution:

In this problem all the compositions are known (in weight percent) as well as one weight—the HI input stream. The unknown quantities are the weights of the CH_3OH (M), the product (P), the waste (W), and the recycle stream. An overall balance about the whole process is all that is needed to calculate the M stream, but it will not include the recycle stream—see the dashed line in Fig. E2.22.

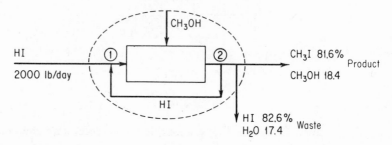

Fig. E2.22.

We first need to convert the product and waste streams to a mole composition in order to make use of the chemical equation. A basis of 100 lb of waste is convenient, although 100 lb of product would also be a satisfactory basis.

Basis: 100 lb waste

comp.	lb = %	MW	lb mole
HI	82.6	128	0.646
H_2O	17.4	18	0.968
	100.0		

Basis: 100 lb product

comp.	lb = %	MW	lb mole	equivalent lb mole of CH_3OH
CH_3I	81.6	142	0.575	0.575
CH_3OH	18.4	32	0.575	0.575
				1.150

(a) From the chemical equation we observe that associated with each pound mole of H_2O in the waste is 1 lb mole of CH_3I. Consequently the entering HI is (per 100 lb waste).

Reacted

$$\frac{0.968 \text{ lb mole } H_2O}{100 \text{ lb waste}} \left| \frac{1 \text{ lb mole } CH_3I}{1 \text{ lb mole } H_2O} \right| \frac{1 \text{ lb mole HI}}{1 \text{ lb mole } CH_3I} \left| \frac{128 \text{ lb HI}}{\text{lb mole HI}} \right. = \frac{124 \text{ lb HI}}{100 \text{ lb waste}}$$

Total HI = reacted + unreacted = $124 + 82.6 = 206.6$ lb HI/100 lb waste

On the same basis the CH_3OH entering is,

$$\frac{0.968 \text{ lb mole } H_2O}{100 \text{ lb waste}} \left| \frac{1 \text{ lb mole } CH_3I}{1 \text{ lb mole } H_2O} \right| \frac{1.150 \text{ lb mole } CH_3OH}{0.575 \text{ lb mole } CH_3I} \left| \frac{32 \text{ lb } CH_3OH}{1 \text{ lb mole } CH_3OH} \right.$$

$$= \frac{61.9 \text{ lb } CH_3OH}{100 \text{ lb waste}}$$

Basis: 1 day

$$\frac{2000 \text{ lb HI}}{1 \text{ day}} \left| \frac{100 \text{ lb waste}}{206.6 \text{ lb HI}} \right| \frac{61.9 \text{ lb } CH_3OH}{100 \text{ lb waste}} = 600 \text{ lb } CH_3OH/\text{day}$$

(b) To obtain the value of the recycle stream R, a balance must be made about a junction point or about just the reactor itself so as to include the stream R. A balance about junction 1 indicates that (all subsequent bases are 1 day)

$$200 + R = \text{reactor feed}$$

but this equation is not of much help since we do not know the rate of feed to the reactor. Similarly, for a balance on junction 2,

$$\text{gross product} = R + P + W$$

we can compute the value of the product and waste but not that of the gross product.

The only additional information available to us that has not yet been used is the fact that the reaction is 40 percent complete on one pass through the reactor. Consequently, 60 percent of the HI in the gross feed passes through the reactor unchanged. We can make an HI balance, then, about the reactor:

$$(2000 + R)(0.60) = R + 800$$

$$400 = 0.40R$$

$$R = 1000 \text{ lb/day}$$

Our material balance around the reactor for the HI is known as a *once-through balance* and says that 60 percent of the HI that enters the reactor (as fresh feed plus recycle) leaves the reactor unchanged in the gross product (recycle plus product).

Two additional commonly encountered types of process streams are shown in Figs. 2.12 and 2.13:

(a) A bypass stream—one which skips one or more stages of the process and goes directly to another stage.

(b) A purge stream—a stream bled off to remove an accumulation of inerts or unwanted material that might otherwise build up in the recycle

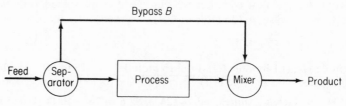

Fig. 2.12. Bypass stream.

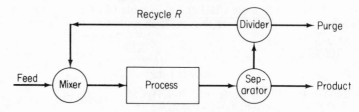

Fig. 2.13. A recycle stream with purge.

stream. The purge rate is adjusted so that the amount of purged material remains below a specified level or so that the

$$\left\{\begin{array}{c}\text{rate of}\\\text{accumulation}\end{array}\right\} = 0 = \left\{\begin{array}{c}\text{rate of entering material}\\\text{and/or production}\end{array}\right\} - \left\{\begin{array}{c}\text{rate of purge}\\\text{and/or loss}\end{array}\right\}$$

Calculations for bypass and purge streams introduce no new principles or techniques beyond those presented so far. An example will make this clear.

EXAMPLE 2.23 Bypass calculations

In the feed-stock preparation section of a plant manufacturing natural gasoline, isopentane is removed from butane-free gasoline. Assume for purposes of simplification that the process and components are as shown in Fig. E2.23. What fraction of the butane-free gasoline is passed through the isopentane tower?

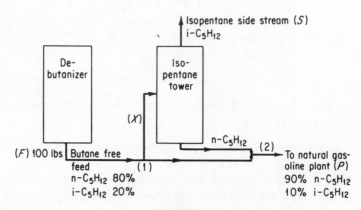

Fig. E2.23.

Solution:

By examining the flow diagram you can see that part of the butane-free gasoline bypasses the isopentane tower and proceeds to the next stage in the natural gasoline plant. All the compositions are known. What kind of balances can we write for this process? We can write the following (each stream is designated by the letter F, S, or P):

Basis: 100 lb feed

(a) *Overall balances* (all compositions are known; a tie component exists):
Total material balance:

in = out

$$100 = S + P \tag{1}$$

Component balance (n-C_5):

in = out

$$100(0.80) = S(0) + P(0.90) \tag{2}$$

Using (2),

$$P = 100\left(\frac{0.80}{0.90}\right) = 89 \text{ lb}$$

$$S = 100 - 89 = 11 \text{ lb}$$

The overall balances will not tell us the fraction of the feed going to the isopentane tower; for this we need another balance.

(b) *Balance around isopentane tower:* Let $x = $ lb of butane-free gas going to isopentane tower.
Total material:

in = out

$$x = 11 + n\text{-}C_5H_{12} \text{ stream (another unknown, } y) \tag{3}$$

Component (n-C_5):

$$x(0.80) = y \quad \text{(a tie component)} \tag{4}$$

Consequently, combining (3) and (4),

$$x = 11 + 0.8x$$

$$x = 55 \text{ lb}, \quad \text{or the desired fraction is 0.55}$$

Another approach to this problem is to make a balance at points (1) or (2) called *mixing points*. Although there are no pieces of equipment at those points, you can see that streams enter and leave the junction.

(c) *Balance around mixing point (2):*

material into junction = material out

Total material:

$$(100 - x) + y = 89 \tag{5}$$

Component (iso-C_5):

$$(100 - x)(0.20) + 0 = 89(0.10) \tag{6}$$

Equation (6) avoids the use of y. Solving,

$$20 - 0.2x = 8.9$$

$$x = 55 \text{ lb, as before}$$

After a little practice you can size up a problem and visualize the simplest types of balances to make.

Up to now we have discussed material balances of a rather simple order of complexity. If you try to visualize all the operations which might be involved in even a moderate-sized plant, as illustrated in Fig. 2.4, the stepwise or simultaneous solution of material balances for each phase of the entire plant is truly a staggering task, but a task that can be eased considerably by the use of computer codes. Keep in mind that a plant can be described by a number of individual, interlocking material balances which, however tedious they are to set up and solve, can be set down according to the principles and techniques discussed in this chapter. In a practical case there is always the problem of collecting suitable information and evaluating its accuracy, but this matter calls for detailed familiarity with any specific process and is not a suitable topic for discussion here. We can merely remark that some of the problems you will encounter have such conflicting data or so little useful data that the ability to perceive what kind of data are needed is the most important attribute you can bring to bear in their solution. By now, from working with simple problems, you should have some insight into the requirements for the solution of more complicated problems. In Chap. 5 you will encounter some of these more complex problems.

WHAT YOU SHOULD HAVE LEARNED FROM THIS CHAPTER

1. You should be able to analyze a material balance problem in order to
   (a) Find out what the problem is.
   (b) Draw a picture of the process.
   (c) Put down all the compositions of the entering and leaving streams.
   (d) Decide which masses (weights) are known and unknown.
   (e) Find a tie component and/or set up algebraic mass balances involving the unknowns.
   (f) Solve for the required values.
2. You should understand the once-through balance for recycle problems involving chemical reactions.

SUPPLEMENTARY REFERENCES

1. Benson, S. W., *Chemical Calculations*, 2nd ed., Wiley, New York, 1963.
2. Henley, E. J., and H. Bieber, *Chemical Engineering Calculations*, McGraw-Hill, New York, 1959.
3. Hougen, O. A., K. M. Watson, and R. A. Ragatz, *Chemical Process Principles*, Part I, 2nd ed., Wiley, New York, 1956.

4. Littlejohn, C. E., and G. F. Meenaghan, *An Introduction to Chemical Engineering*, Van Nostrand Reinhold, New York, 1959.

5. Nash, L., *Stoichiometry*, Addison-Wesley, Reading, Mass., 1966.

6. Ranz, W. E., *Describing Chemical Engineering Systems*, McGraw-Hill, New York, 1970.

7. Schmidt, A. X., and H. L. List, *Material and Energy Balances*, Prentice-Hall, Englewood Cliffs, N.J., 1962.

8. Whitwell, J. C., and R. K. Toner, *Conservation of Mass and Energy*, Ginn/Blaisdell, Waltham, Mass., 1969.

9. Williams, E. T., and R. C. Johnson, *Stoichiometry for Chemical Engineers*, McGraw-Hill, New York, 1958.

PROBLEMS[1]

2.1. The steady-state hydrology of the Great Lakes based on average flow values for the period 1900–1960 is shown in Fig. P2.1. Compute the return into lake St. Clair and the evaporation from Lake Ontario based on the data shown. Would the balance for 1 year be a steady-state balance? For 1 month?

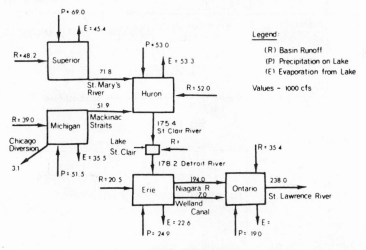

Fig. P2.1.

2.2. By use of an overall steady-state material balance determinine whether or not the petrochemical process indicated in Fig. P2.2 has been properly formulated. The block diagrams represent the steam cracking of naphtha into various products, and all flows are on an annual basis, i.e., per year.

---

[1]An asterisk designates problems appropriate for computer solution. Also refer to the computer problem after Problem 2.68.

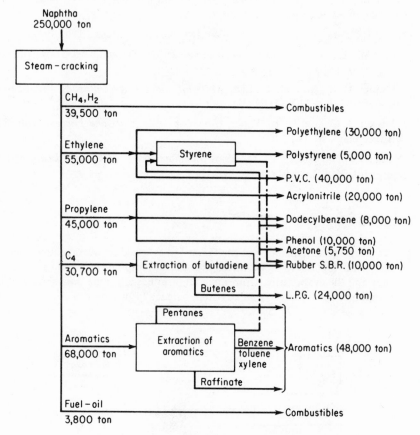

Fig. P2.2.

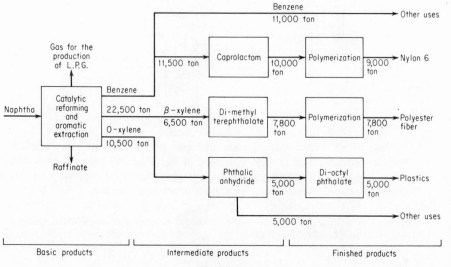

Fig. P2.3.

2.3. Figure P2.3 shows the production of aromatics, synthetic fibers, and plastics using naphtha as a feed stock. Check the overall steady-state material balance to see that the annual flows are correctly represented in the diagram.

2.4. Consider a lake as a system, and discuss the material balance on (a) water, (b) salts, and (c) plant life which might be made for such a lake. What quantities can be measured, and what quantities must be calculated?

2.5. A study of the carbon cycle in the biosphere is fundamentally a study of the overall global interactions of living organisms and their environment. The biosphere contains a complex mixture of carbon compounds in a continuous state of creation, transformation, and decomposition. This dynamic state is maintained through the ability of phytoplankton in the sea and plants on land to capture the energy of sunlight and utilize it to transform carbon dioxide (and water) into organic molecules. A fundamental and interesting problem is the effort to grasp the overall balance and flow of material in the worldwide community of plants and animals that has developed in the few billion years since life began. This is ecology in the broadest sense of the word. Figure P2.5 shows the carbon circulation in the biosphere expressed as billions of metric tons per year and also the carbon in various reservoirs.

(a) Is the carbon cycle as shown in the steady state?

(b) What is the net addition of carbon dioxide to the atmosphere per year?

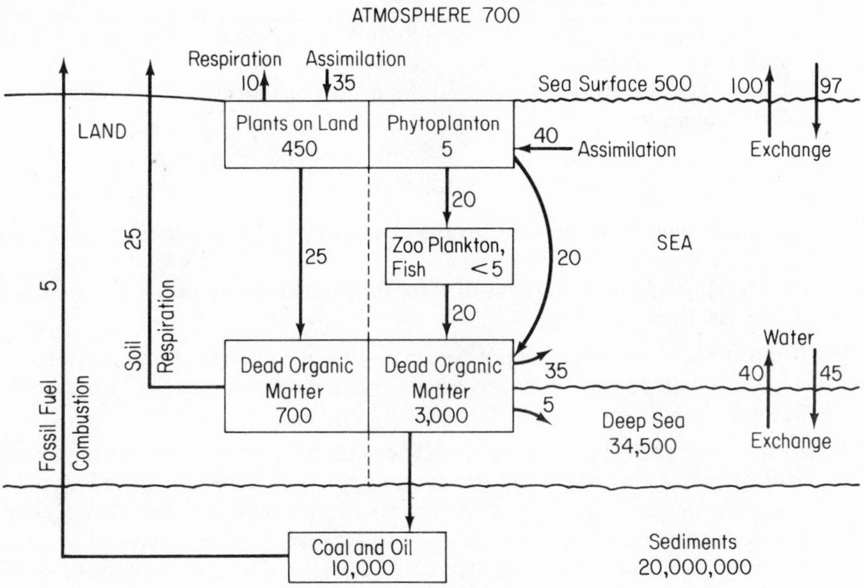

Fig. P2.5.

2.6. Hydrogen-free carbon in the form of coke is burnt (a) with complete combustion using theoretical air, (b) with complete combustion using 50 percent excess air, and (c) using 50 percent excess air but with 10 percent of the carbon burning to CO only. In each case calculate the gas analysis which will be found by testing the flue gases with an Orsat apparatus.

2.7. In the Deacon process for the manufacture of Cl_2, a dry mixture of HCl and air is passed over a heated catalyst that promotes the oxidation of HCl to Cl_2. The air used is 26 percent in excess of that quantitatively required for oxidation according to the reaction

$$4HCl + O_2 \rightleftharpoons 2Cl_2 + 2H_2O$$

Compute the composition of the gas that is fed to the reaction chamber.

2.8. It has been proposed [R. L. Miller and J. D. Winefordner, *Environmental Science and Technology*, v. 5, p. 445 (May 1971)] that a general relation can be developed for the ratio of O_2 to N_2 in combustion. For example, consider the following cases; the fuel might be cellulose $(C_6H_{10}O_5)_n$:

(1) Stoichiometric case: According to the balanced chemical reaction

$$1O_2 + 3.76N_2 + fuels \longrightarrow 0O_2 + 3.76N_2 + products$$

the ratio of O_2 to N_2 is 0.00.

(2) 50 percent excess air case: According to the balanced chemical reaction

$$1.50O_2 + 5.64N_2 + fuel \longrightarrow 0.5O_2 + 5.64N_2 + products$$

the ratio of O_2 to N_2 is $(0.5/5.64) = 0.0886$.

(3) 100 percent excess air case: According to the chemical reaction

$$2O_2 + 7.53N_2 + fuel \longrightarrow 1O_2 + 7.52N_2 + products$$

the O_2 to N_2 ratio is 0.133.

(4) Infinite excess air case: For this limiting case, the O_2 to N_2 ratio is simply $(1/3.76) = 0.0266$.

The ratio of O_2 to N_2, R, and the percentage excess air, P, are related by the equation

$$R = \frac{P}{3.76P + 376}$$

(a) Is the relation given valid for the ratio of O_2 to N_2 for cellulose? For other waste materials? For hexane?

(b) Develop a formula in terms of R for the percent excess air for the combustion of 100 lb of cellulose.

2.9. If moist hydrogen containing 4 percent water by volume is burnt completely in a furnace with 32 percent excess air, calculate the Orsat analysis of the resulting flue gas.

2.10. In the analysis of stack gases from incinerators it is desirable to have a convenient means for measuring the amount of air in excess of the air needed to burn the fuel completely, i.e., stoichiometric combustion. The Orsat analysis provides information that can be used to estimate the percentage of excess air. It is proposed that the ratio of O_2 to N_2 would provide a good means of estimating the percentage of excess air where the fuel is cellulose $(C_6H_{10}O_5)$ and the incinerator temperature is not excessively hot, causing the N_2 in the air to oxidize. Determine what the O_2 to N_2 ratio is in the stack gas of an incinerator for the stoichiometric burning of cellulose and for 20, 50, and 100 percent excess air.

2.11. Aviation gasoline is isooctane, C_8H_{18}. If it is burned with 20 percent excess air and 30 percent of the carbon forms carbon monoxide, what is the flue-gas analysis?

2.12. In the United States the five most common primary air pollutants (in tons emitted annually) are carbon monoxide, sulfur oxides, hydrocarbons, nitrogen oxides, and particulate matter. If a natural gas that analyzes 80 percent CH_4, 10 percent H_2, and 10 percent N_2 is burned with 40 percent excess air and 10 percent of the carbon forms CO, compute the Orsat (dry basis) analysis of the resulting flue gas. Will the nitrogen be oxidized? Recommend a method of eliminating the CO formed.

2.13. A natural gas analyzes CH_4, 80.0 percent and N_2, 20.0 percent. It is burned under a boiler and most of the CO_2 is scrubbed out of the flue gas for the production of dry ice. The exit gas from the scrubber analyzes CO_2, 1.2 percent; O_2, 4.9 percent; and N_2, 93.9 percent. Calculate the
(a) Percentage of the CO_2 absorbed.
(b) Percent excess air used.

2.14. A natural gas that analyzes 80 percent CH_4 and 20 percent N_2 is burned, and the CO_2 is scrubbed out of the resulting products for use in the manufacture of dry ice. The exit gases from the scrubber analyze 6 percent O_2 and 94 percent N_2. Calculate the
(a) Air to gas ratio.
(b) Percent excess air.

2.15. The U.B.T. Development Company is selling a fuel cell which generates electrical energy by direct conversion of coal into electrical energy. The cell is quite simple —it works on the same principle as a storage battery and produces energy free from the Carnot cycle limitation. To make clear the outstanding advantages of the fuel cell, we can consider the reaction

$$\text{fuel} + \text{oxygen} = \text{oxidation products} \tag{1}$$

Reaction (1) is intended to apply to any fuel and always to 1 mole of the fuel. The fuel cell is remarkable in that it can convert chemical energy directly into work and bypass the wasteful intermediate conversion into heat. A typical fuel consists of

C	65%
H	5
O	10
S	4
Ash	16

Assume that no carbon is left in the ash and that the S is oxidized to SO_2. If 20 percent excess air is used for oxidation of the fuel, calculate the compositions of all the oxidation products.

2.16. A steel-annealing furnace burns a fuel oil, the composition of which can be represented as $(CH_2)_n$. It is planned to burn this fuel with 12 percent excess air. Assuming complete combustion, calculate the Orsat analysis of the flue gas. Repeat this problem on the assumption that 5 percent of the carbon in the fuel is burned to CO only.

2.17. A producer gas analyzing CO_2, 4.5 percent; CO, 26 percent; H_2, 13 percent;

CH_2, 0.5 percent; and N_2, 56 percent, is burned in a furnace with 10 percent excess air. Calculate the Orsat analysis of the flue gas.

2.18. The Deacon process can be reversed and HCl formed from Cl_2 and steam by removing the O_2 formed with hot coke as follows:

$$2Cl_2 + 2H_2O + C \longrightarrow 4HCl + CO_2$$

If chlorine cell gas analyzing 90 percent Cl_2 and 10 percent air is mixed with 5 percent excess steam and the mixture is passed through a hot coke bed at 900°C, the conversion of Cl_2 will be 80 percent complete, but all the O_2 in the air will react. Calculate the composition of the exit gases from the converter, assuming no CO formation.

2.19. Solvents emitted from industrial operations can become significant pollutants if not disposed of properly. A chromatographic study of the waste exhaust gas from a synthetic fiber plant has the following analysis in mole percent:

$$
\begin{array}{ll}
CS_2 & 40\% \\
SO_2 & 10 \\
H_2O & 50
\end{array}
$$

It has been suggested that the gas be disposed of by burning with an excess of air. The gaseous combustion products are then emitted to the air through a smokestack. The local air pollution regulations say that no stack gas is to analyze more than 2 percent SO_2 by an Orsat analysis averaged over a 24-hr period. Calculate the minimum percent excess air that must be used to stay within this regulation.

2.20. An industrial gas has the following composition:

$$
\begin{array}{ll}
CO & 35\% \\
H_2 & 45 \\
N_2 & 10 \\
O_2 & 5 \\
CO_2 & 5
\end{array}
$$

Rather than waste the gas, it is more economical to burn it with sufficient air so that the total amount of oxygen available for combustion is that theoretically required for complete combustion. The exit gases, which are sufficiently hot to maintain all water formed in the vapor state, contain 61.4 percent N_2 and 19.4 percent H_2O. Compute the percent completion of the combustion of the H_2.

2.21. Twelve hundred pounds of $Ba(NO_3)_2$ are dissolved in sufficient water to form a saturated solution at 90°C, at which temperature the solubility is 30.6 g/100 g water. The solution is then cooled to 20°C, at which temperature the solubility is 8.6 g/100 g water.

(a) How many pounds of water are required for solution at 90°C, and what weight of crystals is obtained at 20°C?

(b) How many pounds of water are required for solution at 90°C, and what weight of crystals is obtained at 20°C, assuming that 10 percent more water is to be used than necessary for a saturated solution at 90°C?

(c) How many pounds of water are required for solution at 90°C, and what weight of crystals is obtained at 20°C, assuming that the solution is to be made up 90 percent saturated at 90°C?

(d) How many pounds of water are required for solution at 90°C, and what weight of crystals is obtained at 20°C, assuming that 5 percent of the water evaporates on cooling and that the crystals hold saturated solution mechanically in the amount of 5 percent of their dry weight?

2.22. If 100 g of Na_2SO_4 is dissolved in 200 g of H_2O and the solution is cooled until 100 g of $Na_2SO_4 \cdot 10H_2O$ crystallizes out, find

(a) The composition of the remaining solution (*mother liquor*).

(b) The grams of crystals recovered per 100 g of initial solution.

2.23. The solubility of magnesium sulfate at 20°C is 35.5 g/100 g H_2O. How much $MgSO_4 \cdot 7H_2O$ must be dissolved in 100 lb of H_2O to form a saturated solution?

2.24. The solubility of manganous sulfate at 20°C is 62.9 g/100 g H_2O. How much $MnSO_4 \cdot 5H_2O$ must be dissolved in 100 lb of water to give a saturated solution?

2.25. How much water must be evaporated from 100 gal of Na_2CO_3 solution, containing 50 g/l at 30°C, so that 70 percent of the Na_2CO_3 will crystallize out when the solution is cooled to 0°C?

2.26. An evaporator is fed 100 lb/min of 23.1 percent Na_2SO_4 solution and the concentrated product is passed through a crystallizer where it is cooled to 0°C. How much water must be evaporated per minute so that 90 percent of the sodium sulfate (MW = 142) will crystallize out? Note that at 32.4°C Na_2SO_4 becomes a hydrate:

$$Na_2SO_4 + 10H_2O \longrightarrow Na_2SO_4 \cdot 10H_2O$$

The solubility of Na_2SO_4 and the hydrate are

temp., °C	solubility, g/100 g H_2O
100	42.5
32.4	49.5
0	5.0

2.27. The feed to a distillation column is separated into net overhead product containing nothing with a boiling point higher than isobutane and bottoms containing nothing with a boiling point below that of propane. See Fig. P2.27. The composition of the feed is

	mole %
Ethylene	2.0
Ethane	3.0
Propylene	5.0
Propane	15.0
Isobutane	25.0
n-Butane	35.0
n-Pentane	15.0
Total	100.0

The concentration of isobutane in the overhead is 5.0 mole percent, and the concentration of propane in the bottoms is 0.8 mole percent. Calculate the composition of the overhead and bottoms streams per 100 moles of feed.

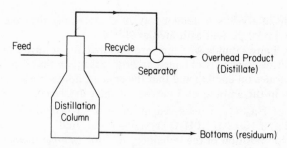

Fig. P2.27.

2.28. A gas containing 80 percent CH_4 and 20 percent He is sent through a quartz diffusion tube (see Fig. P.2.28). to recover the helium. Twenty percent by weight of the original gas is recovered, and its composition is 50 percent He. Calculate the composition of the waste gas if 100 lb moles of gas are processed per minute. The initial gas pressure is 17 psia, and the final gas pressure is 2 psig. The barometer reads 740 mm Hg. The temperature of the process is 70°F.

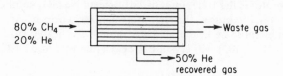

Fig. P2.28.

2.29. A and B are immiscible liquids, but they emulsify in one another to give uniform emulsions. See Fig. P2.29. The uniform emulsion is withdrawn from the lower layer and sent to a settler where the emulsion breaks and is separated. During the addition of 698 kg of A and 1302 kg of B to the extractor, the interface between the layers rises to and stays at a new level of 6 cm higher than the old level. A rise of 1 cm corresponds in volume to 30 kg of A or 40 kg of B. The volumes of A and B are additive in all proportions. The top level in the extrac-

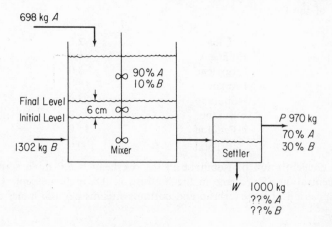

Fig. P2.29.

tor remains constant during the operation. What is the composition of the bottom layer from the settler?

2.30. To prepare a solution of 50.0 percent sulfuric acid, a dilute waste acid containing 28.0 percent H_2SO_4 is fortified with a purchased acid containing 96.0 percent H_2SO_4. How many kilograms of the purchased acid must be bought for each 100 kg of dilute acid?

2.31. A fuel composed of ethane (C_2H_6) and methane (CH_4) in unknown proportions is burned in a furnace with oxygen-enriched air (50.0 mole percent O_2). Your Orsat analysis is: CO_2, 25 percent; N_2, 60 percent; and O_2, 15 percent. Find
(a) The composition of the fuel, i.e., the mole percent methane in the methane-ethane mixture.
(b) The moles of oxygen-enriched air used per mole of fuel.

2.32. In modern U.S. incinerators, refuse burns on moving grates in refractory-lined chambers, and combustible gases and entrained solids burn in secondary combustion chambers or zones. Combustion is 85–90 percent complete for the combustible materials. A large body of technology has grown up and a variety of mechanical designs of incinerators are available. The scientific principles that underlie the technology often are not well defined, partly because of the heterogeneous nature of refuse and its variable moisture. Up to three times as much air is introduced into the incinerator as would be needed to supply the oxygen required to oxidize the refuse completely. The temperature in the bed of burning refuse may reach 2500°F or more, and the excess air is required mainly to hold the temperature in the furnace at 1400°–1800°F. Above 1800°F, slag formation in the furnace can become a problem. The table lists the results from one incinerator.
(a) Based on the gas analysis, what was the percent excess air used in the incinerator?
(b) Does the solid analysis agree with the gas analysis?

Stack Gases	Solid Analysis lb/ton of Refuse	Gas Analysis Fraction by Volume, Dry
Carbon dioxide	1,738	6.05%
Sulfur dioxide	1	22 ppm
Carbon monoxide	10	0.06%
Oxygen	2,980	14.32%
Nitrogen oxides (NO)	3	93 ppm
Nitrogen	14,557	79.57%
Total dry gas	19,289	100%
Water vapor	1,400	
Total	20,689	
Solids, dry basis		
Grate residue	471	
Fly ash	20	
Total, lb/ton of refuse	21,180	

SOURCE: E. R. Kaiser, "Refuse Reduction Processes," in *Proceedings, The Surgeon General's Conference on Solid Waste Management for Metropolitan Washington*, U.S. Public Health Service publication No. 1729, Government Printing Office, Washington, D.C., July 1967, p. 93.

2.33. Ammonia is a gas for which reliable analytical methods are available to determine its concentration in other gases. To measure flow in a natural gas pipeline, pure ammonia gas is injected into the pipeline at a constant rate of 72.3 kg/min for 12 min. Five miles downstream from the injection point, the steady-state ammonia concentration is found to be 0.382 weight percent. The gas upstream from the point of ammonia injection contains no measurable ammonia. How many kilograms of natural gas are flowing through the pipeline per hour?

2.34. A lacquer plant must deliver 1000 lb of an 8 percent nitrocellulose solution. They have in stock a 5.5 percent solution. How much dry nitrocellulose must be dissolved in the solution to fill the order?

2.35. Paper pulp is sold on the basis that it contains 12 percent moisture; if the moisture exceeds this value, the purchaser can deduct any charges for the excess moisture and also deduct for the freight costs of the excess moisture. A shipment of pulp became wet and was received with a moisture content of 22 percent. If the original price for the pulp was $40/ton of air-dry pulp and if the freight is $1.00/100 lb shipped, what price should be paid per ton of pulp delivered?

2.36. To meet certain specifications, a dealer mixes bone-dry glue, selling at 25 cents/lb, with glue containing 22 percent moisture, selling at 14 cents/lb, so that the mixture contains 16 percent moisture. What should be the selling price per pound of the mixed glue?

2.37. A dairy produces casein which when wet contains 23.7 percent moisture. They sell this for $8.00/100 lb. They also dry this casein to produce a product containing 10 percent moisture. Their drying costs are $0.80/100 lb water removed. What should be the selling price of the dried casein to maintain the same margin of profit?

2.38. In 1969 the Federal Water Pollution Control Administration issued a guidance memorandum indicating that the reduction of phosphate in detergents was desirable. The question, of course, was what replacement substances could be used that would not contribute more pollutional effects to receiving waters. A company manufactures Spic and Spotless detergent, which contains 11.1 percent phosphorus (reported as P) on an as purchased (wet) basis and 14.7 percent on a dry basis. Suppose that all the phosphates ($Na_5P_3O_{10}$, $Na_4P_2O_7$, Na_3PO_4, etc.) are to be removed from the soap formulation and replaced by NTA (sodium nitrilotriacetate), which is 1.8 times as effective as a water softener as Na_3PO_4, so that the weight percent NTA is 6.0 percent in the new blend of detergent. If NTA costs $0.75/lb and the phosphates sell for $0.041/lb P, will the new detergent cost more or less than the original detergent? By how much in dollars relative to the original detergent?

2.39. A paper manufacturer contracts for rosin size containing not more than 20 percent water at 12 cents/lb f.o.b. the rosin-size plant, a deduction to be made for water above this amount, and the excess freight on the water is to be charged back to the manufacturer at 80 cents/100 lb. A shipment of 2400 lb is received which analyzes 26.3 percent water. What should the paper manufacturer pay for the shipment?

2.40. In anaerobic decomposition, microbial organisms decompose organic compounds in the absence of oxygen to CO_2, CH_4, H_2S, etc. Anaerobic digestion is used in waste-disposal methods but also occurs naturally in many lakes, swamps, etc., where the characteristic smell of H_2S is easily noted. In sewage purification where the solids decompose on the bottom of basins, to avoid the smell, the upper layers of the water are aerated so that the H_2S is oxidized to less odoriferous compounds. If the Orsat gas analysis above a test basin is SO_2, 7.2 percent; O_2, 10.1 percent; and N_2, 82.7 percent, what percentage of the sulfur in the H_2S was oxidized to SO_3?

2.41. A natural gas which is entirely methane, CH_4, is burned with an oxygen-enriched air so that a higher flame temperature may be obtained. The flue gas analyzes CO_2, 22.2 percent; O_2, 4.4 percent; and N_2, 73.4 percent. Calculate the percentage of O_2 and N_2 in the oxygen-enriched air.

2.42. As superintendent of a lacquer plant, your foreman brings you the following problem: he has to make up 1000 lb of an 8 percent nitrocellulose solution. He has available a tank of a 5.5 percent solution. How much dry nitrocellulose must he add to how much of the 5.5 percent solution in order to fill the order?

2.43. The Clean Air Act of 1970 requires automobile manufacturers to warrant their control systems as satisfying the emission standards for 50,000 miles. It requires owners to have their engine control systems serviced exactly according to manufacturers' specifications and to always use the correct gasoline. In testing an engine exhaust having a known Orsat analysis of 16.2 percent CO_2, 4.8 percent O_2, and 79 percent N_2 at the outlet, you find to your surprise that at the end of the muffler the Orsat analysis is 13.1 percent CO_2. Can this discrepancy be caused by an air leak into the muffler? (Assume the analyses are satisfactory.) If so, compute the moles of air leaking in per mole of exhaust gas leaving the engine.

2.44. Removal of sulfur dioxide from the flue gases of industrial installations is a topic which has attracted worldwide attention during the past decade. Recovery of sulfur dioxide from the flue gases of power plants would reduce air pollution and supplement the recovery of an important industrial chemical. A wide variety of processes is under development throughout the world and selection of the appropriate scheme depends on the size of the installation, the size of the available market for the final commodity produced, transport and handling costs, and the cost of capital plant and operating costs. In one proposal the flue gas from a high-sulfur fuel oil is burned to SO_2 and then further oxidized to SO_3 in a series of catalytic converters and absorbers. Each converter is followed by an absorber. The gas entering the first converter is (on a CO_2 free basis)

SO_2	0.80%
O_2	2.80
N_2	96.40

and the gas leaving the first absorber is

SO_2	0.20%
O_2	2.00
N_2	97.80

Calculate the degree of conversion of SO_2 in the first converter. What is the analysis of the exit gas after complete conversion and absorption has taken place (on a CO_2 free basis)?

$$SO_2 + \tfrac{1}{2}O_2 \longrightarrow SO_3 \qquad SO_3 + H_2O \longrightarrow H_2SO_4$$

2.45. Your boss asks you to calculate the flow through a natural-gas pipeline. Since it is 26 in. in diameter, it is impossible to run the gas through any kind of meter or measuring device. You decide to add 100 lb of CO_2 per minute to the gas through a small $\frac{1}{2}$-in. piece of pipe, collect samples of the gas downstream, and analyze them for CO_2. Several consecutive samples after 1 hr are

time	% CO_2
1 hr, 0 min	2.0
10 min	2.2
20 min	1.9
30 min	2.1
40 min	2.0

(a) Calculate the flow of gas in pounds per minute at the point of injection.

(b) Unfortunately for you, the gas at the point of injection of CO_2 already contained 1.0 percent CO_2. How much was your original flow estimate in error (in percent)?

Note: In part (a) the natural gas is all methane, CH_4.

2.46. A low-grade pyrites containing 32 percent S is mixed with 10 lb of pure sulfur per 100 lb of pyrites so the mixture will burn readily, forming a burner gas that analyzes (Orsat) SO_2, 13.4 percent; O_2, 2.7 percent; and N_2, 83.9 percent. No sulfur is left in the cinder. Calculate the percentage of the sulfur fired that burned to SO_3. (The SO_3 is not detected by the Orsat analysis.)

2.47. Pure barytes, $BaSO_4$, is fused with soda ash, Na_2CO_3, and the fusion mass is then leached with water. The solid residue from the leaching analyzes 33.6 percent $BaSO_4$ and 66.4 percent $BaCO_3$. The soluble salts in solution analyze 41.9 percent Na_2SO_4 and 58.1 percent Na_2CO_3. Calculate the composition of the mixture before fusion.

2.48. A power company operates one of its boilers on natural gas and another on oil (for peak period operation). The analysis of the fuels are as follows:

natural gas	oil
96% CH_4	$(CH_{1.8})_n$
4% CO_2	

When both boilers are on the line, the flue gas shows (Orsat analysis) 10.0 percent CO_2, 4.5 percent O_2, and the remainder N_2. What percentage of the total carbon burned comes from the oil? *Hint:* Do not forget the H_2O in the stack gas.

2.49. A power company operates one of its boilers on natural gas and another on oil. The analyses of the fuels show 96 percent CH_4, 2 percent C_2H_2, and 2 percent CO_2 for the natural gas and $C_nH_{1.8n}$ for the oil. The flue gases from both groups enter the same stack, and an Orsat analysis of this combined flue gas shows 10.0 percent CO_2, 0.63 percent CO, and 4.55 percent O_2. What percentage of the total carbon burned comes from the oil?

2.50. An automobile engine burning a fuel consisting of a mixture of hydrocarbons is found to give an exhaust gas analyzing 10.0 percent CO_2 by the Orsat method. It is known that the exhaust gas contains no oxygen or hydrogen. Careful metering of the air entering the engine and the fuel used shows that 12.4 lb of dry air enter the engine for every pound of fuel used.

(a) Calculate the complete Orsat gas analysis.

(b) What is the weight ratio of hydrogen to carbon in the fuel?

2.51. A fuel oil and a sludge are burned together in a furnace with dry air. Assume the fuel oil contains only C and H.

fuel oil	*sludge*		*flue gas*
	wet	*dry*	
C = ?%	water = 50%	S = 32%	SO_2 = 1.52%
	solids = 50	C = 40	CO_2 = 10.14
H = ?%		H_2 = 4	CO = 2.02
		O_2 = 24	O_2 = 4.65
			N_2 = 81.67

(a) Determine the weight percent composition of the fuel oil.

(b) Determine the ratio of pounds of sludge to pounds of fuel oil.

2.52. A solvent dewaxing unit in an oil refinery is separating 3000 bbl/day of a lubricating distillate into 23 vol percent of slack wax and 77 vol percent of dewaxed oil. The charge is mixed with solvent, chilled, and filtered into wax and oil solution streams. The solvent is then removed from the two streams by two banks of stripping columns, the bottoms from each column in a bank being charged to the next column in the bank. The oil bank consists of four columns, and the wax bank of three. A test on the charge and bottoms from each column gave the following results:

	to 1st column	*percent solvent by volume*			
		no. 1 bottoms	*no. 2 bottoms*	*no. 3 bottoms*	*no. 4 bottoms*
Pressed oil	83	70	27	4.0	0.8
Wax	83	71	23	0.5	—

Calculate the following:

(a) Total solution per day charged to the whole unit.

(b) Percentage of total solvent in oil solution removed by each column in oil bank.

(c) Percentage of total solvent in wax solution removed by each column in wax bank.

(b) Barrels of solvent lost per day (in bottoms from last column of each bank).

2.53. The price of crude oil is based on its API gravity; the highest gravities command the best prices. It is often possible to blend two crudes advantageously so that the price obtained for the mixture is greater than the price of the separate crudes. This is possible because the price vs. API function is discontinuous (like income tax rates). A crude oil with a 34.4°API gravity is to be mixed with 50,000 bbl of a 30.0°API gravity crude to give a mixture which has a gravity of 31.0°API. What is the increased selling price of the mixed crude over that of the separate components?

Data: (1) Crude prices:

API	*$/bbl*
29.0–30.9	$2.60
31.0–32.9	2.65
33.0–34.9	2.70

(2) API gravities are not additive on any basis.

(3) Sp gr at 60°F = 141.5/(131.5 + API).

2.54.* It is desired to mix three L.P.G. (liquified petroleum gas) products in certain proportions in order that the final mixture will meet certain vapor pressure specifications. These specifications will be met by a stream of composition D below. Calculate the proportions in which streams A, B, and C must be mixed to give a product with a composition of D. The values are liquid volume percent.

component	stream A	B	C	D
C_2	5.0			1.4
C_3	90.0	10.0		31.2
iso-C_4	5.0	85.0	8.0	53.4
n-C_4		5.0	80.0	12.6
iso-C_5^+			12.0	1.4
	100.0	100.0	100.0	100.0

2.55. A coal analyzes 74 percent C and 12 percent ash (inert). The flue gas from the combustion of the coal analyzes CO_2, 12.4 percent; CO, 1.2 percent; O_2, 5.7 percent; and N_2, 80.7 percent. Calculate the following:

(a) The pounds of coal fired per 100 moles of flue gas.

(b) The percent excess air.

(c) The pounds of air used per pound of coal.

Assume that there is no nitrogen in the coal.

2.56. Sea water is to be desalinized by reverse osmosis using the scheme indicated in Fig. P2.56. Use the data given in the figure to determine

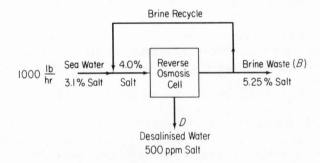

Fig. P2.56.

(a) The rate of waste brine removal (B).

(b) The rate of desalinized water (called potable water) production (D).

(c) The fraction of the brine leaving the osmosis cell (which acts in essence as a separator) that is recycled.

Note: ppm designates parts per million.

2.57. In the production of NH_3, the mole ratio of the N_2 to the H_2 in the feed to the whole process is 1 N_2 to 3 H_2. Of the feed *to the reactor*, 25 percent is converted to NH_3. The NH_3 formed is condensed to a liquid and completely removed from the reactor, while the unreacted N_2 and H_2 are recycled back to mix with the feed to the process. What is the ratio of recycle to feed in pound recycle per pound feed? The feed is at 100°F and 10 atm, while the product is at 40°F and 8 atm.

2.58.* An isomerizer is a catalytic reactor which simply tends to rearrange isomers. The number of moles entering an isomerizer is equal to the number of moles leaving. A process, as shown Fig. P2.58, has been designed to produce a *p-*

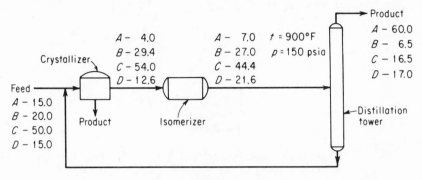

Fig. P2.58.

xylene-rich product from an aromatic feed charge. All compositions on the flow sheet are in mole percent. The components are indicated as follows:

A	ethyl benzene
B	*o*-xylene
C	*m*-xylene
D	*p*-xylene

Eighty percent of the ethyl benzene entering the distillation tower is removed in the top stream from the tower. The ratio of the moles fresh feed to the process as a whole to the moles of product from the crystallizer is 1.63. Find the

(a) Reflux ratio (ratio of moles of stream from the bottom of the distillation tower per mole of feed to the tower).

(b) Composition (in mole percent) of the product from the crystallizer.

(c) Moles leaving the isomerizer per mole of feed.

2.59. In an attempt to provide a means of generating NO cheaply, gaseous NH_3 is burned with 20 per cent excess O_2:

$$4NH_3 + 5O_2 \longrightarrow 4NO + 6H_2O$$

The reaction is 70 percent complete. The NO is separated from the unreacted NH_3, and the latter recycled as shown Fig. P2.59. Compute the

(a) Moles of NO formed per 100 moles of NH_3 fed.

(b) Moles of NH_3 recycled per mole of NO formed.

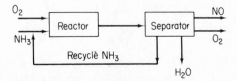

Fig. P2.59.

2.60. Acetic acid is to be generated by the addition of 10 percent excess sulfuric acid to calcium acetate. The reaction $Ca(Ac)_2 + H_2SO_4 \longrightarrow CaSO_4 + 2HAc$ goes to 90 percent completion. The unused $Ca(Ac)_2$ and the H_2SO_4 are separated from the products of the reaction, and the excess $Ca(Ac)_2$ is recycled. The acetic acid is separated from the products. Find the amount of recycle per hour based on 1000 lb of feed per hour, and also the pounds of acetic acid manufactured per hour. See Fig. P2.60.

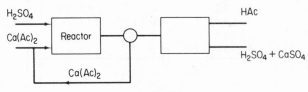

Fig. P2.60.

2.61. Iodine can be obtained commercially from kelp by treating the seaweed with MnO_2 and 20 percent excess H_2SO_4 according to the reaction

$$2NaI + MnO_2 + 2H_2SO_4 \longrightarrow Na_2SO_4 + MnSO_4 + 2H_2O + I_2$$

All the H_2O, Na_2SO_4, $MnSO_4$, I_2, and inerts are removed in the separator. The kelp contains 5 percent NaI, 30 percent H_2O, and the rest can be considered to be inerts. The product contains 54 percent I_2 and 46 percent H_2O. See Fig. P2.61. Assuming the reaction to be 80 percent complete, calculate the following:
(a) The pounds of I_2 produced per ton of kelp used.
(b) The percentage composition of the waste.
(c) The percentage composition of the recycle.

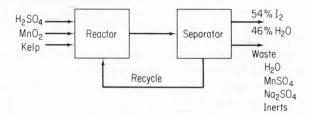

Fig. P2.61.

2.62. A solution containing 10 percent NaCl, 3 percent KCl, and 87 percent water is fed to the process shown in Fig. P2.62 at the rate of 18,400 kg/hr. The compositions of the streams are as follows:

	NaCl	KCl	H$_2$O
Evap. product, *P*	16.8	21.6	61.6
Recycle, *R*	18.9	12.3	68.8

Calculate the kilograms per hour of solution *P* passing from the evaporator to the crystallizer and the kilograms per hour of the recycle *R*.

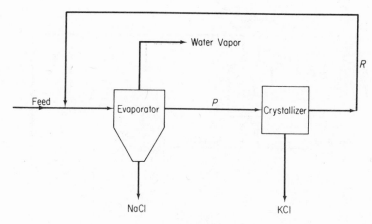

Fig. P2.62.

2.63.* A natural gasoline plant at Short Junction, Oklahoma produces gasoline by removing condensable vapors from the gas flowing out of gas wells. The gas from the well has the following composition:

component	mole %
CH_4	77.3
C_2H_6	14.9
C_3H_8	3.6
iso- and n-C_4H_{10}	1.6
C_5 and heavier	0.5
N_2	2.1
	100.0

This gas passes through an absorption column called a scrubber (see Fig. P2.63), where it is scrubbed with a heavy, nonvolatile oil. The gas leaving the scrubber has the following analysis:

component	mole %
CH_4	92.0
C_2H_6	5.5
N_2	2.5

The scrubbing oil absorbs none of the CH_4 or N_2, much of the ethane, and all of the propane and higher hydrocarbons in the gas stream. The oil stream is then sent to a stripping column, which separates the oil from the absorbed hydrocarbons. The overhead from the stripper is termed *natural gasoline*, while the bottoms from the stripper is called *lean oil*. The lean-oil stream is cooled and returned to the absorption column. Assume that there are no leaks in the system. Gas from the wells is fed to the absorption column at the rate of 52,000 lb moles/day and the flow rate of the rich oil going to the stripper is 1230 lb/min (MW = 140). Calculate the following:

(a) The pounds of CH_4 passing through the absorber per day.

(b) The pounds of C_2H_6 absorbed from the gas stream per day.

(c) The weight percentage of propane in the rich oil stream leaving the scrubber.

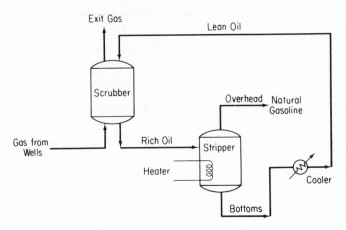

Fig. P2.63.

2.64. Raw water for use in a boiler has the following analysis in ppm (pounds per million pounds of pure water):

Ca^{2+}	90
Mg^{2+}	60
Na^+ and K^+	40
HCO_3^-	270
SO_4^{2-} and Cl^-	160
Total solids	620 ppm

Na_3PO_4 is added in 10 percent excess above that required to form insoluble $Ca_3(PO_4)_2$ and $Mg_3(PO_4)_2$, which are allowed to settle out before the treated water is fed to the boiler. See Fig. P2.64. The reactions are complete, and no other reactions occur. The steam produced entrains 4 lb of liquid per 100 lb of dry steam. Sixty percent of this wet steam returns to the boiler as condensate after

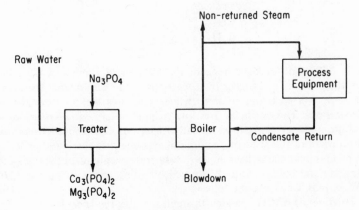

Fig. P2.64.

picking up an additional 40 ppm of dissolved solids during passage through the equipment and return lines. The total solids concentration in the boiler must not exceed 3000 ppm. Therefore water of this concentration is continuously withdrawn, or "blown down." The solids content of the liquid in the steam is the same as that of the liquid in the boiler. Compute the ratio of blowdown to treated-water feed required.

2.65. In the operation of a synthetic ammonia plant, an excess of hydrogen is burned with air so that the burner gas contains nitrogen and hydrogen in a 1:3 mole ratio and no oxygen. Argon is also present in the burner gas since it accounts for 0.94 percent of air. The burner gas is fed to a converter where a 25 percent conversion of the N_2-H_2 mixture to ammonia is produced. The ammonia formed is separated by condensation and the unconverted gases are recycled to the converter. To prevent accumulation of argon in the system, some of the unconverted gases are vented before being recycled to the converter. The upper limit of argon in the converter is to be 4.5 percent of the entering gases. What percentage of the original hydrogen is converted into ammonia?

2.66. A 50 percent NaCl solution is to be concentrated in a triple-effect evaporator as shown in Fig. P2.66. (Each individual evaporator is termed an *effect*.) An equal

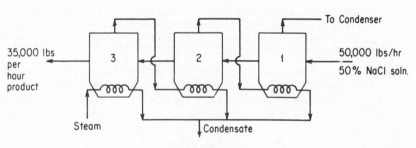

35,000 lbs per hour product

To Condenser

50,000 lbs/hr 50% NaCl soln.

Steam Condensate

Fig. P2.66.

amount of water is evaporated in each effect. Determine the composition of the outlet stream from effect 2 if the internal contents of effect 2 are uniformly mixed so that the outlet stream has the same composition as the internal contents of effect 2. The steam lines in each effect are completely separate from the evaporator contents so that no mixing of the steam with the contents occurs.

2.67. In the extraction process shown in Fig. P2.67, a feed stream, designated A, containing 70 percent by weight of unwanted material, U, is extracted in a countercurrent extraction system with the solvent, chlorex, C. One thousand pounds of A and 3000 lb of C are charged per hour. The paraffin oil, P, exits at one end of the system and the undesirable material exits from the other end. Associated with the paraffin oil, P, is 0.1 lb of chlorex per pound of paraffin oil in the product. The undesired material, U, at the other end of the apparatus will retain the remainder of the chlorex and in addition will retain 0.2 lb of paraffin oil per pound of U (this paraffin oil is not recovered).

(a) What weight of paraffin oil is recovered per hour?

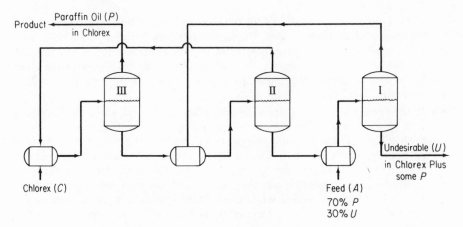

Fig. P2.67.

(b) What weight of undesirable material, including the paraffin oil retained with the *U*, is recovered per hour?

(c) If 700 lb of paraffin oil containing 0.2 lb of *U* per pound of paraffin oil flow per hour from settler I to settler II and 670 lb of paraffin oil containing 0.05 lb of *U* per pound of paraffin oil flow per hour from settler II to settler III, what weights of undesirable material *U* are being transferred per hour from settlers III to II and from II to I?

2.68. Two thousand kilograms per hour of solid industrial waste containing a toxic organic compound are extracted with 20,000 kg/hr of pure solvent in a two-stage countercurrent extraction system. See Fig. P2.68. The solid waste contains 10 percent by weight of extractable compound. After exposure to the solvent, the waste retains 1.5 kg of solvent per kilogram of waste (on a toxic compound-free basis). What fraction of the toxic compound is recovered per hour?

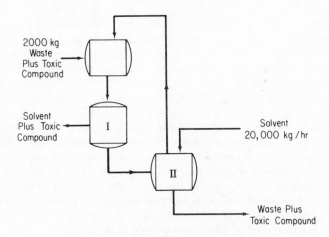

Fig. P2.68.

PROBLEM TO PROGRAM ON THE COMPUTER

2.1. A 15-stage extraction column similar to the one in Example 2.13 is set up to separate acetone and ethanol using two solvents, pure water and pure chloroform. The only added features to Fig. E2.13 in Example 2.13 are (1) that you must consider two solutes in each solvent instead of one solute in the two solvents and (2) the acetone-ethanol feed is introduced into stage 6. You know that the feed composition is 50 mole percent ethanol and 50 mole percent acetone and that the feed rate is 20 lb moles/hr. Assume that the water and chloroform are not soluble in each other. You want to have 1.50 mole percent water in the exit stream from the bottom of the column, and 2×10^{-5} mole percent acetone in the exit stream from the top of the column. Determine the rates of flow of the two solvents. In each stage you are given a relationship between X_i^A and X_i^B ($i = 1, 2, 3$) in the two phases in terms of *mole fractions:*

$$\frac{x_i^B}{x_i^A} = \frac{\gamma_i^B}{\gamma_i^A} \tag{a}$$

Each of the γs in phase A or phase B can be expressed in terms of some known constants and the mole fraction of the other two components. For component i,

$$\ln \gamma_i = 2x_i \sum_{j=1}^{3} x_j a_{ji} + \sum_{j=1}^{3} x_j^2 a_{ij} + \sum_{j=1}^{3} \sum_{\substack{k=1 \\ j \neq i \\ k \neq i \\ j < k}}^{3} x_j x_k a_{ijk}^*$$

$$- 2 \sum_{i=1}^{3} x_i^2 \sum_{j=1}^{3} x_j a_{ij}$$

$$- 2 \sum_{i=1}^{3} \sum_{j=1}^{3} \sum_{k=1}^{3} x_i x_j x_k a_{ijk}^* \tag{b}$$

where $a_{ijk}^* = a_{ij} + a_{ji} + a_{ik} + a_{ki} + a_{jk} + a_{kj}$. The values of the coefficients are

	1: acetone		3: chloroform
	2: ethanol		4: water

a_{11} 0.0	$a_{12} = 0.5446$	$a_{13} = 0.9417$	$a_{14} = 1.872$
$a_{21} = 0.599$	$a_{22} = 0.0$	$a_{23} = 1.61$	$a_{24} = 1.46$
$a_{31} = 0.674$	$a_{32} = 0.501$	$a_{33} = 0.0$	$a_{34} = 5.91$
$a_{41} = 1.338$	$a_{42} = 0.877$	$a_{43} = 4.76$	$a_{44} = 0.0$

The γs are obtained for each component, 1, 2, 3, by permuting the subscripts cyclically $i \longrightarrow j \longrightarrow k \longrightarrow i$. *Hints:* Select some appropriate initial values for the flow rates of A and B. Divide the feed (stage 6) equally between both phases. The initial compositions can be provided by specifying both stage 1 and stage 15 and generating the remaining compositions by linear interpolation. Be sure to include (a) one equation such as Eq. (a) for each component for each stage, (b) the correct number of component material balances for each stage (a total balance for the stage can be substituted for one component balance), and (c) $\sum x_i = 1$ for each stage for one phase (similar equations for the other phase are redundant).

Gases, Vapors, Liquids, and Solids 3

In planning and decision making for our modern technology, the engineer and scientist must know with reasonable accuracy the properties of the fluids and solids with which he deals. If you are engaged in the design of equipment, say the volume required for a process vessel, you need to know the specific volume or density of the gas or liquid that will be in the vessel as a function of temperature and pressure. If you are interested in predicting the possibility or extent of rainfall, you have to know something about the relation between the vapor pressure of water and the temperature. Whatever your current or future job, you need to have an awareness of the character and sources of information concerning the physical properties of fluids.

Clearly it is not possible to have reliable detailed experimental data on all the useful pure compounds and mixtures that exist in the world. Consequently, in the absence of experimental information, we estimate (predict) properties based on modifications of well-established principles, such as the *ideal* gas law, or based on empirical correlations. Thus the foundation of the estimation methods ranges from quite theoretical to completely empirical and their reliability ranges from excellent to terrible.

In this chapter we shall first discuss ideal and real gas relationships including some of the gas laws for pure components and mixtures of ideal gases. You will learn about methods of expressing the *p-V-T* properties of real gases by means of equations of state and, alternatively, by compressibility factors. Next we shall introduce the concepts of vaporization, condensation, and vapor pressure and illustrate how material balances are made for saturated and partially

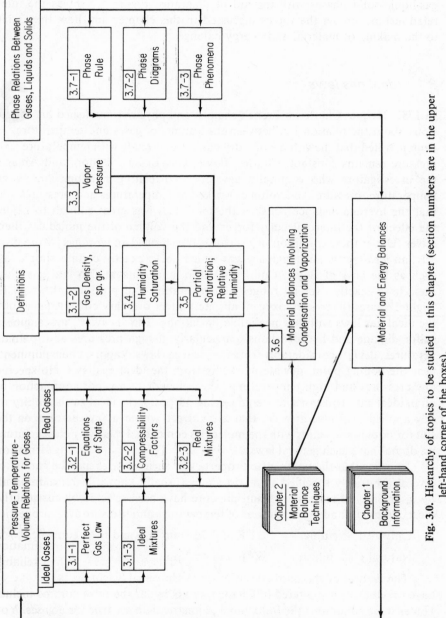

Fig. 3.0. Hierarchy of topics to be studied in this chapter (section numbers are in the upper left-hand corner of the boxes).

147

saturated gases. Finally, we shall examine the qualitative characteristics of gas-liquid-solid phases with the aid of diagrams. Figure 3.0 shows the interrelationships among the topics discussed in this chapter and how they relate to the making of material and energy balances.

3.1 Ideal gas laws

In 1787, Jacques Charles, a French chemist and physicist, published his conclusions about the relationship between the volume of gases and temperature. He demonstrated that the volume of a dry gas varies directly with temperature if the pressure remains constant. Charles, Boyle, Gay-Lussac, Dalton, and Amagat, the investigators who originally developed correlating relations among gas temperature, pressure, and volume, worked at temperatures and pressures such that the average distance between the molecules was great enough to neglect the effect of the intermolecular forces and the volume of the molecules themselves. Under these conditions a gas came to be termed an *ideal* gas. More properly, an *ideal gas* is an imaginary gas which obeys exactly certain simple laws such as the laws of Boyle, Charles, Dalton, and Amagat. No real gas obeys these laws exactly over all ranges of temperature and pressure, although the "lighter" gases (hydrogen, oxygen, air, etc.) under ordinary circumstances obey the ideal gas laws with but negligible deviations. The "heavier" gases such as sulfur dioxide and hydrocarbons, particularly at high pressures and low temperatures, deviate considerably from the ideal gas laws. Vapors, under conditions near the boiling point, deviate markedly from the ideal gas laws. However, at low pressures and high temperatures, the behavior of a vapor approaches that of an ideal gas. Thus for many engineering purposes, the ideal gas laws, if properly applied, will give answers that are correct within a few percent or less. But for liquids and solids with the molecules compacted relatively close together, we do not have such general laws.

In order that the volumetric properties of various gases may be compared, several arbitrarily specified standard states (usually known as *standard conditions*, or S.C.) of temperature and pressure have been selected by custom. The most common standard conditions of temperature and pressure are

Universal Scientific   32°F and 760 mm Hg (and their equivalents)

Natural Gas Industry   60°F and 14.7 psia

The first set of standard conditions, the Universal Scientific, is the one you have previously encountered in Chap. 1 and is by far the most common in use. Under these conditions the following volumetric data are true for any *ideal gas:*

$$1 \text{ g mole} = 22.4 \text{ l at S.C.}$$

$$1 \text{ lb mole} = 359 \text{ ft}^3 \text{ at S.C.}$$

$$1 \text{ kg mole} = 22.4 \text{ m}^3 \text{ at S.C.}$$

The fact that a substance cannot exist as a gas at 32°F and 29.92 in. Hg is immaterial. Thus, as we shall see later, water vapor at 32°F cannot exist at a pressure greater than its saturation pressure of 0.18 in. Hg without condensation occurring. However, the imaginary volume at standard conditions can be calculated and is just as useful a quantity in the calculation of volume-mole relationships as though it could exist. In the following, the symbol V will stand for total volume, and the symbol $\hat{V}$ for volume per mole, or per unit mass.

EXAMPLE 3.1   Use of standard conditions

Calculate the volume, in cubic feet, occupied by 88 lb of CO_2 at standard conditions.

Solution:

$$\text{Basis: 88 lb } CO_2$$

$$\frac{88 \text{ lb } CO_2}{} \left| \frac{1 \text{ lb mole } CO_2}{44 \text{ lb } CO_2} \right| \frac{359 \text{ ft}^3 \text{ } CO_2}{1 \text{ lb mole } CO_2} = 718 \text{ ft}^3 \text{ } CO_2 \text{ at S.C.}$$

Notice how in this problem the information that 359 ft³ at S.C. = 1 lb mole is applied to transform a known number of moles into an equivalent number of cubic feet. Incidentally, whenever you use cubic feet, you must establish the conditions of temperature and pressure at which the cubic feet are measured, since the term "ft³," standing alone, is really not any particular *quantity* of material.

3.1.1 Perfect Gas Law.   Boyle found that the volume of a gas is inversely proportional to the absolute pressure at constant temperature. Charles showed that, at constant pressure, the volume of a given mass of gas varies directly with the absolute temperature. From the work of Boyle and Charles scientists developed the relationship now called the *perfect gas law* (or sometimes *the ideal gas law*):

$$pV = nRT \tag{3.1}$$

In applying this equation to a process going from an initial set of conditions to a final set of conditions, you can set up ratios of similar terms which are dimensionless as follows:

$$\left(\frac{p_1}{p_2}\right)\left(\frac{V_1}{V_2}\right) = \left(\frac{n_1}{n_2}\right)\left(\frac{T_1}{T_2}\right) \tag{3.2}$$

Here the subscripts 1 and 2 refer to the initial and final conditions. This arrangement of the perfect gas law has the convenient feature that the pressures may be expressed in any system of units you choose, such as in. Hg, mm Hg, N/m^2, atm, etc., so long as the same units are used for both conditions of pressure (do not forget that the pressure must be *absolute* pressure in both cases). Similarly, the grouping together of the *absolute* temperature and the volume terms gives ratios that are dimensionless.

Let us see how we can use the perfect gas law both in the form of Eq. (3.2) and Eq. (3.1).

EXAMPLE 3.2   Perfect gas law

An oxygen cylinder used as a standby source of oxygen contains 1.000 ft³ of O_2 at 70°F and 200 psig. What will be the volume of this O_2 in a dry-gas holder at 90°F and 4.00 in. H_2O above atmospheric? The barometer reads 29.92 in. Hg. See Fig. E3.2.

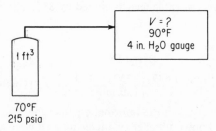

Fig. E3.2.

Solution:

You must first convert the temperatures and pressures into absolute units:

$$460 + 70 = 530°R$$

$$460 + 90 = 550°R$$

atmospheric pressure = 29.92 in. Hg = std atm = 14.7 psia

$$\text{initial pressure} = \frac{200 \text{ psig} + 14.7 \text{ psia}}{} \Bigg| \frac{29.92 \text{ in. Hg}}{14.7 \text{ psia}} = 437 \text{ in. Hg}$$

$$\text{final pressure} = 29.92 \text{ in. Hg} + \frac{4 \text{ in. } H_2O}{12 \text{ in. } H_2O} \Bigg| \frac{29.92 \text{ in. Hg}}{33.91 \text{ ft } H_2O}$$
$$\frac{}{\text{ft } H_2O}$$

$$= 29.92 + 0.29 = 30.21 \text{ in. Hg}$$

The simplest way to proceed, now that the data are in good order, is to apply the laws of Charles and Boyle, and, in effect, to apply the perfect gas law.

From Charles' law, since the temperature *increases*, the volume *increases;* hence the ratio of the temperatures *must* be greater than 1. The pressure *decreases;* therefore from Boyle's law, the volume will *increase*; hence the ratio of pressures will be *greater* than 1.

Basis: 1 ft³ of oxygen at 70°F and 200 psig

$$\text{final volume} = \frac{1.00 \text{ ft}^3}{} \Bigg| \frac{550°R}{530°R} \Bigg| \frac{437 \text{ in. Hg}}{30.21 \text{ in. Hg}}$$

$$= 15.0 \text{ ft at } 90°F \text{ and } 4 \text{ in. } H_2O \text{ gauge}$$

Formally, the same calculation can be made using Eq. (3.2),

$$V_2 = V_1 \left(\frac{p_1}{p_2}\right)\left(\frac{T_2}{T_1}\right) \qquad \text{since } n_1 = n_2$$

EXAMPLE 3.3 Perfect gas law

Probably the most important constituent of the atmosphere that fluctuates is the water. Rainfall, evaporation, fog, and even lightning are associated with water as a vapor or a liquid in air. To obtain some feeling for how little water vapor there is at higher altitudes, calculate the mass of 1.00 m³ of water vapor at 15.5 mm Hg and 23°C. Assume that water vapor is an ideal gas under these conditions.

Solution:

First visualize the information available to you, and then decide how to convert it into the desired mass. If you can convert the original amount of water vapor to S.C. by use of the ideal gas law and then make use of the fact that 22.4 m³ = 1 kg mole, you can easily get the desired mass of water vapor. See Fig. E3.3.

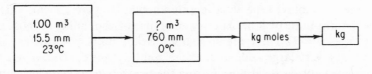

Fig. E3.3.

Basis: 1.00 m³ H_2O vapor at 15.5 mm Hg and 23°C

$$\frac{1.00 \text{ m}^3}{} \left|\frac{15.5 \text{ mm}}{760 \text{ mm}}\right| \frac{(273 + 0)°\text{K}}{(273 + 23)°\text{K}} \left|\frac{1 \text{ kg mole}}{22.4 \text{ m}^3}\right| \frac{18 \text{ kg } H_2O}{1 \text{ kg mole}} = 1.51 \times 10^{-2} \text{ kg } H_2O$$

Notice how the entire calculation can be carried out in a single dimensional equation.

EXAMPLE 3.4 Perfect gas law

You have 10 lb of CO_2 in a 20-ft³ fire extinguisher tank at 30°C. Assuming that the ideal gas laws hold, what will the pressure gauge on the tank read in a test to see if the extinguisher is full? See Fig. E3.4.

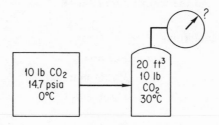

Fig. E3.4.

Solution:

Employing Eq. (3.2), we can write (the subscript 1 stands for standard conditions, 2 for the conditions in the tank)

$$p_2 = p_1 \left(\frac{V_1}{V_2}\right)\left(\frac{T_2}{T_1}\right)$$

$$\overbrace{14.7 \text{ psia}}^{p_1} \underbrace{\left| \frac{10 \text{ lb CO}_2}{} \right| \frac{1 \text{ lb mole CO}_2}{44 \text{ lb CO}_2} \left| \frac{359 \text{ ft}^3}{1 \text{ lb mole}} \right.}_{V_1} \underbrace{\left| \frac{}{20 \text{ ft}^3} \right.}_{V_2} \underbrace{\left| \frac{303°\text{K}}{273°\text{K}} \right.}_{\frac{T_2}{T_1}} = p_2 = 66 \text{ psia}$$

Hence the gauge on the tank will read (assuming that it reads gauge pressure and that the barometer reads 14.7 psia) $66 - 14.7 = 51.3$ psig.

We have not used the gas constant R in the solution of any of the example problems so far, but you can use Eq. (3.1) to solve for one unknown as long as all the other variables in the equation are known. However, such a calculation requires that the units of R be expressed in units corresponding to those used for the quantities p-V-T. There are so many possible units you can use for each variable that a very large table of R values will be required. But, if you want to use R, you can always determine R from the p-V-T data at standard conditions which you have already memorized and used; R merely represents a pV/T relation at some fixed condition of temperature and pressure.

EXAMPLE 3.5   Calculation of R

Find the value for the universal gas constant R for the following combinations of units:

(a) For 1 lb mole of ideal gas when the pressure is expressed in psia, the volume is in ft³/lb mole, and the temperature is in °R.

(b) For 1 g mole of ideal gas when the pressure is in atm, the volume in cm³, and the temperature in °K.

(c) For 1 kg mole of ideal gas when the pressure is in N/m², the volume is in m³/kg mole, and the temperature is in °K.

Solution:

(a) At standard conditions

$$p = 14.7 \text{ psia}$$
$$\hat{V} = 359 \text{ ft}^3/\text{lb mole}$$
$$T = 492°\text{R}$$

Then

$$R = \frac{p\hat{V}}{T} = \frac{14.7 \text{ psia}}{492°\text{R}} \left| \frac{359 \text{ ft}^3}{1 \text{ lb mole}} = 10.73 \frac{(\text{psia})(\text{ft}^3)}{(°\text{R})(\text{lb mole})}$$

(b) Similarly, at standard conditions,

$$p = 1 \text{ atm}$$

$$\hat{V} = 22{,}400 \text{ cm}^3/\text{g mole}$$

$$T = 273°\text{K}$$

$$R = \frac{p\hat{V}}{T} = \frac{1 \text{ atm}}{273°\text{K}} \bigg| \frac{22{,}400 \text{ cm}^3}{1 \text{ g mole}} = 82.06 \frac{(\text{cm}^3)(\text{atm})}{(°\text{K})(\text{g mole})}$$

(c) In the SI system of units standard conditions are

$$p = 1.013 \times 10^5 \text{ N/m}^2$$

$$\hat{V} = 22.4 \text{ m}^3/\text{kg mole}$$

$$T = 273°\text{K}$$

$$R = \frac{p\hat{V}}{T} = \frac{1.013 \times 10^5 \text{ N/m}^2}{273°\text{K}} \bigg| \frac{22.4 \text{ m}^3}{1 \text{ kg mole}} = 8.30 \times 10^3 \frac{(\text{N})(\text{m})}{(°\text{K})(\text{kg mole})}$$

We want to emphasize that R does not have a universal value even though it is called the *universal gas constant*. The value of R depends on the units of p, V, and T. Similarly, you should realize that R is not a dimensionless quantity; i.e., there is no value of $R = 21.9$, but there is a value of

$$R = 21.9 \frac{(\text{in. Hg})(\text{ft}^3)}{(\text{lb mole})(°\text{R})}$$

EXAMPLE 3.6 Perfect gas law

Calculate the volume occupied by 88 lb of CO_2 at a pressure of 32.2 ft of water and at 15°C.

Solution:
See Fig. E3.6.

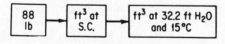

Fig. E3.6.

Solution No. 1 (using Boyle's and Charles' laws):
At S.C.:

$$p = 33.91 \text{ ft H}_2\text{O}$$

$$\hat{V} = 359 \frac{\text{ft}^3}{\text{lb mole}}$$

$$T = 273°\text{K}$$

Basis: 88 lb of CO_2

$$\frac{88 \text{ lb CO}_2}{} \bigg| \frac{1 \text{ lb mole CO}_2}{44 \text{ lb CO}_2} \bigg| \frac{359 \text{ ft}^3}{1 \text{ lb mole}} \bigg| \frac{288}{273} \bigg| \frac{33.91}{32.2} = 798 \text{ ft}^3 \text{ CO}_2 \text{ at 32.2 ft H}_2\text{O and 15°C}$$

Solution No. 2 (using gas constant R):

First, the value of R must be obtained in the same units as the variable p, $\hat{V}$, and T. For 1 lb mole,

$$R = \frac{p\hat{V}}{T}$$

and at S.C.,

$$p = 33.91 \text{ ft } H_2O$$

$$\hat{V} = 359 \text{ ft}^3/\text{lb mole}$$

$$T = 273°K$$

$$R = \frac{33.91 \mid 359}{\mid 273} = 44.6 \frac{(\text{ft } H_2O)(\text{ft}^3)}{(\text{lb mole})(°K)}$$

Now, using Eq. (3.1), inserting the given values, and performing the necessary calculations, we get

Basis: 88 lb CO_2

$$V = \frac{nRT}{p} = \frac{88 \text{ lb } CO_2}{\dfrac{44 \text{ lb } CO_2}{\text{lb mole } CO_2}} \left| \frac{(44.6 \text{ ft } H_2O)(\text{ft}^3)}{(\text{lb mole})(°K)} \right| \frac{288°K}{32.2 \text{ ft } H_2O}$$

$$= 798 \text{ ft}^3 \ CO_2 \text{ at } 32.2 \text{ ft } H_2O \text{ and } 15°C$$

If you will inspect both solutions closely, you will observe that in both cases the same numbers appear and that both are identical except that in the second solution using R two steps are required to obtain the solution.

EXAMPLE 3.7 Perfect gas law

One important source of emissions from gasoline-powered automobile engines that causes smog is the nitrogen oxides NO and NO_2. They are formed whether combustion is complete or not as follows: At the high temperatures which occur in an internal combustion engine during the burning process, oxygen and nitrogen combine to form nitric oxide (NO). The higher the peak temperatures and the more oxygen available, the more NO is formed. There is insufficient time for the NO to decompose back to O_2 and N_2 because the burned gases cool too rapidly during the expansion and exhaust cycles in the engine. Although both NO and nitrogen dioxide (NO_2) are significant air pollutants (together termed NO_x), the NO_2 is formed in the atmosphere as NO is oxidized.

Suppose that you collect a sample of a NO-NO_2 mixture (after having removed the other combustion gas products by various separations procedures) in a 100-cm^3 standard cell at 30°C. Certainly some of the NO will have been oxidized to NO_2

$$2NO + O_2 \longrightarrow 2NO_2$$

during the collection, storage, and processing of the combustion gases so that measurement of NO alone will be misleading. If the standard cell contains 0.291 g of NO_2 plus NO and the pressure measured in the cell is 1265 mm Hg, what percent of the NO + NO_2 is in the form of NO? See Fig. E3.7.

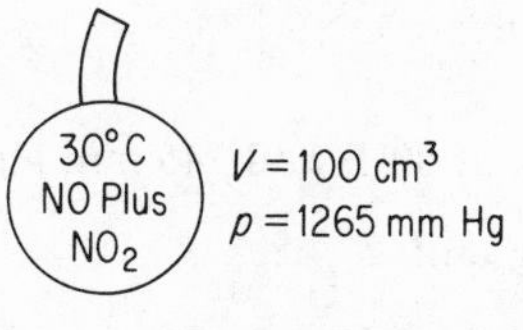

Fig. E3.7.

Solution:

The gas in the cell is composed partly of NO and partly of NO_2. We can use the ideal gas law to calculate the total gram moles present in the cell and then, by using the chemical equation and the principles of stoichiometry, compute the composition in the cell.

Basis: 100 cm³ gas at 1265 mm Hg and 30°C

$$R = \frac{760 \text{ mm Hg}}{273°K} \bigg| \frac{22.4\,1}{1 \text{ g mole}} \bigg| \frac{1000 \text{ cm}^3}{1\,1} = 6.24 \times 10^4 \frac{(\text{mm Hg})(\text{cm}^3)}{(°K)(\text{g mole})}$$

$$n = \frac{pV}{RT} = \frac{1265 \text{ mm Hg}}{6.24 \times 10^4 \dfrac{(\text{mm Hg})(\text{cm}^3)}{(°K)(\text{g mole})}} \bigg| \frac{100 \text{ cm}^3}{303°K} = 0.00670 \text{ g mole}$$

If the mixture is composed of NO (MW = 30) and NO_2 (MW = 46), because we know the total mass in the cell we can compute the fraction of, say, NO. Let $x =$ grams of NO; then $(0.291 - x) = $ g NO_2.

Basis: 0.291 g total gas

The sum of the moles is

$$\frac{x \text{ g NO}}{} \bigg| \frac{1 \text{ g mole NO}}{30 \text{ g NO}} + \frac{(0.291 - x) \text{ g } NO_2}{} \bigg| \frac{1 \text{ g mole } NO_2}{46 \text{ g } NO_2} = 0.00670$$

$$0.0333x + (0.291 - x)(0.0217) = 0.00670$$

$$x = 0.033 \text{ g}$$

The weight percent NO is

$$\frac{0.033}{0.291}(100) \cong 11\%$$

and the mole percent NO is

$$\frac{(0.0333)(0.033)}{(0.00670)}(100) \cong 17\%$$

3.1.2 Gas Density and Specific Gravity. The density of a gas is defined as the mass per unit volume and can be expressed in pounds per cubic foot, grams per liter, or other units. Inasmuch as the mass contained in a unit volume varies with the temperature and pressure, as we have previously mentioned, you should always be careful to specify these two conditions. Unless otherwise specified, the volumes are presumed to be at S.C. Density can be calculated by selecting a unit volume as the basis and calculating the mass of the contained gas.

EXAMPLE 3.8   Gas density

What is the density of N_2 at 80°F and 745 mm Hg expressed in

(a) American engineering units?
(b) cgs units?
(c) SI units?

Solution:
(a)

Basis: 1 ft³ N_2 at 80°F and 745 mm Hg

$$\frac{1 \text{ ft}^3 \mid 492°R \mid 745 \text{ mm Hg} \mid 1 \text{ lb mole} \mid 28 \text{ lb}}{\mid 540°R \mid 760 \text{ mm Hg} \mid 359 \text{ ft}^3 \mid 1 \text{ lb mole}} = 0.0696 \text{ lb}$$

density = 0.0696 lb/ft³ of N_2 at 80°F and 745 mm Hg

(b)

Basis: 1 *l* N_2 at 80°F and 745 mm Hg

$$\frac{1 \text{ } l \mid 492°R \mid 745 \text{ mm Hg} \mid 1 \text{ g mole} \mid 28 \text{ g}}{\mid 540°R \mid 760 \text{ mm Hg} \mid 22.4 \text{ } l \mid 1 \text{ g mole}} = 1.115 \text{ g}$$

density = 1.115 g/*l* of N_2 at 80°F (27°C) and 745 mm Hg

(c)

Basis: 1 m³ N_2 at 80°F and 745 mm Hg

$$\frac{1 \text{ m}^3 \mid 492°R \mid 745 \text{ mm Hg} \mid 1 \text{ kg mole} \mid 28 \text{ kg}}{\mid 540°R \mid 760 \text{ mm Hg} \mid 22.4 \text{ m}^3 \mid 1 \text{ kg mole}} = 1.115 \text{ kg}$$

density = 1.115 kg/m³ of N_2 at 80°F (300°K) and 745 mm Hg (9.94×10^4 N/m²)

The *specific gravity* of a gas is usually defined as the ratio of the density of the gas at a desired temperature and pressure to that of air (or any specified reference gas) at a certain temperature and pressure. The use of specific gravity occasionally may be confusing because of the manner in which the values of specific gravity are reported in the literature. You must be very careful in using literature values of specific gravity. Be particularly careful to ascertain that the conditions of temperature and pressure are known both for the gas in question and for the reference gas. Among the examples below, several represent inadequate methods of expressing specific gravity.

(a) *What is the specific gravity of methane?* Actually this question may have the same answer as the question, How many grapes are in a bunch? Unfortunately, occasionally one may see this question and the best possible answer is

$$\text{sp gr} = \frac{\text{density of methane at S.C.}}{\text{density of air at S.C.}}$$

(b) *What is the specific gravity of methane* ($H_2 = 1.00$)? Again a poor question. The notation of ($H_2 = 1.00$) means that H_2 at S.C. is used as the reference gas, but the question does not even give a hint regarding the conditions of temperature and pressure of the methane. Therefore the best interpretation is

$$\text{sp gr} = \frac{\text{density of methane at S.C.}}{\text{density of } H_2 \text{ at S.C.}}$$

(c) *What is the specific gravity of ethane* (air = 1.00)? Same as question (b) except that in the petroleum industry the following is used:

$$\text{sp gr} = \frac{\text{density of ethane at } 60°F \text{ and } 760 \text{ mm Hg}}{\text{density of air at S.C. } (60°F, 760 \text{ mm Hg})}$$

(d) *What is the specific gravity of butane at* 80°F *and* 740 mm Hg? No reference gas nor state of reference gas is mentioned. However, when no reference gas is mentioned, it is taken for granted that air is the reference gas. In the case at hand the best thing to do is to assume that the reference gas and the desired gas are under the same conditions of temperature and pressure:

$$\text{sp gr} = \frac{\text{density of butane at } 80°F \text{ and } 740 \text{ mm Hg}}{\text{density of air at } 80°F \text{ and } 740 \text{ mm Hg}}$$

(e) *What is the specific gravity of* CO_2 *at* 60°F *and* 740 mm Hg (air = 1.00)?

$$\text{sp gr} = \frac{\text{density of } CO_2 \text{ at } 60°F \text{ and } 740 \text{ mm Hg}}{\text{density of air at S.C.}}$$

(f) *What is the specific gravity of* CO_2 *at* 60°F *and* 740 mm Hg (ref. air at S.C.)?

$$\text{sp gr} = \text{same as question (e)}$$

EXAMPLE 3.9   Specific gravity of a gas

What is the specific gravity of N_2 at 80°F and 745 mm Hg compared to

(a) Air at S.C. (32°F and 760 mm Hg)?

(b) Air at 80°F and 745 mm Hg?

Solution:

First you must obtain the density of the N_2 and the air at their respective conditions of temperature and pressure, and then calculate the specific gravity by taking a ratio of their densities. Example 3.8 covers the calculation of the density of a gas, and therefore, to save space, no units will appear in the following calculations:

(a)

Basis: 1 ft³ N_2 at 80°F and 745 mm Hg

$$\frac{1}{} \left| \frac{492}{540} \right| \frac{745}{760} \left| \frac{}{359} \right| \frac{28}{} = 0.0696 \text{ lb } N_2/\text{ft}^3 \text{ at } 80°F, 745 \text{ mm Hg}$$

Basis: 1 ft³ of air at 32°F and 760 mm Hg

$$\frac{1}{} \left| \frac{492}{492} \right| \frac{760}{760} \left| \frac{}{359} \right| \frac{29}{} = 0.0808 \text{ lb air/ft}^3 \text{ at } 32°F, 760 \text{ mm Hg}$$

Therefore

$$\text{sp gr} = \frac{0.0696}{0.0808} = 0.863 \frac{\text{lb N}_2/\text{ft}^3 \text{ N}_2 \text{ at } 80°\text{F, 745 mm Hg}}{\text{lb air/ft}^3 \text{ air at S.C.}}$$

Note: specific gravity is not a dimensionless number.

(b)

Basis: 1 ft³ air at 80°F and 745 mm Hg

$$\frac{1}{}\bigg|\frac{492}{540}\bigg|\frac{745}{760}\bigg|\frac{29}{359}\bigg| = 0.0721 \text{ lb/ft}^3 \text{ at } 80°\text{F and 745 mm Hg}$$

$$(\text{sp gr})_{\text{N}_2} = \frac{0.0696}{0.721} = 0.965 \frac{\text{lb N}_2/\text{ft}^3 \text{ N}_2 \text{ at } 80°\text{F, 745 mm Hg}}{\text{lb air/ft}^3 \text{ air at } 80°\text{F, 745 mm Hg}}$$

$$= 0.965 \text{ lb N}_2/\text{lb air}$$

Note: You can work part (b) by dividing the unit equations instead of dividing the resulting densities:

$$\text{sp gr} = \frac{d_{\text{N}_2}}{d_{\text{air}}} = \frac{\dfrac{1}{}\bigg|\dfrac{492}{540}\bigg|\dfrac{745}{760}\bigg|\dfrac{}{359}\bigg|\dfrac{28}{}}{\dfrac{1}{}\bigg|\dfrac{492}{540}\bigg|\dfrac{745}{760}\bigg|\dfrac{}{359}\bigg|\dfrac{29}{}} = \frac{28}{29} = 0.965 \text{ l b N}_2/\text{lb air}$$

The latter calculation shows that the specific gravity is equal to the *ratio of the molecular weights* of the gases when the densities of *both* the desired gas and the reference gas are at the same temperature and pressure. This, of course, is true only for ideal gases and should be no surprise to you, since Avogadro's law in effect states that at the same temperature and pressure 1 mole of any ideal gas is contained in identical volumes.

3.1.3 Ideal Gaseous Mixtures.

In the majority of cases, as an engineer you will deal with mixtures of gases instead of individual gases. There are three ideal gas laws which can be applied successfully to gaseous mixtures:

(a) Dalton's law of partial pressures.

(b) Amagat's law of partial volumes.

(c) Dalton's law of the summation of partial pressures.

(a) *Dalton's laws.* Dalton postulated that the total pressure of a gas is equal to the sum of the pressures exerted by the individual molecules of each component gas. He went one step further to state that each individual gas of a gaseous mixture can hypothetically be considered to exert a *partial pressure*. A partial pressure is the pressure that would be obtained if this *same mass* of individual gas were alone in the *same total volume* at the *same temperature*. The sum of these partial pressures for each component in the gaseous mixture would be equal to the total pressure, or

$$p_1 + p_2 + p_3 + \cdots + p_n = p_t \tag{3.3}$$

Equation (3.3) is Dalton's law of the summation of the partial pressures.

To illustrate the significance of Eq. (3.3) and the meaning of partial pressure, suppose that you carried out the following experiment with ideal gases. Two tanks of 150 ft³ volume, one containing gas A at 300 mm Hg and the other gas B at 400 mm of Hg (both gases being at the same temperature of 70°F), are connected together. All the gas in B is forced into tank A isothermally. Now you have a 150-ft³ tank of A + B at 700 mm of Hg. For this mixture (in the 150-ft³ tank at 70°F and a total pressure of 700 mm Hg) you could say that gas A exerts a partial pressure of 300 mm and gas B exerts a partial pressure of 400 mm. Of course you cannot put a pressure gauge on the tank and check this conclusion because the pressure gauge will read only the total pressure. These partial pressures are hypothetical pressures that the individual gases would exert and are equivalent to the pressures they actually would have if they were put into the same volume at the same temperature all by themselves.

You can surmise that, at constant volume and at constant temperature, the pressure is a function only of the number of molecules of gas present. If you divide the perfect gas law for component 1, $p_1 V_1 = n_1 R T_1$, by that for component 2, $p_2 V_2 = n_2 R T_2$, at *constant temperature and volume*, you can obtain

$$\frac{p_1}{p_2} = \frac{n_1}{n_2} \tag{3.4}$$

which shows that the ratio of the partial pressures is exactly the same numerically as the ratio of the moles of components 1 and 2. Similarly, dividing the ideal gas law for component 1 by the gas law for all the molecules, $p_t V_t = n_t R T_t$, you will get Dalton's law of partial pressures:

$$\frac{p_1}{p_t} = \frac{n_1}{n_t} = \text{mole fraction} = y_1 \tag{3.5}$$

Equation (3.5) shows that the ratio of the partial pressure of an individual component to the total pressure is exactly the same numerically as the ratio of the moles of the individual component to the total moles. With this principle under your belt, if the mole fraction of an individual gaseous component in a gaseous mixture is known and the total pressure is known, then you are able to calculate the partial pressure of this component gas by generalizing Eq. (3.5):

$$p_i = y_i p_t \tag{3.5a}$$

where *i* stands for any component.

(b) *Amagat's law.* Amagat's law of additive volumes is analogous to Dalton's law of additive pressures. Amagat stated that the total volume of a gaseous mixture is equal to the sum of the volumes of the individual gas components if they were to be measured at the same temperature and at the total pressure of all the molecules. The individual volumes of these individual components at the same temperature and pressure are called the *partial volumes* (or sometimes *pure component* volumes) of the individual components:

$$V_1 + V_2 + V_3 + \cdots + V_n = V_t \tag{3.6}$$

Reasoning in the same fashion as in our explanation of partial pressures, *at the same temperature and pressure*, the partial volume is a function only of number of molecules of the individual component gas present in the gaseous mixture, or

$$\frac{V_1}{V_2} = \frac{n_1}{n_2} \tag{3.7}$$

and

$$\frac{V_1}{V_t} = \frac{n_1}{n_t} = y_1 = \text{mole fraction} \tag{3.8}$$

which shows that the ratio of the partial volumes is exactly the same, numerically, as the ratio of the moles of components 1 and 2, or the ratio of the moles of component 1 to the total moles. Equation (3.8) states the principle, presented without proof in Chap. 1, that

$$\text{volume fraction} = \text{mole fraction} = y_i \tag{3.9}$$

for an ideal gas.

EXAMPLE 3.10 Partial pressures and volumes

A gas-tight room has a volume of 10,000 ft³. This room contains air (considered to be 21 percent O_2 and 79 percent N_2) at 70°F and a total pressure of 760 mm Hg.

(a) What is the partial volume of O_2 in the room?
(b) What is the partial volume of N_2 in the room?
(c) What is the partial pressure of O_2 in the room?
(d) What is the partial pressure of N_2 in the room?
(e) If all of the O_2 were removed from the room by some method, what would be the subsequent total pressure in the room?

Solution:
See Fig. E3.10(a)

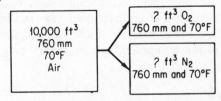

Fig. E3.10(a).

Basis: 10,000 ft³ of air at 70°F and 760 mm Hg

Partial volumes can be calculated by multiplying the total volume by the respective component mole fractions [Eq. (3.8)]:

(a) $V_{O_2} = (0.21)(10,000) = 2,100$ ft³ O_2 at 70°F, 760 mm
(b) $V_{N_2} = (0.79)(10,000) = 7,900$ ft³ N_2 at 70°F, 760 mm
total volume = 10,000 ft³ air at 70°F, 760 mm

Note how the temperature and pressure have to be specified for the partial volumes to make them meaningful.

Partial pressures can be calculated by multiplying the total pressure by the respective component mole fractions [Eq. (3.5a)]; the basis is still the same:

(c) $p_{O_2} = (0.21)(760) = 160$ mm Hg when $V = 10,000$ ft^3 at 70°F

(d) $p_{N_2} = (0.79)(760) = \underline{600}$ mm Hg when $V = 10,000$ ft^3 at 70°F

            total pressure $= 760$ mm Hg when $V = 10,000$ ft^3 at 70°F

(e)   If a tight room held dry air at 760 mm Hg and all the oxygen were removed from the air by a chemical reaction, the pressure reading would fall to 600 mm Hg. This is the partial pressure of the nitrogen and inert gases in the air. Alternatively, if it were possible to remove the nitrogen and inert gases and leave only the oxygen, the pressure reading would fall to 160 mm Hg. In either case, you would have left in the room 10,000 ft^3 of gas at 70°F. See Fig. E3.10(b).

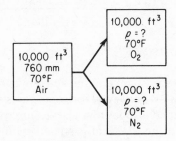

Fig. E3.10(b).

For use in our subsequent calculations you should clearly understand now that the original room contained:

1.   7,900 ft^3 dry N_2 at 760 mm Hg and 70°F
2.   <u>2,100</u> ft^3 dry O_2 at 760 mm Hg and 70°F
3.   10,000 ft^3 dry air at 760 mm Hg and 70°F
                            (add 1 and 2)

or

1. 10,000 ft^3 dry O_2 at 160 mm Hg and 70°F
2. 10,000 ft^3 dry N_2 at <u>600</u> mm Hg and 70°F
3. 10,000 ft^3 dry air at <u>760</u> mm Hg and 70°F
                            (add 1 and 2)

EXAMPLE 3.11   Ideal gas mixtures

A flue gas analyzes CO_2, 14.0 percent; O_2, 6.0 percent; and N_2, 80.0 percent. It is at 400°F and 765.0 mm Hg pressure. Calculate the partial pressure of each component.

Solution:

Basis: 1.00 kg mole flue gas

comp.	kg moles	p (mm Hg)
CO_2	0.140	107.1
O_2	0.060	45.9
N_2	0.800	612.0
	1.000	765.0

On the basis of 1.00 kg mole of flue gas, the mole fraction y of each component, when multiplied by the total pressure, gives the partial pressure of that component. Thus, V ft³ of flue gas represents four different things:

V ft³ flue gas	at 765.0 mm Hg and 400°F
V ft³ carbon dioxide	at 107.1 mm Hg and 400°F
V ft³ oxygen	at 45.9 mm Hg and 400°F
V ft³ nitrogen	at 612.0 mm Hg and 400°F

3.2   Real gas relationships

We have said that at room temperature and pressure many gases can be assumed to act as ideal gases. However, for some gases under normal conditions, and for most gases under conditions of high pressure, values of the gas properties that you might obtain using the ideal gas law would be at wide variance with the experimental evidence. You might wonder exactly how the behavior of real gases does compare with that calculated from the ideal gas laws. In Fig. 3.1 you can see how the $p\hat{V}$ product of several gases deviates from that predicted by the ideal gas laws as the pressure increases substantially. Thus it is clear that we need some way of computing the p-V-T properties of a gas that is not ideal.

   Essentially there are four methods of handling real gas calculations:

(a) Equations of state.
(b) Compressibility charts.
(c) Estimated properties.[1]
(d) Actual experimental data.

Even if experimental data is available, the other three techniques still may be quite useful for certain types of calculations. Of course, under conditions such that part of the gas liquefies, the gas laws apply only to the gas phase portion of the system—you cannot extend these real gas laws into the liquid region any more than you can apply the ideal gas laws to a liquid.

---

   [1]Computer programs are available to estimate physical properties of compounds and mixtures based on structural group contributions and other basic parameters; refer to E. L. Meadows, *Chem. Eng. Progr.*, vol. 61, p. 93 (1965).

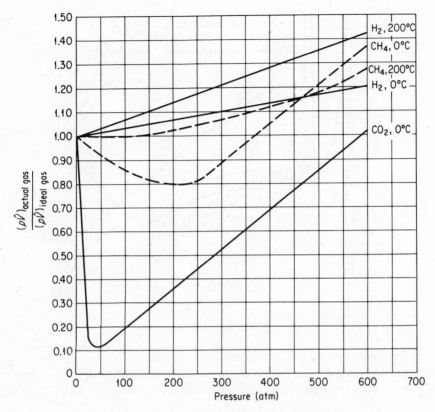

Fig. 3.1. Deviation of real gas from ideal gas laws at high pressures.

3.2.1 Equations of State. Equations of state relate the p-V-T properties of a pure substance (or mixtures) by semitheoretical or empirical relations. By *property* we shall mean any measurable characteristic of a substance, such as pressure, volume, or temperature, or a characteristic that can be calculated or deduced, such as internal energy, to be discussed in Chap 4. Experimental data for a gas, carbon dioxide, are illustrated in Fig. 3.2, which is a plot of pressure, versus molal volume with absolute temperature (converted to °C) as the third parameter. For a constant temperature the perfect gas law, $p\hat{V} = RT$ (for 1 mole), is, on this plot, a rectangular hyperbola because $p\hat{V} = $ constant. The experimental data at high temperatures and low pressures is not shown but closely approximate the perfect gas law. However, as we approach the point C, called the critical point (discussed in Sec. 3.2.2), we see that the traces of the experimental data are quite different from the lines which would represent the relation $p\hat{V} = $ constant.

The problem, then, is to predict these experimental points by means of some equation which can, with reasonable accuracy, represent the experimental data throughout the gas phase. Over 100 equations of state have been proposed

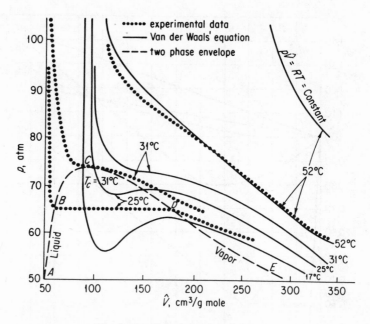

Fig. 3.2. *p-V-T* properties of CO_2.

to accomplish this task, of which a few are shown in Table 3.1. We would not, of course, expect the function to follow the experimental data within the two-phase region outlined by the area inside the envelope *A, B, C, D,* and *E.*

(a) van der Waal's equation.

van der Waals' equation, first proposed in 1893, does not yield satisfactory predictions at low gas temperatures (see Fig. 3.2), but it does have historical interest and is one of the simplest equations of state. Thus it illustrates some of the typical theoretical development that has taken place as well as some of the computational problems involved in using equations of state. Keep in mind that some of the equations listed in Table 3.1 are completely empirical while others, such as van der Waals' equation, are semiempirical; that is, although they were developed from theory, the constants in the equation or portions of the equation are determined by empirical methods.

van der Waals tried to include in the ideal gas law the effect of the attractive forces among the molecules by adding to the pressure term, in the ideal gas law, the term

$$\frac{n^2 a}{V^2} \tag{3.10}$$

where

n = number of moles
V = volume
a = a constant, different for each gas

TABLE 3.1 EQUATIONS OF STATE (for 1 mole)

van der Waals:

$$\left(p + \frac{a}{\hat{V}^2}\right)(\hat{V} - b) = RT$$

Lorentz:

$$p = \frac{RT}{\hat{V}^2}(\hat{V} + b) - \frac{a}{\hat{V}^2}$$

Dieterici:

$$p = \frac{RT}{(\hat{V} - b)}e^{-a/\hat{V}\,RT}$$

Berthelot:

$$p = \frac{RT}{(\hat{V} - b)} - \frac{a}{T\hat{V}^2}$$

Redlich-Kwong:

$$\left[p + \frac{a}{T^{1/2}\hat{V}(\hat{V} + b)}\right](\hat{V} - b) = RT$$

$$a = 0.4278\frac{R^2 T_c^{2.5}}{p_c}$$

$$b = 0.0867\frac{RT_c}{p_c}$$

Kammerlingh-Onnes:

$$p\hat{V} = RT\left[1 + \frac{B}{\hat{V}} + \frac{C}{\hat{V}^2} + \cdots\right]$$

Holborn:

$$p\hat{V} = RT[1 + B'p + C'p^2 + \cdots]$$

Beattie-Bridgeman:

$$p\hat{V} = RT + \frac{\beta}{\hat{V}} + \frac{\gamma}{\hat{V}^2} + \frac{\delta}{\hat{V}^3}$$

$$\beta = RTB_0 - A_0 - \frac{Rc}{T^2}$$

$$\gamma = -RTB_0b + aA_0 - \frac{RB_0c}{T^2}$$

$$\delta = \frac{RB_0bc}{T^2}$$

Benedict-Webb-Rubin:

$$p\hat{V} = RT + \frac{\beta}{\hat{V}} + \frac{\sigma}{\hat{V}^2} + \frac{\eta}{\hat{V}^4} + \frac{w}{\hat{V}^5}$$

$$\beta = RTB_0 - A_0 - \frac{C_0}{T^2}$$

$$\sigma = bRT - a + \frac{c}{T^2}\exp\left(-\frac{\gamma}{\hat{V}^2}\right)$$

$$\eta = c\gamma\exp\left(-\frac{\gamma}{\hat{V}^2}\right)$$

$$w = a\alpha$$

He also tried to take into account the effect of the volume occupied by the molecules themselves by subtracting a term from the volume in the ideal gas law. These corrections led to the following equation:

$$\left(p + \frac{n^2 a}{V^2}\right)(V - nb) = nRT \tag{3.11}$$

where a and b are constants determined by fitting van der Waals' equation to experimental p-V-T data, particularly the values at the critical point. Values of the van der Waals constants for a few gases are given in Table 3.2.

Let us look at some of the computational problems that may arise when using equations of state to compute p-V-T properties by using van der Waals' equation as an illustration. van der Waals' equation can easily be explicitly solved for p as follows:

$$p = \frac{nRT}{(V - nb)} - \frac{n^2 a}{V^2} \tag{3.12}$$

However, if you want to solve for V (or n), you can see that the equation becomes cubic in V (or n), or

$$V^3 - \left(nb + \frac{nRT}{p}\right)V^2 + \frac{n^2 a}{p}V - \frac{n^3 ab}{p} = 0 \tag{3.13}$$

TABLE 3.2 VAN DER WAALS' CONSTANTS FOR GASES
(From the data of the International Critical Tables)

	$a*$ atm $\left(\dfrac{cm^3}{g\ mole}\right)^2$	$b\dagger$ $\left(\dfrac{cm^3}{g\ mole}\right)$
Air	1.33×10^6	36.6
Ammonia	4.19×10^6	37.3
Carbon dioxide	3.60×10^6	42.8
Ethylene	4.48×10^6	57.2
Hydrogen	0.245×10^6	26.6
Methane	2.25×10^6	42.8
Nitrogen	1.347×10^6	38.6
Oxygen	1.36×10^6	31.9
Water vapor	5.48×10^6	30.6

*To convert to psia $\left(\dfrac{ft^3}{lb\ mole}\right)^2$, multiply table value by 3.776×10^{-3}.

†To convert to $\dfrac{ft^3}{lb\ mole}$, multiply table value by 1.60×10^{-2}.

Consequently to solve for V (or n) you would have to

(a) Resort to a trial-and-error method,
(b) Plot the equation assuming various values of V (or n), and see at what value of V (or n) the curve you plot intersects the abscissa, or
(c) Use Newton's method or some other method for solving nonlinear equations (refer to Appendix L).

A computer program can be of material help in technique (c). You can obtain a close approximation to V (or n) in many cases from the ideal gas law, useful at least for the first trial, and then you can calculate a more exact value of V (or n) by using van der Waals' equation.

EXAMPLE 3.12 van der Waals' equation

A 5.0-ft³ cylinder containing 50.0 lb of propane (C_3H_8) stands in the hot sun. A pressure gauge shows that the pressure is 665 psig. What is the temperature of the propane in the cylinder? Use van der Waals' equation. See Fig. E3.12.

Solution:

Basis: 50 lb of propane

The van der Waals' constants obtained from any suitable handbook are

$$a = 3.27 \times 10^4 \text{ psia} \left(\frac{ft^3}{lb\ mole}\right)^2$$

$$b = 1.35 \frac{ft^3}{lb\ mole}$$

$$\left(p + \frac{n^2a}{V^2}\right)(V - nb) = nRT$$

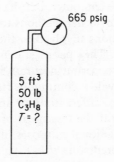

Fig. E3.12.

All the additional information you need is as follows:

$$p = 665 \text{ psig} + 14.7 = 679.7 \text{ psia}$$

$$R \text{ in proper units is} = 10.73 \frac{\text{(psia)(ft}^3)}{\text{(lb mole)(}^\circ\text{R)}}$$

$$n = \frac{50 \text{ lb}}{44 \text{ lb/lb mole}} = 1.137 \text{ lb moles propane}$$

$$\left[679.7 + \frac{(1.137)^2(3.27 \times 10^4)}{(5)^2}\right][5 - (1.137)(1.35)] = 1.137(10.73)\,T$$

$$T = 673^\circ\text{R} = 213^\circ\text{F}$$

(b) Other equations of state.

Among the numerous other equations of state which have been proposed in addition to van der Waals' equation are those shown in Table 3.1. The form of these equations is of interest inasmuch as they are attempts to fit the experimental data with as few constants in the equation as possible. One form of the equation known as the *virial* form is illustrated by the equations of Kammerlingh-Onnes and Holborn. These are essentially power series in $1/\hat{V}$ or in p, and the quantities B, C, D, etc., are known as *virial coefficients*. These equations reduce to $p\hat{V} = RT$ at low pressures. The virial form of an equation of state is the best we have today; however, it is not possible to solve theoretically by statistical thermodynamics for the constants beyond the first two. Therefore, fundamentally, these equations are semiempirical, the constants being determined by fitting the equation to experimental data. The Beattie-Bridgeman equation, which has five constants (exclusive of R), and the Benedict-Webb-Rubin equation, which has eight constants, are among the best we have at the present time. The two-constant Redlich-Kwong equation also appears to be quite good according to a study of Thodos and Shah.[2] Naturally, the use of these equations involves time-consuming calculations, particularly when carried

[2]G. Thodos and K. K. Shah, *Ind. Eng. Chem.*, v. 57, p. 30 (1965).

out by hand rather than on the computer, but the results will frequently be more accurate than those obtained by other methods to be described shortly.

In spite of the complications involved in their use, equations of state are important for several reasons. They permit a concise summary of a large mass of experimental data and also permit accurate interpolation between experimental data points. They provide a continuous function to facilitate thermodynamic calculations involving differentiation and integration. Finally, they provide a point of departure for the treatment of thermodynamic properties of mixtures. However, for instructional purposes, and for many engineering calculations, the techniques of predicting *p-V-T* values discussed in the next section are considerably more convenient to use and are usually just as accurate as equations of state.

3.2.2 Compressibility Factors.
In the attempt to devise some truly universal gas law for high pressures, the idea of corresponding states was developed. Early experimenters found that at the critical point all substances are in approximately the same state of molecular dispersion. Consequently, it was felt that their thermodynamic and physical properties should be similar. The *law of corresponding states* expresses the idea that in the critical state all substances should behave alike.

Now that we have mentioned *critical state* several times, let us consider exactly what the term means. You can find many definitions, but the one most suitable for general use with pure component systems as well as with mixtures of gases is the following:

The *critical state* for the gas-liquid transition is the set of physical conditions at which the density and other properties of the liquid and vapor become identical.

Referring to Fig. 3.3, note how as the pressure increases the specific volumes of the liquid and gas approach each other and finally become the same at the point *C*. This point, for a pure component (only), is the highest temperature at which liquid and vapor can exist in equilibrium. From a slightly different view a critical point is a limiting point that marks the disappearance of a state.

Consider Fig. 3.2. When gaseous carbon dioxide is compressed at 25°C its pressure rises until a certain density is reached (*D*). On further compression the pressure remains constant and the gas condenses, liquid and vapor being in coexistence. Only after all vapor is condensed to liquid (*B*) does the pressure rise again. If the temperature is raised a little, the vapor reaches a higher density before it starts condensing; on the other hand, the condensation is completed at lower liquid density. Thus the coexisting phases have a smaller difference in density at the higher temperature. Above 30°C this density range diminishes very rapidly with temperature and finally, just above 30°C, the two densities at which condensation is initiated and completed coincide. Vapor and liquid are no longer distinguishable; the separating meniscus disappears. Below the

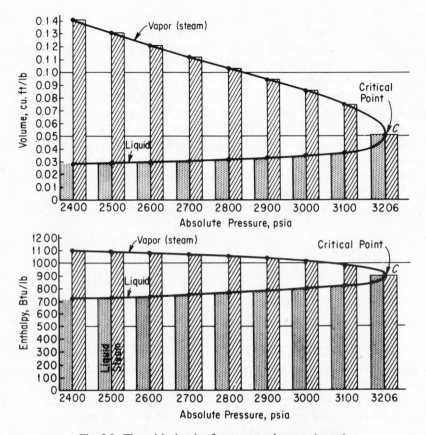

Fig. 3.3. The critical point for a pure substance (water).

critical temperature, the system can exhibit two coexisting phases; above, it goes from the dilute to the dense state without a phase transition. The critical point marks the highest temperature on the coexistence curve.

Experimental values of the critical temperature (T_c) and the critical pressure (p_c) for various compounds will be found in Appendix D. If you cannot find a desired critical value in this text or in a handbook, you can always consult Reid and Sherwood[3] or Hakata and Hirata,[4] which describe and evaluate methods of estimating critical constants for various compounds.

The gas-liquid transition described above is only one of several possible transitions exhibiting a critical point. Critical phenomena are observed in liquids and solids as well. For example, at low temperatures water is only partially miscible in phenol and vice versa. Thus, if equal amounts of water and phenol

[3] R. C. Reid and T. K. Sherwood, *The Properties of Gases and Liquids*, 2nd ed., McGraw-Hill, New York, 1965.

[4] T. Hakata and M. Hirata, *J. Chem. Engr. of Japan*, 3, p. 5 (1970).

are shaken together, they will separate in two phases, one consisting of water with some phenol and the other of phenol with some water dissolved in it. If the temperature is increased, the concentration of phenol in water increases and also that of water in phenol. Finally (if the initial composition is chosen correctly) a critical temperature of the binary liquid mixture is reached at which the two phases become equal in composition, so that the interface disappears and the phases can no longer be distinguished from one another. Above the critical temperature, the system cannot coexist in two phases. A plot of the co-existing compositions against temperature would give a coexistence curve closely analogous to the coexistence curves for the gas-liquid transition.

Another set of terms with which you should immediately become familiar are the *reduced* conditions. These are *corrected*, or *normalized*, conditions of temperature, pressure, and volume and are expressed mathematically as

$$T_r = \frac{T}{T_c} \tag{3.14}$$

$$p_r = \frac{p}{p_c} \tag{3.15}$$

$$V_r = \frac{V}{V_c} \tag{3.16}$$

The idea of using the reduced variables to correlate the p-V-T properties of gases, as suggested by van der Waals, is that all substances behave alike in their reduced, i.e., their corrected, states. In particular, any substance would have the same *reduced* volume at the same *reduced* temperature and pressure. If a relationship does exist involving the reduced variables that can be used to predict p_r, T_r, and V_r, what would the equation be? Certainly the simplest form of such an equation would be to imitate the ideal gas law, or

$$p_r V_r = \gamma T_r \tag{3.17}$$

where γ is some constant.

Now, how does this concept work out in practice? If we plot p_r vs. T_r with a third parameter of constant V_r,

$$\frac{p_r}{T_r} = \frac{\gamma}{V_r} \tag{3.18}$$

we should get straight lines for constant V_r. That this is actually what happens can be seen from Fig. 3.4 for various hydrocarbons. Furthermore, this type of correlation works in the region where $pV = nRT$ is in great error. Also, at the critical point ($p_r = 1.0$, $T_r = 1.0$), V_r should be 1.0 as it is in the diagram.

However, the ability of Eq. (3.18) to predict gas properties breaks down for most substances as they approach the perfect gas region (the area of low pressures somewhat below $p_r = 1.0$). If this concept held in the perfect gas region

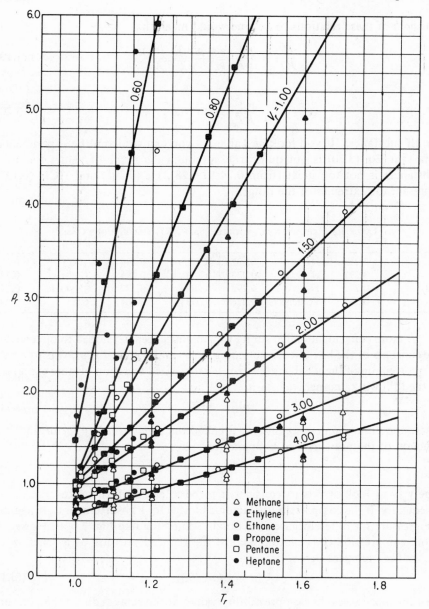

Fig. 3.4. A test of the relation $p_r/T_r = \gamma/V_r$.

for all gases, then for 1 mole

$$p\hat{V} = RT \tag{3.1a}$$

should hold as well as Eq. (3.17)

$$p_r V_r = \gamma T_r$$

For *both* of these relations to be true would mean that

$$\frac{p\hat{V}}{RT} = \frac{p_r V_r}{\gamma T_r} = \frac{(p/p_c)(\hat{V}/\hat{V}_c)}{\gamma(T/T_c)} \tag{3.19}$$

or

$$\frac{p_c \hat{V}_c}{RT_c} = \text{constant} \tag{3.20}$$

for all substances. If you look at Table 3.3 you can see from the experimental data that Eq. (3.20) is not quite true, although the range of values for most substances is not too great (from 0.24 to 0.28). He (0.305) and HCN (0.197) represent the extremes of the range.

TABLE 3.3 EXPERIMENTAL VALUES OF $p_c \hat{V}_c/RT_c$ FOR VARIOUS GASES

NH_3	0.243	HCN	0.197	CH_4	0.290	C_2H_4	0.270
Ar	0.291	H_2O	0.233	C_2H_6	0.285	CH_3OH	0.222
CO_2	0.279	N_2	0.292	C_3H_8	0.277	C_2H_5OH	0.248
He	0.305	Toluene	0.270	C_6H_{14}	0.264	CCl_4	0.272

A more convenient and accurate way that has been developed to tie together the concepts of the law of corresponding states and the ideal gas law is to revert to a correction of the ideal gas law, i.e., a generalized equation of state expressed in the following manner:

$$pV = znRT \tag{3.21}$$

where the dimensionless quantity z is called the *compressibility factor* and is a function of the pressure and temperature:

$$z = \psi(p, T) \tag{3.22}$$

One way to look at z is to consider it to be a factor which makes Eq. (3.21) an equality. If the compressibility factor is plotted for a given temperature against the pressure for different gases, we obtain something like Fig. 3.5(a). However, if the compressibility is plotted against the reduced pressure as a function of the reduced temperature,

$$z = \psi(p_r, T_r) \tag{3.23}$$

then for most gases the compressibility values at the same reduced temperature and reduced pressure fall at about the same point, as illustrated in Fig. 3.5(b).

This permits the use of what is called a generalized compressibility factor, and Figs. 3.6(a) and (b) through 3.10 are the *generalized compressibility charts* or *z*-factor charts prepared by Nelson and Obert.[5] These charts are based on

[5] L. C. Nelson and E. F. Obert, *Chem. Eng.*, vol. 61, no. 7, pp. 203–208 (1954). O. A. Hougen and K. M. Watson, *Chemical Process Principles*, J. Wiley, New York, 1943, also presents *z*-factor charts, as do many articles in the literature for specialized gases, such as natural gases.

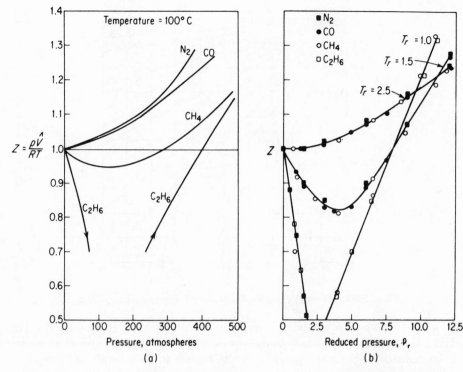

Fig. 3.5(a). Compressibility factor as a function of temperature and pressure.

Fig. 3.5(b). Compressibility as a function of reduced temperature and reduced pressure.

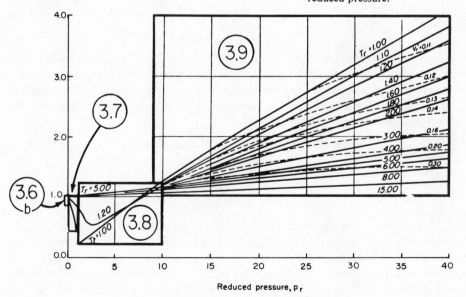

Fig. 3.6(a). General compressibility chart.

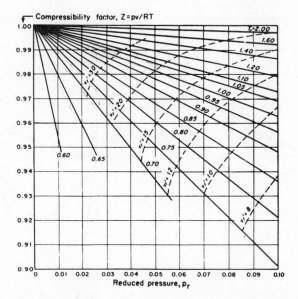

Fig. 3.6(b). General compressibility chart, very low temperatures.

30 gases. Figures 3.6(b) and 3.7 represent z for 26 gases (excluding H_2, He, NH_3, H_2O) with a maximum deviation of 1 percent, and H_2 and H_2O within a deviation of 1.5 percent. Figure 3.8 is for 26 gases and is accurate to 2.5 percent, while Fig. 3.9 is for nine gases and errors can be as high as 5 percent. For H_2 and He only, Newton's corrections to the actual critical constants are used to give pseudocritical constants

$$T'_c = T_c + 8°K \tag{3.24}$$

$$p'_c = p_c + 8 \text{ atm} \tag{3.25}$$

which enable you to use Figs. 3.6 through 3.10 for these two gases as well with minimum error. Figure 3.10 is a unique chart which, by having several parameters plotted simultaneously on it, helps you avoid trial-and-error solutions or graphical solutions of real gas problems. One of these helpful factors is the ideal reduced volume defined as

$$V_{r_i} \text{ (or } V'_r \text{ in the Nelson and Obert charts)} = \frac{\hat{V}}{\hat{V}_{c_i}} \tag{3.26}$$

$\hat{V}_{c_i}$ is the ideal critical volume, or

$$\hat{V}_{c_i} = \frac{RT_c}{p_c} \tag{3.27}$$

Both V_{r_i} and $\hat{V}_{c_i}$ are easy to calculate since T_c and p_c are presumed known. The development of the generalized compressibility chart is of considerable practical as well as pedagogical value because it enables engineering calculations to be made with considerable ease and also permits the development of thermodynamic functions for gases for which no experimental data are available. All you need to know to use these charts are the critical temperature and the critical

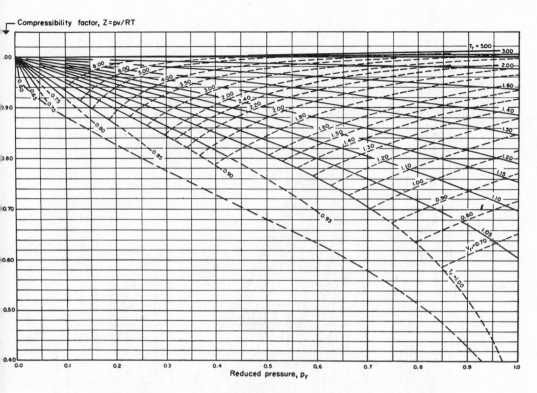

Fig. 3.7. General compressibility chart, low pressures.

pressure for a pure substance (or the pseudovalues for a mixture, as we shall see later). The value $z = 1$ represents ideality, and the value $z = 0.27$ is the compressibility factor at the critical point.

Quite a few authors have suggested that the generalized compressibility relation, Eq. (3.23), could provide better accuracy if another parameter for the gas were included in the argument on the right-hand side (in addition to p_r and T_r). Clearly, if a third parameter is used, adding a third dimension to p_r and T_r, you have to employ a *set* of tables or charts rather than a single table or chart. Lydersen et al.[6] developed tables of z that included as the third parameter the critical compressibility factor, $z_c = p_c \hat{V}_c / RT_c$. Pitzer[7] used a different third parameter termed the *acentric* factor, defined as being equal to $-\ln p_{rs} - 1$, where p_{rs} is the value of the reduced vapor pressure at $T_r = 0.70$.

Viswanath and Su[8] compared the z factors from the Nelson and Obert charts, the Lydersen-Greenkorn-Hougen charts, and the two-parameter (T_r and p_r) Viswanath and Su chart with the experimental z's for T_r's of 1.00–15.00

[6]A. L. Lydersen, R. A. Greenkorn, and O. A. Hougen, *University of Wisconsin Engineering Experimental Station Report no. 4*, Madison, Wis., 1955.
[7]K. S. Pitzer, *J. Amer. Chem. Soc.*, v. 77, p. 3427 (1955).
[8]D. S. Viswanath and G. J. Su, *A.I.Ch.E. J.*, v. 11, p. 202 (1965).

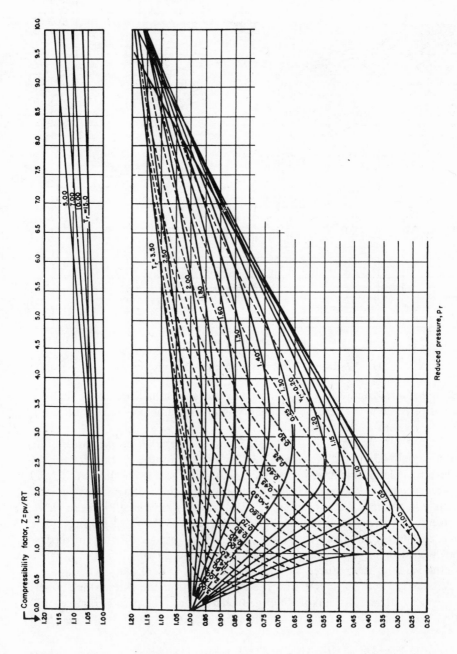

Fig. 3.8. General compressibility chart, medium pressures.

Compressibility factor, $Z = pv/RT$

Reduced pressure, p_r

176

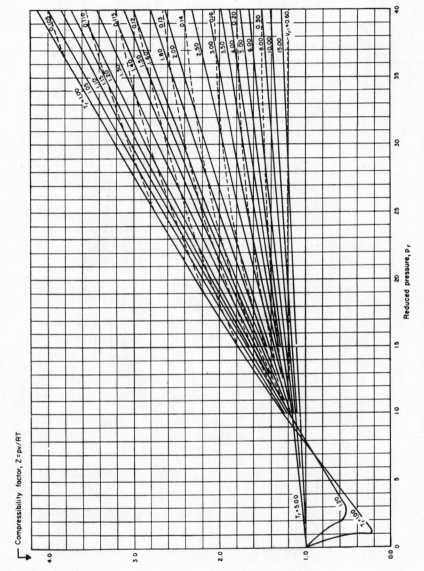

Fig. 3.9. Generalized compressibility chart, high pressures.

177

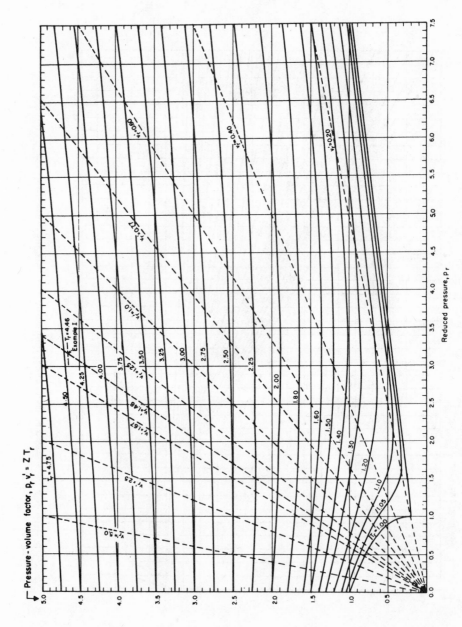

Fig. 3.10. Generalized compressibility chart, with special scales.

178

and p_r's of 0–40 for 19 gases with the following results:

	Viswanath- Su	Nelson- Obert	Lyderson- Greenkorn- Hougen
Number of points	415	355	339
Average deviation (%) in z	0.61	0.83	1.26

It is seen that all these charts yield quite reasonable values for engineering purposes. Because the use of a third parameter does not necessarily yield gas calculations of greater accuracy and because the presentation of z values so as to include the third parameter is considerably more cumbersome, we shall not show the three-parameter tables here.

EXAMPLE 3.13   Use of the compressibility factor

In spreading liquid ammonia fertilizer, the charges for the amount of NH_3 are based on the time involved plus the pounds of NH_3 injected into the soil. After the liquid has been spread, there is still some ammonia left in the source tank (volume = 120 ft³), but in the form of a gas. Suppose that your weight tally, which is obtained by difference, shows a net weight of 125 lb of NH_3 left in the tank as a gas at 292 psig. Because the tank is sitting in the sun, the temperature in the tank is 125°F.

Your boss complains that his calculations show that the specific volume of the gas is 1.20 ft³/lb and hence that there are only 100 lb of NH_3 in the tank. Could he be correct? See Fig. E3.13.

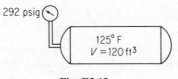

Fig. E3.13.

Solution:

Basis: 1 lb of NH_3

Apparently your boss used the ideal gas law in calculating his figure of 1.20 ft³/lb of NH_3 gas:

$$R = 10.73 \frac{(\text{psia})(\text{ft}^3)}{(\text{lb mole})(°R)}$$

$$T = 125°F + 460 = 585°R$$

$$p = 292 + 14.7 = 307 \text{ psia}$$

$$n = \frac{1 \text{ lb}}{17 \text{ lb/lb mole}}$$

$$\hat{V} = \frac{nRT}{p} = \frac{(1/17)(10.73)(585)}{307} = 1.20 \text{ ft}^3/\text{lb}$$

However, he should have used the compressibility factor because NH_3 does not behave as an ideal gas under the observed conditions of temperature and pressure. Let us again compute the mass of gas in the tank this time using

$$pV = znRT$$

What is known and unknown in the equation?

$$p = 307 \text{ psia}$$
$$V = 120 \text{ ft}^3$$
$$z = ?$$
$$n = \tfrac{1}{17} \text{ lb mole}$$
$$T = 585°R$$

The additional information needed (taken from Appendix D) is

$$T_c = 405.4°K \cong 729°R$$
$$p_c = 111.3 \text{ atm} \cong 1640 \text{ psia}$$

Then, since z is a function of T_r and p_r,

$$T_r = \frac{T}{T_c} = \frac{585°R}{729°R} = 0.803$$

$$p_r = \frac{p}{p_c} = \frac{307 \text{ psia}}{1640 \text{ psia}} = 0.187$$

From the Nelson and Obert chart, Fig. 3.7, you can read $z \cong 0.845$. Now $\hat{V}$ can be calculated as

$$\hat{V} = \frac{znRT}{p}$$

or

$$\hat{V} = \frac{1.20 \text{ ft}^3 \text{ ideal}}{\text{lb}} \bigg| \frac{0.845 \text{ ft}^3 \text{ actual}}{1 \text{ ft}^3 \text{ ideal}} = 1.01 \text{ ft}^3/\text{lb } NH_3$$

Note in the calculation above that we have added some hypothetical units to z to make the conversion from ideal cubic feet to actual cubic feet clear:

$$\frac{1 \text{ lb } NH_3}{1.01 \text{ ft}^3} \bigg| \frac{120 \text{ ft}^3}{} = 119 \text{ lb } NH_3$$

Certainly 119 lb is a more realistic figure than 100 lb, and it is easily possible to be in error by 6 lb if the residual weight of NH_3 in the tank is determined by difference. As a matter of interest you might look up the specific volume of NH_3 at the conditions in the tank in a handbook—you would find that $\hat{V} = 0.973 \text{ ft}^3/\text{lb}$, and hence the compressibility factor calculation yielded a volume with an error of only about 4 percent.

EXAMPLE 3.14 Use of the compressibility factor

A single-stage-to-orbit vehicle with space-shuttle capabilities may be feasible if the mechanical strength of frozen propellants can be utilized—particularly frozen oxygen, if it could be vaporized at the required rate. During the thrust period, the solid oxygen cylinder would be melted from its bottom surface and the resulting liquid propellant conveyed to a conventional pump-fed rocket engine where the oxygen would be vaporized and combined with vaporized fuel.

Suppose that 3.500 kg of O_2 is vaporized into a tank of 0.0284-m^3 volume at $-25°C$. What will the pressure in the tank be; will it exceed 100 atm?

Solution:
See Fig. E3.14.

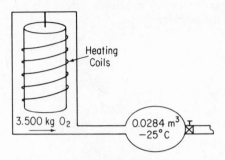

Fig. E3.14.

Basis: 3.500 kg O_2

We do know from Appendix D that

$$T_c = -118.8°C \quad \text{or} \quad -118.8 + 273.1 = 154.3°K$$

$$p_c = 49.7 \text{ atm}$$

However, this problem cannot be worked exactly the same way as the previous problem because we do not know the pressure of the O_2 in the tank. Thus we need to use the other parameter, V_{r_i} (i.e., V'_r on the chart), that is available on the Nelson and Obert charts:

$$\hat{V} \text{ (molal volume)} = \frac{0.0284 \text{ m}^3}{3.500 \text{ kg}} \left| \frac{32 \text{ kg}}{1 \text{ kg mole}} \right. = 0.259 \text{ m}^3/\text{kg mole}$$

Note that the *molal volume must* be used in calculating V_{r_i} since $\hat{V}_{c_i}$ is a volume per mole. Next,

$$\hat{V}_{c_i} = \frac{RT_c}{p_c} = \frac{0.08206 \text{(m}^3\text{)(atm)}}{\text{(kg mole)(}°K\text{)}} \left| \frac{154.3°K}{49.7 \text{ atm}} \right. = 0.255 \frac{\text{m}^3}{\text{kg mole}}$$

Then

$$V_{r_i} = \frac{\hat{V}}{\hat{V}_{c_i}} = \frac{0.259}{0.255} = 1.02$$

Now we know two parameters, V_{r_i} and, with $T = -25°C + 273 = 248°K$,

$$T_r = \frac{T}{T_c} = \frac{248°K}{154.3°K} = 1.61$$

From the Nelson and Obert chart, Figs. 3.8 or 3.10,

$$p_r = 1.43$$

Then

$$p = p_r p_c$$
$$= 1.43(49.7) = 71.0 \text{ atm}$$

The pressure of 100 atm is not exceeded.

EXAMPLE 3.15 Use of compressibility factor

Repeat Example 3.12, this time using the compressibility factor, i.e., the equation of state $pV = znRT$. "A 5-ft³ cylinder containing 50.0 lb of propane (C_3H_8) stands in the hot sun. A pressure gauge shows that the pressure is 665 psig. What is the temperature of the propane in the cylinder?"

Solution:
See Fig. E3.15.

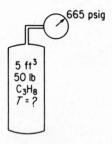

665 psig

5 ft³
50 lb
C_3H_8
$T = ?$

Fig. E3.15.

In the equation $pV = znRT$, we know that

$$V = 5 \text{ ft}^3$$

$$n = \frac{50 \text{ lb}}{44 \text{ lb/lb mole}} = 1.135 \text{ lb moles } C_3H_8$$

$$p = 665 \text{ psig} + 14.7 \cong 680 \text{ psia}$$

However, z as well as T is unknown. Since z is a function of T_r and p_r, we should first calculate the latter two quantities ($T_c = 370°K$, or $666°R$; $p_c = 42.1$ atm, or 617 psia, from Appendix D) but are stopped by the fact that T is unknown. Another parameter available for use, to avoid a trial-and-error or graphical solution, is the factor zT_r (or V'_r could be used, as in the last example):

$$zT = \frac{pV}{nR} \quad \text{or} \quad zT_r = \frac{pV}{nRT_c}$$

$$zT_r = \frac{680 \text{ psia} \mid 5 \text{ ft}^3 \mid}{\mid 1.135 \text{ lb mole} \mid 10.73 \dfrac{(\text{psia})(\text{ft}^3)}{(\text{lb mole})(°R)} \mid 666°R}$$

$$= 0.42$$

From the Nelson and Obert chart, Fig. 3.10, at

$$\left. \begin{array}{l} zT_r = 0.42 \\ \\ p_r = \dfrac{680}{617} = 1.10 \end{array} \right\}$$

you can find $T_r = 1.02$ and

$$T = T_r T_c = (370)(1.02) = 378°K$$

$$= (666)(1.02) = 680°R$$

3.2.3 Gaseous Mixtures. So far we have discussed only pure compounds and their p-V-T relations. Most practical problems involve gaseous mixtures and there is very little experimental data available for gaseous mixtures. Thus the question is, How can we predict p-V-T properties with reasonable accuracy for gaseous mixtures? We treated ideal gas mixtures in Sec. 3.1.3, and saw, as we shall see later (in Chap. 4 for the thermodynamic properties), that for ideal mixtures the properties of the individual components can be added together to give the desired property of the mixture. But this technique does not prove to be adequate for real gases. The most desirable technique would be to develop methods of calculating p-V-T properties for mixtures based solely on the properties of the pure components. Possible ways of doing this for real gases are discussed below.

(a) Equations of state.

Take, for example, van der Waals' equation, Eq. (3.12). You can compute the partial pressure of each component by van der Waals' equation and then add the individual partial pressures together according to Dalton's law to give the total pressure of the system,

for component A,

$$p_A = \frac{n_A RT}{V - n_A b_A} - \frac{n_A^2 a_A}{V^2} \tag{3.28}$$

for component B,

$$p_B = \frac{n_B RT}{V - n_B b_B} - \frac{n_B^2 a_B}{V^2} \tag{3.29}$$

    etc.,
in total,

$$p_T = RT\left[\frac{n_A}{V - n_A b_A} + \frac{n_B}{V - n_B b_B} + \cdots\right]$$
$$- \frac{1}{V^2}[n_A^2 a_A + n_B^2 a_B + \cdots] \tag{3.30}$$

This technique provides a direct solution for p_T if everything else is known; however, the solution for V or n_A from Eq. (3.30) is quite cumbersome.

(b) Average constants.

A simpler and usually equally effective way to solve gas mixture problems is to use average constants in the equation of state. The problem resolves itself into this question, How should the constants be averaged to give the least error on the whole? For van der Waals' equation, you should proceed as follows. For:

b (Use linear mole fraction weight):

$$b_{\text{mixture}} = b_A y_A + b_B y_B + \cdots \tag{3.31}$$

a (Use linear square root average weight):

$$a^{1/2}_{\text{mixture}} = a^{1/2}_A y_A + a^{1/2}_B y_B + \cdots \tag{3.32}$$

If you are puzzled as to the reasoning behind the use of the square root weighting for *a*, remember that the term in van der Waals' equation involving *a* is

$$\frac{n^2 a}{V^2}$$

in which the number of moles is squared. The method of average constants can be successfully applied to most of the other equations of state listed in Table 3.1. The average constants can also be calculated from the pseudoreduced constants described below. For example, for van der Waals' equation,

$$a = \frac{27}{64}\frac{R^2 T_c^2}{p_c} \quad \text{and} \quad b = \frac{RT_c}{8p_c}$$

(c) Mean compressibility factor.

Another approach toward the treatment of gaseous mixtures is to say that $pV = z_m nRT$, where z_m can be called the *mean compressibility factor*. With such a relationship the only problem is how to evaluate the mean compressibility factor satisfactorily. One obvious technique that might occur to you is to make z_m a mole average as follows:

$$z_m = z_A y_A + z_B y_B + \cdots \tag{3.33}$$

Since z is a function of both the reduced temperature and the reduced pressure, it is necessary to decide what pressure will be used to evaluate p_r.

(1) *Assume Dalton's law of partial pressures.* For each component z is evaluated at T_r and the reduced *partial pressure* for each gaseous component. The reduced partial pressure is defined as

$$p_{T_A} = \frac{p_A}{p_{c_A}} = \frac{(p_T)y_A}{p_{c_A}} \tag{3.34}$$

(2) *Assume Amagat's law of pure component volumes.* For each component z is evaluated at T_r and the reduced *total pressure* on the system.

(d) Pseudocritical properties.

Many weighting rules have been proposed to combine the critical properties of the components of a real gas mixture in order to get an effective set of critical properties (*pseudocritical*) that enable *pseudoreduced* properties of the mixture to be computed.[9] The pseudoreduced properties in turn are used in exactly the same way as the reduced properties for a pure compound, and in effect the mixture can be treated as a pure compound. In instances where you know no-

[9]You should still add Newton's corrections for H_2 and He.

thing about the gas mixture this technique is preferable to any of those discussed previously.

In Kay's method pseudocritical values for mixtures of gases are calculated on the assumption that each component in the mixture contributes to the pseudocritical value in the same proportion as the number of moles of that component. Thus the pseudocritical values are computed as follows:

$$p'_c = p_{c_A}y_A + p_{c_B}y_B + \cdots \tag{3.35}$$

$$T'_c = T_{c_A}y_A + T_{c_B}y_B + \cdots \tag{3.36}$$

where p'_c = pseudocritical pressure, T'_c = pseudocritical temperature. (It has also been found convenient in some problems to calculate similarly a weighted pseudo-ideal-critical volume V'_{c_i}.) You can see that these are linearly weighted mole average pseudocritical properties. (In Sec. 3.8 we shall compare the true critical point of gaseous mixture with the pseudocritical point). Then the respective pseudoreduced values are

$$p'_r = \frac{p}{p'_c} \tag{3.37}$$

$$T'_r = \frac{T}{T'_c} \tag{3.38}$$

If you are faced with a complicated mixture of gases whose composition is not well known, you still can estimate the pseudocritical constants from charts[10] such as shown in Figs. 3.11(a) and (b) if you know the gas specific gravity. Figure

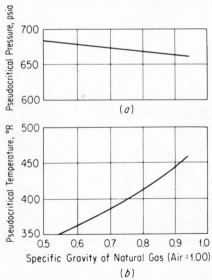

Fig. 3.11. Estimation of critical properties of natural gases.

[10]Natural Gasoline Supply Men's Association, *Engineering Data Book*, Tulsa, Okla., 1957, p. 103.

3.11 is good only for natural gases composed mainly of methane which contain less than 5 percent impurities (CO_2, N_2, H_2S, etc.).

Kay's method is known as a two-parameter rule since only p_c and T_c for each component are involved in the calculation of z. If a third parameter such as z_c, the Pitzer acentric factor, or $\hat{V}_{c_i}$ is included in the determination of the mean compressibility factor, then we would have a three-parameter rule. Reid and Leland[11] have shown that all the weighting rules can be obtained from a common base, and all reduce to Kay's rule as limiting cases with the proper assumptions. All the pseudocritical methods do not provide equal accuracy in predicting

TABLE 3.4 PREDICTION OF p-V-T VALUES OF GAS DENSITY
BY PSEUDOCRITICAL METHODS

Method	% Root-mean-square deviation of density, 35 systems	% Root-mean-square deviation of density, CO_2 and H_2S free systems
(1) *Three-parameter Kay's rule:*		
$T'_c = \sum T_{c_i} y_i$		
$p'_c = \sum p_{c_i} y_i$	10.74	6.82
$z'_c = \sum z_{c_i} y_i$		
(2) *Empirical:*		
$T'_c = \sum T_{c_i} y_i$		
$\dfrac{T'_c}{p'_c} = \sum \dfrac{T'_{c_i} y_i}{p_{c_i}}$	6.06	5.85
$\hat{V}'_c = \sum \hat{V}_{c_i} y_i$		
(3) *Virial approach (Joffe's method III):*		
$\dfrac{T'_c}{p'_c} = \dfrac{1}{8} \sum_i \sum_j y_i y_j \left[\left(\dfrac{T_{c_i}}{p_{c_i}} \right)^{1/3} + \left(\dfrac{T_{c_j}}{p_{c_j}} \right)^{1/3} \right]^3$		
$\dfrac{T'_c}{\sqrt{p'_c}} = \sum y_i \dfrac{T_{c_i}}{\sqrt{p_{c_i}}}$	4.98	4.24
$\hat{V}'_c = \sum \hat{V}_{c_i} y_i$		
(4) *"Recommended" method:*		
$\dfrac{T'_c}{p'_c} = \dfrac{1}{3} \sum y_i \dfrac{T_{c_i}}{p_{c_i}} + \dfrac{2}{3} \left[\sum y_i \left(\dfrac{T_{c_i}}{p_{c_i}} \right)^{1/2} \right]^2$		
$\dfrac{T'_c}{\sqrt{p'_c}} = \sum y_i \dfrac{T_{c_i}}{\sqrt{p_{c_i}}}$	4.32	3.26
$z'_c = \sum z_{c_i} y_i$		

[11]R. C. Reid and T. W. Leland, *A.I.Ch.E. J.*, vol. 11, p. 228 (1965).

p-V-T properties, however, but most suffice for engineering work. Stewart[12] et al. reviewed 21 different methods of determining the pseudoreduced parameters by three-parameter rules (see Table 3.4). Although Kay's method was not the most accurate, it was easy to use and not considerably poorer than some of the more complex techniques of averaging critical properties.

In summary, to evaluate all the various methods which have been presented for treating *p-V-T* relationships of gaseous mixtures, we would have to state that no one method will consistently give the best results. Kay's pseudocritical method, on the average, will prove reliable, although other methods of greater accuracy (and of greater complexity) are available in the literature, as brought out in Table 3.4. Some of the longer equations of state, with average constants, are quite accurate. All these methods begin to break down near the true critical point for the mixture.

EXAMPLE 3.16   *p-V-T* **Relations for gas mixtures**

A gaseous mixture has the following composition (in mole percent):

Methane	CH_4	20
Ethylene	C_2H_4	30
Nitrogen	N_2	50

at 90 atm pressure and 100°C. Compare the molal volume as computed by the methods of

(a) Perfect gas law.
(b) van der Waals' equation plus Dalton's law.
(c) van der Waals' equation using averaged constants.
(d) Mean compressibility factor and Dalton's law.
(e) Mean compressibility factor and Amagat's law.
(f) Pseudoreduced technique (Kay's method).

Solution:

Basis: 1 g mole of gas mixture

Additional data needed are:

	$a(atm)\left(\dfrac{cm^3}{g\ mole}\right)^2$	$b\left(\dfrac{cm^3}{g\ mole}\right)$	$T_c(°K)$	$p_c(atm)$
CH_4	2.25×10^6	42.8	191	45.8
C_2H_4	4.48×10^6	57.2	283	50.9
N_2	1.35×10^6	38.6	126	33.5

$$R = 82.06 \frac{(cm^3)(atm)}{(g\ mole)(°K)}$$

---

[12]W. E. Stewart, S. F. Burkhart, and David Voo, Paper given at the A.I.Ch.E. meeting in Kansas City, Mo., May 18, 1959. Also refer to A. Satter and J. M. Campbell, *Soc. Petrol. Engrs. J.*, p. 333 (Dec. 1963); and H. E. Barner and W. C. Quinlan, *I&EC Process Design and Development*, vol. 8, p. 407 (1969).

(a) Perfect gas law:

$$V = \frac{nRT}{p} = \frac{1(82.06)(373)}{90} = 340 \text{ cm}^3 \text{ at 90 atm and } 373°K$$

(b) Combine van der Waals' equation and Dalton's law according to Eq. (3.30):

$$p_T = RT\left[\frac{n_{CH_4}}{V - n_{CH_4}b_{CH_4}} + \frac{n_{C_2H_4}}{V - n_{C_2H_4}b_{C_2H_4}} + \frac{n_{N_2}}{V - n_{N_2}b_{N_2}}\right]$$
$$- \frac{1}{V^2}[n_{CH_4}^2 a_{CH_4} + n_{C_2H_4}^2 a_{C_2H_4} + n_{N_2}^2 a_{N_2}]$$

Substitute the numerical values (in the proper units):

$$90 = 82.06(373)\left[\frac{0.2}{V - 8.56} + \frac{0.3}{V - 17.1} + \frac{0.5}{V - 19.3}\right]$$
$$- \frac{1}{V^2}[9 \times 10^4 + 40.5 \times 10^4 + 33.8 \times 10^4] \tag{1}$$

The solution to this equation can be obtained by trial and error or graphical means. For the first trial assume ideal conditions. By plotting the right-hand side of Eq. (1) against V until the total is 90, we find the equation converges at about $V = 332$ cm^3 at 90 atm and 373°K.

(c) To use average constants in van der Waals' equation, write it in the following fashion:

$$V^3 - \left(\bar{b} + \frac{RT}{p}\right)nV^2 + \left(\frac{\bar{a}}{p}\right)n^2V - \frac{\bar{a}\bar{b}}{p}n^3 = 0$$

where $\bar{a}$ and $\bar{b}$ are the average constants.

$$(\bar{a})^{1/2} = 0.2a_{CH_4}^{1/2} + 0.3a_{C_2H_4}^{1/2} + 0.5a_{N_2}^{1/2}$$
$$\bar{a} = 2.30 \times 10^6 \text{ atm}\left(\frac{\text{cm}^3}{\text{g mole}}\right)^2$$
$$\bar{b} = 0.2b_{CH_4} + 0.3b_{C_2H_4} + 0.5b_{N_2} = 45.0\frac{\text{cm}^3}{\text{g mole}}$$
$$n = 1 \text{ g mole (the basis)}$$

$$V^3 - \left[45.0 + \frac{(82.06)(373)}{90}\right]V^2 + \left[\frac{2.30 \times 10^6}{90}\right]V - \left[\frac{(2.30 \times 10^6)(45.0)}{90}\right] = 0 \tag{2}$$

Let us work out a method of solving Eq. (2) for V different from the method used in part (b):

$$f(V) = V^3 - 385V^2 + 2.56 \times 10^4 V - 1.15 \times 10^6 = 0$$
$$-V^3 + 385V^2 = 2.56 \times 10^4 V - 1.15 \times 10^6 \tag{3}$$

Splitting this equation into two parts,

$$f_1(V) = 2.56 \times 10^4 V - 1.15 \times 10^6 \tag{4}$$
$$f_2(V) = -V^3 + 385V^2 = V^2(385 - V) \tag{5}$$

and plotting these parts (see Fig. E3.16) to see where they cross is one method of obtaining the real root of the equation. You know V is in the vicinity of 340 cm^3.

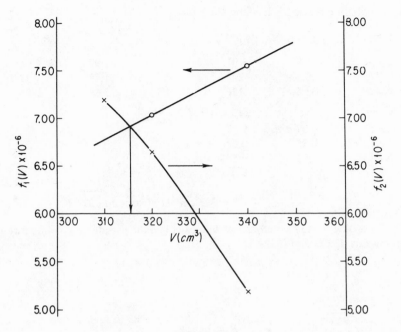

Fig. E3.16.

$V(cm^3)$	$f_1(V)$	$f_2(V)$
340	7.55×10^6	5.20×10^6
320	7.04×10^6	6.65×10^6
310	(only 2 points required for straight line)	7.20×10^6

From the graph $V \cong 316\ cm^3$ at 90 atm and 373°K.

(d) The table below shows how Dalton's law can be used to estimate a mean compressibility factor.

Comp.	$p_c(atm)$	$T_c(°K)$	y	$(90)y = p$	$p_r = \dfrac{p}{p_c}$	$T_r = \dfrac{373}{T_c}$
CH_4	45.8	191	0.2	18	0.394	1.95
C_2H_4	50.9	283	0.3	27	0.535	1.32
N_2	33.5	126	0.5	45	1.340	2.95

Comp.	z	$(z)(y)$
CH_4	0.99	0.198
C_2H_4	0.93	0.279
N_2	1.00	0.500
	z_{mean} =	0.977

Then

$$V = \frac{z_{mean}\,RT}{p} = \frac{(0.977)(82.06)(373)}{90}$$

$$= 332\ cm^3 \text{ at 90 atm and 373°K}$$

(e) Combining Amagat's law and the z factor,

Comp.	$p_c(atm)$	$T_c(°K)$	y	$p_r = \dfrac{90}{p_c}$	$T_r = \dfrac{373}{T_c}$
CH_4	45.8	191	0.2	1.97	1.95
C_2H_4	50.9	283	0.3	1.78	1.32
N_2	33.5	126	0.5	2.68	2.95

Comp.	z	$(z)(y)$
CH_4	0.97	0.194
C_2H_4	0.75	0.225
N_2	1.01	0.505
z_{mean} =		0.924

$$V = \frac{0.924(82.06)(373)}{90} = 313 \text{ cm}^3 \text{ at 90 atm and 373°K}$$

(f) According to Kay's method, we first calculate the pseudocritical values for the mixture by Eqs. (3.35) and (3.36).

$$\qquad\qquad\qquad\qquad\qquad \mathbf{CH_4} \qquad\quad \mathbf{C_2H_4} \qquad\quad \mathbf{N_2}$$

$$p'_c = p_{c_A}y_A + p_{c_B}y_B + p_{c_C}y_C = (45.8)(0.2) + (50.9)(0.3) + (33.5)(0.5)$$
$$= 41.2 \text{ atm}$$

$$T'_c = T_{c_A}y_A + T_{c_B}y_B + T_{c_C}y_C = (191)(0.2) + (283)(0.3) + (126)(0.5)$$
$$= 186°K$$

Then we calculate the pseudoreduced values for the mixture by Eqs. (3.37) and (3.38):

$$p'_r = \frac{p}{p'_c} = \frac{90}{41.2} = 2.18, \qquad T'_r = \frac{T}{T'_c} = \frac{373}{186} = 2.00$$

With the aid of these two parameters we can find from Fig. 3.8 that $z = 0.965$. Thus

$$V = \frac{zRT}{p} = \frac{(0.965)(82.06)(373)}{90} = 328 \text{ cm}^3 \text{ at 90 atm and 373°K}$$

EXAMPLE 3.17 Use of pseudoreduced ideal molal volume

In instances where the temperature or pressure of a gas mixture is unknown, it is convenient, in order to avoid a trial-and-error solution using the generalized compressibility charts, to compute a pseudocritical ideal volume and a pseudoreduced ideal volume as illustrated below. Suppose we have given that the molal volume of the gas mixture in the previous problem was 326 cm³ at 90.0 atm. What was the temperature?

Solution:

Basis: 326 cm³ gas at 90.0 atm and $T°K$

Comp.	$T_c(°K)$	$p_c(atm)$	y	yT_c	yp_c
CH_4	190.7	45.8	0.20	38.14	9.16
C_2H_4	283.1	50.9	0.30	84.93	15.27
N_2	126.2	33.5	0.50	63.10	16.75
				186.2	41.2

Note we have used mole fractions as weighting factors to find an average $T'_c = 186.2$ and $p'_c = 41.2$ as in the previous example. Next we compute V'_{c_i}:

$$\hat{V}'_{c_i} = \frac{RT'_c}{p'_c} = \frac{(82.06)(186.2)}{(41.2)} = 371.0 \text{ cm}^3/\text{g mole}$$

$$V'_r = \frac{\hat{V}}{\hat{V}'_{c_i}} = \frac{326}{371} = 0.879$$

$$p'_r = \frac{p}{p'_c} = \frac{90.0}{41.2} = 2.19$$

From Fig. 3.8 or Fig. 3.10, $T'_r = 1.98$. Then

$$T = T'_c T'_r = (186.2)(1.98) = 369°\text{K}$$

3.3 *Vapor pressure*

The terms *vapor* and *gas* are used very loosely. A gas which exists below its critical temperature is usually called a vapor because it can condense. If you continually compress a pure gas at constant temperature, provided that the temperature is below the critical temperature, some pressure is eventually reached at which the gas starts to condense into a liquid. Further compression does not increase the pressure but merely increases the fraction of gas that condenses. A reversal of the procedure just described will cause the liquid to be transformed into the gaseous state again. From now on, the word *vapor* will be reserved to describe a gas below its critical point in a process in which the phase change is of primary interest, while the word *gas* or *noncondensable gas* will be used to describe a gas above the critical point or a gas in a process in which it cannot condense.

Vaporization and condensation at constant temperature and pressure are *equilibrium* processes, and the equilibrium pressure is called the *vapor pressure*. At a given temperature there is only one pressure at which the liquid and vapor phases of a pure substance may exist in equilibrium. Either phase alone may exist, of course, over a wide range of conditions. By equilibrium we mean a state in which there is no tendency toward spontaneous change. Another way to say the same thing is to say equilibrium is a state in which all the rates of attaining and departing from the state are balanced.

You can visualize vapor pressure, vaporization, and condensation more easily with the aid of Fig. 3.12. Figure 3.12 is an expanded *p-T* diagram for pure water. For each temperature you can read the corresponding pressure at which water vapor and water liquid exist in equilibrium. You have encountered this condition of equilibrium many times—for example, in boiling. Any substance has an infinite number of boiling points, but by custom we say the "normal" boiling point is the temperature at which boiling takes place under a pressure of 1 atm (760 mm Hg). Unless another pressure is specified, 1 atm is assumed,

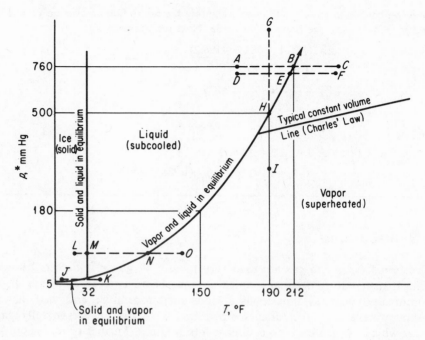

Fig. 3.12. Vapor pressure curve for water.

and the term *boiling point* is taken to mean the "normal boiling point." The normal boiling point for water occurs when the vapor pressure of the water equals the pressure of the atmosphere on top of the water. A piston with a force of 14.7 psia could just as well take the place of the atmosphere, as shown in Fig. 3.13. For example, you know that at 212°F water will boil (vaporize) and the pressure

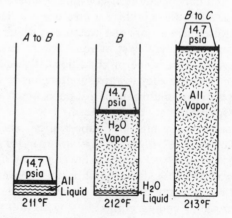

Fig. 3.13. Transformation of liquid water into water vapor at constant pressure.

will be 760 mm Hg, or 1 atm (point *B*). Suppose that you heat water starting at 170°F (point *A*) in an open pan—what happens? We assume that the water vapor above the pan is at all times in equilibrium with the liquid water. This is a constant-pressure process since the air around the water in the pan acts similarly to a piston in a cylinder to keep the pressure at atmospheric pressure. As the temperature rises and the confining pressure stays constant, nothing particularly noticeable occurs until 212°F is reached, at which time the water begins to boil, i.e., evaporate. It pushes back the atmosphere and will completely change from liquid into vapor. If you heated the water in an enclosed cylinder, and if after it had all evaporated at point *B* you continued heating the water vapor formed at constant pressure, you could apply the gas laws in the region *B-C* (and at higher temperatures). A reversal of this process would cause the vapor to condense at *B* to form a liquid. The temperature at point *B* would in these circumstances represent the *dew point*.

Suppose that you went to the top of Pikes Peak and repeated the experiment—what would happen then? Everything would be the same (points *D-E-F*) with the exception of the temperature at which the water would begin to boil, or condense. Since the pressure of the atmosphere at the top of Pikes Peak would presumably be lower than 760 mm Hg, the water would start to displace the air, or boil, at a lower temperature. However, water still would exert a vapor pressure of 760 mm Hg at 212°F. You can see that (a) at any given temperature water exerts its vapor pressure (at equilibrium); (b) as the temperature goes up, the vapor pressure goes up; and (c) it makes no difference whether water vaporizes into air, into a cylinder closed by a piston, or into an evacuated cylinder—at any temperature it still exerts the same vapor pressure as long as the water is in equilibrium with its vapor.

A process of vaporization or condensation at constant temperature is illustrated by the lines *G-H-I* or *I-H-G*, repectively, in Fig. 3.12. Water would vaporize or condense at constant temperature as the pressure reached point *H* on the vapor-pressure curve (also look at Fig. 3.14.)

The *p-T* conditions at which ice (in its common form) and water vapor are in equilibrium are also seen in Fig. 3.12. When the solid passes directly into the vapor phase without first becoming a liquid (line *J-K* as opposed to line *L-M-N-O*), it is said to *sublime*. Iodine crystals do this at room temperature; water sublimes only below 32°F, as when the frost disappears in the winter when the thermometer reads 20°F.

The vapor-pressure line indicates the separation of both solid and liquid regions from the vapor region and extends well past the conditions shown in Fig. 3.12 all the way to the critical temperature and pressure (not shown). Above the critical temperature, water can exist only as a gas. A term commonly applied to the vapor-liquid portion of the vapor-pressure curve is the word *saturated*. It means the same thing as vapor and liquid in equilibrium with each other. If a gas is just ready to start to condense its first drop of liquid, the gas is called a saturated gas; if a liquid is just about to vaporize, it is called a saturated liquid.

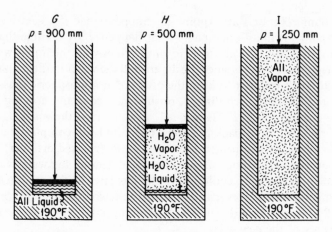

Fig. 3.14. Transformation of liquid water into water vapor at constant temperature.

These two conditions are also known as the *dew point* and *bubble point*, respectively.

If you have a mixture of liquid and vapor at equilibrium (called a *wet gas*), both the liquid and vapor are said to be saturated at the specified conditions. The vapor-pressure line in Fig. 3.12 thus represents the state of a pure component designated by a number of special terms depending on what aspect of the state is of primary importance:

(a) The saturated liquid.
(b) The saturated vapor.
(c) The bubble point.
(d) The dew point.

The region to the right of the vapor-pressure curve in Fig. 3.12 is called the *superheated* region and the one to the left of the vapor-pressure curve is called the *subcooled* region. The temperatures in the superheated region, if measured as the difference $(O - N)$ between the actual temperature of the superheated vapor and the saturation temperature for the same pressure, are called *degrees of superheat*. For example, steam at 500°F and 100 psia (the saturation temperature for 100 psia is 327.8°F) has $(500 - 327.8) = 172.2$°F of superheat. Another new term you will find used frequently is the word *quality*. A "wet" vapor consists of saturated vapor and saturated liquid in equilibrium. The weight fraction of vapor is known as the quality.

EXAMPLE 3.18   Properties of wet vapors

The properties of a mixture of vapor and liquid in equilibrium (for a single component) can be computed from the individual properties of the saturated vapor and saturated liquid. The steam tables (Appendix C) are a good source of data to illustrate such computations. At 400°F and 247.3 psia the specific volume of a wet steam mixture is 1.05 ft³/lb. What is the quality of the steam?

Solution:
From the steam tables the specific volumes of the saturated vapor and liquid are

$$\hat{V}_l = 0.0186 \text{ ft}^3/\text{lb}, \qquad \hat{V}_g = 1.8633 \text{ ft}^3/\text{lb}$$

Basis: 1 lb wet steam mixture

Let x = weight fraction vapor.

$$\frac{0.0186 \text{ ft}^3}{1 \text{ lb liquid}} \bigg| \frac{(1-x) \text{ lb liquid}}{} + \frac{1.8633 \text{ ft}^3}{1 \text{ lb vapor}} \bigg| \frac{x \text{ lb vapor}}{} = 1.05 \text{ ft}^3$$

$$0.0186 - 0.0186x + 1.8633x = 1.05$$

$$1.845x = 1.03$$

$$x = 0.56$$

Other properties of wet mixtures can be treated in the same manner.

3.3.1 Change of Vapor Pressure with Temperature.

A large number of experiments on many substances have shown that a plot of the vapor pressure (p^*) of a compound against temperature does not yield a straight line but a curve, as you saw in Fig. 3.12. Many types of correlations have been proposed to transform this curve to a linear form ($y = mx + b$); a plot of log (p^*) vs. ($1/T$), for moderate temperature intervals, is reasonably linear:

$$\underbrace{\log (p^*)}_{y} = \underbrace{m\left(\frac{1}{T}\right) + b}_{mx + b} \qquad (3.39)$$

Equation (3.39) is derived from the Clausius-Clapeyron equation (see Chap. 4). Empirical correlations of vapor pressure are frequently given in the following form (refer to Appendix G for values of the constants):

$$\log (p^*) = -\frac{A}{t + C} + B \qquad (3.40)$$

where A, B, C = constants different for each substance
t = temperature in °C

Over wide temperature intervals the experimental data are not exactly linear as indicated by Eq. (3.39), but have a slight tendency to curve. This curvature can be straightened out by using a special plot known as a Cox chart.[13] The log of the vapor pressure of a compound is plotted against a special nonlinear temperature scale constructed from the vapor-pressure data for water (called a *reference substance*).

[13]E. R. Cox, *Ind. Eng. Chem.*, v. 15, p. 592 (1923).

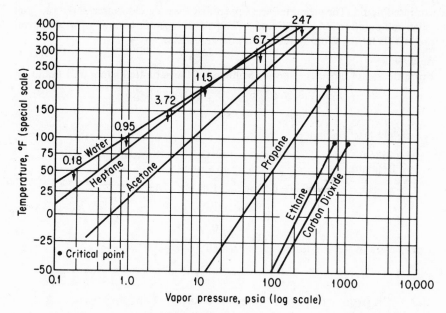

Fig. 3.15. Cox chart.

As illustrated in Fig. 3.15, the temperature scale is established by recording the temperature at a given vapor pressure of water for a number of vapor pressures. The numbers on the line for water indicate the vapor pressures of water for selected values (50, 100, 150, etc.) of temperature. The vapor pressures of other substances plotted on this same graph will yield straight lines over extensive temperature ranges and thus facilitate the extrapolation and interpolation of vapor-pressure data. It has been found that lines so constructed for closely related compounds, such as hydrocarbons, all meet at a common point. Since straight lines can be obtained on a Cox chart only two sets of vapor-pressure data are needed to provide complete information about the vapor pressure of a substance over a considerable temperature range. We shall discuss in Chap. 4 under the topic of the Clausius-Clapeyron equation other information that can be obtained from vapor-pressure plots.

3.3.2 Change of Vapor Pressure with Pressure.

The equation for the change of vapor pressure with total pressure at constant temperature in a system is

$$\left(\frac{\partial(p^*)}{\partial p_T}\right)_T = \frac{\hat{V}_l}{\hat{V}_g} \tag{3.41}$$

where $\hat{V}$ = molal volume of saturated liquid or gas

p_T = total pressure on the system

Under normal conditions the effect is negligible.

EXAMPLE 3.19 Extrapolation of vapor-pressure data

The control of existing solvents is described in the Federal Register, v. 36, no. 158, dated August 14, 1971, under Title 42, Chapter 4, Appendix B, Section 4.0, Control of Organic Compound Emissions. Section 4.6 indicates that reductions of at least 85 percent can be accomplished by (a) incineration or (b) carbon adsorption.

Chlorinated solvents and many other solvents used in industrial finishing and processing, dry-cleaning plants, metal degreasing, printing operations, and so forth, can be recycled and reused by the introduction of carbon adsorption equipment. To predict the size of the adsorber, you first need to know the vapor pressure of the compound being adsorbed at the process conditions.

The vapor pressure of chlorobenzene is 400 mm Hg at 110.0°C and 5 atm at 205°C. Estimate the vapor pressure at 245°C and at the critical point (359°C).

Solution:

The vapor pressures will be estimated by use of a Cox chart. See Fig. E3.19. The temperature scale is constructed by using the following data from the steam tables:

p^* H$_2$O (*psia*)	t (°F)
0.95	100
3.72	150
11.5	200
29.8	250
67.0	300
247	400
680	500
1543	600
3094	700

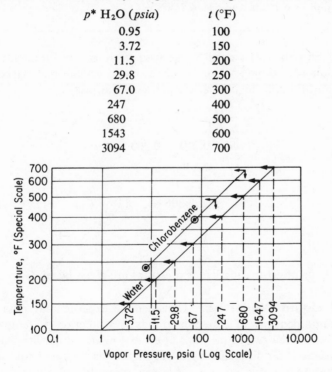

Fig. E3.19.

The vapor pressures from 1 to 3100 psia are marked on the horizontal logarithmic scale. Next, draw a line representing the vapor pressure of water at any suitable angle on the graph so as to stretch the desired temperature range from the bottom to the top of the vertical axis. For each vapor pressure, the temperature is marked and a horizontal line drawn to the ordinate. This establishes the temperature scale (which looks almost logarithmic in nature).

Now convert the two vapor pressures of chlorobenzene into psia,

$$\frac{400 \text{ mm}}{} \left| \frac{14.7 \text{ psia}}{760 \text{ mm}} = 7.74 \text{ psia} \right| 110°C = 230°F$$

$$\frac{5 \text{ atm}}{} \left| \frac{14.7 \text{ psia}}{1 \text{ atm}} = 73.5 \text{ psia} \right| 205°C = 401°F$$

and plot these two points on the graph paper. Next draw a straight line between them, and extrapolate to 471°F (245°C) and 678°F (359°C). At these two temperatures read off the estimated vapor pressures:

	471°F (245°C)	678°F (359°C)
Estimated	150 psia	700 psia
Experimental	147 psia	666 psia

Experimental values given for comparison.

An alternative way to extrapolate vapor pressure data would be to use the Othmer[14] plot as explained in Sec. 4.4.

3.3.3 Estimating Vapor Pressures.

Miller[15] recommends the following equation to estimate vapor pressures based on the normal boiling-point temperature (T_b) and the critical temperature and pressure:

$$\log p_r = -\frac{G}{T_r}[1 - T_r^2 + k(3 + T_r)(1 - T_r)^3]$$

where

$$G = 0.210 + 0.200a$$

$$a = \frac{T_{rb} \text{ lb } p_c}{1 - T_{rb}}$$

$$k = \frac{(a/2.306\ G) - (1 + T_{rb})}{(3 + T_{rb})(1 - T_{rb})^2}$$

$$T_{rb} = \frac{T_b}{T_c}$$

Other methods of estimating vapor pressures are given by McGowan.[16]

3.3.4 Liquid Properties.

Considerable experimental data are available for liquid densities of pure compounds as a function of temperature and pressure. Gold[17] reviews 13 methods of estimating liquid densities and provides recommendations for different classes of compounds if you cannot locate experimental data. As to liquid mixtures, it is even more difficult to predict the *p-V-T*

[14]D. F. Othmer and E. S. Yu, *Indus. Engr. Chem.* v. 60, p. 20 (1968).
[15]D. G. Miller, *J. Phys. Chem.*, v. 69, p. 3209 (1965).
[16]J. C. McGowan, *Rec. Trav. Chim.*, v. 84, p. 99 (1965).
[17]P. I. Gold, *Chem. Engr.*, p. 170 (Nov. 18, 1968).

properties of liquid mixtures than of real gas mixtures. Probably more experimental data (especially at low temperatures) are available than for gases, but less is known about the estimation of the p-V-T properties of liquid mixtures. Corresponding state methods are usually used, but with less assurance than for gas mixtures.

3.4 Saturation

How can you predict the properties of a mixture of a pure vapor (which can condense) and a noncondensable gas? A mixture containing a vapor behaves somewhat differently than does a pure component by itself. A typical example with which you are quite familiar is that of water vapor in air.

Water vapor is a gas, and like all gases its molecules are free to migrate in any direction. They will do so as long as they are not stopped by the walls of a container. Furthermore, the molecules will distribute themselves evenly throughout the entire volume of the container.

When any pure gas (or a gaseous mixture) comes in contact with a liquid, the gas will acquire vapor from the liquid. If contact is maintained for a considerable length of time, equilibrium is attained, at which time the *partial pressure of the vapor* (vaporized liquid) *will equal the vapor pressure* of the liquid at the temperature of the system. Regardless of the duration of contact between the liquid and gas, after equilibrium is reached no more net liquid will vaporize into the gas phase. The gas is then said to be *saturated* with the particular vapor at the given temperature. We can also say that the gas mixture is at its dew point.

As an illustration, assume that you put dry air at 147°F into a container in which liquid water is present. The initial total pressure on the air and water is to be 760 mm Hg. If you keep the total pressure on the container constant at 760 mm Hg (Figs. 3.16 and 3.17), eventually the water will vaporize, and water vapor will enter and mix with the air until the partial pressure of the water in

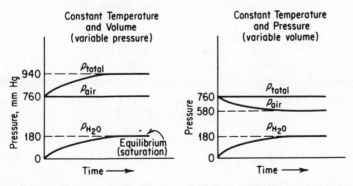

Fig. 3.16. Change of partial and total pressures on vaporization of water into air at constant temperature.

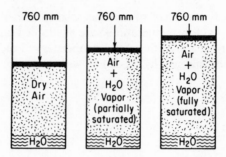

Fig. 3.17. Evaporation of water at constant pressure and temperature.

the air reaches approximately 180 mm Hg (the vapor pressure of water at 147°F). Regardless of the duration of contact between the water and the air (after this partial pressure of water vapor is reached), no more water vapor can enter the air. The air is saturated with respect to water vapor and cannot contain additional water. Of course the volume of the system will change if the total pressure is maintained at 760 mm Hg, or else some air will have to leave the system if the volume and pressure are to stay constant. If the volume of the system is fixed, the total pressure will rise to $760 + 180 = 940$ mm Hg.

Assuming that the ideal gas laws apply to both air and water vapor, as they do with excellent precision, you can calculate the partial pressure of the air as follows *at saturation:*

$$p_{air} + p_{H_2O} = p_{total}, \qquad p_{air} + 180 \text{ mm} = 760 \text{ mm}$$

$$p_{air} = 760 - 180 = 580 \text{ mm Hg}$$

From Eq. (3.4), we know that

$$\frac{p_{air}}{p_{H_2O}} = \frac{n_{air}}{n_{H_2O}} \qquad \text{at constant temperature}$$

or, from Eq. (3.7),

$$\frac{V_{air}}{V_{H_2O}} = \frac{n_{air}}{n_{H_2O}} \qquad \text{at constant temperature}$$

Then

$$\frac{p_{air}}{p_{H_2O}} = \frac{p_{air}}{p_{total} - p_{air}} = \frac{V_{air}}{V_{total} - V_{air}} \tag{3.42}$$

We can generalize these expressions for any two components as follows:

$$\frac{p_1}{p_2} = \frac{p_1}{p_t - p_1} = \frac{n_1}{n_2} = \frac{V_1}{V_t - V_1} = \frac{V_1}{V_2} \tag{3.43}$$

and

$$p_1 = \frac{V_1}{V_2} p_2 \qquad \text{or} \qquad V_1 = \frac{p_1}{p_2} V_2 \tag{3.44}$$

where the subscripts indicate component 1, component 2, and total.

If the temperature rises in an air-water mixture and the vapor phase is initially saturated with water vapor, the change in the partial pressure of the water and the air and in the total pressure, as a function of temperature, can be graphically shown as in Fig. 3.18.

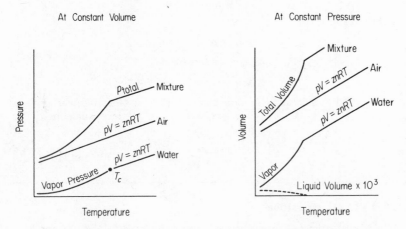

At Constant Volume   At Constant Pressure

Fig. 3.18. The change of total pressure (volume) and partial pressures (volumes) of air and water in a saturated mixture with increasing temperature.

EXAMPLE 3.20 Saturation

What is the minimum number of cubic feet of dry air at 20°C and 738 mm Hg that are necessary to evaporate 13.1 lb of alcohol if the total pressure remains constant at 738 mm Hg? Assume that the air is blown over the alcohol to evaporate it in such a way that the exit pressure of the air-alcohol mixture is at 738 mm Hg.

Solution:

See Fig. E3.20. Assume that the process is isothermal. The additional data needed are

$$p^*_{\text{alcohol}} \text{ at } 20°C \ (68°F) = 44.5 \text{ mm Hg}$$

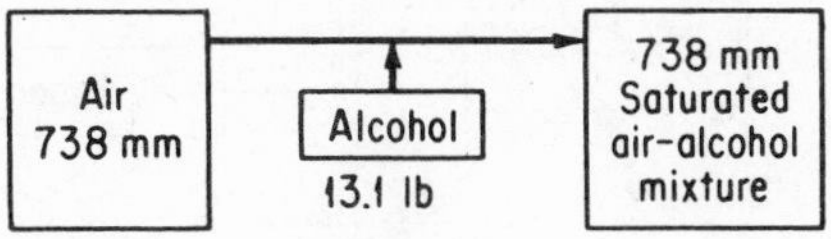

Fig. E3.20.

The most alcohol the air can pick up is a saturated mixture; any condition less than saturated would require more air.

Basis: 13.1 lb alcohol

The ratio of moles of alcohol to moles of air in the final gaseous mixture is the same as the ratio of the partial pressures of these two substances. Since we know the moles of alcohol, we can find the number of moles of air:

$$\frac{p_{alcohol}}{p_{air}} = \frac{n_{alcohol}}{n_{air}}$$

From Dalton's law,

$$p_{air} = p_{total} - p_{alcohol}$$

$$p_{alcohol} = 44.5 \text{ mm}$$

$$p_{air} = (738 - 44.5) \text{ mm}$$

$$\frac{13.1 \text{ lb alcohol}}{} \left| \frac{1 \text{ lb mole alcohol}}{46 \text{ lb alcohol}} \right| \frac{(738 - 44.5) \text{ lb moles air}}{44.5 \text{ lb moles alcohol}}$$

$$\frac{359 \text{ ft}^3}{1 \text{ lb mole}} \left| \frac{293°K}{273°K} \right| \frac{760 \text{ mm}}{738 \text{ mm}} = 1800 \text{ ft}^3 \text{ air}$$
$$\text{at } 20°C \text{ and } 738 \text{ mm Hg}$$

EXAMPLE 3.21 Saturation

If 1250 cm³ of wet H_2 are saturated with water at 30°C and 742 mm Hg, what is the volume of dry gas at standard conditions? The vapor pressure water at 30°C is 32 mm Hg.

Solution:
(a) *Long Solution:*
Your first impulse in selecting a basis is probably to ask yourself the question, What do I have or what do I want? The answer might be that you have 1250 cm³ of saturated H_2 at 30°C and 742 mm Hg. See Fig. E3.21.

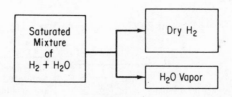

Fig. E3.21.

Basis: 1250 cm³ H_2 saturated with water at 30°C and 742 mm Hg

$$\frac{1250 \text{ cm}^3 \text{ wet gas}}{} \left| \frac{(742 - 32) \text{ cm}^3 \text{ dry } H_2}{742 \text{ cm}^3 \text{ wet gas}} \right| \frac{273°K}{303°K} \left| \frac{742 \text{ mm}}{760 \text{ mm}} \right.$$

$$= 1051 \text{ cm}^3 \text{ dry } H_2 \text{ at S.C.}$$

(b) *Shorter Solution:*
You can save a little time by recalling that

$$p_{total} = p_{H_2O} + p_{H_2}$$
$$742 = 32 + 710$$

In the given gas mixture you have *any* of the following:

(a) 1250 cm³ wet H_2 at 30°C and 742 mm Hg.
(b) 1250 cm³ dry H_2 at 30°C and 710 mm Hg.
(c) 1250 cm³ water vapor at 30°C and 32 mm Hg.

Therefore, since the problem asks for the volume of dry H_2, you could take as a basis the dry H_2, and then apply the ideal gas laws to the dry H_2.

Basis: 1250 cm³ dry H_2 at 30°C and 710 mm Hg

$$\frac{1250 \text{ cm}^3 \text{ dry } H_2}{} \left| \frac{273°K}{303°K} \right| \frac{710 \text{ mm}}{760 \text{ mm}} = 1051 \text{ cm}^3 \text{ dry } H_2 \text{ at S.C.}$$

You can see that by choosing the second basis you can eliminate one step in the solution.

EXAMPLE 3.22   Saturation

A telescopic gas holder contains 10,000 ft³ of saturated gas at 80°F and a pressure of 6.0 in. H_2O above atmospheric. The barometer reads 28.46 in. Hg. Calculate the weight of water vapor in the gas.

Solution:
The total pressure of the saturated gas must be calculated first in in. Hg. See Fig. E3.22.

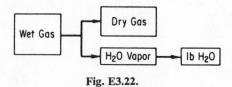

Fig. E3.22.

$$\frac{6.0 \text{ in. } H_2O}{} \left| \frac{1 \text{ ft}}{12 \text{ in.}} \right| \frac{29.92 \text{ in. Hg}}{33.91 \text{ ft } H_2O} = 0.44 \text{ in. Hg}$$

$$\text{barometer} = 28.46 \text{ in. Hg}$$
$$p_{total} = 28.90 \text{ in. Hg}$$

p_w = vapor pressure of H_2O at 80°F = 1.03 in. Hg

$$p_t = p_g + p_w$$
$$28.90 = 27.87 + 1.03$$

There exists

$$10{,}000 \text{ ft}^3 \text{ wet gas} \quad \text{at} \quad 28.90 \text{ in. Hg and } 80°\text{F}$$
$$10{,}000 \text{ ft}^3 \text{ dry gas} \quad \text{at} \quad 27.87 \text{ in. Hg and } 80°\text{F}$$
$$10{,}000 \text{ ft}^3 \text{ water vapor} \quad \text{at} \quad 1.03 \text{ in. Hg and } 80°\text{F}$$

$$°R = 80 + 460 = 540°R$$

From the volume of water vapor present at known conditions, you can calculate the volume at S.C. and hence the pound moles and pounds of water.

Basis: 10,000 ft³ water vapor at 1.03 in. Hg and 80°F

$$\frac{10{,}000 \text{ ft}^3 \text{ H}_2\text{O vapor}}{} \left| \frac{492°R}{540°R} \right| \frac{1.03 \text{ in. Hg}}{29.92 \text{ in. Hg}} \left| \frac{1 \text{ lb mole}}{359 \text{ ft}^3} \right| \frac{18 \text{ lb H}_2\text{O}}{1 \text{ lb mole H}_2\text{O}}$$

$$= 15.7 \text{ lb H}_2\text{O}$$

EXAMPLE 3.23   Smokestack emission and pollution

A local pollution-solutions group has reported the Simtron Co. boiler plant as being an air polluter and has provided as proof photographs of heavy smokestack emissions on 20 different days in January and February. As the chief engineer for the Simtron Co., you know that your plant is not a source of pollution because you burn natural gas (essentially methane) and your boiler plant is operating correctly. Your boss believes the pollution-solutions group has made an error in identifying the stack—it must belong to the company next door that burns coal. Is he correct? Is the pollution-solutions group correct? See Fig. E3.23.

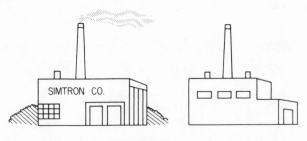

Fig. E3.23.

Solution:

Methane (CH_4) contains 2 lb moles of H_2 per pound mole of C, while coal (see Example 4.24) contains 71 lb of C per 5.6 lb of H_2 in 100 lb of coal. The coal analysis is equivalent to

$$\frac{71 \text{ lb C}}{} \left| \frac{1 \text{ lb mole C}}{12 \text{ lb C}} = 5.9 \text{ lb moles C} \qquad \frac{5.6 \text{ lb H}_2}{} \left| \frac{\text{lb mole H}_2}{2.02 \text{ lb H}_2} = 2.8 \text{ lb mole H}_2 \right.$$

or a ratio of $2.8/5.9 = 0.47$ lb mole H_2/lb mole C. Suppose that each fuel burns with 40 percent excess air and that combustion is complete. We can compute the mole fraction of water vapor in each stack gas.

Basis: 1 lb mole C

Natural gas

$$CH_4 + 2O_2 \longrightarrow CO_2 + 2H_2O$$

fuel comp.	*lb moles*	*composition of combustion gases, lb moles*			
		CO_2	*excess* O_2	N_2	H_2O
C	1.0	1.0			
H_2	2.0				2.0
Air			0.80	10.5	
		1.0	0.80	10.5	2.0

Excess O_2: $2(0.40) = 0.80$

N_2: $(2)(1.40)(79/21) = 10.5$

The total pound moles of gas produced are 14.3 and the mole fraction H_2O is

$$\frac{2.0}{14.3} \cong 0.14$$

Coal

$$C + O_2 \longrightarrow CO_2 \qquad H_2 + \tfrac{1}{2}O_2 \longrightarrow H_2O$$

fuel comp.	*lb moles*	*composition of gas produced, lb moles*			
		CO_2	*excess* O_2	N_2	H_2O
C	1	1			
H_2	0.47				0.47
Air			0.49	6.5	
		1	0.49	6.5	0.47

Excess O_2: $\underset{1(0.40)\ =\ 0.40}{\text{C}} \qquad \underset{(0.47)(1/2)(0.40)\ =\ 0.094}{\text{H}_2}$

N_2: $1.40(79/21)(1 + 0.47(1/2)) = 6.5$

The total pound moles of gas produced are 8.46 and the mole fraction H_2O is

$$\frac{0.47}{8.46} \cong 0.056$$

If the barometric pressure is 14.7 psia, then the stack gas becomes saturated at

	natural gas	*coal*
Pressure	$14.7(0.14) = 2.06$ psia	$14.7(0.056) = 0.083$ psia
Equivalent temperature:	127°F	96°F

By mixing the stack gas with air, the mole fraction water vapor is reduced, and hence the condensation temperature is reduced. However, for equivalent dilution, the coal-burning plant will always have a lower condensation temperature. Thus, on cold winter days, the condensation of water vapor will occur more often and to a greater extent from power plants firing natural gas than from those using other fuels.

    The public, unfortunately, sometimes concludes that all the emissions they perceive are pollution. Natural gas could appear to the public to be a greater pollutant than either oil or coal when, in fact, the emissions are just water vapor. The sulfur content of coal and oil can be released as sulfur dioxide to the atmosphere, and the polluting capacities of coal and oil are much greater than natural gas when all three are being burned properly. The sulfur contents as delivered to the consumers are as follows: natural gas, 4×10^{-4} percent (as added mercaptans); number 6 fuel oil, up to 2.6 percent; and coal, from 0.5 to 5 percent.

What additional steps would you take to resolve the questions that were originally posed?

3.5 Partial saturation and humidity

In the foregoing section we dealt with mixtures of gas and vapor in which the gas was saturated with the vapor. More often, the contact time required between the gas and liquid for equilibrium (or saturation) to be attained is too long, and the gas is not completely saturated with the vapor. Then the vapor is not in equilibrium with a liquid phase, and the partial pressure of the vapor is less than the vapor pressure of the liquid at the given temperature. This condition is called *partial saturation*. What we have is simply a mixture of two or more gases which obey the real gas laws. What distinguishes this case from the previous examples for gas mixtures is that under suitable conditions it is possible to condense part of one of the gaseous components. In Fig. 3.19 you can see how the partial

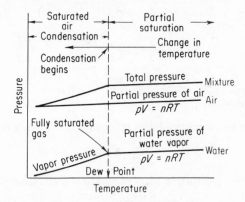

Fig. 3.19. Transformation of a partially saturated water vapor-air mixture into a saturated mixture as the temperature is lowered (volume = constant).

pressure of the water vapor in a gaseous mixture at constant volume obeys the ideal gas laws as the temperature drops until saturation is reached, at which time the water vapor starts to condense. Until these conditions are achieved, you can confidently apply the gas laws to the mixture.

Several ways exist to express the concentration of a vapor in a gas mixture. You sometimes encounter weight or mole fraction (or percent), but more frequently one of the following:

(a) Relative saturation (relative humidity); Sec. 3.5.1.
(b) Molal saturation (molal humidity); Sec. 3.5.2.

(c) "Absolute" saturation ("absolute" humidity) or percent saturation (percent humidity); Secs. 3.5.3 and 5.3.

(d) Humidity; Sec. 3.5.2 and 5.3.

When the vapor is water vapor and the gas is air, the special term *humidity* applies. For other gases or vapors, the term saturation is used.

3.5.1 Relative Saturation. Relative saturation is defined as

$$\mathcal{RS} = \frac{p_{\text{vapor}}}{p_{\text{satd}}} = \text{relative saturation} \tag{3.45}$$

where p_{vapor} = partial pressure of the vapor in the gas mixture

p_{satd} = partial pressure of the vapor in the gas mixture *if* the gas were saturated at the given temperature of the mixture, i.e., the vapor pressure of the vapor

Then, for brevity, if the subscript 1 denotes vapor,

$$\mathcal{RS} = \frac{p_1}{p_1^*} = \frac{p_1/p_t}{p_1^*/p_t} = \frac{V_1/V_t}{V_{\text{satd}}/V_t} = \frac{n_1}{n_{\text{satd}}} = \frac{\text{lb}_1}{\text{lb}_{\text{satd}}} \tag{3.46}$$

You can see that relative saturation, in effect, represents the fractional approach to total saturation, as shown in Fig. 3.20. If you listen to the radio or TV and

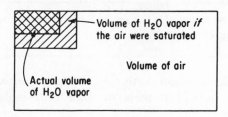

Fig. 3.20. A gas partially saturated with water vapor.

hear the announcer say that the temperature is 70°F and the *relative humidity* is 60 percent, he means that

$$\frac{p_{\text{H}_2\text{O}}}{p_{\text{H}_2\text{O}}^*}(100) = \% \,\mathcal{RH} \tag{3.47}$$

with both the $p_{\text{H}_2\text{O}}$ and the $p_{\text{H}_2\text{O}}^*$ being measured at 70°F. Zero percent relative saturation means no vapor in the gas; 100 percent relative saturation means that the partial pressure of the vapor is the same as the vapor pressure of the substance that is the vapor.

EXAMPLE 3.24 Relative humidity

The weather man on the radio this morning reported that the temperature this afternoon would reach 94°F, the relative humidity would be 43 percent, the barometer

29.67 in. Hg, partly cloudy to clear, with the wind from SSE at 8 mi/hr. How many pounds of water vapor would be in 1 mi³ of afternoon air? What would be the dew point of this air?

Solution:

The vapor pressure of water at 94°F is 1.61 in. Hg. We can calculate the partial pressure of the water vapor in the air from the given percent relative humidity; from this point forward, the problem is the same as the examples in the previous section.

$$p_w = (1.61 \text{ in. Hg})(0.43) = 0.69 \text{ in. Hg}$$

$$p_{air} = p_t - p_w = 29.67 - 0.69 = 28.98 \text{ in. Hg}$$

Basis: 1 mi³ water vapor at 94°F and 0.69 in. Hg

$$\frac{1 \text{ mi}^3}{} \left| \left(\frac{5280 \text{ ft}^3}{1 \text{ mi}}\right)^3 \right| \frac{492°\text{R}}{554°\text{R}} \left| \frac{0.69 \text{ in. Hg}}{29.92 \text{ in. Hg}} \right| \frac{1 \text{ lb mole}}{359 \text{ ft}^3} \left| \frac{18 \text{ lb H}_2\text{O}}{1 \text{ lb mole}} \right.$$

$$= 1.51 \times 10^8 \text{ lb H}_2\text{O}$$

Now the dew point is the temperature at which the water vapor in the air will first condense on cooling at *constant total pressure and composition*. As the gas is cooled you can see from Eq. (3.47) that the percent relative humidity increases since the partial pressure of the water vapor is constant while the vapor pressure of water decreases with temperature. When the percent relative humidity reaches 100 percent,

$$100 \frac{p_1}{p_1^*} = 100\% \quad \text{or} \quad p_1 = (p_1^*)$$

the water vapor will start to condense. This means that at the dew point the vapor pressure of water will be 0.69 in. Hg. From the steam tables you can see that this corresponds to a temperature of about 69°F.

3.5.2 Molal Saturation.   Another way to express vapor concentration in a gas is to use the ratio of the moles of vapor to the moles of vapor-free gas:

$$\frac{n_{\text{vapor}}}{n_{\text{vapor-free gas}}} = \text{molal saturation} \tag{3.48}$$

If subscripts 1 and 2 represent the vapor and the dry gas, respectively, then for a binary system,

$$p_1 + p_2 = p_t \tag{3.49}$$

$$n_1 + n_2 = n_t \tag{3.50}$$

$$\frac{n_1}{n_2} = \frac{p_1}{p_2} = \frac{V_1}{V_2} = \frac{n_1}{n_t - n_1} = \frac{p_1}{p_t - p_1} = \frac{V_1}{V_t - V_1} \tag{3.51}$$

By multiplying by the appropriate molecular weights, you can find the weight of vapor per weight of dry gas:

$$\frac{(n_{\text{vapor}})(\text{mol. wt}_{\text{vapor}})}{(n_{\text{dry gas}})(\text{mol. wt}_{\text{dry gas}})} = \frac{\text{wt}_{\text{vapor}}}{\text{wt}_{\text{dry gas}}} \tag{3.52}$$

The special term *humidity* ($\mathcal{H}$) refers to the pounds of water vapor per pound of bone-dry air and is used in connection with the humidity charts in Sec. 5.3.

3.5.3 "Absolute" Saturation; Percentage Saturation. "Absolute" saturation is defined as the ratio of the moles of vapor per mole of *vapor-free* gas to the moles of vapor *which would be present* per mole of *vapor-free* gas if the mixture were completely saturated at the existing temperature and total pressure:

$$\mathcal{CS} = \text{``absolute saturation''} = \frac{\left(\dfrac{\text{moles vapor}}{\text{moles vapor-free gas}}\right)_{\text{actual}}}{\left(\dfrac{\text{moles vapor}}{\text{moles vapor-free gas}}\right)_{\text{saturated}}} \qquad (3.53)$$

Using the subscripts 1 for vapor and 2 for vapor-free gas,

$$\text{percent absolute saturation} = \frac{\left(\dfrac{n_1}{n_2}\right)_{\text{actual}}}{\left(\dfrac{n_1}{n_2}\right)_{\text{saturated}}}(100) = \frac{\left(\dfrac{p_1}{p_2}\right)_{\text{actual}}}{\left(\dfrac{p_1}{p_2}\right)_{\text{saturated}}}(100) \qquad (3.54)$$

Since p_1 saturated $= p_1^*$ and $p_t = p_1 + p_2$,

$$\text{percent absolute saturation} = 100\frac{\dfrac{p_1}{p_t - p_1}}{\dfrac{p_1^*}{p_t - p_1^*}} = \frac{p_1}{p_1^*}\left(\frac{p_t - p_1^*}{p_t - p_1}\right)100 \qquad (3.55)$$

Now you will recall that $p_1/p_1^* =$ relative saturation. Therefore,

$$\text{percent absolute saturation} = (\text{relative saturation})\left(\frac{p_t - p_1^*}{p_t - p_1}\right)100 \qquad (3.56)$$

Percent absolute saturation is always less than relative saturation except at saturated conditions (or at zero percent saturation) when percent absolute saturation = percent relative saturation.

EXAMPLE 3.25 Partial saturation

Helium contains 12 percent (by volume) of ethyl acetate. Calculate the percent relative saturation and the percent absolute saturation of the mixture at a temperature of 30°C and a pressure of 740 mm of Hg.

Solution:
The additional data needed are

p_{EtAc}^* at 30°C = 119 mm Hg   (from any suitable handbook)

Using Dalton's laws,

$$p_{\text{EtAc}} = p_t y_{\text{EtAc}} = p_t\left(\frac{n_{\text{EtAc}}}{n_t}\right) = p_t\left(\frac{V_{\text{EtAc}}}{V_t}\right)$$
$$= (740)(0.12) = 88.9 \text{ mm Hg}$$

$$p_{\text{He}} = p_t - p_{\text{EtAc}}$$
$$= 740 - 88.9 = 651.1 \text{ mm Hg}$$

At 30°C the

(a) *Percent relative saturation =*

$$100\frac{p_{EtAc}}{p^*_{EtAc}} = 100\frac{88.9}{119} = 74.6\%$$

(b) *Percent absolute saturation =*

$$100\frac{\dfrac{p_{EtAc}}{p_t - p_{EtAc}}}{\dfrac{p^*_{EtAc}}{p_t - p^*_{EtAc}}} = \frac{\dfrac{88.9}{740 - 88.9}}{\dfrac{119}{740 - 119}}100 = \frac{\dfrac{88.9}{651.1}}{\dfrac{119}{621}}100$$

$$= 71.1\%$$

EXAMPLE 3.26 Partial saturation

A mixture of ethyl acetate vapor and air has a relative saturation of 50 percent at 30°C and a total pressure of 740 mm Hg. Calculate the analysis of the vapor and the molal saturation.

Solution:

The vapor pressure of ethyl acetate at 30°C from the previous problem is 119 mm Hg.

$$\% \text{ relative saturation} = 50 = \frac{p_{EtAc}}{p^*_{EtAc}}100$$

From the above relation, the p_{EtAc} is

$$p_{EtAc} = 0.50(119) = 59.5 \text{ mm Hg}$$

(a)
$$\frac{n_{EtAc}}{n_t} = \frac{p_{EtAc}}{p_t} = \frac{59.5}{740} = 0.0805$$

Hence the vapor analyzes EtAc, 8.05 percent; air, 91.95 percent.

(b) Molal saturation is

$$\frac{n_{EtAc}}{n_{air}} = \frac{p_{EtAc}}{p_{air}} = \frac{p_{EtAc}}{p_t - p_{EtAc}} = \frac{59.5}{740 - 59.5}$$

$$= 0.0876 \frac{\text{mole EtAc}}{\text{mole air}}$$

EXAMPLE 3.27 Partial saturation

The percent absolute humidity of air at 86°F and a total pressure of 750 mm of Hg is 20 percent. Calculate the percent relative humidity and the partial pressure of the water vapor in the air. What is the dew point of the air?

Solution:

Data from the steam tables are

$$p^*_{H_2O} \text{ at } 86°F = 31.8 \text{ mm Hg}$$

To get the relative humidity, $p_{H_2O}/p^*_{H_2O}$, we need to find the partial pressure of the water vapor in the air. This may be obtained from

$$\mathcal{RH} = 20 = \frac{\dfrac{p_{H_2O}}{p_t - p_{H_2O}}}{\dfrac{p^*_{H_2O}}{p_t - p^*_{H_2O}}} 100 = \frac{\dfrac{p_{H_2O}}{750 - p_{H_2O}}}{\dfrac{31.8}{750 - 31.8}} 100$$

This equation can be solved for p_{H_2O}

$$0.00885 = \frac{p_{H_2O}}{750 - p_{H_2O}}$$

$$6.65 - 0.00885 p_{H_2O} = p_{H_2O}$$

(a)
$$p_{H_2O} = 6.6 \text{ mm Hg}$$

(b)
$$\% \mathcal{RH} = 100 \frac{6.6}{31.8} = 20.7\%$$

(c) The dew point is the temperature at which the water vapor in the air would commence to condense. This would be at the vapor pressure of 6.6 mm, or about 41°F.

3.6   Material balances involving condensation and vaporization

The solution of material balance problems involving partial saturation, condensation, and vaporization will now be illustrated. Remember the drying problems in Chap. 2? They included water and some bone-dry material, as shown in the top of Fig. 3.21. To complete the diagram, we can now add the air that is used to remove the water from the material being dried.

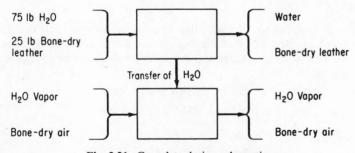

Fig. 3.21. Complete drying schematic.

You can handle material balance problems involving water vapor in air in exactly same fashion as you handled the material balance problems for the drying of leather (or paper, etc.), depending on the information provided and sought. (You can find additional humidity and saturation problems which include the use of energy balances and humidity charts in Chap. 5.)

We should again stress in connection with the examples to follow that if you know the dew point of a gas mixture, you automatically know the partial pressure of the water vapor in the gas mixture. When a partially saturated gas is cooled at constant pressure, as in the cooling of air containing some water vapor at atmospheric pressure, the volume of the mixture may change slightly, but the partial pressures of the air and water vapor remain constant until the dew point is reached. At this point water begins to condense; the air remains saturated as the temperature is lowered. All that happens is that more water goes from the vapor into the liquid phase. At the time the cooling is stopped, the air is still saturated, and at its dew point.

EXAMPLE 3.28 Material balance with condensation

If the atmosphere in the afternoon during a humid period is at 90°F and 80 percent ℛℋ (barometer reads 738 mm Hg) while at night it is at 68°F (barometer reads 745 mm Hg), what percent of the water in the afternoon air is deposited as dew?

Solution:
The data are shown in Fig. E3.28.

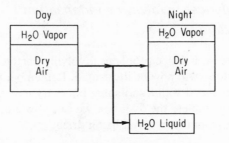

Fig. E3.28.

If any dew is deposited at night, the air must be fully saturated at 68°F. You can easily check the dew point of the day air and find it is 83°F. Thus the air first becomes saturated at 83°F. The partial pressures of air and water vapor during the day and night are

$$p_w \text{ day} = 36(0.80) = 28.8 \text{ mm}$$

$$p_w \text{ night} = 17.5(1.00) = 17.5 \text{ mm}$$

$$p_t = p_{air} + p_w$$

day: $\quad p_{air} = 738 - 28.8 = 709.2 \text{ mm}$

night: $\quad p_{air} = 745 - 17.5 = 727.5 \text{ mm}$

This gives us in effect the compositions of all the streams.

As a basis you could select 1 ft³ of wet air, 1 lb mole wet (or dry) air, or many other suitable bases. The simplest basis to take is

Basis: 738 lb moles moist day air

because then the moles = the partial pressures.

	initial mixture (day)		*final mixture (night)*
	lb moles =	tie	*lb moles =*
comp.	*partial press.*		*partial press.*
Air	709.2	←—— element ——→	727.5
H_2O	28.8		17.5
Total	738.0		745.0

On the basis of 738 lb moles of moist day air we have the following water in the night air:

$$\frac{17.5 \text{ lb mole } H_2O \text{ in night air}}{727.5 \text{ lb mole air}} \left| \frac{709.2 \text{ lb mole air}}{} \right. = 17.0 \text{ lb mole } H_2O$$

$$28.8 - 17.0 = 11.8 \text{ lb moles } H_2O \text{ deposited as dew}$$

$$100\frac{11.8}{28.8} = 41\% \text{ of water in day air deposited as dew}$$

This problem could also be solved on the basis of

Basis: 1 lb mole of bone-dry air (BDA)

$$\begin{array}{ccc} \textbf{initial} & - & \textbf{final} & = \textbf{change} \\ \dfrac{28.8 \text{ lb moles } H_2O}{709.2 \text{ lb moles BDA}} & - & \dfrac{17.5 \text{ lb moles } H_2O}{727.5 \text{ lb moles BDA}} & = \text{change} \end{array}$$

$$0.0406 \quad - \quad 0.0241 \quad = 0.0165 \frac{\text{lb mole } H_2O}{\text{lb mole DBA}}$$

$$\frac{0.0165}{0.0406}100 = 41\% \text{ of water vapor deposited as dew}$$

Note especially that the operation

$$\begin{array}{cc} \text{initial} & \text{final} \\ \dfrac{28.8}{738} & - \dfrac{17.5}{745} \end{array}$$

is meaningless since two different bases are involved—the wet initial air and the wet final air.

EXAMPLE 3.29   Dehydration

By absorption in silica gel you are able to remove all (0.72 lb) of the H_2O from moist air at 60°F and 29.2 in. Hg. The same air measures 1000 ft³ at 70°F and 32.0 in. Hg when dry. What was the relative humidity of the moist air?

Solution:
See Fig. E3.29.

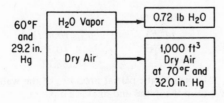

Fig. E3.29.

Basis: 1000 ft³ bone-dry air (BDA) at 70°F and 32.0 in. Hg

$$\frac{1000 \text{ ft}^3 \text{ BDA} \mid 492°R \mid 32.0 \text{ in. Hg} \mid 1 \text{ lb mole}}{\mid 530°R \mid 29.92 \text{ in. Hg} \mid 359 \text{ ft}^3} = 2.77 \text{ lb moles dry air}$$

The wet air now appears to have the following composition:

	lb moles
BDA	2.77
H₂O	
0.72/18 =	0.040
Wet air	2.81

The partial pressure of the water in the moist air was the total pressure times the mole fraction water vapor:

$$\frac{29.2 \text{ in. Hg} \mid 0.04}{\mid 2.81} = 0.415 \text{ in. Hg}$$

Saturated air at 60°F has a vapor pressure of water of 0.52 in. Hg. Consequently, the relative humidity was

$$\frac{p_{H_2O}}{p_{H_2O}^*} = \frac{0.415}{0.52} 100 = 80\%$$

EXAMPLE 3.30 Humidification

One thousand cubic feet of moist air at 760 mm Hg and 72°F and with a dew point of 53°F enter a process. The air leaves the process at 740 mm Hg with a dew point of 137°F. How many pounds of water vapor are added to each pound of wet air entering the process?

Solution:
See Fig. E3.30.

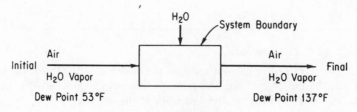

Fig. E3.30.

Additional data are

dew point	
t (°F)	$p_{H_2O}^*$ (*mm Hg*)
53°F	10.3*
137°F	138.2*

*These give the partial pressures of the water vapor in the initial and final gas mixtures.

We know the initial and final compositions of the air-water vapor mixture, since the partial pressure of the water vapor in the moist air is the pressure of water at the dew point for each case.

Basis: 760 moles initial wet air

component	*initial* moles		*final* moles	*pressure (mm)*
Vapor	10.3			138.2
BDA*	749.9	←—tie element—→	749.9	601.8
Total	760			740

*Bone-dry air.

In the final gas we can say that there are

$$138.2 \text{ moles vapor}/601.8 \text{ moles BDA}$$

$$138.2 \text{ moles vapor}/740 \text{ moles moist air}$$

$$601.8 \text{ moles BDA}/740 \text{ moles moist air}$$

Since we have a tie element of 750 moles of bone-dry air and want to know how many moles of vapor are present in the final air on our basis of 760 moles initial wet air,

$$\frac{749.7 \text{ moles BDA}}{} \left| \frac{138.2 \text{ moles vapor in final}}{601.8 \text{ moles BDA in final}} \right. = 172 \text{ moles vapor in final}$$

$$172 \text{ moles final vapor} - 10.3 \text{ moles initial vapor} = 162 \text{ moles of vapor added}$$

But these calculations were all based on 760 moles of initial wet gas. Therefore our answer should be

$$\frac{162 \text{ moles vapor added}}{760 \text{ moles wet gas in}} = \frac{0.213 \text{ moles vapor added}}{1 \text{ mole wet gas in}}$$

Now the molecular weight of water is 18; the average molecular weight of the original wet air has to be calculated:

comp.	moles	mol. wt	lb
BDA	750	29	21,750
H_2O	10	18	180
	760		21,930

$$\frac{21,930 \text{ lb}}{760 \text{ lb mole}} = 28.8$$

$$\frac{0.213 \text{ lb mole } H_2O}{1 \text{ lb mole wet gas}} \left| \frac{18 \text{ lb}}{1 \text{ lb mole } H_2O} \right| \frac{1 \text{ lb mole wet gas}}{28.8 \text{ lb}} = 0.133 \frac{\text{lb } H_2O}{\text{lb wet gas}}$$

Comments:

(a) If the original problem had asked for the moles of final wet gas, we could have used the tie element in the following manner:

$$\frac{750 \text{ moles BDA}}{} \left| \frac{740 \text{ moles wet gas out}}{602 \text{ moles BDA out}} \right. = 921 \text{ moles wet gas out}$$

(b) If a basis of 1 mole of bone-dry air had been used, the following calculation would apply:

Basis: 1 mole BDA

in:
$$\frac{\text{moles vapor}}{\text{mole BDA}} = \frac{10.3}{749.7} = 0.0138$$

out:
$$\frac{\text{moles vapor}}{\text{mole BDA}} = \frac{138.2}{601.8} = 0.229$$

Since there are more moles of vapor in the final air (per mole of bone-dry air), we know that vapor must have been added:

$$\text{moles vapor added} = 0.229 - 0.0138 = 0.215 \text{ moles}$$

The remainder of the calculations would be the same.

We have gone over a number of examples of condensation and vaporization, and you have seen how a given amount of air at atmospheric pressure can hold only a certain maximum amount of water vapor. This amount depends on the temperature of the air, and any decrease in the temperature will lower the water-bearing capacity of the air. An increase in pressure also will accomplish the same effect. If a pound of saturated air at 75°F is isothermally compressed (with a reduction in volume, of course), liquid water will be deposited out of the air just like water being squeezed out of a wet sponge (Fig. 3.22). This process has been previously described in the *p-T* diagram for water (Fig. 3.12), by line *G-H-I*.

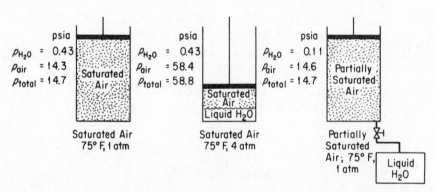

Fig. 3.22. Effect of an increase of pressure on saturated air and a return to the initial pressure.

For example, if a pound of saturated air at 75°F and 1 atm is compressed isothermally to 4 atm (58.8 psia), almost three-quarters of the original content of water vapor now will be in the form of liquid, and the air has a dew point of 75°F at 4 atm. Remove the liquid water, expand the air isothermally back to 1 atm, and you will find the dew point has been lowered to about 36°F. Mathematically (1 = state at 1 atm, 4 = state at 4 atm) with $z = 1.00$ for both components:

For saturated air at 75°F and 4 atm,

$$\left(\frac{n_{H_2O}}{n_{air}}\right)_4 = \left(\frac{p^*_{H_2O}}{p_{air}}\right)_4 = \frac{0.43}{58.4}$$

For the same air saturated at 75°F and 1 atm,

$$\left(\frac{n_{H_2O}}{n_{air}}\right)_1 = \left(\frac{p^*_{H_2O}}{p_{air}}\right)_1 = \frac{0.43}{14.3}$$

Since the air is the tie element in the process,

$$\left(\frac{n_4}{n_1}\right)_{H_2O} = \frac{\dfrac{0.43}{58.4}}{\dfrac{0.43}{14.3}} = \frac{14.3}{58.4} = 0.245$$

24.5 percent of the original water will remain as vapor. After the air-water vapor mixture is returned to a total pressure of 1 atm, the following two familiar equations now apply:

$$p_{H_2O} + p_{air} = 14.7$$

$$\frac{p_{H_2O}}{p_{air}} = \frac{n_{H_2O}}{n_{air}} = \frac{0.43}{58.4} = 0.00737$$

From these two relations you can find that

$$p_{H_2O} = 0.108 \text{ psia}$$
$$p_{air} = 14.6$$
$$p_{total} = \overline{14.7} \text{ psia}$$

The pressure of the water vapor represents a dew point of about 36°F, and a relative humidity of

$$100\frac{p_{H_2O}}{p^*_{H_2O}} = \frac{0.108}{0.43}100 = 25\%$$

3.7 Phase phenomena

We shall now consider briefly some of the qualitative characteristics of vapors, liquids, and solids.

3.7.1 The Phase Rule.

You will find the phase rule a useful guide in establishing how many properties, such as pressure and temperature, have to be specified to definitely fix all the remaining properties and number of phases that can coexist for any physical system. The rule can be applied only to systems in *equilibrium*. It says

$$F = C - \mathcal{P} + 2 \tag{3.57}$$

where F = number of degrees of freedom, i.e., the number of independent properties which have to be specified to determine all the intensive properties of each phase of the system of interest.

C = number of components in the system. For circumstances involving chemical reactions, C is *not* identical to the number of chemical compounds in the system but is equal to the number of chemical compounds less the number of independent-reaction and other equilibrium relationships among these compounds.

$\mathcal{P}$ = number of phases that can exist in the system. A phase is a homogeneous quantity of material such as a gas, a pure liquid, a solution, or a homogeneous solid.

Variables of the kind with which the phase rule is concerned are called *phase-rule variables*, and they are *intensive* properties of the system. By this we mean properties which do not depend on the quantity of material present. If you think about the properties we have employed so far in this text, you have the feeling that pressure and temperature are independent of the amount of material present. So is concentration, but what about volume? The total volume of a system is called an *extensive* property because it does depend on how much material you have; the specific volume, on the other hand, the cubic feet per pound, for example, is an intensive property because it is independent of the amount of material present. In the next chapter we shall take up additional properties such as internal energy and enthalpy; you should remember that the specific (per unit mass) values of these quantities are intensive properties; the total quantities are extensive properties.

An example will clarify the use of these terms in the phase rule. You will remember for a pure gas that we had to specify three of the four variables in the ideal gas equation $pV = nRT$ in order to be able to determine the remaining one unknown. For a single phase $\mathcal{P} = 1$, and for a pure gas $C = 1$, so that

$$F = C - \mathcal{P} + 2 = 1 - 1 + 2 = 2 \qquad \text{variables to be specified}$$

How can we reconcile this apparent paradox with our previous statement? Since the phase rule is concerned with intensive properties only, the following are phase-rule variables in the ideal gas law:

$$\left.\begin{array}{l} P \\ \hat{V} \text{ (specific molar volume)} \\ T \end{array}\right\} \quad \text{3 intensive properties}$$

Thus, the ideal gas law should be written as follows:

$$p\hat{V} = RT$$

and once two of the intensive variables are fixed, the third is also automatically fixed.

An *invariant* system is one in which no variation of conditions is possible without one phase disappearing. An example with which you may be familiar is the ice-water-water vapor system which exists at only one temperature (0.01°C):

$$F = C - \mathcal{P} + 2 = 1 - 3 + 2 = 0$$

With the three phases present, none of the physical conditions can be varied without the loss of one phase. As a corollary, if the three phases are present, the temperature, the specific volume, etc., must always be fixed at the same values. This phenomenon is useful in calibrating thermometers and other instruments.

A complete discussion of the significance of the term C and the other terms in the phase rule is beyond the scope of this text; for further information you can read one of the general references on the phase rule listed at the end of this chapter or consult an advanced book on thermodynamics or physical chemistry. We are now going to consider how phase phenomena can be illustrated by means of diagrams.

3.7.2 Phase Phenomena of Pure Components. We can show the properties of a single-phase pure component, such as a gas, on two-dimensional plots. According to the phase rule,

$$F = C - \mathcal{P} + 2 = 1 - 1 + 2 = 2$$

only two physical variables have to be specified to fix the system definitely; i.e., any point on the two-dimensional diagram will be enough to fix the other properties of the system at that point.

Actually, if we are to clearly understand phase phenomena, the properties of a substance should be shown in three dimensions, particularly if there can be more than one phase in the region for which the p-V-T properties are to be presented. We previously overcame the handicap of using two dimensions to present three-dimensional data by showing, on the two-dimensional graph, lines of constant value for the other properties of the system (see Fig. 3.2, which has p and $\hat{V}$ as axes, and lines of constant temperature as the third parameter). In this section we are going to present a more elaborate treatment of these same p-V-T properties. Since the data on water, a substance that expands on freezing (to ice I), and carbon dioxide, a substance that contracts on freezing, are so well known, we have illustrated in Figs 3.23 and 3.24 the relationships between the three-dimensional presentations of p-V-T data for these two compounds and the conventional two-dimensional diagrams you usually encounter. Be certain that you view Figs. 3.23 and 3.24 as *surfaces* and not as solid figures; only the p-V-T points on the surface exist, and no point exists above or below the surface.

You can see that the two-dimensional diagrams are really projections of the three-dimensional figure onto suitable axes. On the p-$\hat{V}$ diagram, you will observe lines of constant temperature called *isothermal lines*, and on the p-T diagram you will see lines of constant volume, called *isometric lines* or *isochores*. On the T-$\hat{V}$ diagram, we might have put lines of constant pressure known as *isobaric lines*, but these are not shown since they would obscure the more impor-

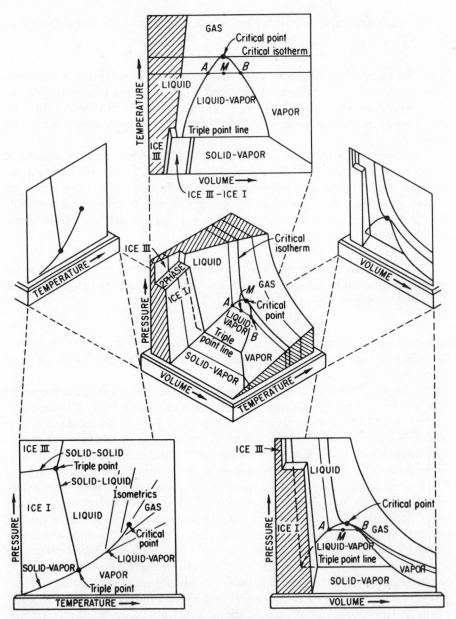

Fig. 3.23. *p-V-T* surface and projections for H_2O.

tant features which are shown. The two-phase vapor-liquid region is represented by the heavy envelope in the *p-$\hat{V}$* and the *T-$\hat{V}$* diagrams. On the *p-T* diagram this two-phase region appears only as a single line because you are looking at the three-dimensional diagram from the side. The *p-T* diagram shows what we have

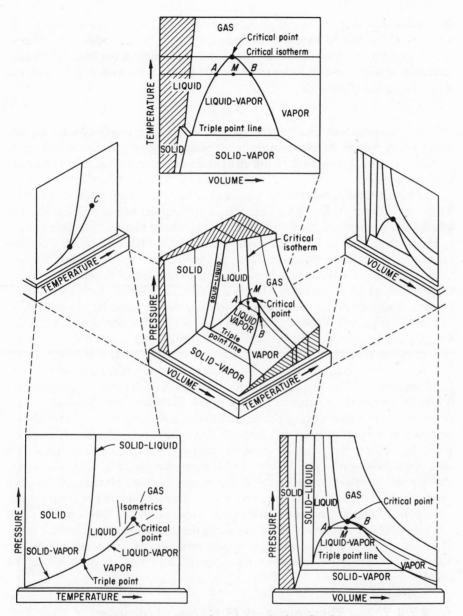

Fig. 3.24. *p-V-T* surface and projections for CO_2.

so far called the *vapor-pressure curve*, while the associated diagrams show that the "curve" is really a surface on which two phases coexist, each having the properties found at the appropriate bounding line for the surface. A point on the surface shows merely the overall composition of one phase in equilibrium with another, as, for example, at point *A* we have liquid and at point *B* we have vapor and in between them no single phase exists. If we added the liquid at *A* to the vapor at *B*, we would have a two-phase mixture with the gross properties as shown by point *M*. You will notice that in the *p*-$\hat{V}$ diagram the lines of constant temperature and pressure become horizontal in the two-phase region, because if you apply the phase rule

$$F = C - \mathcal{P} + 2 = 1 - 2 + 2 = 1$$

only one variable can change, the specific volume. In the two-phase region, where condensation or evaporation takes place, the pressure and temperature remain constant. Think back to our discussion of water boiling in this connection.

Other features illustrated on the diagrams are the critical point; the critical isotherm; the triple point, which becomes a triple-point line actually in the *p*-$\hat{V}$ and in the *T*-$\hat{V}$ diagrams; and the existence of two solid phases and a liquid phase in equilibrium, such as (Fig. 3.23) solid ice I, solid ice III, and liquid. The temperature at the freezing point for water, i.e., the point where ice, water, air, and water vapor are in equilibrium, is not quite the same temperature as that at the triple point, which is the point where ice, water, and water vapor are in equilibrium. The presence of air lowers the freezing temperature by about 0.001°C so that the triple point is at 0.00602 atm and 0.01°C. The corresponding values for CO_2 are 5 atm and -57°C, respectively. Another triple point for the water system exists at 2200 atm and 20°C between ice III, ice I, and liquid water.

The critical point can be determined from a *p*-$\hat{V}$ diagram by plotting constant-temperature lines for the experimental data. As you proceed to higher and higher pressures, the horizontal portion of the isothermal line becomes shorter and shorter until finally the isotherm has only an inflection (point of zero slope). Any lines at subsequently higher pressures do not have this inflection. If carefully done, the critical temperature, pressure, and volume can be determined this way. Another way is to use the law of *rectilinear diameter*. Fig. 3.25, which says that the arithmetic average of the densities (or specific volumes) of the pure components in the liquid and the vapor state is a linear function of temperature, if the pairs of measurements are made at the same pressure. This is not an exact relationship but is extremely useful in getting the critical density or critical specific volume, since the linear average density relation can be extrapolated until it crosses the two-phase envelope where experimental data are quite difficult to collect.

3.7.3 Phase Phenomena of Mixtures.

Presentation of phase phenomena for mixtures which are completely miscible in the liquid state involves some rather complex reduction of three-, four-, and higher-dimensional diagrams into two dimensions. We shall restrict ourselves to two-component systems because, although the ideas discussed here are applicable to any number of components, the graphical presentation of more complex systems in an elementary text such as this is probably more confusing than helpful. We shall

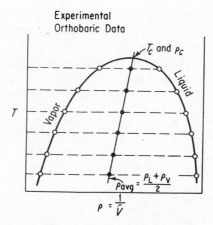

Fig. 3.25. Law of rectilinear diameters used to find critical temperature and density.

use hydrocarbons for most of our examples of two-component systems since a vast amount of experimental work has been reported for these compounds.

It would be convenient if the critical temperature of a mixture were the mole weight average of the critical temperatures of its pure components, and the critical pressure of a mixture were simply a mole weight average of the critical pressures of the pure components (according to Kay's rule), but these maxims simply are not true, as shown in Fig. 3.26. The pseudocritical temperature falls on the dashed line between the critical temperatures of CO_2 and SO_2, while

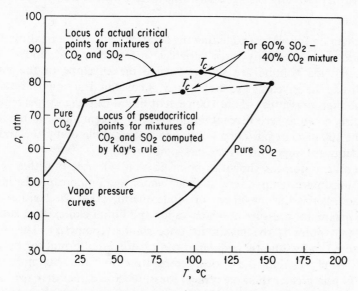

Fig. 3.26. Critical and pseudocritical points for a mixture.

the actual critical point for the mixture lies somewhere else. The dashed line as in Fig. 3.31 illustrates (for another system) the three-dimensional aspects of the locus of the actual critical points.

Figure 3.27 is a *p-T* diagram for a 25 percent methane-75 percent butane mixture. You can compare this figure with the one for a pure substance, water, as in Fig. 3.12; the *p-T* curve for pure butane or pure methane would look just

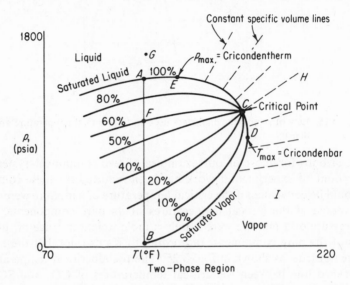

Fig. 3.27. *p-T* diagram for binary system: 25 percent CH_4 — 75 percent C_4H_{10}.

like the diagram for water (excluding the solid phases). For a mixture of a fixed composition, we have only three variables to consider, p, $\hat{V}$, and T. Lines of constant specific volume are shown outside the envelope of the two-phase region. Inside the envelope are shown lines of constant fractions of liquid, start-ing at the high pressures, where 100 percent liquid exists (zero percent vapor), and ranging down to zero percent liquid and 100 percent vapor at the low pres-sures. The 100 percent saturated liquid line is the bubble-point line and the 100 percent saturated vapor line is the dew-point line.

The critical point is shown as point *C*. You will note that this is not the point of maximum temperature at which vapor and liquid can exist in equilib-rium; the latter is the maximum cricondentherm, point *D*. Neither is it the point of maximum pressure at which vapor and liquid can exist in equilibrium because that point is the maximum cricondenbar (point *E*). The maximum temperature at which liquid and vapor can exist in equilibrium for this mixture is about 30° higher than the critical temperature, and the maximum pressures is 60 or 70 psia greater than the critical pressure. You can clearly see, therefore, that the best definition of the critical point was the one which stated that the density and other properties of the liquid and vapor become identical at the critical point. This definition holds for single components as well as mixtures.

With the aid of Fig. 3.27 we can indicate some of the phenomena that can take place when you change the pressure on a two-component mixture at constant temperature or change the temperature of a mixture at constant pressure. If you proceed from *A* in the liquid region to *F* and then to *B* in the vapor region by means of an isothermal process of some nature, you can go from 100 percent liquid to 100 percent vapor and notice a phase change. A similar type of analysis could have been applied to the *p-T* water diagram (Fig. 3.12) and the same conclusion reached, although the fraction of vapor and liquid cannot be found from Fig. 3.12. It is only when you cross the two-phase region that a change of phase will be noticed. If you go from *A* to *G* to *H* to *B*, the liquid will change from 0 to 100 percent vapor, but there will be no noticeable phase change. Figure 3.28 illustrates this transition for a mixture by a change of shading, and a similar diagram could be drawn for a pure component.

A very interesting phenomenon called *retrograde condensation* (or vaporization) can occur in two-phase systems, a phenomenon which initially appears to be quite contrary to your normal expectations. If you start at point 1 in Fig. 3.29 (an enlarged portion of Fig. 3.27 in the vicinity of the critical region) you

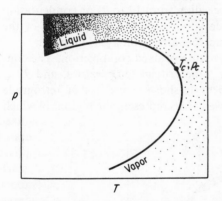

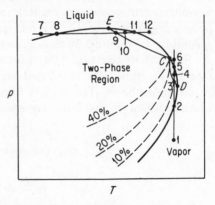

Fig. 3.28. Change in phase characteristics for a mixture of two components.

Fig. 3.29. The two types of retrograde condensation.

are in the vapor region. As you proceed to raise the pressure at constant temperature, you will arrive at point 2, where you are on the saturated-vapor line, or the dew point line. Continuing to increase the pressure will, as you might expect, increase the amount of liquid formed. However, soon you will reach a point where a maximum amount of liquid has formed (3), and increasing the pressure still further to points 4 and 5 will reduce the amount of liquid present, and eventually you will wind up back in the vapor region at 6. This phenomenon of retrograde condensation can happen, of course, only if the critical point, *C*, is located above and to the left of D as shown on the diagram so that there is a "bulge" in the vicinity of D for the mixture. The interesting thing about this phenomenon is that you would expect, as the pressure increased at constant temperature, to continually increase the amount of liquid formed, whereas the opposite is actually true after point 3 is reached.

Another type of retrograde phenomenon exists in a few systems when you change the temperature at constant pressure. This would be illustrated by the line 7, 8, 9, 10, 11, and 12, where first, as the temperature increases to point 8 on the bubble-point curve, some liquid starts to vaporize. The liquid continues to vaporize until a maximum of vapor is obtained at 9, and then the quantity of vapor starts to decrease until you reach a point where you have all liquid again at 11. Although the points of maximum percent liquid (or vapor) enveloping the regions of retrograde phenomena are shown in the diagram by solid lines, there are actually no phase boundaries at these points. The solid lines merely represent the locus of the region in which the maximum amount of liquid or vapor can exist (the point of infinite or zero slope on the lines indicating the fraction liquid), exaggerated on the diagram for the purpose of illustration. Formally, retrograde phenomena of these two types are known as (a) isothermal retrograde vaporization (proceeding from 1 to 6) or condensation (proceeding from 6 to 1), or retrograde vaporization or condensation, respectively, of the "first type," and (b) isobaric retrograde vaporization (proceeding from 12 to 7) or condensation (proceeding from 7 to 12), or retrograde vaporization (condensation) of the "second type." Figure 3.30 shows the same phenomena on a p-$\hat{V}$ diagram.

Figure 3.30, like Fig. 3.29, is for a mixture of fixed composition. The constant-temperature lines illustrated are T_c, the critical temperature, and T_{max}, the cricondentherm. The middle line, 6 to 1, shows a process of retrograde condensation of the first kind. The shadowed areas represent the regions in which

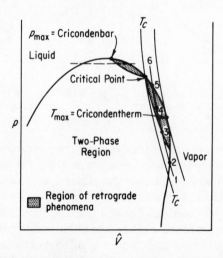

Fig. 3.30. Retrograde condensation on a p-V diagram.

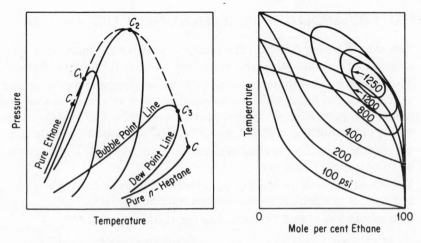

Fig. 3.31. *p-T* and *T-x* diagrams for ethane-heptane mixtures of different compositions.

retrograde phenomena take place. A $T\text{-}\hat{V}$ diagram is not illustrated since it is not particularly enlightening.

The diagrams shown so far represent two-component systems with a fixed overall composition. You might wonder what a diagram would look like if we were to try to show systems of several compositions on one page. This has been done in Fig. 3.31. Here we have a composite *p-T* diagram, which is somewhat awkward to visualize, but represents the bubble-point and dew-point curves for various mixtures of ethane and heptane. These curves in essence are intersections of planes in the composition coordinate sliced out of a three-dimensional system and are stacked one in front of the other, although in two dimensions it appears that they overlie one another. The vapor-pressure curves for the two pure components are at the extreme sides of the diagrams as single lines (as you might expect). Each of the loops represents the two-phase area for a system of any specific composition. An infinite number of these slices are possible, of course. The dotted line indicates the envelope of the critical points for each possible composition. Although this line appears to be two-dimensional in Fig. 3.31, it actually is a three-dimensional line of which only the projection is shown in the figure.

Another way to handle and illustrate the phase phenomena for the two-component systems we have been discussing is to use pressure-composition diagrams at constant temperature or, alternatively, to use temperature-composition diagrams at constant pressure. A temperature-composition diagram with pressure as the third parameter is illustrated in Fig. 3.31 for the ethane-heptane system.

Discussion of the phase diagrams for multicomponent systems (ternary and higher) is beyond our scope here.

WHAT YOU SHOULD HAVE LEARNED FROM THIS CHAPTER

1. You should be familiar with the ideal gas laws, know under what circumstances they are applicable, and know how to apply them in problems.
2. You should be able to distinguish under which circumstances a gas acts as an ideal gas and under which circumstances it acts as a real gas, and be able to use, as appropriate, (a) the ideal gas equation, $pV = nRT$; (b) the compressibility factor together with $pV = znRT$; or (c) an equation of state.
3. You should know how to treat ideal and real gas mixtures, and in particular be able to apply Kay's rule.
4. You should be able to use the generalized compressibility chart with ease, and be able to calculate the parameters required for its use.
5. You should know what the following terms mean:

vapor	saturated vapor
gas	bubble point
vapor pressure	dew point
critical point	superheat
critical temperature	subcooled
critical pressure	quality
pseudocritical temperature	saturated air
pseudocritical pressure	relative humidity
condensation	absolute humidity
vaporization	isometric
sublimation	isobaric
equilibrium	isothermal
normal boiling point	retrograde condensation
saturated liquid	

6. You should be able to estimate vapor pressures as a function of temperature.
7. You should know how to solve material balance problems involving vaporization, condensation, saturation, drying, and humidification.
8. You should be able to apply the phase rule to simple systems.
9. You should be familiar with the qualitative phase behavior of pure components and mixtures and also be able to draw typical diagrams illustrating this behavior.

SUPPLEMENTARY REFERENCES

1. American Petroleum Institute, Division of Refining, *Technical Data Book—Petroleum Refining*, API, New York, 1970.
2. Edmister, Wayne C., *Applied Hydrocarbon Thermodynamics*, Gulf Publications Co., Houston, 1961, Chaps. 1–4.

3. Henley, E. J., and H. Bieber, *Chemical Engineering Calculations*, McGraw-Hill, New York, 1959.

4. Hougen, O. A., K. M. Watson, and R. A. Ragatz, *Chemical Process Principles*, Part I, 2nd ed., Wiley, New York, 1956.

5. Natural Gas Processors Suppliers Association, *Engineering Data Book*, 10th ed., Tulsa, Okla., 1970.

6. Reid, R. C., and T. K. Sherwood, *The Properties of Gases and Liquids*, 2nd ed., McGraw-Hill, New York, 1966.

7. Wexler, A., *Humidity and Moisture*, 4 vols., Van Nostrand Reinhold, New York, 1965.

8. Williams, E. T., and R. C. Johnson, *Stoichiometry for Chemical Engineers*, McGraw-Hill, New York, 1958.

References on Phase Phenomena

1. Ricci, J. E., *The Phase Rule and Heterogeneous Equilibrium*, Van Nostrand Reinhold, New York, 1951.

2. Stanley, H. E., *Introduction to Phase Transitions and Critical Phenomena*, Oxford University Press, London, 1971.

PROBLEMS[18]

3.1. A given amount of N_2 has a volume of $200 \, cm^3$ at $15.7°C$. Find its volume at $0°C$, the pressure remaining constant.

3.2. A gas measured $300 \, cm^3$ at $27.5°C$, and because of a change in temperature, the pressure remaining constant, the volume decreased to $125 \, cm^3$. What was the new temperature in $°C$?

3.3. Under standard conditions, a gas measures 10.0 liters in volume. What is its volume at $92°F$ and 29.4 in. Hg?

3.4. Since 1917, Goodyear has built over 300 airships, most of them military. One of the newest is the America, which has a bag 192 ft long, is 50 ft in diameter, and holds about $202,000 \, ft^3$ of helium. Her twin 210-hp engines produce a cruising speed of 30–35 mi/hr and a top speed of 50 mi/hr. The 23-ft gondola will hold the pilot and six passengers. Blimps originally were made of rubber-impregnated cotton, but the bag of the America is made of a two-ply fabric of Dacron polyester fiber coated with neoprene. The bag's outer surface is covered with an aluminized coat of Hypalon synthetic rubber. Assuming that the bag size cited is at 1 atm and $25°C$, estimate the temperature increase or decrease in the bag at a height of 1000 m (where the pressure is 740 mm Hg) if the bag volume does not change. If the temperature remains $25°C$, explain how you might estimate the volume change in the bag.

[18]An asterisk designates problems appropriate for computer solution. Also refer to the computer problems after Problem 3.129.

3.5. The pressure on a confined gas, at 100°F, was 792 mm Hg. If the pressure later registered 820 mm Hg, what was the temperature then, the volume remaining unchanged?

3.6. Ten cubic feet of an ideal gas are under a pressure of 27.3 in. Hg. What is the volume of the gas at 29.92 in. Hg if there is no change in temperature?

3.7. Argon gas occupies a volume of 200 cc under a pressure of 780 mm Hg. The temperature remaining constant, what pressure must be applied to reduce the volume to 400 cc?

3.8. If a cubic meter of gas at standard conditions has its temperature raised to 100°C, what is the final pressure on the gas in Newtons per square meter if the volume of the gas remains unchanged?

3.9. A recent newspaper report states, "Home meters for fuel gas measure the volume of gas usage based on a standard temperature, usually 60 degrees. But gas contracts when it's cold and expands when warm. East Ohio Gas Co. figures that in chilly Cleveland, the homeowner with an outdoor meter gets more gas than the meter says he does, so that's built into the company's gas rates. The guy who loses is the one with an indoor meter: If his home stays at 60 degrees or over, he'll pay for more gas than he gets. (Several companies make temperature-compensating meters, but they cost more and aren't widely used. Not surprisingly, they are sold mainly to utilities in the North.)"

Suppose that the outside temperature drops from 60 to 10°F. What is the percentage increase in the mass of the gas passed by a noncompensated outdoor meter that operates at constant pressure? Assume that the gas is CH_4.

3.10. The highest temperature of a gas holder in summer is 42°C, and the lowest in winter is −38°C. Calculate how many more kilograms of methane may be contained by the gas holder of volume equal to 2000 m^3 at the lowest winter temperature compared to the highest summer temperature. Assume that the pressure in the gas holder is maintained at 780 mm Hg in both cases.

3.11. One cubic meter of gas at 7.90 psia and 20°C is heated to 100°C. What is the pressure in newtons per square meter applied on the gas if its volume is unaltered?

3.12. Five hundred cubic centimeters of gas are at a pressure of 70.0 kN/m^2 and a temperature of 40°C. What is the volume of the gas at standard conditions (in cubic meters)?

3.13. A steel tank having a capacity of 3 m^3 contains helium at 35°C and 205 kN/m^2. Calculate the mass, in kilograms, of the helium.

3.14. *Chem. Eng.*[19]—The problem all sulfuric acid and oleum plant operators face is that, if there is any water present after SO_2 has been oxidized, the trioxide will combine to form extremely small droplets of mist. The particles, most of which are below 3 microns, pass right through the rest of the plant and produce the familiar bluish plume on the stack outlet. Such a plume, points out York, is at once a health hazard, an equipment destroyer, and an economic loss. Stack losses can be estimated from the following typical data:

Acid tonnage	1,000 tons/day
Stack gas flow rate	73,000 ft^3/min
Entrainment	100 mg/ft^3

[19] *Chem. Eng.*, p. 112, (Oct. 25, 1965).

These figures, admittedly from a poorly operating plant, mean that 11.56 tons of sulfuric acid are sent up the stack each day. Catching it would save between \$52,000 and \$86,000/yr.[19] Verify the stack loss based on the given figure of 11.56 tons of H_2SO_4.

3.15. A boiler of unknown volume contains air at 70°F and atmospheric pressure. One hundred pounds of dry ice (solid CO_2) are thrown into the tank and the manhole is closed. When all the CO_2 is vaporized, the pressure in the boiler becomes 25.4 psig. What is the volume of the boiler?

3.16. A cylinder of nitrogen contains 1.00 m³ of gas at 20°C and atmospheric pressure. If the valve on the cylinder is opened and the cylinder is heated to 100°C, calculate the fraction of the nitrogen that leaves the cylinder.

3.17. Fill in the blanks for the following statements:
(a) As oil in the tank is pumped out of a tank 100 ft in diameter, the air vent plugs and the pressure inside the tank drops to 1 psia less than atmospheric pressure. This difference of pressure will cause an external force on the roof of _____ lb_f.
(b) The two tanks, 10 ft in diameter and 20 ft in diameter, respectively, are both the same height and full of water. Which one exerts the greater weight against its foundation? _____
(c) For the same tanks of part (b), which one exerts the greater pressure against its foundation? _____

3.18. One of the experiments in the fuel-testing laboratory has been giving some trouble because a particular barometer gives erroneous readings owing to the presence of a small amount of air above the mercury column. At a pressure of 755 mm Hg the barometer reads 748 Hg, and at 740 the reading is 736. What will the barometer read when the actual pressure is 760 mm Hg?

3.19. Hyperbaric oxygen therapy means placing a patient in a pressure chamber and subjecting him to two or more times normal air pressure. The patient is then given pure oxygen to breathe by means of a regular face mask. Under such circumstances the patient's blood carries about 15 times its normal quota of oxygen, enabling his system to be "drenched" with oxygen at a time when oxygen deficiency may be critical. Such critical situations may occur for patients suffering from gas gangrene or carbon monoxide poisoning. It may be valuable for "blue" babies, shock, strokes, etc. For a cylindrical chamber $41\frac{1}{2}$ ft long and 7 ft in diameter, the air must be changed and filtered 8 times/hr. If the chamber operates at 2.5 atm and 70°F, how many cubic feet of air at STP (standard temperature and pressure) must be compressed by the air compressor?

3.20. Explain whether or not the statement is correct, and, if not, modify the statement to make it correct:
(a) Pressure is how much a fluid weighs.
(b) If more of a gas is pumped into a closed drum, the volume of gas in the drum increases.
(c) If more of a gas is pumped into a closed drum, the weight of gas in the drum decreases.
(d) A pump has a maximum discharge pressure of 43.3 psi. The maximum height that water can be pumped with this pump for 0 psig suction is equivalent to 43.3 psia.

3.21. Georgian[20] has suggested that the gas constant arises from an incorrect selection of the temperature scale. Starting with $p\hat{V} = RT$, he proposes to select the temperature scale so that $R = 1.0$ and the gas law becomes $p\hat{V} = T$. In the mks system of units, temperature will have the units of Joules per kilomole and the dimensions of length squared per time squared. The scale of temperature can be established from the fundamental constant $p_0\hat{V}_0 = T_0 = 2,271,160$ for an ideal gas at 0°C and $p_0 \longrightarrow 0$. The Boltzmann constant becomes the reciprocal of Avogadro's number. Comment on the usefulness of such a temperature scale. What is the conversion factor from degrees Rankine to Joules per kilomole?

3.22. Two tanks are initially sealed off from one another by means of valve A. Tank I initially contains 1.00 ft³ of air at 100 psia and 150°F. Tank II initially contains a nitrogen-oxygen mixture containing 95 mole percent nitrogen at 200 psia and 200°F. Valve A is then opened allowing the contents of the two tanks to mix. After complete mixing has been effected, the gas was found to contain 85 mole percent nitrogen. Calculate the volume of tank II. See Fig. P3.22.

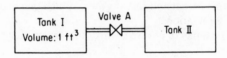

Fig. P3.22.

3.23. Calculate the molar gas constant R in the following units:
   (a) (atm)(cm³)/(g mole)(°K).
   (b) (psia)(ft³)/(lb mole)(°R).
   (c) (atm)(ft³)/(lb mole)(°K).
   (d) kilowatt-hr/(lb mole)(°R).
   (e) hp-hr/(lb mole)(°R).

3.24. A gas-fired furnace uses 3000 ft³/hr of gas at a pressure of 30.0 in. Hg and at a temperature of 100°F. The gas is purchased on a metered basis referred to 30.0 in. Hg and 60°F. Your accounting department wants to know if there is any difference between these two measurements. What percentage increase or decrease, if any, in volume is there? Base your percentage on the conditions of the gas as used in the furnace.

3.25. How much mercuric oxide must be heated to liberate a liter of oxygen measured at 25°C and 765 mm Hg pressure?

3.26. How much $KClO_2$ is required to be decomposed to produce 5.00 l of O_2 at 20°C and 740 mm.

3.27. An industrial fuel gas consists of 40 percent CO and 60 percent CH_4; 121 ft³/min of the gas, measured at 500 psia and 90°F, is completely burned with 25 percent excess air that enters at 70°F and 770 mm Hg. The resultant gases leave the furnace at 570°F and 770 mm Hg. Compute the composition and number of cubic feet per minute of the hot gases leaving the furnace.

[20]J. C. Georgian, *Nature*, v. 201, p. 695 (1964).

3.28. Turbojet aircraft operating under most conditions produce lower concentrations of pollutants than do motor vehicles. At idle, CO and unburned hydrocarbons are higher than in motor vehicles at idle, but in the operating mode the emission index (pounds of pollutant per 1000 lb of fuel) is

	CO	Hydrocarbons	NO_x
Turbojet	8.7	0.16	2.7
Automotive piston	300	55	27

Given that the Orsat analysis from a turbojet shows CO_2, 12.2 percent; CO, 0.4 percent; and O_2, 6.2 percent, compute the net hydrogen to carbon ratio in the fuel (the fuel contains negligible sulfur and nitrogen). Also compute the cubic feet of air used at 80°F and 29.56 in. Hg/lb of fuel burned.

3.29. A natural gas from an oil well has the following composition:

Component	Vol %
CH_4 (methane)	80.0
C_2H_6 (ethane)	6.0
C_3H_8 (propane)	4.0
C_4H_{10} (butane)	3.0
C_5H_{12} (pentane)	1.5
CO_2 (carbon dioxide)	3.5
N_2 (nitrogen)	2.0

(a) What is the composition in percent by weight?

(b) What is the composition in mole percent?

(c) It is planned to run the gas through an absorber plant so that the lighter fractions can be injected back into the porous media to maintain the well pressure. If all the pentane, all the butane, and 90 percent by weight of the propane is removed from the gas in the absorber, what is the composition of the gas leaving the absorber?

(d) An alternative use for the gas leaving the absorber is to burn it under the boilers of a power plant with the theoretical amount of air. What is the composition of the flue gases (wet) and how many cubic feet at 600°F and 29.10 in. Hg of the flue gases (dry) would be produced per pound of natural gas taken from the well?

3.30. Pine wood has the following composition: C, 50.31 percent; H_2, 6.20 percent; O_2, 43.08 percent; and ash, 0.41 percent.

(a) Calculate the cubic feet of air at 76°F and 29.4 in. Hg necessary for complete combustion per pound of wood.

(b) If 30 percent excess air were used, calculate the cubic feet of dry gas at 600°F and 29.4 in. Hg produced per pound of wood.

(c) Calculate the Orsat analysis of the flue gas for parts (a) and (b).

3.31. In a test on an oil-fired boiler, it is not possible to measure the amount of oil burned, but the air used is determined by inserting a venturi meter in the air line. It is found that 5000 ft³/min of air at 80°F and 10 psig is used. The dry gas analyzes CO_2, 10.7 percent; CO, 0.55 percent; O_2, 4.75 percent; and N_2, 84.0 percent. If the oil is assumed to be all hydrocarbon, calculate the gallons per hour of oil burned. The sp gr of the oil is 0.94.

3.32. A hydrogen-free coke analyzes moisture, 4.2 percent; ash, 10.3 percent; and carbon, 85.5 percent. It is burned giving a stack gas which analyzes CO_2, 13.6 percent; CO, 1.5 percent; O_2, 6.5 percent; and N_2, 78.4 percent on a dry basis. Calculate the following:
   (a) The percentage of excess air used.
   (b) The cubic feet of air at 80°F and 740 mm entering per pound carbon burned.
   (c) Same as part (b) per pound coke burned.
   (d) The cubic feet of dry flue gas at 690°F per pound coke.
   (e) The cubic feet of wet flue gas at S.C. per pound coke.

3.33. A gas analyzes CO_2, 5 percent; CO, 40 percent; H_2, 36 percent; CH_4, 4 percent; and N_2, 15 percent. It is burned with 60 percent excess air; combustion is complete. Gas and air enter at 60°F, and the stack gas leaves at 500°F. Calculate the following:
   (a) The Orsat gas analysis.
   (b) The cubic feet of air per cubic foot of gas.
   (c) The cubic feet of wet stack gas per cubic foot of gas.

3.34. A furnace produces a flue gas that contains 16.8 percent CO_2. The flue gas is drawn through a waste heat boiler and on exiting it is found to contain 15.2 percent CO_2. Calculate the cubic feet of air that have leaked into the system per cubic foot of flue gas.

3.35. In the oxidation of ammonia to oxides of nitrogen, the reaction is

$$4NH_3 + 5O_2 \longrightarrow 4NO + 6H_2O$$

What air-ammonia ratio (cubic feet of air per cubic feet of ammonia) should be used to secure 18 percent excess air? If the reaction is 93 percent complete, calculate the composition of the products.

3.36. A rigid closed vessel having a volume of 1 ft³ contains NH_3 gas at 330°C and 30 psia. Into the closed vessel is pumped 0.35 ft³ of HCl gas measured at 200°F and 20 psia. NH_4Cl is formed according to the reaction

$$NH_3 + HCl \longrightarrow NH_4Cl$$

Assume that the reaction goes to completion and that the vapor pressure of NH_4Cl at 330°C is 610.6 mm Hg.
   (a) How much NH_4Cl will be formed?
   (b) Assuming that some NH_4Cl exists as a solid, what will be the final pressure in the closed vessel if the final temperature is 330°C?

3.37. Gas at 60°F and 31.2 in. Hg is flowing through an irregular duct. To determine the rate of flow of the gas, CO_2 is passed into the gas stream. The gas analyzes 1.2 percent CO_2 (by volume) before, and 3.4 percent CO_2 after, addition. The CO_2 tank is placed on a scale and found to lose 15 lb in 30 min. What is the rate of flow of the entering gas in cubic feet per minute?

3.38. Gas at 60°F and 42.1 in. Hg is flowing through an irregular duct. To determine the rate of flow of the gas, He is passed into the gas stream. The gas analyzes 1.0 percent He (by volume) before, and 1.4 percent He after, addition. The He tank is placed on a scale and is observed to lose 10 lb in 30 min. What is the rate of flow of the gas in the duct in cubic feet per minute?

3.39. For the manufacture of dry ice a furnace produces 1.31×10^5 ft³/hr (at 750°F and 1 atm) of a flue gas that contains 16.8 percent CO_2. It passes through a waste heat boiler to the absorbers, at which point it contains 15.2 percent CO_2. Calculate the cubic feet of air that have leaked into the system per hour if the air is at 70°F and 1 atm.

3.40. Automobiles built in 1975 and thereafter must be capable of markedly lower emissions than any built heretofore in the United States. To measure the exhaust emissions of a vehicle, it is driven through a prescribed speed-time pattern on a dynamometer. A fraction of the exhaust is collected in a bag, and at the end of the test, pollutant concentrations in the bag are measured with special electronic gas-analyzing equipment to give the emissions in grams per mile of travel. A tank presumably containing propane is placed on a scale and fed to a new type of engine that burns propane instead of gasoline. The gas collected in the bag analyzes (on a dry basis) CO_2, 10.1 percent; CO, 1.1 percent; O_2, 4.7 percent; and N_2, 84.1 percent. During the test the cylinder looses 10 lb in weight while the bag picks up 61 ft³ of gas at 60°F and 32.00 in. Hg. Because of some confusion in the stockroom, the tank was not labeled, and there is some question as to whether the hydrocarbon burned was really butane or perhaps another hydrocarbon. Determine whether or not an error has been made.

3.41. A volatile compound of chlorine and oxygen has the following weight percent composition: Cl_2, 38.77 percent; O_2, 61.23 percent. At 27°C and 760 mm Hg, 250 cc of its vapor weighs 1.86 g. Determine the following:
  (a) The molecular weight.
  (b) The molecular formula of this compound.

3.42. What is the density of propane gas (C_3H_8) in pounds per cubic foot at 100°F and 30.0 psig? What is the sp gr of propane?

3.43. What is the sp gr of gas (C_3H_8) at 100°F and 800 mm Hg relative to air at 60°F and 760 mm Hg?

3.44. A natural gas is composed entirely of methane, CH_4. Calculate its density at 80°F and 32.0 in. Hg. What is its sp gr relative to air at the same temperature and pressure?

3.45. A mixture of bromine vapor in air contains 1 percent bromine by volume.
  (a) What weight percent bromine is present?
  (b) What is the average molecular weight of the mixture?
  (c) What is its sp gr?
  (d) What is its sp gr compared to bromine?
  (e) What is its sp gr at 100°F and 100 psig compared to air at 60°F and 30 in. Hg?

3.46. What is the weight of 1 ft³ of H_2 at 0°F and 29.6 in. Hg? What is the specific gravity of this H_2 compared to air at 0°F and 29.6 in. Hg?

3.47. A glass weighing cell is used to determine the density of a gas. For a certain determination, the data are as follows:

> Weight of the cell full of air in air = 18.602 g
> Weight of the evacuated cell in air = 18.294 g
> Weight of the cell filled with sample gas in air = 18.345 g

The density of the air is 0.0750 lb/ft³. Calculate the density of the sample gas in pounds per cubic foot.

3.48. Gas from the Padna Field, Louisiana, is reported to have the components and volume percent composition given below. What is (a) the mole percent, (b) the weight percent of each component in the gas, (c) the apparent molecular weight of the gas, and (d) its sp gr?

	%		%
Methane	87.09	Pentanes	0.46
Ethane	4.42	Hexanes	0.29
Propane	1.60	Heptanes	0.06
Isobutane	0.40	Nitrogen	4.76
Normal butane	0.5	Carbon dioxide	0.40
			100.00

3.49. A natural gas from a gas well has the following composition:

component	%	mol. wt
CH_4	60	16
C_2H_6	16	30
C_3H_8	10	44
C_4H_{10}	14	58

(a) Whât is the composition in weight percent?
(b) What is the composition in mole percent?
(c) How many cubic feet will be occupied by 100 lb of the gas at 70°F and 74 cm Hg?
(d) What is the density of the gas in pounds per cubic foot at 70°F and 740 mm Hg?
(e) What is the sp gr of the gas?

3.50. In the manufacture of dry ice, a fuel is burned to a flue gas which contains 16.2 percent CO_2, 4.8 percent O_2, and the remainder N_2. This flue gas passes through a heat exchanger and then goes to an absorber. The data show that the analysis of the flue gas entering the absorber is 13.1 percent CO_2 with the remainder O_2 and N_2. Apparently something has happened. To check your initial assumption that an air leak has developed in the heat exchanger, you collect the following data with a wet-test meter on the heat exchanger:

Entering flue gas in a 2-min period   47,800 ft³ at 600°F and 740 mm of Hg
Exit flue gas in a 2-min period   30,000 ft³ at   60°F and 720 mm of Hg

Was your assumption about an air leak a good one, or was perhaps the analysis of the gas in error? Or both?

3.51. A gaseous mixture consisting of 50 mole percent hydrogen and 50 mole percent acetaldehyde (C_2H_4O) is initially contained in a rigid vessel at a total pressure of 760 mm Hg abs. The formation of ethanol (C_2H_6O) occurs according to

$$C_2H_4O + H_2 \longrightarrow C_2H_6O$$

After a time it was noted that the total pressure in the rigid vessel had dropped to 700 mm Hg abs. Calculate the degree of completion of the reaction using the following assumptions:

All reactants and products are in the gaseous state.
The vessel and its contents were at the same temperature when the two pressures were measured.

3.52. The gaseous products from a catalytic cracker amount to 27,400,000 std ft³/day (27.4 M²SCFD). The composition is as follows:

CH_4	72.5%	$n\text{-}C_4H_{10}$	1.6%
C_2H_4	4.0	$n\text{-}C_4H_8$	2.3
C_2H_6	8.8	iso-C_5H_{12}	0.5
C_3H_6	3.1	C_5H_{10}	0.4
C_3H_8	4.0	$n\text{-}C_5H_{12}$	0.7
iso-C_4H_{10}	1.8	iso-C_6^+	0.3
			100.0

The average molecular weight of the iso-C_6^+ is 96, and its sp gr in the liquid state at 60°F is 0.89.

(a) Calculate the gallons of each condensable constituent present per 1000 SCF of gas (G/M). Condensables are those hydrocarbons which can be liquefied at 100°F at any pressure.

(b) Express the composition of the condensable portion as volume percent and weight percent.

(c) Using the information determined in part (a), calculate:
 (1) The total G/M of condensables present.
 (2) The total B/D (bbl/day) of condensables available.
 (3) The B/D of the material available for alkylation-plant feed stock. (Alkylation plants use isobutane (iso-C_4) and condensable unsaturants for feed.)

3.53. Butane heating fuel contains 50 percent propane (C_3H_8), 48 percent normal butane ($n\text{-}C_4H_{10}$), and 2 percent isobutane (iso-C_4H_{10}) by liquid volume.

(a) If 10 gal of this mixture is vaporized, how many cubic feet are produced at the standard conditions of the petroleum industry (60°F and 14.7 psia, dry gas)?

(b) What is the API gravity of the fuel?

3.54. The contents of a gas cylinder are found to contain 20 percent CO_2, 60 percent O_2, and 20 percent N_2 at a pressure of 740 mm Hg and at 20°C. What are the partial pressures of each of the components? If the temperature is raised to 40°C, will the partial pressures change? If so, what will they be?

3.55. A natural gas has the composition of

CH_4	80%
C_2H_4	10
N_2	10

If the pressure in the line is 100 kN/m², what are the partial pressures of the three components?

3.56. A liter of oxygen at 760 mm is forced into a vessel containing a liter of nitrogen at 760 mm. What will be the resulting pressure? What assumptions are necessary for your answer?

3.57. A mixture of 15 lb N_2 and 20 lb H_2 is at a pressure of 50 psig and a temperature of 60°F. Determine the following:

(a) The partial pressure of each component.

(b) The partial (or pure component) volumes.

(c) The specific volume of the mixture.

3.58. Lead nitrate decomposes on heating according to the following reaction:

$$Pb(NO_3)_2(s) \longrightarrow PbO(s) + N_2O_4 + \tfrac{1}{2}O_2$$

When the gaseous products are cooled to 50°C, 45 percent of the nitrogen tetroxide (N_2O_4) is dissociated into nitrogen dioxide (NO_2) and the partial pressure of the oxygen is 0.184 atm. Calculate the following:
(a) The partial pressures of the NO_2 and N_2O_4.
(b) The density of the gaseous mixture at 50°C.

3.59. Using data from the steam tables for the specific volume of water vapor, plot percentage deviation curves for the actual specific volume of water vapor from that calculated by the ideal gas laws:

$$\% \text{ deviation} = 100\frac{V_{ideal} - V_{actual}}{V_{ideal}}$$

where V_{ideal} is calculated from the ideal gas law. For the abscissa use temperatures from 32° to 1200°F.
(a) Show the deviations for the saturated vapor, i.e., the saturated steam up to the critical point.
(b) Show also how the deviations would change at constant pressure for 14.7 psia, 1000 psia, and 2000 psia, from the saturated steam curve to 1200°F.

3.60. A tank with a volume of 3 ft³ contains 104 lb of carbon monoxide at 75°C. What is the pressure in the tank?

3.61. An 80-lb block of ice is put into a 10-ft³ container and is heated to 900°K. What is the pressure in the container?

3.62. Nineteen cubic feet of CO_2 gas is held at a constant pressure of 36.9×10^6 N/m² and is heated from 16° to 77°C. Compare the volumes calculated for the gas after heating by considering it (a) as a nonideal gas and (b) as an ideal gas.

3.63. A block of dry ice weighing 50 lb is dropped into an empty steel bomb, the volume of which is 5.0 ft³. The bomb is closed and heated until the pressure gauge reads 1600 psig. What was the temperature of the gas in the bomb?

3.64. A cylinder of ethylene is standing in the sun so that its temperature becomes 38°C (100°F). The cylinder has a volume of 1.2 ft³ and contains 2.3 lb ethylene. What is the pressure in the cylinder?

3.65. Ethylene at 500 atm pressure and a temperature of 100°C is contained in a cylinder of internal volume of 1.0 ft³. How many pounds of C_2H_4 are in the cylinder?

3.66. Calculate the volume occupied by 2.0 lb air at 735 psia and 392°F.

3.67. A high-pressure line carries natural gas (all methane) at 1500 psia and 110°F. What volume under these conditions is equivalent to M ft³ at S.C. (S.C. natural gas industry = 14.7 psia, 60°F, and dry).

3.68. What weight of ethane is contained in a gas cylinder of 1.0-ft³ volume if the gas is at 100°F and 2000 psig?

3.69. A reactor containing 12.00 ft³ of propane (25.0 lb of propane) is at 450 psig. What is the temperature of the propane?

3.70. A steel cylinder contains ethylene (C_2H_4) at 200 psig. The cylinder and gas weigh 222 lb. The supplier refills the cylinder with ethylene until the pressure reaches 1000 psig, at which time the cylinder and gas weigh 250 lb. The temperature is constant at 25°C. Calculate the charge to be made for the ethylene if the ethylene is sold at $0.11/lb, and what the weight of the cylinder is for use in billing the freight charges. Also find the volume of the cylinder in cubic feet.

3.71. Two identical gas cylinders contain different gases. One contains ethylene and the other contains nitrogen. Both are at 25°C and the pressure in each is 1000 psig.

(a) Which cylinder contains the largest volume of gas measured at standard conditions? Verify your answer by giving the ratio of volumes at standard conditions.

(b) Which cylinder weighs the most? Verify your answer by an appropriate calculation. What is the ratio of the weights of the gases contained in the cylinders?

3.72. Polyethylene can be made from ethylene in a very high-pressure process. The compression of ethylene is carried out by piston-type compressors, as sketched in Fig. P3.72. The ethylene enters at 1000 psia and is compressed to 20,000

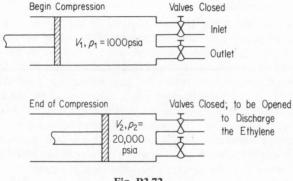

Fig. P3.72.

psia. (Assume that the temperature is constant at 25°C.) Calculate the following:

(a) The compression ratio, i.e., V_1/V_2.

(b) The density of the high-pressure ethylene.

3.73. "After 58 years of lying two miles under the Atlantic Ocean, the luxury liner *Titanic* may once more ride the ocean surface. A 21-man international team led by Douglas Woolley of Baldock, England, hopes to refloat the 66,000-ton wreck using plastic pontoons filled with hydrogen produced by electrolysis of sea water at the site.

"The team will use remote controlled equipment to circumvent the main problem—a pressure of more than 5000 p.s.i.g. at a depth of 2 miles. In Mr. Woolley's plan, the television-controlled deep-sea device will be used to break in some of the ship's more than 2000 portholes to attach lines connected to large bags or pontoons. Each of the 200 nylon-reinforced polyvinyl chloride pontoons will be linked to the ship by 10 2-inch-diameter nylon ropes. Each rope will support about 200 tons, for a total lifting capacity of about 400,000 tons.

"A ship-borne 200 volt 20-megawatt generator will provide power to electrolyze the sea water inside the cylindrical pontoons. Current will flow from an external anode through an aperture on the pontoon's underside to the internal cathodes. Mr. Woolley estimates it will take a week to generate the 85,000 cubic yards of hydrogen required." [*Chem. Eng. News*, p. 22 (Oct. 5, 1970).]

(a) How many cubic feet of hydrogen will be produced at 5000 psig and 35°F in 1 week by the generator if all the 20 MW of electricity are used to produce gas?

(b) Comment on the proposed lifting method. *Hint:* Write the equation for the generation of the hydrogen. Also, what happens as the pontoons rise?

3.74. Gauges used on cylinders containing compressed gases frequently are calibrated to show both the pressure in the cylinder and the volumetric contents of the cylinder. You have a Matheson Co. size 1A cylinder containing methane gas. You have a gauge for a 1A hydrogen cylinder, but not a gauge for methane. When you put the hydrogen gauge on the methane cylinder, it registers 2000 psig at 77°F and a capacity of 200 ft^3 (as hydrogen at 77°F and 1 atm pressure). Prepare a calibration so that the cubic feet as hydrogen can be converted to cubic feet as methane.

3.75. Current thinking about the smog-formation process is that reactive hydrocarbons from several sources combine with oxides of nitrogen stemming largely from automobiles and combustion processes to produce "oxidants" under certain conditions of meteorology and sunlight. Hydrocarbons can leak into the atmosphere from reservoirs, gasoline storage, natural gas storage, and so forth. It is proposed to store a gas composed of 60 percent methane and 40 percent ethylene by volume when measured at 1 atm in a high-pressure underground storage reservoir. The amount of gas to be stored is 2.79×10^5 lb and the reservoir volume is 1.81×10^4 ft^3. Assume that the temperature in the reservoir is 108°F. Calculate the pressure in the reservoir based on the following:

(a) Assuming that the mixture obeys the ideal gas laws.

(b) Assuming that the gases are nonideal; use the compressibility factor determined from the pseudocritical point of the mixture.

(c) Is the use of the storage reservoir feasible as planned if the maximum allowable pressure in the reservoir is only 1250 psia? How many pounds of the mixture can be stored at 1250 psia?

3.76. A sample of natural gas taken at 500 psig and 250°F is separated by chromatography at standard conditions. It was found by calculation that the grams of each component in the gas were

component	g
Methane (CH$_4$)	100
Ethane (C$_6$H$_2$)	240
Propane (C$_3$H$_8$)	150
Nitrogen (N$_2$)	50
Total	540

What was the density of the original gas sample?

3.77.* A gaseous mixture has the following composition (in moles percent):

$$C_2H_4 \quad 57$$
$$Ar \quad 40$$
$$He \quad 3$$

at 120 atm pressure and 25°C. Compare the experimental volume of 0.14 l/g
mole with that computed by the following:
(a) van der Waals' equation plus Dalton's law.
(b) van der Waals' equation using averaged constants.
(c) Mean compressibility factor and Dalton's law.
(d) Mean compressibility factor and Amagat's law.
(e) Compressibility factor using pseudoreduced conditions (Kay's method).
(f) Perfect gas law.

3.78. You are in charge of a pilot plant using an inert atmosphere composed of
60 percent ethylene (C_2H_4) and 40 percent argon (A). How big a cylinder (or
how many) must be purchased if you are to use 300 ft^3 of gas measured at the
pilot plant conditions of 100 atm and 300°F?

cylinder type	cost	pressure (psig)	lb gas
1A	$40.25	2000	62
2	32.60	1500	47
3	25.50	1500	35

State any additional assumptions. You can buy only one type of cylinder.

3.79. A gas has the following composition by volume:

$$CO_2 \quad 10\%$$
$$CH_4 \quad 40$$
$$C_2H_4 \quad 50$$

It is desired to distribute 33.6 lb of this gas per cylinder. Cylinders are to be
designed so that the maximum pressure will not exceed 2400 psig when the
temperature is 180°F. Calculate the volume of the cylinder required:
(a) Assuming the mixture to be an ideal gas.
(b) Using the pseudocritical point of the mixture and the compressibility
factor.
What pressure would exist in the cylinder if a cylinder designed from data
in part (b) and charged with 33.6 lb of this gas mixture were stored at 57°F?
(Use the pseudocritical method.)

3.80.* Prepare a Cox chart for
(a) Acetic acid vapor,
(b) Heptane,
(c) Ammonia, and
(d) Methanol
from 32°F to the critical point (for each substance). Compare the estimated
vapor pressure at the critical point with the critical pressure.

3.81.* Estimate the vapor pressure of benzene at 125°C from vapor pressure data
taken from a handbook or a journal.

3.82.* Estimate the vapor pressure of aniline at 350°C from the following vapor pressure data (the experimental vapor pressure is 40 atm):

t (°C)	184.4	212.8	254.8	292.7
p^* (atm)	1.00	2.00	5.00	10.00

3.83. Suppose that a vessel of dry nitrogen at 70.0°F and 29.90 in. Hg is saturated thoroughly with water. What will be the pressure in the vessel after saturation if the temperature is still 70.0°F?

3.84. An 8.00-liter cylinder contains a gas saturated with water vapor at 25.0°C and a pressure of 770 mm Hg. What is the volume of the gas when dry at standard conditions?

3.85. One cubic meter of propane saturated with water vapor is collected at 15°C and a total pressure of 750 mm Hg. If the propane were collected in a dry state at the same temperature and pressure, how many cubic meters would there be? What would the volume be, dry, under standard conditions?

3.86. A laboratory technician collected a sample of 40.0 cm³ of dangerous gas saturated with water at 20°C and 7.0×10^4 N/m². The gas is to be dried and transferred to a 50.0-cm³ standard cell at 1.00 atm pressure. What temperature must be maintained in the cell if it is to be exactly filled?

3.87. Fifty cubic feet of air saturated with water 90.0°F and 29.80 in. Hg is dehydrated. Compute the volume of the dry air and the pounds of moisture removed.

3.88. Fifty pounds of propane from a cylinder is gasified and collected over water in a gas holder at 25°C and a pressure of 6.0 in. H_2O above the normal barometer. What volume of gas is collected?

3.89. A mixture of acetylene (C_2H_2) with an excess of oxygen measured 350 ft³ at 25°C and 745 mm pressure. After explosion the volume of the dry gaseous product was 300 ft³ at 60°C and its partial pressure of 745 mm. Calculate the volume of acetylene and of oxygen in the original mixture. Assume that the final gas is saturated and that only enough water is formed to saturate the gas.

3.90. Uranium dioxide (UO_2) powder suitable for forming and sintering into high-density ceramic fuel may be prepared by the hydrogen reduction of ammonium diuranate. What size of the commercial hydrogen cylinder is needed to provide 205 ft³ of H_2 saturated with water vapor at a total pressure of 29.8 in. Hg and at 70°F? A commercial hydrogen cylinder contains dry hydrogen at 60°F and a pressure of 2000 psig.

3.91. Carbon disulfide (CS_2) at 20°C has a vapor pressure of 352 mm Hg. Dry air is bubbled through the CS_2 at 20°C until 4.45 lb of CS_2 are evaporated. What was the volume of the dry air required to evaporate this CS_2 (assuming that the air is saturated) if the air was initially at 20°C and 10 atm and the final pressure on the air-CS_2 vapor mixture is 750 mm Hg?

3.92. Toluene is used as a diluent in lacquer formulas. Its vapor pressure is 40 mm Hg at 31.8°C. You are painting as a "low" moves in, and the barometer drops from 780 to 720 mm Hg in 1 hr. Is there any change in the volume of dry air required to evaporate the toluene? Is there any change in the weight of dry air required? Express any change as percentage change from the conditions before the front moved in.

3.93. A gaseous mixture of 50 percent nitrogen and 50 percent water vapor at 200°F and 1 atm is cooled at constant pressure to a temperature low enough to condense 50 mole percent of the water. The remaining gases are compressed and leave the system at 250°F and 2 atm. Compute the volume of gases leaving per cubic foot of the original mixture.

3.94. Oxalic acid ($H_2C_2O_2$) is burned with 248 percent excess air, 65 percent of the carbon burning to CO. Calculate the following:
(a) The flue gas analysis.
(b) The volume of air at 90°F and 785 mm Hg used per pound of oxalic acid burned.
(c) The volume of stack gases at 725°F and 785 mm Hg per pound of oxalic acid burned.
(d) The dew point of the stack gas.

3.95. A solution containing 12 wt percent of dissolved nonvolatile solid is fed to a flash distillation unit. The molecular weight of the solid is 123.0. The effective vapor pressure of the solution is equal to the mole fraction of water, i.e.,

$$p = p^*x$$

where p = effective vapor pressure of the solution
$\quad x$ = mole fraction of water
$\quad p^*$ = vapor pressure of pure water

The pressure in the flash distillation unit is 1.121 psia and the temperature is 110°F. Calculate the pounds of pure water obtained in the vapor stream per 100 lb of feed solution and the weight percent of the dissolved nonvolatile solid leaving in the liquid stream.

3.96. A rigid vessel which is 1 ft^3 in volume contains 1 lb of N_2 and 1 lb of H_2O at 100°F.
(a) What is the pressure (psia) in the vessel?
(b) What is the molal humidity in the vapor phase?
(c) What mass fraction of the water is liquid?

3.97. Leather containing 100 percent of its own weight of water (i.e., if the dry leather is 1 lb, the water is 1 lb) is dried by means of air. The dew point of the entering air is 40°F, and in the exit air it is 55°F. If 2000 lb of the entering wet air are forced through the dryer per hour, how many pounds of water are removed per hour? The barometer reads 750 mm Hg.

3.98. Under what circumstances can the relative humidity and percent absolute humidity be equal?

3.99. The Environmental Protection Agency has promulgated a national ambient air quality standard for hydrocarbons: 160 μg/m^3 is the maximum 3-hr concentration not to be exceeded more than once a year. It was arrived at by considering the role of hydrocarbons in the formation of photochemical smog. Suppose that in an exhaust gas benzene vapor is mixed with air at 25°C such that the partial pressure of the benzene vapor is 2.20 mm Hg. The total pressure is 800 mm Hg. Calculate the following:
(a) The moles of benzene vapor per mole of gas (total).
(b) The moles of benzene per mole of benzene free gas.
(c) The weight of benzene per unit weight of benzene-free gas.

(d) The relative saturation.

(e) The percent saturation.

(f) The micrograms of benzene per cubic meter.

(g) The grams of benzene per cubic foot.

Does the exhaust gas concentration exceed the national quality standard?

3.100. A chemist determines the humidity of the air by absorbing the water vapor in P_2O_5 and weighing the water found in a definite volume of air. His setup looks like that in Fig. P3.100. A total of 1.153 ft³ of air is drawn through the wet-test

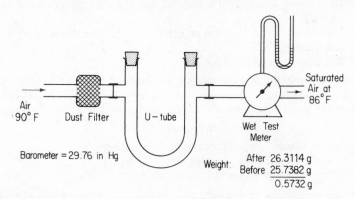

Fig. P3.100.

meter. The thermometer on the wet-test meter reads 86°F and the manometer shows that the pressure within is 0.4 in. H_2O less than the outside pressure. What is the molal humidity of the air? What is its percent absolute humidity? Assume that all the water in the air is removed in the P_2O_5 tube.

3.101. Toluene is mixed with air at 70°F in such proportions that the partial pressure of the vapor is 10 mm Hg. The total pressure is 745 mm Hg. Calculate the following:

(a) The relative saturation.

(b) The moles of toluene per mole of vapor-free gas.

(c) The weight of toluene per unit weight of vapor-free gas.

(d) The percent saturation.

(e) The percentage of toluene by volume.

(f) The grains of toluene per cubic foot of mixture.

3.102. Acetone (C_3H_6O) at 25°C has a vapor pressure of 229.2 mm Hg and a sp gr of $d_4^{25} = 0.780$. Dry air at 25°C and an absolute pressure of 715 mm Hg is bubbled through the acetone at 25°C; 15 gal of acetone evaporate. How many pounds of dry air are required?

3.103. On Thursday the temperature was 90°F, and the dew point was 70°F. At 2 P.M. the barometer read 29.83 in. Hg, but owing to an approaching storm it dropped by 5 P.M. to 29.08 in. Hg, with no other changes. What change occurred in (a) the relative humidity and (b) the percent absolute humidity, between 2 and 5 P.M.?

3.104. Air is saturated with water at 140°F and a barometer reading of 29.68 mm Hg.

If the temperature is reduced to 100°F and the gas compressed to 25 psia, what percentage of the water separates out?

3.105. A "wet" natural gas consists essentially of hexane (C_6H_{14}) and methane (CH_4). A natural gasoline plant of the pressure-refrigeration type is designed to recover the hexane from the gas, which is flowing at a pressure of 75 psi gauge and 30°C in a gas pipeline. One hundred grams of hexane is adsorbed and 100 ft³ (at 760 mm and 30°C) of CH_4 remains. To what temperature must the wet gas at 75 psi be cooled to recover 80 percent of the hexane?

3.106. Air saturated with water vapor is at 140°F and a pressure of 29.68 in. Hg.
(a) To what temperature must the air be cooled to separate 68 percent of the water in it (pressure is constant)?
(b) To what pressure must the air be compressed to separate 68 percent of the water in it (temperature is constant)?
(c) If the temperature is reduced to 100°F and the gas is compressed to 25 psia, what percentage of the water separates out?

3.107. Moist day air at 90°F, dew point 70°F, and barometer 29.8 in. Hg cools at night to a temperature of 50°F, barometer unchanged. Calculate the following:
(a) The cubic feet of night air per 1000 ft³ of day air.
(b) The pounds of water deposited per 1000 ft³ of day air.

3.108. A flue gas that analyzes CO_2, 12.0 percent; O_2, 8.0 percent; and N_2, 80.0 percent, by volume, has a dew point of 90°F. Calculate the density of the stack gas (flue gas + water vapor with it) at 600°F and 29.4 in. Hg. What is its sp gr compared with air at 80°F and 29.92 in. Hg?

3.109. Moist air is partially dehydrated and cooled before it is passed through a refrigerator room maintained at 0°F, to prevent excessive ice formation on the cooling coils. The cool air is passed through the room at the rate of 20,000 ft³/24 hr measured at the entrance temperature (40°F) and pressure. At the end of 30 days the refrigerator room must be warmed in order to remove the ice from the coils. How many pounds of water are removed? See Fig. P3.109.

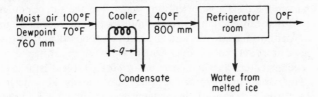

Fig. P3.109.

3.110. Air at 80°F and 1 atm has a dew point of 60°F.
(a) What is the percent relative humidity?
(b) To what pressure must this air be compressed to cause condensation to start (the temperature remains at 80°F)?
(c) To what temperature must this air be cooled to remove 25 percent of the moisture (the pressure stays constant at 1 atm)?
(d) What would be the percent relative humidity of this air after heating to 150°F (at constant pressure)?

(e) Did the molal saturation change during the heating indicated in part (d)?

3.111. Around airports jet aircraft can become major contributors to pollution, and as aircraft activity increases and public concern brings other sources under control, the relative contribution of aircraft to pollution could go up. Recently, federal-, state-, and local-government pressure has speeded the introduction of new combustors in aircraft. In a test for a supersonic aircraft fuel with the average composition $C_{1.20}H_{4.40}$, the fuel is completely burned with the exact stochiometric amount of air required. The air is supplied at 75°F and 760 mm Hg, with a humidity of 80 percent. The combustion products leave at 900°F and 800 mm Hg pressure and are passed through a heat exchanger from which they emerge at 135°F and 760 mm Hg pressure.
(a) For the entering air, compute the following:
  (1) The dew point.
  (2) The molal humidity.
  (3) The relative humidity.
(b) How much water is condensed in the heat exchanger per pound of gas burned, and hence must be removed as liquid water?

3.112. In Problem 3.40, a test approach is described for automobile exhaust emission studies. It is important that water be prevented from condensing in the bag. Pure hexane, C_6H_{14}, is burned in an internal combustion engine with 31.8 percent excess air. Of the carbon in the fuel, 95.0 percent goes to CO_2 and 5.00 percent goes to CO. The pressure in the bag is 722 mm Hg, and the exhaust gases have a temperature of 800°F. Assume that the exhaust gases obey the perfect gas law. Calculate the following:
(a) The Orsat analysis of the exhaust gas.
(b) The volume of exhaust gas produced per pound of hexane.
(c) The dew point of the exhaust gas to the nearest degree Fahrenheit. Will water condense out at room temperature?

3.113. Thermal pollution is the introduction of waste heat into the environment in such a way as to adversely affect environmental quality. Most thermal pollution results from the discharge of cooling water into the surroundings. It has been suggested that power plants use cooling towers and recycle water rather than dump water into the streams and rivers. In a proposed cooling tower, air enters and passes through baffles over which warm water from the heat exchanger falls. The air enters at a temperature of 80°F and leaves at a temperature of 70°F. The partial pressure of the water vapor in the air entering is 5 mm Hg and the partial pressure of the water vapor in the air leaving the tower is 18 mm Hg. The total pressure is 740 mm Hg. Calculate the following:
(a) The relative humidity of the air-water vapor mixture entering and of the mixture leaving the tower.
(b) The percentage composition by volume of the moist air entering and of that leaving.
(c) The percentage composition by weight of the moist air entering and of that leaving.
(d) The percent absolute humidity of the moist air entering and leaving.
(e) The pounds of water vapor per 1000 ft³ of mixture both entering and leaving.

(f) The pounds of water vapor per 1000 ft³ of vapor-free air both entering and leaving.

(g) The weight of water evaporated if 800,000 ft³ of air (at 740 mm and 80°F) enters the cooling tower per day.

3.114. A wet sewage sludge contains 50 percent by weight of water. A centrifuging step removes water at a rate of 100 lb/hr. The sludge is dried further by air. Use the data in Fig. P3.114 to determine how much moist air (in cubic feet per hour) is required for the process shown.

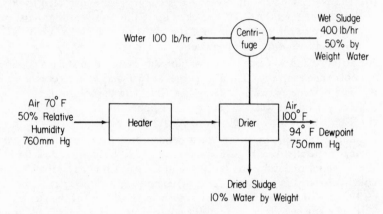

Fig. P3.114.

3.115. A process must be supplied with air and water vapor in small amounts. It is calculated that 0.04 lb of water vapor should accompany each pound of dry air into the process. The air enters at 1.00 atm. The engineers decide to accomplish this by saturating the air with water. The atmospheric air is at 77°F, 1 atm, and 55 percent humidity. The air is heated and then bubbled through water in an insulated tank.

(a) To what temperature was the air heated?

(b) How many cubic feet of entering air were required to evaporate 5 gal of water?

3.116. Air is humidified in the spray chamber shown in Fig. P3.116. Calculate how much water must be added per hour to the tower to process 10,000 ft³/hr of air metered at the entrance conditions.

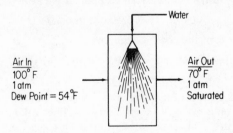

Fig. P3.116.

3.117. Soybean flakes from an extraction process are reduced from 0.96 lb of C_2HCl_3 per pound of dry flakes to 0.05 lb of C_2HCl_3 per pound of dry flakes in a desolventizer by a stream of N_2 which vaporizes the C_2HCl_3. The entering N_2 contains C_2HCl_3 such that its dew point is 30°C. The N_2 leaves at 90°C with a relative saturation of 60 percent. The pressure in the desolventizer is 760 mm, and 1000 lb/hr of dry flakes pass through the drier.
   (a) Compute the volume of N_2 plus C_2HCl_3 leaving the desolventizer at 90°C and 760 mm Hg in ft³/min.
   (b) The N_2 leaving the desolventizer is compressed and cooled to 40°C, thus condensing out the C_2HCl_3 picked up in the desolventizer. What must the pressure in the condenser be if the gas is to have a dew point of 30°C at the pressure of the desolventizer?

3.118. Air at 100°F and 20 percent relative humidity is forced through a cooling tower to cool condenser water. The air emerges at 75°F and 70 percent relative humidity. The barometer reads 29.48 in. Hg. Calculate the following:
   (a) The pounds of water evaporated per pound of dry air.
   (b) The cubic feet of air out per 1000 ft³ of air in.

3.119. A natural gas contains 5.0 percent condensable hydrocarbons, of which the average molecular weight is 86 and the density in the liquid state is 0.659 g/cm³. What volume of natural gas at 120°F and 100 psig contains 1 gal of the condensable hydrocarbons? What volume of completely stripped gas would result at 60°F and 30.0 in. Hg? (Give your answer in cubic feet.) If the gas is cooled to 32°F where the vapor pressure of the condensate is 43 mm Hg, what fraction of the condensable material is recovered?

3.120. Methane gas contains CS_2 vapor in an amount such that the relative saturation at 35°C is 85 percent. To what temperature must the gas mixture be cooled to condense out 60 percent of the CS_2 by volume? The total pressure is constant at 750 mm Hg, and the vapor pressure of CS_2 (mol. wt 76.12) is given by the following relation:

$$p^* = 15.4T + 130$$

   where p^* = vapor pressure of CS_2, mm Hg
   T = temperature, °C

3.121.* A certain gas contains moisture, and you have to remove this by compression and cooling so that the gas will finally contain not more than 1 percent moisture (by volume). You decide to cool the final gas down to 70°F.
   (a) Determine the minimum final pressure needed.
   (b) If the cost of the compression equipment is

$$\text{cost in \$} = (\text{pressure in psia})^{1.40}$$

   and the cost of the cooling equipment is

$$\text{cost in \$} = (350 - \text{temp. °K})^{1.9}$$

   is 70°F the best temperature to use?

3.122. One thousand pounds of a slurry containing 10 percent by weight of $CaCO_3$ is to be filtered on a rotary vacuum filter. The filter cake from the filter contains 60 percent water. This cake is then placed in a drier and dried to a moisture content of 9.09 percent (9.09 lb of H_2O per 100 lb of $CaCO_3$) on the dry basis.

If the humidity of the air entering the drier is 0.005 lb of water per pound of dry air and the humidity of the air leaving the drier is 0.015 lb of water per pound of dry air, calculate the following:

(a) The pounds of water removed by the filter.

(b) The pounds of dry air needed in the drier.

3.123. An absorber receives a mixture of air containing 12 percent carbon disulfide (CS_2). The absorbing solution is benzene, and the gas exits from the absorber with a CS_2 content of 3 percent and a benzene content of 3 percent (because some benzene evaporates). What fraction of the CS_2 was recovered?

3.124. In the manufacture of paper pulp by the sulfite process, a gas containing SO_2 is passed through an absorption system containing a milk-of-lime suspension to produce a calcium bisulfite cooking liquor for the wood chips. The gas enters the absorption system at 150°F, at a total pressure of 36.4 in. Hg and a partial pressure of SO_2 of 3.8 in. Hg. The other gases are inert gases. The exit gas from the absorption system is at 80°F, at a total pressure of 30.2 in. Hg and a partial pressure of SO_2 of 0.4 in. Hg. Calculate the following:

(a) The cubic feet of dry exit gas per 1000 ft³ of dry entering gas.

(b) The percent loss of SO_2/1000 ft³ dry entering gas.

3.125. Refer to the process flow diagram (Fig. P3.125) for a process which produces maleic anhydride by the partial oxidation of benzene. The moles of O_2 fed to the reactor per mole of pure benzene fed to the reactor is 18.0. All the maleic acid produced in the reactor is removed with water in the bottom stream from the water scrubber. All the C_6H_6, O_2, CO_2, and N_2 leaving the reactor leave in the stream from the top of the water scrubber, saturated with H_2O. Originally, the benzene contains trace amounts of a nonvolatile contaminant

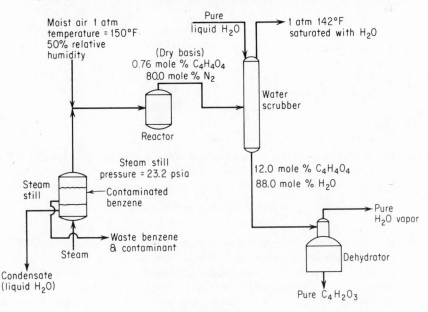

Fig. P3.125.

which would inhibit the reaction. This contaminant is removed by steam distillation in the steam still. The steam still contains liquid phases of both benzene and water (benzene is completely insoluble in water). The benzene phase is 80 wt percent, and the water phase is 20 wt percent of the two liquid phases in the still. Other process conditions are given in the flow sheet. Use the vapor pressure data given below. Calculate the following:

(a) The moles of benzene undergoing reaction (2) per mole of benzene feed to the reactor.

(b) The pounds of H_2O removed in the top stream from the dehydrator per pound mole of benzene feed to the reactor.

(c) The composition (mole percent, wet basis) of the gases leaving the top of the water scrubber.

(d) The pounds of pure liquid H_2O added to the top of the water scrubber per pound mole of benzene feed to the reactor.

Vapor pressure data:

temperature, °F	benzene, psia	water, psia
110	4.045	1.275
120	5.028	1.692
130	6.195	2.223
140	7.570	2.889
150	9.178	3.718
160	11.047	4.741
170	13.205	5.992
180	15.681	7.510
190	18.508	9.339
200	21.715	11.526

The reactions are

$$(1)\ C_6H_6 + 4\tfrac{1}{2}O_2 \longrightarrow \underset{\substack{\\ CH-C}}{\overset{\substack{CH-C}}{\|}} \begin{matrix} O \\ OH \\ O \\ OH \end{matrix} + 2CO_2 + H_2O$$

$$(2)\ C_6H_6 + 7\tfrac{1}{2}O_2 \longrightarrow 6CO_2 + 3H_2O$$

3.126. In the Girbotol process to remove hydrogen sulfide from natural and refinery gases, monoethanolamine or other ethanolamines are allowed to react with the hydrogen sulfide, forming compounds which may be broken down by heat:

$$RNH_2 + H_2S \rightleftharpoons RNH_3HS$$
(R represents an organic radical)

This plant uses monoethanolamine (MEA) in aqueous solution which circulates through an absorber and a reactivator. As shown in Fig. P3.126, the products of this process are acid gas containing 40 percent hydrogen sulfide and purified gas of 0.2 percent hydrogen sulfide content. Five hundred cubic feet of impure gas at 15 psia and 60°F are fed with water into the absorber. The impure

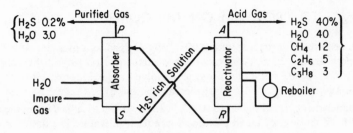

Fig. P3.126.

gas has the following composition:

Methane	65%
Ethane	20
Propane	5
Hydrogen sulfide	10

Compositions of the acid gas and purified gas are as shown.

(a) How many pounds of purified gas are obtained?

(b) Based on the entire amount of hydrogen sulfide entering the absorber, what weight of MEA is required for absorption, assuming 30 percent excess?

(c) If the reactivator is only 90 percent effective in removing the components of the acid gas from stream S, determine the pounds of liquid in streams R and S and determine their compositions.

3.127. Examine the following statements:

(a) The vapor pressure of gasoline is about 14 psia at 130°F.

(b) The vapor pressure of the system, water-furfural diacetate, is 760 mm Hg at 99.96°C.

Are the statements correct? If not, correct them. Assume that the numerical values are correct.

3.128. A constant-volume bomb contains air at 66°F and 21.2 psia. One pound of liquid water is introduced into the bomb. The bomb is then heated to a constant temperature of 180°F. After equilibrium is reached, the pressure in the bomb is 33.0 psia. The vapor pressure of water at 180°F is 7.51 psia.

(a) Did all the water evaporate?

(b) Compute the volume of the bomb in cubic feet.

(c) Compute the humidity of the air in the bomb at the final conditions in pounds of water per pound of air.

3.129. A laboratory accident recently occurred in a refinery when a sample container of butane ruptured and flashed. The container was an Air Force surplus O_2 container (completely flushed with N_2 to avoid explosion) of a design pressure of 500 psia and a working pressure of 400 psia. It had been filled in the field at 26°F and a line pressure of 220 psig and was brought into the laboratory which was at 75°F. Explain the cause of the rupture; show calculations.

PROBLEMS TO PROGRAM ON THE COMPUTER

3.1. You need to fill a reactor with 8.00 lb of NH_3 gas to a pressure of 150 psia at a temperature of 250°F. Apply van der Waals' equation to find the volume of the reactor. Prepare a computer program to solve Eq. (3.13) for V. How do you select the initial estimate of V to start the calculations?

3.2. An experiment was designed to test the validity of Charles' law using the expansion of a fixed amount of air in a balloon in a flask of water. The data taken were as follows at 751 mm Hg constant pressure:

Temp., °C	Incremental vol of balloon, cm³	Temp., °C	Incremental vol of balloon, cm³
22	0.0	44	4.2
23	0.3	47	4.9
24	0.45	50	5.5
25	0.8	53	6.5
26	0.8	55	7.0
27	1.2	59	9.0
29	2.0	61	10.0
34	3.0	63	11.9
37	3.2	65	13.5
39	3.8	67	14.7
43	3.8		

Ascertain how well the ideal gas law (Charles' law) is obeyed. A least-squares computer code should be used to evaluate the experimental data. What was the initial volume of gas (at 22°C)? Assume that the pressure on the air in the balloon is constant.

3.3. Use the Benedict-Webb-Rubin (BWR) equation of state (see Table 3.1)

$$p = RT\rho + \left(B_0 RT - A_0 - \frac{C_0}{T^2}\right)\rho^2 + (bRT - a)\rho^3$$
$$+ a\alpha\rho^6 + \frac{c\rho^3}{T^2}(1 + \gamma\rho^2)\exp(-\gamma\rho^2)$$

where ρ is the density, to predict the density of a mixture of two nonideal gases. Let the pressure be in atmospheres and calculate the density in both gram moles per liter and pound moles per cubic foot for selected pairs of p and T. Use the following rule illustrated for A_0 to obtain the coefficients A_0, B_0, C_0, a, b, c, α, and γ for the mixture

$$A_0 = \left[\sum_{i=1}^{n} x_i A_{0i}^{1/2}\right]^2$$

where n is the number of components in the mixture and x_i is the mole fraction. Values of the BWR constants for 38 pure components have been tabulated by H. W. Cooper and J. C. Goldfrank, *Hydrocarbon Processing*, v. 46, no. 12, p. 141 (1967). For reduced temperatures above 0.6 and reduced specific volumes less than 0.5, the BWR equation should be accurate to within 1 percent.

Energy Balances 4

How much power will be used in the world 10 years from now?—100 years from now? Forecasts[1] yield a considerable range of disagreement, but one common feature of all the forecasts is that the trend in energy usage is up, and up sharply, as illustrated in Fig. 4.1, for the United States. The combination of growing energy consumption per capita with population growth calls for substantial increases in total energy consumption that dwarfs our current productive facilities. If public expectations make these forecasts come true, the economic allocation of the sources of power and the design of new types of power production facilities will require that engineers continue to play a vital role in our technology.

Another facet of power generation is its relation to our environment. The generation of power by combustion or by nuclear fission by its very nature results in some form of pollution. Fossil- and nuclear-fueled power plants have an impact on land, water, and air resources during the production, processing, utilization, and disposal of the products of the generation of energy. Table 4.1 lists some of these effects. With electrical power usage doubling every 9 years, it is no surprise that the effect on the environment of the generation of energy becomes of considerable concern.

To provide publicly acceptable, effective, and yet economical conversion of our resources into energy and to properly utilize the energy so generated, you must understand the basic principles underlying the generation, uses, and transformation of energy in its different forms. The answers to questions such as, Is

[1] For example, see *An Energy Model for the United States*, the U.S. Department of the Interior, Government Printing Office, Washington, D.C., 1968.

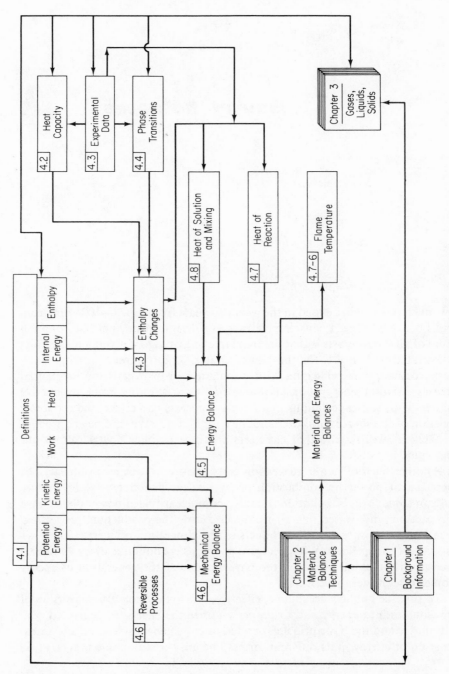

Fig. 4.0. Heirarchy of topics to be studied in this chapter (section numbers are in the upper left-hand corner of the boxes).

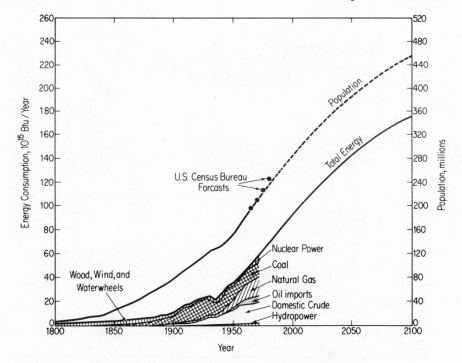

Fig. 4.1. Forecast of future energy consumption.

thermal pollution inherently necessary, What is the most economic source of fuel, What can be done with waste heat, and other related questions can only arise from an understanding of the treatment of energy transfer by natural processes or machines.

In this chapter we shall discuss energy balances together with the accessory background information needed to understand and apply them correctly. Before taking up the energy balance itself, we shall discuss the types of energy that are of major interest to engineers and scientists and some of the methods that are used to measure and evaluate these forms of energy. Our main attention will be devoted to heat, work, enthalpy, and internal energy. Next, the energy balance will be described and applied to practical and hypothetical problems. Finally, we shall discuss the generation of energy through the mechanisms of reaction and how energy generation fits into industrial process calculations. Figure 4.0 shows the relationships among the topics discussed in this chapter and also in previous chapters.

4.1 Concepts and units

When you think about energy, what sort of image comes to mind? Perhaps it is an electric generator whirring at the bottom of a dam, or perhaps a bullet speed-

TABLE 4.1 IMPACT OF POWER PLANTS ON THE ENVIRONMENT

Source of Energy	Effect on Land			Effect on Water			Effect on Air		
	Production	Processing	Utilization	Production	Processing	Utilization	Production	Processing	Utilization
Coal	Disturbed land, mine wastes	Solid waste disposal	Ash, slag disposal	Acid mine drainage, silting of streams	Leaching of waste piles	Increased water temps.	Mine fires	Dust	Sulfur oxides, nitrogen oxides, particulate matter
Oil	Pipelines, spills			Oil spills, brines		Increased water temps.			Carbon monoxide, nitrogen oxides, hydrocarbons
Natural gas	Pipelines					Increased water temps.	Waste flaring		
Uranium	Disturbed land		Disposal of radioactive material		Disposal of radioactive material	Increased water temps., accidents			Accidents

ing through the air. If it is the latter, you instinctively feel that the bullet in motion is different from one at rest. You know a ball of iron in the sun is in a different state than one in the shade. If a gas confined in a cylinder at high pressure is allowed to expand, it can do work against a rétarding force, and at the same time the temperature of the gas may fall. However, the same change in the properties of the gas can be brought about in other ways which seem to have very little to do with gas expansion; for example, the temperature of the gas might be reduced by placing the cylinder on a block of ice.

We are able to interpret and explain all these and many other phenomena in terms of the physical quantity *energy*. In Sec. 4.5 we shall relate different types of energy by means of an energy balance. In this section we shall discuss the units used to express energy and shall define several common types of energy. Unfortunately, many of the terms described below are used loosely in our ordinary conversation and writing, whereas others are so fundamental to our thinking that they defy exact definition. You may feel that you understand them from long acquaintance—be sure that you really do.

Table 1.1 lists the units used to express energy in various systems; you can find the conversion factors for these and other units in Appendix A. The calorie is roughly defined as the amount of energy required to raise the temperature of 1 g of water 1° C at a pressure of 1 atm. Similarly, the British thermal unit (Btu) is the amount of energy required to raise 1 lb of water 1° F. However, since the heat capacity of water varies with temperature, it is necessary to be a little more precise in specifying what type of calorie and British thermal unit (Btu) is under consideration (there are four or five common kinds). The type of calorie that forms the basis of the thermochemical properties of substances in this text is the *thermochemical calorie* (equal to 4.184 J). A second type of calorie is the *International Steam Table (I.T.) calorie* (equal to 4.1867 J). The British thermal unit is defined in terms of the calorie:

$$\frac{1 \text{ Btu}}{\text{lb}_m} = \frac{1.8 \text{ cal}}{\text{g}}$$

You can see that these calories differ by less than one part per thousand, so that for calculations made to slide-rule accuracy (but not for precise measurements) there is hardly any reason for you to consider the differences between them.

Certain terms that have been described in earlier chapters occur repeatedly in this chapter; these terms are summarized below with some elaboration in view of their importance.

(a) *System.* Any arbitrarily specified mass of material or segment of apparatus to which we would like to devote our attention. A system must be defined by surrounding it with a system boundary. A system enclosed by a boundary which prohibits the transfer of mass across the boundary is termed a *closed* system, or *nonflow* system, in distinction to an *open* system, or *flow* system, in which the exchange of mass is permitted. All the mass or apparatus external to the defined system is termed the *surroundings*. Reexamine some of the example problems in Chap. 2 for illustrations of the location of system boundaries. You should always draw similar boundaries in the solution of your problems, since this will fix clearly the system and surroundings.

(b) *Property*. A characteristic of material which can be measured, such as pressure, volume, or temperature—or calculated, if not directly measured, such as certain types of energy. The properties of a system are dependent on its condition at any given time and not on what has happened to the system in the past.

An *extensive property* (variable, parameter) is one whose value is the sum of the values of each of the subsystems comprising the whole system. For example, a gaseous system can be divided into two subsystems which have volumes or masses different from the original system. Consequently, mass or volume is an extensive property.

An *intensive property* (variable, parameter) is one whose value is not additive and does not vary with the quantity of material in the subsystem. For example, temperature, pressure, density (mass per volume), etc., do not change if the system is sliced in half or if the halves are put together.

Two properties are *independent* of each other if at least one variation of state for the system can be found in which one property varies while the other remains fixed. The set of independent intensive properties necessary and sufficient to fix the state of the system can be ascertained from the phase rule of Sec. 3.7.1.

(c) *State*. Material with a given set of properties at a given time. The state of a system does not depend on the shape or configuration of the system but only on its intensive properties.

Next we shall take up some new concepts, although they may not be entirely new, because you perhaps have encountered some of them before.

(a) *Heat*. In a discussion of *heat* we enter an area in which our everyday use of the term may cause confusion, since we are going to use heat in a very restricted sense when we apply the laws governing energy changes. Heat (Q) is commonly defined as that part of the total energy flow across a system boundary that is caused by a temperature difference between the system and the surroundings. Heat may be exchanged by conduction, convection, or radiation. A more effective general qualitative definition is given by Callen[2]:

> A macroscopic observation [of a system] is a kind of "hazy" observation which discerns gross features but not fine detail. . . . Of the enormous number of atomic coordinates [which can exist], a very few with unique symmetric properties survive the statistical averaging associated with a transition to the macroscopic description [and are macroscopically observable]. Certain of these surviving coordinates are mechanical in

[2]H. B. Callen, *Thermodynamics*, Wiley, New York, 1960, p. 7.

nature [such as volume]. Others are electrical in nature [such as dipole moments]. . . . It is equally possible to transfer energy to the hidden atomic modes of motion [of the atom] as well as to those modes which happen to be macroscopically observable. An energy transfer to the hidden atomic modes is called heat.

To evaluate heat transfer *quantitatively*, unless given a priori, you must apply the energy balance that is dicussed in Sec. 4.5, and evaluate all the terms except Q. Heat transfer can be *estimated* for engineering purposes by many empirical relations, which can be found in books treating heat transfer or transport processes.

(b) *Work*. Work (W) is commonly defined as energy transferred between the system and its surroundings by means of a vector force acting through a vector displacement on the system boundaries

$$W = \int F \, dl$$

where F is the direction of dl. However this definition is not exact inasmuch as

(1) The displacement may not be easy to define.
(2) The product of $F \, dl$ does not always result in an equal amount of work.
(3) Work can be exchanged without a force acting on the system boundaries (such as through magnetic or electric effects).

Since heat and work are by definition mutually exclusive exchanges of energy between the system and the surroundings, we shall qualitatively classify work as energy that can be transferred to or from a mechanical state, or mode, of the system, while heat is the transfer of energy to atomic or molecular states, or modes, which are not macroscopically observable. To measure or evaluate work *quantitatively* by a mechanical device is difficult, so that, unless one of the energy balances in Sec. 4.5 or 4.6 can be applied, in many instances the value of the work done must be given a priori.

(c) *Kinetic Energy*. Kinetic energy (K) is the energy a system possesses because of its velocity relative to the surroundings. Kinetic energy may be calculated from the relation

$$K = \tfrac{1}{2}mv^2 \tag{4.1}$$

or

$$\hat{K} = \tfrac{1}{2}v^2 \tag{4.1a}$$

where the superscript caret ($^\wedge$) refers to the energy per unit mass (or sometimes per mole) and not the total kinetic energy as in Eq. (4.1)

(d) *Potential Energy*. Potential energy (P) is energy the system possesses because of the body force exerted on its mass by a gravitational field with respect to a reference surface. Potential energy can be calculated from

$$P = mgh \tag{4.2}$$

or

$$\hat{P} = gh \tag{4.2a}$$

where the symbol ($\wedge$) again means potential energy per unit mass (or sometimes per mole).

EXAMPLE 4.1   Energy changes

A 100-lb ball initially on the top of a 15-ft ladder is dropped and hits the ground. See Fig. E4.1. With reference to the ground, determine the following:

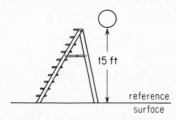

Fig. E4.1.

(a) What is the initial kinetic and potential energy of the ball in (ft)(lb$_f$)?

$$\text{initial } K = \tfrac{1}{2}mv^2 = 0$$

$$\text{initial } P = mgh = \frac{100 \text{ lb}_m \left| \; g\dfrac{\text{ft}}{\text{sec}^2} \; \right| 15 \text{ ft}}{\left| \; g_c\dfrac{(\text{ft})(\text{lb}_m)}{(\text{sec}^2)(\text{lb}_f)} \; \right|} = +1500 \text{ (ft)(lb}_f)$$

(b) What is the final kinetic and potential energy of the ball?

$$\text{final } K = 0$$

$$\text{final } P = 0$$

(c) What is the change in kinetic and potential energy for the process?

$$K_2 - K_1 = \tfrac{1}{2}m(v_2^2 - v_1^2) = 0$$

$$P_2 - P_1 = mg(h_2 - h_1) = 0 - (+1500) = -1500 \text{ (ft)(lb}_f)$$

A decrease in potential energy has occurred.

(d) If all the initial potential energy of the ball were somehow converted into heat, how many Btu would this amount to?

$$\frac{1500 \text{ (ft)(lb}_f) \; \left| \; 1 \text{ Btu} \right.}{\left| \; 778 \text{ (ft)(lb}_f) \right.} = 1.93 \text{ Btu}$$

The system has not been specified in this example. Where would you place the system boundary?

(e) *Internal Energy.* Internal energy (U) is a macroscopic measure of the molecular, atomic, and subatomic energies, all of which follow definite microscopic conservation rules for dynamic systems. Because no instruments exist with which to measure internal energy directly on a macroscopic scale, internal energy is calculated from certain other variables that can be measured macroscopically, such as pressure, volume, temperature, and composition.

To calculate the internal energy per unit mass ($\hat{U}$) from the variables that can be measured, we make use of a special property of internal energy, namely, that it is an exact differential (because it is a *point* or *state* property, a matter to be described shortly) and, for a pure component, can be expressed in terms of the temperature and specific volume alone. If we say that

$$\hat{U} = \hat{U}(T, \hat{V})$$

by taking the total derivative, we find that

$$d\hat{U} = \left(\frac{\partial \hat{U}}{\partial T}\right)_{\hat{V}} dT + \left(\frac{\partial \hat{U}}{\partial \hat{V}}\right)_T d\hat{V} \tag{4.3}$$

By definition $(\partial\hat{U}/\partial T)_{\hat{V}}$ is the heat capacity at constant volume, given the special symbol C_v. For most practical purposes in this text the term $(\partial\hat{U}/\partial\hat{V})_T$ is so small that the second term on the right-hand side of Eq. (4.3) can be neglected. Consequently, changes in the internal energy can be computed by integrating Eq. (4.3) as follows:

$$\hat{U}_2 - \hat{U}_1 = \int_{T_1}^{T_2} C_v \, dT \tag{4.4}$$

Note that you can only calculate *differences* in internal energy, or calculate the internal energy relative to a reference state, but not absolute values of internal energy. Instead of using Eq. (4.4), internal energy changes are usually calculated from enthalpy values, the next topic of discussion.

(f) *Enthalpy.* In applying the energy balance you will encounter a variable which is given the symbol H and the name *enthalpy* (pronounced en'-thal-py). This variable is defined as the combination of two variables which will appear very often in the energy balance

$$H = U + pV \tag{4.5}$$

where p is the pressure and V the volume. [The term enthalpy has replaced the now obsolete terms "heat content" or "total heat," to eliminate any connection whatsoever with heat as defined above in (a).]

To calculate the enthalpy per unit mass, we use the property that the enthalpy is an exact differential. For a pure substance, the enthalpy can be expressed in terms of the temperature and pressure (a more convenient variable for enthalpy than volume). If we let

$$\hat{H} = \hat{H}(T, p)$$

by taking the total derivative of $\hat{H}$, we can form an expression corresponding to Eq. (4.3):

$$d\hat{H} = \left(\frac{\partial \hat{H}}{\partial T}\right)_p dT + \left(\frac{\partial \hat{H}}{\partial p}\right)_T dp \tag{4.6}$$

By definition $(\partial \hat{H}/\partial T)_p$ is the heat capacity at constant pressure, given the special symbol C_p. For most practical purposes $(\partial \hat{H}/\partial p)_T$ is so small at modest pressures that the second term on the right-hand side of Eq. (4.6) can be neglected. Changes in enthalpy can then be calculated by integration of Eq. (4.6) as follows:

$$\hat{H}_2 - \hat{H}_1 = \int_{T_1}^{T_2} C_p \, dT \tag{4.7}$$

However, in high-pressure processes the second term on the right-hand side of Eq. (4.6) cannot necessarily be neglected, but must be evaluated from experimental data. One property of ideal gases that should be noted is that their enthalpies and internal energies are functions of temperature only and are not influenced by changes in pressure or volume.

As with internal energy, enthalpy has no absolute value; only changes in enthalpy can be calculated. Often you will use a reference set of conditions (perhaps implicitly) in computing enthalpy changes. For example, the reference conditions used in the steam tables are liquid water at 32°F and its vapor pressure. This does not mean that the enthalpy is actually zero under these conditions but merely that the enthalpy has arbitrarily been assigned a value of zero at these conditions. In computing enthalpy changes, the reference conditions cancel out as can be seen from the following:

initial state of system   *final state of system*
enthalpy $= \hat{H}_1 - \hat{H}_{\text{ref}}$ enthalpy $= \hat{H}_2 - \hat{H}_{\text{ref}}$
Net enthalpy change $= (\hat{H}_2 - \hat{H}_{\text{ref}}) - (\hat{H}_1 - \hat{H}_{\text{ref}}) = \hat{H}_2 - \hat{H}_1$

(g) *Point, or State, Functions.* The variables enthalpy and internal energy are called *point functions*, or *state variables*, which means that in their differential form they are exact differentials.[3] Figure 4.2 illustrates geometrically the concept

---

[3]To test if a differential dS is exact, where

$$dS = R \, dx + Z \, dy$$

form the partial derivatives $(\partial R/\partial y)_x$ and $(\partial Z/\partial x)_y$, and verify whether

$$\left(\frac{\partial R}{\partial y}\right)_x = \left(\frac{\partial Z}{\partial x}\right)_y$$

If so, dS is exact; if not, dS is not exact. The statement that dS is an exact differential is completely equivalent to saying that S is independent of path. As an example, applying the test to Eq. (4.6) yields

$$\frac{\partial}{\partial p}\left(\frac{\partial \hat{H}}{\partial T}\right)_p \overset{?}{=} \frac{\partial}{\partial T}\left(\frac{\partial \hat{H}}{\partial p}\right)_T$$

or

$$\left(\frac{\partial^2 \hat{H}}{\partial p \, \partial T}\right) = \left(\frac{\partial^2 \hat{H}}{\partial T \, \partial p}\right)$$

as expected.

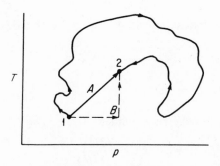

Fig. 4.2. Point function.

of state variable. In proceeding from state 1 to state, 2, the actual process is shown by the wiggly line. However, you may calculate $\Delta \hat{H}$ by route A or B, or any other route, and still obtain the same net enthalpy change as for the route shown by the wiggly line. The change of enthalpy depends only on the initial and final states of the system. A process that proceeds first at constant pressure and then at constant temperature from 1 to 2 will yield exactly the same $\Delta \hat{H}$ as one that takes place first at constant temperature and then at constant pressure as long as the end point is the same. The concept of the point function is the same as that of an airplane passenger who plans to go straight to Chicago from New York but is detoured because of bad weather by way of Cincinnati. His trip costs him the same whatever way he flies, and he eventually arrives at his destination. The gasoline consumption of the plane may vary considerably, and in analogous fashion heat (Q) or work (W), the two "path" functions with which we deal, may vary depending on the specific path chosen, while $\Delta \hat{H}$ is the same regardless of path. If the plane were turned back by bad weather and landed at New York, the passenger might be irate, but at least he could get his money back. Thus $\Delta \hat{H} = 0$ if a cyclical process is involved which goes from state 1 to 2 and back to state 1 again, or

$$\oint d\hat{H} = 0$$

All the intensive properties we shall work with, such as $P, T, \hat{U}, p, \hat{H}$, etc., are point functions and depend only on the state of the substance of interest, so that we can say, for example, that

$$\oint dT = 0$$

$$\oint d\hat{U} = 0$$

Always keep in mind that the values for a difference in a point function can be calculated by taking the value in the final state and subtracting the value in the initial state, regardless of the actual path.

Before formulating the general energy balance, we shall discuss in some detail the calculation of enthalpy changes and provide some typical examples of such calculations. The discussion will be initiated with consideration of the heat capacity C_p.

4.2 Heat capacity

The two heat capacities have been defined as

(a)
$$C_p = \left(\frac{\partial \hat{H}}{\partial T}\right)_p$$

(b)
$$C_v = \left(\frac{\partial \hat{U}}{\partial T}\right)_{\hat{v}}$$

To give these two quantities some physical meaning, you can think of them as representing the amount of energy required to increase the temperature of a substance by 1 degree, energy which might be provided by heat transfer in certain specialized processes. To determine from experiment values of C_p (or C_v), the enthalpy (or internal energy) change must first be calculated from the general energy balance, Eq. (4.24), and then the heat capacity evaluated from Eq. (4.7) [or (4.4)] for small enthalpy changes.

We can provide a rough illustration of the calculation of C_p using data from the steam tables. We shall take the enthalpies for water vapor at 1 psia and at 300° and 350°F. Clearly the temperature difference is far too large to precisely approximate a derivative, but it should suffice under the selected conditions to give a C_p valid to two significant figures:

$$\Delta \hat{H} = (1217.3 - 1194.4) = 22.9 \text{ Btu 1 lb}$$

$$C_p = \left(\frac{\partial \hat{H}}{\partial T}\right)_p \cong \left(\frac{\Delta \hat{H}}{\Delta T}\right)_p = \frac{22.9}{50} = 0.46 \frac{\text{Btu}}{(\text{lb})(\Delta °F)}$$

In line with the definition of the calorie or Btu presented in the previous section, we can see that the heat capacity can be expressed in various systems of units and still have the same numerical value; for example, heat capacity may be expressed in the units of

$$\frac{\text{cal}}{(\text{g mole})(°C)} = \frac{\text{kcal}}{(\text{kg mole})(°C)} = \frac{\text{Btu}}{(\text{lb mole})(°F)}$$

Alternatively, heat capacity may be in terms of

$$\frac{\text{cal}}{(\text{g})(°C)} = \frac{\text{Btu}}{(\text{lb})(°F)} \quad \text{or} \quad \frac{\text{J}}{(\text{kg})(°K)}$$

Note that

$$\frac{1 \text{ Btu}}{(\text{lb})(°F)} = \frac{4.184 \text{ J}}{(\text{g})(°K)}$$

and that the heat capacity of water in the SI system is 4184 J/(kg)(°K). These relations are worth memorizing. Note that in each system of units the heat capacity consists of energy divided by the product of the mass times the temperature difference.

The heat capacity over a wide temperature range for a pure substance is represented figuratively in Fig. 4.3, where the heat capacity is plotted as a function of the absolute temperature. We see that at zero degrees absolute the heat capacity is zero. As the temperature rises, the heat capacity also increases until a certain point is reached at which a phase transition takes place. The phase transitions are shown also on a related p-T diagram in Fig. 4.4 for water. The phase transi-

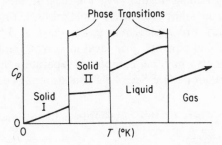

Fig. 4.3. Heat capacity as a function of temperature for a pure substance.

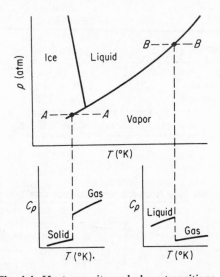

Fig. 4.4. Heat capacity and phase transitions.

tion may take place between two solid states, or between a solid and a liquid state, or between a solid and a gaseous state, or between a liquid and a gaseous state. Figure 4.3 shows first a transition between solid state I and solid state II, then the transition between solid state II and the liquid state, and finally the transition between the liquid and the gaseous state. Note that the heat capacity is a continuous function *only* in the region between the phase transitions; consequently it is not possible to have a heat capacity equation for a substance which will go from absolute zero up to any desired temperature. What the engineer does is to determine experimentally the heat capacity between the temperatures at which the phase transitions occur, fit the data with an equation, and then stop and determine a new heat capacity equation for the next range of temperatures between the succeeding phase transitions.

Experimental evidence indicates that the heat capacity of a substance is not constant with temperature, although at times we may assume that it is constant in order to get approximate results. For the ideal monoatomic gas, of course, the heat capacity at constant pressure is constant even though the temperature varies (see Table 4.2). For typical real gases, see Fig. 4.5; the heat capacities shown are for pure components. For ideal mixtures, the heat capacities of the individual components may be computed separately and each component handled as if it were alone (see Sec. 4.8 for additional details).

TABLE 4.2 HEAT CAPACITIES OF IDEAL GASES

Type Molecule	Approximate Heat Capacity (C_p)	
	High Temperature (Translational, Rotational, and Vibrational Degrees of Freedom)	Room Temperature (Translational and Rotational Degrees of Freedom Only)
Monoatomic	$\frac{5}{2}R$	$\frac{5}{2}R$
Polyatomic, linear	$(3n - \frac{3}{2})R$	$\frac{7}{2}R$
Polyatomic, nonlinear	$(3n - 2)R$	$4R$

n = number of atoms per molecule
R = gas constant defined in Sec. 3.1.

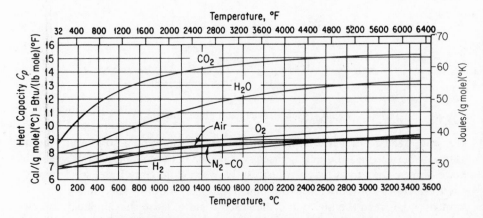

Fig. 4.5. Heat capacity curves for the combustion gases.

Most of the equations for the heat capacities of solids, liquids, and gases are empirical. We usually express the heat capacity at constant pressure C_p as a function of temperature in a power series, with constants a, b, c, etc.; for example,

$$C_p = a + bT$$

or

$$C_p = a + bT + cT^2$$

where the temperature may be expressed in degrees Centigrade, degrees Fahrenheit, degrees Rankine, or degrees Kelvin. If C_p is expressed in the form of

$$C_p = a + bT + cT^{-1/2}$$
$$C_p = a + bT - cT^{-2}$$

or a form such that we are dividing by T, then it is necessary to use degrees Kelvin or degrees Rankine in the heat capacity equations, because if degrees Centigrade or Fahrenheit were to be used, we might be dividing at some point in the temperature range by zero. Since these heat capacity equations are valid only over moderate temperature ranges, it is possible to have equations of different types represent the experimental heat capacity data with almost equal accuracy. The task of fitting heat capacity equations to heat capacity data is greatly simplified these days by the use of digital computers, which can determine the constants of best fit by means of a standard prepared program and at the same time determine how precise the predicted heat capacities are. Heat capacity information can be found in Appendix E. The change of C_p with pressure at high pressures is beyond the scope of our work here. Details can be found in several of the references listed at the end of the chapter.

Specific heat is a term similar to *specific gravity* in that it is a ratio of the heat capacity of one substance to the heat capacity of a reference substance.[4] The common reference substance for solids and liquids is water, which is assigned the heat capacity of 1.0 at about 17°C. Since the heat capacity of water is approximately unity in the cgs and American engineering systems, the numerical values of specific heats and heat capacities are about the same although their units are not. For example, the units of the ratio of the heat capacity of substance A to that of water are

$$\frac{C_{p_A}}{C_{p_{H_2O}}} = \frac{\text{Btu}/(\text{lb}_A)(^\circ\text{F})}{\text{Btu}/(\text{lb}_{H_2O})(^\circ\text{F})}$$

EXAMPLE 4.2 Heat capacity equation

The heat capacity equation for CO_2 gas is

$$C_p = 6.393 + 10.100T \times 10^{-3} - 3.405T^2 \times 10^{-6}$$

---

[4]In some fields of engineering and science, specific heat means heat capacity based on a pound or gram.

with C_p expressed in cal/(g mole)(°K) and T in °K. Convert this equation into a form so that the heat capacity will be expressed over the entire temperature range in

(a) Cal/(g mole)(°C) with T in °C and ΔT in Δ°C.

(b) Btu/(lb mole)(°F) with T in °F and ΔT in Δ°F.

(c) Cal/(g mole)(°K) with T in °F and ΔT in Δ°K.

(d) J/(kg mole)(°K) with T in °K and ΔT in Δ°K.

Solution:

Changing a heat capacity equation from one set of units to another is merely a problem in the conversion of units. To avoid confusion in the conversion, you must remember to distinguish between the temperature symbols that represent temperature and the temperature symbols that represent temperature difference even though the same symbol often is used for each concept. In the conversions below we shall distinguish between the temperature and the temperature difference for clarity.

(a) The heat capacity equation with T in °C and ΔT in Δ°C is

$$C_p \frac{\text{cal}}{(\text{g mole})(\Delta°C)} = 6.393 \frac{\text{cal}}{(\text{g mole})(\Delta°K)} \left| \frac{1\,\Delta°K}{1\,\Delta°C} \right.$$

$$+ 10.1000 \times 10^{-3} \frac{\text{cal}}{(\text{g mole})(\Delta°K)(°K)} \left| \frac{1\,\Delta°K}{1\,\Delta°C} \right| \frac{(T_{°C} + 273)°K}{}$$

$$- 3.405 \times 10^{-6} \frac{\text{cal}}{(\text{g mole})(\Delta°K)(°K)^2} \left| \frac{1\,\Delta°K}{1\,\Delta°C} \right| \frac{(T_{°C} + 273)^2°K^2}{}$$

$$= 6.393 + 10.100 \times 10^{-3} T_{°C} + 2.757 - 3.405 \times 10^{-6} T_{°C}^2$$

$$- 1.860 \times 10^{-3} T_{°C} - 0.254$$

$$= 8.896 + 8.240 \times 10^{-3} T_{°C} - 3.405 \times 10^{-6} T_{°C}^2$$

(b)

$$C_p \frac{\text{Btu}}{(\text{lb mole})(\Delta°F)} = 6.393 \frac{\text{cal}}{(\text{g mole})(\Delta°K)} \left| \frac{454\,\text{g moles}}{1\,\text{lb mole}} \right| \frac{1\,\text{Btu}}{252\,\text{cal}} \left| \frac{1\,\Delta°K}{1.8\,\Delta°F} \right.$$

$$+ 10.100 \times 10^{-3} \frac{\text{cal}}{(\text{g mole})(\Delta°K)(°K)} \left| \frac{454\,\text{g moles}}{1\,\text{lb mole}} \right| \frac{1\,\text{Btu}}{252\,\text{cal}} \left| \frac{1\,\Delta°K}{1.8\,\Delta°F} \right| \frac{\left[273 + \dfrac{T_{°F} - 32}{1.8}\right]°K}{}$$

$$- 3.405 \times 10^{-6} \frac{\text{cal}}{(\text{g mole})(\Delta°K)(°K)^2} \left| \frac{454\,\text{g moles}}{1\,\text{lb mole}} \right| \frac{1\,\text{Btu}}{252\,\text{cal}} \left| \frac{1\,\Delta°K}{1.8\,\Delta°F} \right| \frac{\left[273 + \dfrac{T_{°F} - 32}{1.8}\right]^2°K^2}{}$$

$$= 6.393 + 10.100 \times 10^{-3}[273 + (T_{°F} - 32)/1.8]$$

$$- 3.405 \times 10^{-6}[273 + (T_{°F} - 32)/1.8]^2$$

$$= 6.393 + 2.575 + 5.61 \times 10^{-3} T_{°F} - 0.222$$

$$- 0.964 \times 10^{-3} T_{°F} - 1.05 \times 10^{-6} T_{°F}^2$$

$$= 8.746 + 4.646 \times 10^{-3} T_{°F} - 1.05 \times 10^{-6} T_{°F}^2$$

Note: In Sec. 1.4 we have developed and made use of the expression

$$T_{°K} = 273 + \frac{T_{°F} - 32}{1.8}$$

If the equation for part (a) were available, the solution to part (b) would be simplified by employing it. For example, we use

$$C_p \frac{\text{cal}}{(\text{g mole})(\Delta°C)} = 8.896 + 8.240 \times 10^{-3} T_{°C} - 3.405 \times 10^{-6} T_{°C}$$

and convert it into

$$C_p \frac{\text{Btu}}{(\text{lb mole})(\Delta°\text{F})} = \frac{8.896 \text{ cal}}{(\text{g mole})(\Delta°\text{C})} \left| \frac{454 \text{ g moles}}{\text{lb mole}} \right| \frac{1 \text{ Btu}}{252 \text{ cal}} \left| \frac{1 \Delta°\text{C}}{1.8 \Delta°\text{F}} \right.$$

$$+ 8.240 \times 10^{-3} \frac{\text{cal}}{(\text{g mole})(\Delta°\text{C})(°\text{C})} \left| \frac{454 \text{ g moles}}{1 \text{ lb mole}} \right| \frac{1 \text{ Btu}}{252 \text{ cal}} \left| \frac{1 \Delta°\text{C}}{1.8 \Delta°\text{F}} \right| \frac{[(T°_\text{F} - 32)/1.8]°\text{C}}{}$$

$$- 3.405 \times 10^{-6} \frac{\text{cal}}{(\text{g mole})(\Delta°\text{C})(°\text{C}^2)} \left| \frac{454 \text{ g moles}}{1 \text{ lb mole}} \right| \frac{1 \text{ Btu}}{252 \text{ cal}} \left| \frac{1 \Delta°\text{C}}{1.8 \Delta°\text{F}} \right| \frac{[(T°_\text{F} - 32)/1.8]^2°\text{C}^2}{}$$

$$= 8.746 + 4.646 \times 10^{-3} T°_\text{F} - 1.05 \times 10^{-6} T°^2_\text{F}$$

(c) Since

$$\frac{\text{Btu}}{(\text{lb mole})(\Delta°\text{F})} = \frac{\text{cal}}{(\text{g mole})(\Delta°\text{K})}$$

the equation developed in part (b) gives the heat capacity in the desired units.

(d)

$$C_p \frac{\text{J}}{(\text{kg mole})(\Delta°\text{K})} = 6.93 \frac{\text{cal}}{(\text{g mole})(\Delta°\text{K})} \left| \frac{4.184 \text{ J}}{1 \text{ cal}} \right| \frac{1000 \text{ g}}{1 \text{ kg}}$$

$$+ 10.1000 \times 10^{-3} \frac{\text{cal}}{(\text{g mole})(\Delta°\text{K})(°\text{K})} \left| \frac{4.184 \text{ J}}{1 \text{ cal}} \right| \frac{1000 \text{ g}}{1 \text{ kg}} \left| T°_\text{K} \right.$$

$$- 3.405 \times 10^{-6} \frac{\text{cal}}{(\text{g mole})(°\text{K})^2(\Delta°\text{K})} \left| \frac{4.184 \text{ J}}{1 \text{ cal}} \right| \frac{1000 \text{ g}}{1 \text{ kg}} \left| T°^2_\text{K} \right.$$

$$= 2.90 \times 10^4 + 42.27 T°_\text{K} - 1.425 \times 10^{-2} T°^2_\text{K}$$

EXAMPLE 4.3   Heat capacity of the ideal gas

Show that $C_p = C_v + \hat{R}$ for the ideal monoatomic gas.

Solution:

The heat capacity at constant volume is defined as

$$C_v = \left(\frac{\partial \hat{U}}{\partial T} \right)_{\hat{V}} \tag{a}$$

For any gas,

$$C_p = \left(\frac{\partial \hat{H}}{\partial T} \right)_p = \left[\frac{\partial \hat{U} + \partial (p\hat{V})}{\partial T} \right]_p = \left[\frac{\partial \hat{U} + p\partial \hat{V}}{\partial T} \right]_p$$

$$= \left(\frac{\partial \hat{U}}{\partial T} \right)_p + p \left(\frac{\partial \hat{V}}{\partial T} \right)_p \tag{b}$$

For the *ideal gas*, since $\hat{U}$ is a function of temperature only,

$$\left(\frac{\partial \hat{U}}{\partial T} \right)_p = \left(\frac{\partial \hat{U}}{\partial T} \right)_{\hat{V}} = C_v \tag{c}$$

and from $p\hat{V} = \hat{R}T$ we can calculate

$$\left(\frac{\partial \hat{V}}{\partial T} \right)_p = \frac{\hat{R}}{p} \tag{d}$$

so that

$$C_p = C_v + \hat{R}$$

4.2.1 Estimation of Heat Capacities. We now mention a few ways by which to *estimate* heat capacities of solids, liquids and gases. For the most accurate results, you should employ actual experminental heat capacity data or equations derived from such data in your calculations However, if experimental data are not available, there are a number of equations that you may use which give estimates of values for the heat capacities.

(a) *Solids*. Only very rough approximations of solid heat capacities can be made. *Kopp's rule* (1864) should only be used as a last resort when experimental data cannot be located or new experiments carried out. Kopp's rule states that at room temperature the sum of the heat capacities of the individual elements is approximately equal to the heat capacity of a solid compound. For elements below potassium, numbers have been assigned from experimental data for the heat capacity for each element as shown in Table 4.3. For liquids Kopp's rule can be applied with a modified series of values for the various elements, as shown in Table 4.3

TABLE 4.3 VALUES FOR MODIFIED KOPP'S RULE

Atomic Heat Capacity at 20°C in cal/(g atom)(°C)		
Element	Solids	Liquids
C	1.8	2.8
H	2.3	4.3
B	2.7	4.7
Si	3.8	5.8
O	4.0	6.0
F	5.0	7.0
P or S	5.4	7.4
All others	6.2	8.0

EXAMPLE 4.4   Use of Kopp's rule

Determine the heat capacity at room temperature of $Na_2SO_4 \cdot 10H_2O$.

Basis: 1 g mole $Na_2SO_4 \cdot 10H_2O$

$$
\begin{array}{lll}
\text{Na} & 2 \times 6.2 = & 12.4 \text{ cal/(g atom)(°C)} \\
\text{S} & 1 \times 5.4 = & 5.4 \\
\text{O} & 14 \times 4.0 = & 56.0 \\
\text{H} & 20 \times 2.3 = & \underline{46.0} \\
& & 119.8 \text{ cal/(g mole)(°C)}
\end{array}
$$

The experimental value is about 141 cal/(g mole)(°C), so you can see that Kopp's rule does not yield very accurate estimates.

(b) *Liquids*

(1) *Aqueous solutions.* For the special but very important case of aqueous solutions, a rough rule in the absence of experimental data is to use the heat capacity of the water only. For example, a 21.6 percent solution of NaCl is assumed to have a heat capacity of 0.784 cal/(g)(°C); the experimental value at 25°C is 0.806 cal/(g)(°C).

(2) *Hydrocarbons.* An equation for the heat capacity of liquid hydrocarbons and petroleum products recommended by Fallon and Watson[5] is

$$C_p = [(0.355 + 0.128 \times 10^{-2} \, °API) + (0.503 + 0.117 \times 10^{-2} \, °API)$$
$$\times \, 10^{-3}t][0.05K + 0.41]$$

where °API $= \dfrac{141.5}{\text{sp gr } (60°F/60°F)} - 131.5$ and is a measure of specific gravity.

$t = °F$

$K =$ The Universal Oil Products characterization factor which has been related to six easily applied laboratory tests. This factor is not a fundamental characteristic but is easy to determine experimentally. Values of K range from 10.1 to 13.0 and are discussed in Appendix K.

$C_p = $ Btu/(lb)(°F)

(3) *Organic liquids.* A simple and reasonably accurate relation between C_p and molecular weight is

$$C_p = kM^a$$

where M is the molecular weight and k and a are constants. Pachaiyappan et al.[6] give a number of values of the set (k, a). For example,

compounds	k	a
Alcohols	0.85	−0.1
Acids	0.91	−0.152
Ketones	0.587	−0.0135
Esters	0.60	−0.0573
Hydrocarbons, aliphatic	0.873	−0.113

Gold[7] reviews a number of estimation methods for liquids and indicates their precision.

(c) *Gases and Vapors*

(1) *Petroleum vapors.* The heat capacity of petroleum vapors can be estimated from[8]

$$C_p = \frac{(4.0 - s)(T + 670)}{6450}$$

[5] J. F. Fallon and K. M. Watson, *Natl. Petrol. News, Tech. Sec.* (June 7, 1944).
[6] V. Pachaiyappan, S. H. Ibrahim, and N. R. Kuloor, *Chem. Eng.*, p. 241 (Oct 9, 1967).
[7] P. I., Gold, *Chem. Eng.*, p. 130 (April 7, 1969).
[8] Bahlke and Kay, *Ind. Eng. Chem.*, v. 21, p. 942 (1929).

where C_p is in Btu/(lb)(°F), T is in °F, and s is the sp gr at 60°F/60°F, with air as the reference gas.

(2) Kothari and Doraiswamy[9] recommend plotting

$$C_p = a + b \log_{10} T_r$$

Given two values of C_p at known temperatures, values of C_p can be predicted at other temperatures with good precision. The most accurate methods of estimating heat capacities of vapors are those of Dobratz[10] based on spectroscopic data and generalized correlations based on reduced properties[11].

Additional methods of estimating solid and liquid heat capacities may be found in Reid and Sherwood,[11] who compare various techniques we do not have the space to discuss and make recommendations as to their use.

4.3 Calculation of enthalpy changes without change of phase

Now that we have examined sources for heat capacity values, we can turn to the details of the calculation of enthalpy changes. If we omit for the moment consideration of phase changes and examine solely the problem of how to calculate enthalpy changes from heat capacity data, we can see that in general if we use Eq. (4.7), $\Delta \hat{H}$ is the area under the curve in Fig. 4.6:

$$\int_{H_1}^{H_2} d\hat{H} = \Delta \hat{H} = \int_{T_1}^{T_2} C_p \, dT$$

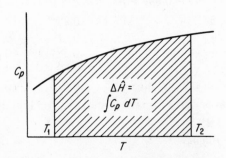

Fig. 4.6. Calculation of enthalpy change.

[9]M. S. Kothari and L. K. Doraiswamy, *Hydrocarbon Processing and Petroleum Refiner*, v. 43, no. 3, p. 133 (1964).

[10]C. J. Dobratz, *Ind. Eng. Chem.*, v. 33, p. 759 (1941).

[11]R. C. Reid and T. K. Sherwood, *The Properties of Gases and Liquids*, 2nd ed., McGraw-Hill, New York, 1966.

(The technique of graphical integration is illustrated in Example 6.6.) If our heat capacity is expressed in the form $C_p = a + bT + cT^2$, then

$$\Delta \hat{H} = \int_{T_1}^{T_2} (a + bT + cT^2)dT = a(T_2 - T_1) + \frac{b}{2}(T_2^2 - T_1^2)$$
$$+ \frac{c}{3}(T_2^3 - T_1^3) \tag{4.8}$$

If a different functional form of the heat capacity is available, the integration can still be handled readily.

It is also possible to define a mean heat capacity, which, if known, provides a quick and convenient way of calculating enthalpy changes. The *mean heat capacity*, C_{p_m}, is defined as the enthalpy change divided by the temperature difference for this change, or

$$C_{p_m} = \frac{\hat{H}_2 - \hat{H}_1}{T_2 - T_1} \tag{4.9}$$

Then it is possible to compute an enthalpy change, if C_{p_m} is available, as

$$\Delta \hat{H} = C_{p_m} \Delta T = C_{p_m}(T_2 - T_1) \tag{4.10}$$

where C_{p_m} is between T_1 and T_2.

TABLE 4.4a MEAN HEAT CAPACITIES OF COMBUSTION GASES
[cal/(g mole)(°C)]
(To convert to J/(kg mole)(°K), multiply by 4184.)
Reference Temperature: 0°C; Pressure: 1 atm

°C	N_2	O_2	Air	H_2	CO	CO_2	H_2O
0	6.959	6.989	6.946	6.838	6.960	8.595	8.001
18	6.960	6.998	6.949	6.858	6.961	8.706	8.009
25	6.960	7.002	6.949	6.864	6.962	8.716	8.012
100	6.965	7.057	6.965	6.926	6.973	9.122	8.061
200	6.985	7.154	7.001	6.955	7.050	9.590	8.150
300	7.023	7.275	7.054	6.967	7.057	10.003	8.256
400	7.075	7.380	7.118	6.983	7.120	10.360	8.377
500	7.138	7.489	7.190	6.998	7.196	10.680	8.507
600	7.207	7.591	7.266	7.015	7.273	10.965	8.644
700	7.277	7.684	7.340	7.036	7.351	11.221	8.785
800	7.350	7.768	7.414	7.062	7.428	11.451	8.928
900	7.420	7.845	7.485	7.093	7.501	11.68	9.070
1000	7.482	7.916	7.549	7.128	7.570	11.85	9.210
1100	7.551	7.980	7.616	7.165	7.635	12.02	9.348
1200	7.610	8.039	7.674	7.205	7.688	12.17	9.482
1300	7.665	8.094	7.729	7.227	7.752	12.32	9.613
1400	7.718	8.146	7.781	7.260	7.805	12.44	9.740
1500	7.769	8.192	7.830	7.296	7.855	12.56	9.86

SOURCE: Page 30 of Reference 12 in Table 4.6.

If the heat capacity can be expressed as a power series, $C_p = a + bT + cT^2$, then C_{p_m} is

$$C_{p_m} = \frac{\int_{T_1}^{T_2} C_p \, dT}{\int_{T_1}^{T_2} dT} = \frac{\int_{T_1}^{T_2} (a + bT + cT^2) \, dT}{T_2 - T_1} \tag{4.11}$$

$$= \frac{a(T_2 - T_1) + (b/2)(T_2^2 - T_1^2) + (c/3)(T_2^3 - T_1^3)}{T_2 - T_1}$$

By choosing 0°F or 0°C as the reference temperature for C_{p_m}, you can simplify the expression for C_{p_m}. Tables 4.4a and 4.4b lists mean heat capacities for combustion gases based on the same reference point (0°C) for two different temperature scales.

TABLE 4.4b MEAN HEAT CAPACITIES OF COMBUSTION GASES
[Btu/(lb mole)(°F)]
Reference Temperature: 32°F; Pressure: 1 atm

°F	N_2	O_2	Air	H_2	CO	CO_2	H_2O
32	·6.959	6.989	6.946	6.838	6.960	8.595	8.001
60	6.960	6.996	6.948	6.855	6.961·	8.682	8.008
77	6.960	7.002	6.949	6.864	6.962	8.715	8.012
100	6.961	7.010	6.952	6.876	6.964	8.793	8.019
200	6.964	7.052	6.963	6.921	6.972	9.091	8.055
300	6.970	7.102	6.978	6.936	6.987	9.362	8.101
400	6.984	7.159	7.001	6.942	7.007	9.612	8.154
500	7.002	7.220	7.028	6.961	7.033	9.844	8.210
600	7.026	7.283	7.060	6.964	7.065	10.060	8.274
700	7.055	7.347	7.096	6.978	7.101	10.262	8.341
800	7.087	7.409	7.134	6.981	7.140	10.450	8.411
900	7.122	7.470	7.174	6.984	7.182	10.626	8.484
1000	7.158	7.529	7.214	6.989	7.224	10.792	8.558
1100	7.197	7.584	7.256	6.993	7.268	10.948	8.634
1200	7.236	7.637	7.298	7.004	7.312	11.094	8.712
1300	7.277	7.688	7.341	7.013	7.355	11.232	8.790
1400	7.317	7.736	7.382	7.032	7.398	11.362	8.870
1500	7.356	7.781	7.422	7.054	7.439	11.484	8.950
1600	7.395	7.824	7.461	7.061	7.480	11.60	9.029
1700	7.433	7.865	7.500	7.073	7.519	11.71	9.107
1800	7.471	7.904	7.537	7.081	7.558	11.81	9.185
1900	7.507	7.941	7.573	7.093	7.595	11.91	9.263
2000	7.542	7.976	7.608	7.114	7.631	12.01	9.339

SOURCE: Page 31 of Reference 12 in Table 4.6.

EXAMPLE 4.5 Calculation of $\Delta \hat{H}$ using mean heat capacity

Calculate the enthalpy change for 1 lb mole of nitrogen (N_2) which is heated at constant pressure (1 atm) from 60°F to 2000°F.

Solution:

The enthalpy data can be taken from Table 4.4b; reference conditions, 32°F:

$$C_{pm} = 7.542 \text{ Btu/(lb mole)(°F)} \quad \text{at 2000°F}$$

$$C_{pm} = 6.960 \text{ Btu/(lb mole)(°F)} \quad \text{at 60°F}$$

Basis: 1 lb mole of N_2

$$\Delta \hat{H}_{2000-60} = \Delta \hat{H}_{2000} - \Delta \hat{H}_{60}$$

$$= \frac{7.542 \text{ Btu}}{\text{(lb mole)(°F)}} \bigg| (2000 - 32°F) - \frac{6.960 \text{ Btu}}{\text{(lb mole)(°F)}} \bigg| (60 - 32°F)$$

$$= 14,740 \text{ Btu/lb mole}$$

As shown above, if the reference conditions are not zero (0°K, 0°F etc.), the $\Delta \hat{H}_{2000-60}$ is

$$C_{pm2000}(T_{2000} - T_{ref}) - C_{pm60}(T_{60} - T_{ref})$$

and *not* $C_{pm2000} T_{2000} - C_{pm60} T_{60}$.

EXAMPLE 4.6 Calculation of $\Delta \hat{H}$ using heat capacity equations

The conversion of solid wastes to innocuous gases can be accomplished in incinerators in an environmentally acceptable fashion. However, the hot exhaust gases must be cooled or diluted with air. An economic feasibility study indicates that solid municipal waste can be burned to a gas of the following composition (on a dry basis):

CO_2	9.2
CO	1.5
O_2	7.3
N_2	82.0
	100.0

What is the enthalpy difference for this gas between the bottom and the top of the stack if the temperature at the bottom of the stack is 550°F and the temperature at the top is 200°F? Ignore the water vapor in the gas. Because these are ideal gases, you can neglect any energy effects resulting from the mixing of the gaseous components.

Solution:

Heat capacity equations from Table E.1 in Appendix E (T in °F; C_p = Btu/(lb mole)(°F)) are

N_2: $C_p = 6.895 + 0.7624 \times 10^{-3} T - 0.7009 \times 10^{-7} T^2$

O_2: $C_p = 7.104 + 0.7851 \times 10^{-3} T - 0.5528 \times 10^{-7} T^2$

CO_2: $C_p = 8.448 + 5.757 \times 10^{-3} T - 21.59 \times 10^{-7} T^2 + 3.049 \times 10^{-10} T^3$

CO: $C_p = 6.865 + 0.8024 \times 10^{-3} T - 0.7367 \times 10^{-7} T^2$

Basis: 1.00 lb mole gas

By multiplying these equations by the respective mole fraction of each component, and then adding them together, you can save time in the integration.

N_2: $0.82(6.895 + 0.7624 \times 10^{-3} T - 0.7009 \times 10^{-7} T^2)$

O_2: $0.073(7.104 + 0.7851 \times 10^{-3} T - 0.5528 \times 10^{-7} T^2)$

CO_2: $\quad 0.092(8.448 + 5.757 \times 10^{-3}T - 21.59 \times 10^{-7}T^2 + 3.059 \times 10^{-10}T^3)$

CO: $\quad 0.015(6.865 + 0.8024 \times 10^{-3}T - 0.7367 \times 10^{-7}T^2)$

$C_{p_{net}} = 7.049 + 1.2243 \times 10^{-3}T - 2.6164 \times 10^{-7}T^2 + 0.2815 \times 10^{-10}T^2$

$$\Delta \hat{H} = \int_{550}^{200} C_p \, dT = \int_{550}^{200} (7.049 + 1.2243 \times 10^{-3}T - 2.6164 \times 10^{-7}T^2$$
$$+ 0.2815 \times 10^{-10}T^3) \, dT$$

$$= 7.049[(200) - (550)] + \frac{1.2243 \times 10^{-3}}{2}[(200)^2 - (550)^2]$$

$$- \frac{2.6164 \times 10^{-7}}{3}[(200)^3 - (550)^3]$$

$$+ \frac{0.2815 \times 10^{-10}}{4}[(200)^4 - (550)^4]$$

$$= -2465 - 160.6 + 13.8 - 0.633$$

$$= -2612 \text{ Btu/lb mole gas}$$

We shall now briefly discuss the use of enthalpy tables and charts to calculate enthalpy changes. The simplest and quickest method of computing enthalpy

TABLE 4.5a ENTHALPIES OF COMBUSTION GASES
(Btu/lb mole)
Pressure: 1 atm

°R	N_2	O_2	Air	H_2	CO	CO_2	H_2O
492	0.0	0.0	0.0	0.0	0.0	0.0	0.0
500	55.67	55.93	55.57	57.74	55.68	68.95	64.02
520	194.9	195.9	194.6	191.9	194.9	243.1	224.2
537	313.2	315.1	312.7	308.9	313.3	392.2	360.5
600	751.9	758.8	751.2	744.4	752.4	963	867.5
700	1450	1471	1450	1433	1451	1914	1679
800	2150	2194	2153	2122	2154	2915	2501
900	2852	2931	2861	2825	2863	3961	3336
1000	3565	3680	3579	3511	3580	5046	4184
1100	4285	443	4306	4210	4304	6167	5047
1200	5005	5219	5035	4917	5038	7320	5925
1300	5741	6007	5780	5630	5783	.8502	6819
1400	6495	6804	6540	6369	6536	9710	7730
1500	7231	6712	7289	7069	7299	10942	8657
1600	8004	8427	8068	7789	8072	12200	9602
1700	8774	9251	8847	8499	8853	13470	10562
1800	9539	10081	9623	9219	9643	14760	11540
1900	10335	10918	10425	9942	10440	16070	12530
2000	11127	11760	11224	10689	11243	17390	13550
2100	11927	12610	12030	11615	12050	18730	14570
2200	12730	13460	12840	12160	12870	20070	15610
2300	13540	14320	13660	12890	13690	21430	16660
2400	14350	15180	14480	13650	14520	22800	17730
2500	15170	16040	15300	14400	15350	24180	18810

SOURCE: Page 30 of Reference 12 in Table 4.6.

changes is to use tabulated enthalpy data available in the literature or in reference books. Thus, rather than using integrated heat capacity equations or mean heat capacity data, you should first look to see if you can find enthalpy data listed in tables as a function of temperature. Tables 4.5a and 4.5b list typical enthalpy data for the combustion gases. Some sources of enthalpy data are listed in Table 4.6. The most common source of enthalpy data for water is the steam tables; see Appendix C.[12] The reference by Kobe and associates (Table 4.6, 12) lists heat capacity equation, enthalpy values, and many other thermodynamic functions for over 100 compounds of commercial importance. Some of this information can be found in the Appendix. Computer tapes can be purchased providing information on the physical properties of large numbers of compounds as described in Reference 13 in Table 4.6, or from the Institution of Chemical Engineers, London.

Remembering that enthalpy values are all relative to some reference state, you can make enthalpy difference calculations merely by subtracting the final enthalpy from the initial enthalpy for any two sets of conditions. Of course, if

TABLE 4.5b ENTHALPIES OF COMBUSTION GASES
(cal./g mole)

°K	N_2	O_2	Air	H_2	CO	CO_2	H_2O
273	0.0	0.0	0.0	0.0	0.0	0.0	0.0
291	125.3	126.0	125.1	123.4	125.5	156.7	144.2
298	174.0	175.0	173.7	171.6	174.1	217.9	200.3
300	187.9	189.0	187.6	182.5	188.0	235.7	216.4
400	883.2	896.9	883.6	873.6	884.2	1172	1024
500	1588	1628	1592	1575	1590	2200	1853
600	2301	2383	2312	2275	2310	3300	2707
700	3024	3161	3044	2978	3047	4459	3589
800	3766	3959	3795	3684	3800	5667	4499
900	4532	4773	4569	4394	4571	6916	5440
1000	5299	5601	5346	5112	5357	8200	6411
1100	6088	6439	6142	5838	6157	9513	7412
1200	6888	7288	6950	6575	6968	10852	8440
1300	7700	8145	7768	7320	7790	12211	9494
1300	8518	9009	8593	8076	8621	13590	10573
1500	9356	9880	9434	8842	9459	14980	11675
1750	11458	12083	11550	10821	11582	18510	14520
2000	13600	14320	13700	12830	13740	22100	17480
2250	15770	16600	15880	14900	15910	25750	20520
2500	17940	18910	18080	17020	18110	29440	23630
2750	20140	21250	20300	19190	20320	33150	26790
3000	22350	23620	22530	21380	22530	36890	30000
3500	26800	28450	27040	25820	27000	44430	36520
4000	31280	33380	31590	30480	31500	52050	43120

[12]See R. W. Haywood, *Thermodynamic Tables in SI (Metric) Units*, Cambridge University Press, London, 1968, for steam tables in metric units.

the enthalpy data are not available, you will have to rely on calculations involving heat capacities or mean heat capacities.

If a pure material exists in solid, liquid, and gaseous form, a chart or diagram can be prepared showing its properties if sufficient reliable experimental information is available. Examples of such charts are shown in Fig. 4.7 and in Appendix J; sources of such charts are listed in Table 4.7. Many forms of charts

TABLE 4.6 SOURCES OF ENTHALPY DATA

1. American Petroleum Institute, Division of Refining, *Technical Data Book—Petroleum Refining*, 2nd ed., API, Washington D.C., 1970.
2. American Petroleum Institute, Research Project 44, *Selected Values of Physical and Thermodynamic Properties of Hydrocarbons and Related Compounds*, API, Washington, D.C., 1953.
3. Chermin, H. A. G., *Hydrocarbon Processing & Petroleum Refining*, Parts 26–32, 1961. (Continuation of Kobe compendium for gases; see item 12 below.)
4. Egloff, G., *Physical Constants of Hydrocarbons*, 5 vols., Van Nostrand Reinhold, New York, 1939–1953.
5. Hamblin, F. D., *Abridged Thermodynamic and Thermochemical Tables* (*S.I. Units*), Pergamon Press, Elmsford, N.Y. (paperback edition available), 1972.
6. Hilsenrath, J., and others, "Tables of Thermal Properties of Gases," *National Bureau of Standards* (*U.S.*), *Circ.* 564, 1955. (Additional single sheets issued as NBS-NACA sponsored tables.)
7. International Atomic Energy Agency, *Thermodynamics of Nuclear Materials*, Vienna, Austria, 1962.
8. *Journal of Physical and Chemical Reference Data*, American Chemical Society, 1972 and following.
9. Keenan, J. H., and F. G. Keyes, *Thermodynamic Properties of Steam*, Wiley, New York, 1936.
10. Keenan, J. H., and J. Kay, *Gas Tables*, Wiley, New York, 1948.
11. Kelley, K. K., various *U.S. Bureau of Mines Bulletins* on the Thermodynamic Properties of Substances, particularly minerals; 1935–1962.
12. Kobe, K. A., et al. "Thermochemistry of Petrochemicals," Reprint no. 44 from the *Petroleum Refiner*. A collection of a series of 24 articles covering 105 different substances that appeared in the *Petroleum Refiner*.
13. Meadows, E. L., "Estimating Physical Properties," *Chem. Eng. Progr.*, v. 61, p. 93 (1965). (A.I.Ch.E. Machine Computation Committee report on the preparation of thermodynamic tables on tapes.)
14. Perry, J. H., *Chemical Engineers' Handbook*, 4th ed., McGraw-Hill, New York, 1964.
15. Ross, L. W., *Chem. Eng.*, p. 205 (Oct. 10, 1966).
16. Rossini, F. K., and others, "Tables of Selected Values of Chemical Thermodynamic Properties," *National Bureau of Standards* (*U.S.*), *Circ.* 500, 1952. (Revisions are being issued periodically under the Technical Note 270 series by other authors).
17. Sage, B. H., and W. N. Lacey, *Some Properties of the Lighter Hydrocarbons, Hydrogen Sulfide, and Carbon Dioxide*, American Petroleum Institute, New York, 1955.
18. Stull, D. R., and H. Prophet, *JANAF Thermochemical Tables*, 2nd ed., Government Printing Office No. 0303–0872, C13, 48: 37 (AD-732 043), 1971. (Data for over 1000 compounds.)
19. Zeise, H., *Thermodynamik*, Band III/I Taballen, S. Hirzel Verlag, Leipzig, 1954. (Tables of heat capacities, enthalpies, entropies, free energies, and equilibrium constants.)

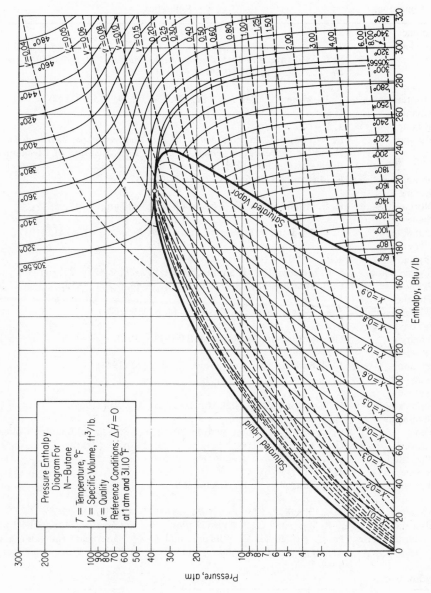

Fig. 4.7. Pressure enthalpy diagram for *n*-butane.

279

TABLE 4.7 THERMODYNAMIC CHARTS SHOWING ENTHALPY DATA FOR PURE COMPOUNDS
[For mixtures, see V. F. Lesavage et al. *Ind. Eng. Chem.*, v. 59, no. 11, p. 35 (1967).]

Compound	Reference
Acetone	2
Acetylene	1
Air	V. C. Williams, *Am. Inst. Chem. Engrs. Trans.*, v. 39, p. 93 (1943); *A.I. Ch. E. J.*, v. 1, p. 302 (1955).
Benzene	1
1, 3-Butadiene	C. H. Meyers, *J. Res. Natl. Bur. Std., A.*, v. 39, p. 507 (1947).
i-Butane	1, 3
n-Butane	1, 3, 4
n-Butanol	L. W. Shemilt, in *Proceedings of the Conference on Thermodynamic Transport Properties of Fluids*, London, 1957, Institute of Mechanical Engineers London (1958).
Butanol, *tert*.	F. Maslan, *A.I.Ch.E. J.*, v. 7, p. 172 (1961).
n-Butene	1
Chlorine	R. E. Hulme and A. B. Tilman, *Chem. Eng.* (Jan. 1949).
Ethane	1, 3, 4
Ethanol	R. C. Reid and J. M. Smith, *Chem. Eng. Progr.*, v. 47, p. 415 (1951).
Ethyl ether	2
Ethylene	1, 3
Ethylene oxide	J. E. Mock and J. M. Smith, *Ind. Eng. Chem.*, v. 42, p. 2125 (1950).
n-Heptane	E. B. Stuart, et al., *Chem. Eng. Progr.*, v. 46, p. 311 (1950).
n-Hexane	1
Hydrogen sulfide	J. R. West, *Chem. Eng. Progr.*, v. 44, p. 287 (1948).
Isopropyl ether	2
Mercury	General Electric Co. Report GET-1879A, 1949.
Methane	1, 3, 4
Methanol	J. M. Smith, *Chem. Eng. Progr.*, v. 44, p. 52 (1948).
Methyl ethyl ketone	2
Monomethyl hydrazine	F. Bizjak, and D. F. Stai, *A.I.A.A. J.*, v. 2, p. 954 (1964).
Neon	Cryogenic Data Center, National Bureau of Standards, Boulder, Colo.
Nitrogen	G. S. Lin, *Chem. Eng. Progr.*, v. 59, no. 11, p. 69 (1963).
n-Pentane	1, 3, 4
Propane	1, 3, 4
n-Propanol	L. W. Shemilt, (see ref. for butanol).
Propylene	1, 3
Refrigerant 245	R. L. Shank, *J. Chem. Eng. Data*, v. 12, p. 474 (1967).
Sulfur dioxide	J. R. West, and G. P. Giusti, *J. Phys. Colloid Chem.*, v. 54, p. 601 (1950).
Combustion gases	H. C. Hottel, G. C. Williams, and C. N. Satterfield, *Thermodynamic Charts for Combustion Processes* (I) Text, (II) Charts, Wiley, New York, 1949.
Hydrocarbons	3

1. L. N. Cajar, et al., *Thermodynamic Properties and Reduced Correlations for Gases*, Gulf Publishing Co., Houston, Texas, 1967. (Series of articles which appeared in the magazine *Hydrocarbon Processing* from 1962 to 1965.)
2. P. T. Eubank and J. M. Smith, *J. Chem. Eng. Data*, v. 7, p. 75 (1962).
3. W. C. Edmister, *Applied Hydrocarbon Thermodynamics*, Gulf Publishing Co., Houston, Texas, 1961.
4. K. E. Starling et al., *Hydrocarbon Processing*, 1971 and following; to be published as a book. Charts available from Gulf Publishing Co., Houston, Texas, separately.

are available. Some types that may occur to you would have as coordinate axes

$$p \text{ vs. } \hat{H}$$
$$p \text{ vs. } \hat{V}$$
$$p \text{ vs. } T$$
$$\hat{H} \text{ vs. } \hat{S}^{[13]}$$

Since a chart has only two dimensions, the other parameters of interest have to be plotted as lines of constant value across the face of the chart. Recall, for example, that in the $p\hat{V}$ diagram for CO_2, Figs. 3.2 and 3.24, lines of constant temperature were shown as parameters. Similarly, on a chart with pressure and enthalpy as the axes, lines of constant volume and/or temperature might be drawn.

You should have noted for pure gases in the illustrated problems that we had to specify two physical properties such as temperature and pressure to definitely fix the *state* of the gas. Since we know from our previous discussion in Chap. 3 that specifying two properties for a pure gas will ensure that all the other properties will have definite values, any two properties can be chosen at will. The ones usually selected are the ones easiest to measure (such as temperature and pressure); then the others (specific volume, density, specific enthalpy, specific internal energy, etc.) are fixed. Since a particular state for a gas can be defined by any two independent properties, a two-dimensional thermodynamic chart can be seen to be a handy way to present combinations of physical properties.

Although diagrams and charts are helpful in that they portray the relations among the phases and give approximate values for the physical properties of a substance, to get more precise data (unless the chart is quite large), usually you will find it best to see if tables are available which list the physical properties you want.

Enthalpies and other thermodynamic properties can be *estimated* by generalized methods based on the theory of corresponding states or additive bond contributions.[14,15,16]

EXAMPLE 4.7   **Calculation of enthalpy change using tabulated enthalpy values**

Recalculate Example 4.5 using data from enthalpy tables.

Solution:
For the same change as in Example 4.5, the data from Table 4.5a are

$$
\begin{aligned}
\text{at } 2000°F &\circ 2460°R: \quad \Delta\hat{H} = 14,880 \text{ Btu/lb mole} \\
\text{at } 60°F &\circ 520°R: \quad \Delta\hat{H} = 194.9 \text{ Btu/lb mole}
\end{aligned}
\Biggr\} \text{ ref. temp.} = 32°F
$$

Basis: 1 lb mole N_2

$$\Delta\hat{H} = 14,880 - 194.9 = 14,865 \text{ Btu/lb mole}$$

---

[13]S is *entropy*—this diagram is called a *Mollier* diagram.

[14]Lyderson, Greenkorn, and Hougen, *op. cit.*

[15]E. J. Janz, *Estimation of Thermodynamic Properties of Organic Compounds*, Academic Press, New York, 1967.

[16]E. K. Landis, M. T. Cannon, and L. N. Canjar, *Hydrocarbon Processing*, v. 44, p. 154 (1965).

EXAMPLE 4.8   Use of the steam tables

What is the enthalpy change in British thermal units when 1 gal of water is heated from 60° to 1150°F and 240 psig?

Solution:

$$\text{Basis: 1 lb H}_2\text{O at 60°F}$$

From steam tables (ref. temp. = 32°F),

$$\hat{H} = 28.07 \text{ Btu/lb} \qquad \text{at 60°F}$$

$$\hat{H} = 1604.5 \text{ Btu/lb} \qquad \text{at 1150°F and 240 psig (254.7 psia)}$$

$$\Delta\hat{H} = (1604.5 - 28.07) = 1576.4 \text{ Btu/lb}$$

$$\Delta H = 1576(8.345) = 13,150 \text{ Btu/gal}$$

Note: The enthalpy values which have been used for liquid water as taken from the steam tables are for the saturated liquid under its own vapor pressure. Since the enthalpy of liquid water changes negligibly with pressure, no loss of accuracy is encountered for engineering purposes if the initial pressure on the water is not stated.

4.4   Enthalpy changes for transitions

In making enthalpy calculations, we noted in Fig. 4.3 that the heat capacity data are discontinuous at the points of a phase transition. The name usually given to the enthalpy changes for these phase transitions is *latent heat* changes, latent meaning "hidden" in the sense that the substance (for example, water) can absorb a large amount of heat without any noticeable increase in temperature. Unfortunately, the word *heat* is still associated with these enthalpy changes for historical reasons, although they have nothing directly to do with heat as defined in Sec. 4.1. Ice at 0°C can absorb energy amounting to 80 cal/g without undergoing a temperature rise or a pressure change. The enthalpy change from a solid to a liquid is called the *heat of fusion*, and the enthalpy change for the phase transition between solid and liquid water is thus 80 cal/g. The *heat of vaporization* is the enthalpy change for the phase transition between liquid and vapor, and we also have the *heat of sublimation*, which is the enthalpy change for the transition directly from solid to vapor. Dry ice at room temperature and pressure sublimes. The $\Delta\hat{H}$ for the phase change from gas to liquid is called the *heat of condensation*. You can find experimental values of latent heats in the references in Table 4.6, and a brief tabulation is shown in Appendix D. The symbols used for latent heat changes vary, but you usually find one or more of the following employed: $\Delta\hat{H}, L, \lambda, \Lambda$.

To calculate the enthalpy change for the vaporization of water using steam table data, the liquid or gaseous water should be heated or cooled to the dew point and vaporized at the dew point. Keep in mind that the enthalpy changes for vaporization given in the steam tables are for water under its vapor pressure at the indicated temperature. *National Bureau of Standards Circular* 500 provides the data at 25°C that we can use to compare the heat of vaporization of water at its vapor pressure and at 1 atm:

°C	mm Hg	kcal/g mole	Btu/lb mole
25	23.75	10.514	18,925
25	760.	10.520	18,936

You can see that the effect of pressure is quite small and may be neglected for ordinary engineering calculations.

In the absence of experimental values for the latent heats of transition, the following approximate methods will provide a rough *estimate* of the *molar latent heats*. Reid and Sherwood, *The Properties of Gases and Liquids*, give more sophisticated methods.[17]

4.4.1 Heat of Fusion. The heat of fusion for many elements and compounds can be expressed as

$$\frac{\Delta \hat{H}_f}{T_f} = \text{constant} = \begin{cases} 2\text{--}3 & \text{for elements} \\ 5\text{--}7 & \text{for inorganic compounds} \\ 9\text{--}11 & \text{for organic compounds} \end{cases} \qquad (4.12)$$

where $\Delta \hat{H}_f$ = molar heat of fusion in cal/g mole
     T_f = melting point in °K

4.4.2 Heat of Vaporization. (a) *Kistyakowsky's equation.* Kistyakowsky[18] developed Eq. (4.13) for nonpolar liquids, and it provides quite accurate values of the molar heat of vaporization for these liquids:

$$\frac{\Delta \hat{H}_{v_b}}{T_b} = 8.75 + 4.517 \log_{10} T_b \qquad (4.13)$$

where $\Delta \hat{H}_{v_b}$ = molar heat of vaporization at T_b in cal/g mole
     T_b = normal boiling point in °K

(b) *Clausius-Clapeyron Equation.* The Clapeyron equation itself is an exact thermodynamic relationship between the slope of the vapor-pressure curve and the molar heat of vaporization and the other variables listed below:

$$\frac{dp^*}{dT} = \frac{\Delta \hat{H}_v}{T(\hat{V}_g - \hat{V}_l)} \qquad (4.14)$$

where p^* = the vapor pressure
     T = absolute temperature
    $\Delta \hat{H}_v$ = molar heat of vaporization at T
    $\hat{V}_i$ = molar volume of gas or liquid as indicated by the subscript g or l

[17]Also see S. H. Fishtine, *Hydrocarbon Processing*, v. 45, no. 4, p. 173 (1965).
[18]W. Kistyakowsky, *Z. Physik u. Chem.*, v. 107, p. 65 (1923).

Any consistent set of units may be used. If we assume that
  (1) $\hat{V}_l$ is negligible in comparison with $\hat{V}_g$,
  (2) The ideal gas law is applicable for the vapor:

$$\hat{V}_g = RT/p^*$$

$$\frac{dp^*}{p^*} = \frac{\Delta\hat{H}_v \, dT}{RT^2}$$

Rearranging the form,

$$\frac{d \ln p^*}{d(1/T)} = 2.303\frac{d \log_{10} p^*}{d(1/T)} = -\frac{\Delta\hat{H}_v}{R} \tag{4.15}$$

You can plot the $\log_{10} p^*$ vs. $1/T$ and obtain the slope $-(\Delta\hat{H}_v/2.303R)$.

  (3) $\Delta\hat{H}_v$ is constant over the temperature range of interest, integration of Eq. (4.15) yields an indefinite integral

$$\log_{10} p^* = -\frac{\Delta\hat{H}_v}{2.3RT} + B \tag{4.16}$$

or a definite one

$$\log_{10} \frac{p_1^*}{p_2^*} = \frac{\Delta\hat{H}_v}{2.3R}\left(\frac{1}{T_2} - \frac{1}{T_1}\right) \tag{4.17}$$

Either of these equations can be used graphically or analytically to obtain $\Delta\hat{H}_v$ for a short temperature interval.

EXAMPLE 4.9   Heat of vaporization from the Clausius-Clapeyron equation

Estimate the heat of vaporization of isobutyric acid at 200°C.

Solution:

The vapor-pressure data for isobutyric acid (from Perry) are

pressure (mm)	temp. (°C)	pressure (atm)	temp. (°C)
100	98.0	1	154.5
200	115.8	2	179.8
400	134.5	5	217.0
760	154.5	10	250.0

Basis: 1 g mole isobutyric acid

Since $\Delta\hat{H}_v$ remains essentially constant for short temperature intervals, the heat of vaporization can be estimated from Eq. (4.17) and the vapor-pressure data at 179.8°C and 217.0°C.

$$179.8°C \doteqdot 452.8°K; \qquad 217°C \doteqdot 490°K$$

$$\log_{10} \frac{2}{5} = \frac{\Delta\hat{H}_v}{(2.303)(1.987)}\left[\frac{1}{490} - \frac{1}{452.8}\right]$$

$$\Delta\hat{H}_v = 10,700 \text{ cal/g mole} = 19,500 \text{ Btu/lb mole}$$

A reduced form of the Clapeyron equation, i.e., an equation in terms of the reduced pressure and temperature which gives good results is described in several of the references at the end of this chapter.

(c) *Reference substance plots.* Several methods have been developed to estimate the molal heat of vaporization of a liquid at any temperature (not just at its boiling point) by comparing the $\Delta \hat{H}_v$ for the unknown liquid with that of a known liquid such as water.

(1) *Duhring plot.* The temperature of the wanted compound A is plotted against the temperature of the known (reference) liquid at *equal vapor pressure*. For example, if the temperatures of A (isobutyric acid) and the reference substance (water) are determined at 760, 400, and 200 mm pressure, then a plot of the temperatures of A vs. the temperatures of the reference substance will be approximately a straight line over a wide temperature range.

To indicate how to apply the Duhring plot to estimate the heat of vaporization of compound A, we apply the Clapeyron equation to each substance:

$$\frac{d(\ln p_A^*)}{d(\ln p_{\text{ref}}^*)} = \frac{\dfrac{-\Delta \hat{H}_{v_A}}{R\, d(1/T_A)}}{\dfrac{-\Delta \hat{H}_{v_{\text{ref}}}}{R\, d(1/T_{\text{ref}})}} \tag{4.18}$$

If the vapor pressures of A and the reference substance are the same, rearrangement of Eq. (4.18) gives

$$\frac{d\left(\dfrac{1}{T_{\text{ref}}}\right)}{dT_A} = \frac{\dfrac{dT_{\text{ref}}}{T_{\text{ref}}^2}}{\dfrac{dT_A}{T_A^2}} = \frac{\Delta \hat{H}_{v_A}}{\Delta \hat{H}_{v_{\text{ref}}}}$$

or in terms of the slope of the Duhring plot,

$$\frac{dT_{\text{ref}}}{dT_A} = \frac{\Delta \hat{H}_{v_A}}{\Delta \hat{H}_{v_{\text{ref}}}} \frac{T_{\text{ref}}^2}{T_A^2} \tag{4.19}$$

Given the slope of the Duhring plot, $\Delta \hat{H}_{v_{\text{ref}}}$, and T_{ref} and T_A, Eq. (4.19) can be used to calculate $\Delta \hat{H}_{v_A}$.

(2) *Othmer plot.* The Othmer plot[19] is based on the same concepts as the Duhring plot except that the *logarithms* of vapor pressures are plotted against each other at *equal temperature*. As illustrated in Fig. 4.8, a plot of the $\log_{10}(p_A^*)$ against $\log_{10}(p_{\text{ref}}^*)$ chosen at the same temperature yields a straight line over a very wide range. By choosing values of vapor pressure at equal

[19]D. F. Othmer, *Ind. Eng. Chem.*, v. 32, p. 841 (1940). For a complete review of the technique and a comparative statistical analysis among various predictive methods, refer to D. F. Othmer and H. N. Huang, *Ind. Eng. Chem.*, v. 57, p. 40 (1965); D. F. Othmer and E. S. Yu, *Ind. Eng. Chem.*, v. 60, no. 1, p. 22 (1968); and D. F. Othmer and H. T. Chem, *Ind. Eng. Chem.*, v. 60, no. 4, p. 39 (1968).

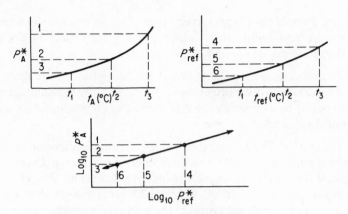

Fig. 4.8. Othmer plot and vapor-pressure curves.

temperature ($T_A = T_{ref}$) in Eq. (4.18), we obtain

$$\frac{(d \ln p_A^*)}{(d \ln p_{ref}^*)} = \frac{\Delta \hat{H}_{v_A}}{\Delta \hat{H}_{v_{ref}}} = m = \text{slope of Othmer plot as in Fig. 4.8.} \quad (4.20)$$

The Othmer plot works well because the errors inherent in the assumptions made in deriving Eq. (4.20) cancel out to a considerable extent. At very high pressures the Othmer plot is not too effective. Incidentally, this type of plot has been applied to a wide variety of thermodynamic and transport relations, such as equilibrium constants, diffusion coefficients, solubility relations, ionization and dissociation constants, etc., with considerable success.

Othmer recommended another type of relationship that can be used to estimate the heats of vaporization:

$$\ln p_{r_A}^* = \frac{\Delta \hat{H}_{v_A}}{\Delta H_{v_{ref}}} \frac{T_{c_{ref}}}{T_{c_A}} \ln p_{r_{ref}}^* \quad (4.21)$$

where p_r^* is the reduced vapor pressure. Equation (4.21) is effective and gives a straight line through the point $p_{c_A}^* = 1$ and $T_{c_A} = 1$. The Gordon method plots the log of the vapor pressure of A versus that of the reference substance at equal *reduced temperature*.

EXAMPLE 4.10 Use of Othmer plot

Repeat the calculation for the heat of vaporization of isobutyric acid at 200°C using an Othmer plot.

Solution:
Data are as follows:

temp. (°C)	p_{iso}^* (atm)	log p*	$p_{H_2O}^*$ (atm)	log p*
154.5	1	0	5.28	0.723
179.8	2	0.3010	9.87	0.994
217.0	5	0.698	21.6	1.334
250.0	10	1.00	39.1	1.592

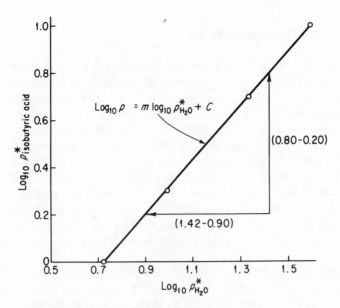

Fig. E4.10.

From Othmer plot, Fig. E4.10

$$\text{slope} = \frac{\Delta \hat{H} v_{\text{iso}}}{\Delta \hat{H} v_{\text{H}_2\text{O}}} = \frac{0.80 - 0.20}{1.42 - 0.90} = 1.15$$

At 200°C (392°F), from the steam tables,

$$\Delta \hat{H} v_{\text{H}_2\text{O}} = 15,000 \text{ Btu/lb mole} \qquad (833.8 \text{ Btu/lb})$$

Then

$$\Delta \hat{H} v_{\text{iso}} = (15,000)(1.15) = 17,250 \text{ Btu/lb mole}$$

The answer in this case is lower than that in Example 4.9. Without the experimental value, it is difficult to say what the proper answer is. However, at the normal boiling point (154°C), $\Delta \hat{H} v_{\text{iso}}$ is known to be 17,700 Btu/lb mole, and $\Delta \hat{H} v_{\text{iso}}$ should be lower at 200°C so that 17,250 Btu/lb mole seems to be the more satisfactory value.

(d) *Empirical relation of* Watson.[20] Watson found empirically that

$$\frac{\Delta \hat{H}_{v_2}}{\Delta \hat{H}_{v_1}} = \left(\frac{1 - T_{r_2}}{1 - T_{r_1}} \right)^{0.38}$$

where $\Delta \hat{H}_{v_2}$ = heat of vaporization of a pure liquid at T_2

$\Delta \hat{H}_{v_1}$ = heat of vaporization of the same liquid at T_1

Gold[21] summarizes numerous estimation methods and delineates their ranges of application.

[20]K. M. Watson, *Ind. Eng. Chem.*, v. 23, p. 360 (1931); v. 35, p. 398 (1943).
[21]P. I. Gold, *Chem. Eng.*, p. 109. (Feb. 24, 1969).

4.5 The general energy balance

Scientists did not begin to write energy balances for physical systems prior to the latter half of the nineteenth century. Before 1850 they were not sure what energy was or even if it was important. But in the 1850s the concepts of energy and the energy balance became clearly formulated. We do not have the space here to outline the historical development of the energy balance and of special cases of it, but it truly makes a most interesting story and can be found elsewhere.[22,23,24,25,26] Today we consider the energy balance to be so fundamental a physical principle that we invent new kinds of energy to make sure that the equation indeed does balance. Equation (4.22) as written below is a generalization of the results of numerous experiments on relatively simple special cases. We universally believe the equation is valid because we cannot find exceptions to it in practice, taking into account the precision of the measurements.

It is necessary to keep in mind two important points as you read what follows. First, we shall examine only systems that are homogeneous, not charged, and without surface effects, in order to make the energy balance as simple as possible. Second, the energy balance will be developed and applied from the macroscopic viewpoint (overall about the system) rather than from a microscopic viewpoint, i.e., an elemental volume within the system.

The concept of the macroscopic energy balance is similar to the concept of the macroscopic material balance, namely,

$$
\begin{Bmatrix} \text{Accumulation of} \\ \text{energy within the} \\ \text{system} \end{Bmatrix} = \begin{Bmatrix} \text{transfer of energy} \\ \text{into system through} \\ \text{system boundary} \end{Bmatrix} - \begin{Bmatrix} \text{transfer of energy out} \\ \text{of system through} \\ \text{system boundary} \end{Bmatrix}
$$
$$
+ \begin{Bmatrix} \text{energy genera-} \\ \text{tion within} \\ \text{system} \end{Bmatrix} - \begin{Bmatrix} \text{energy con-} \\ \text{sumption} \\ \text{within system} \end{Bmatrix}
\tag{4.22}
$$

While the formulation of the energy balance in words as outlined in Eq. (4.22) is easily understood and rigorous, you will discover in later courses that to express each term of (4.22) in mathematical notation may require certain simplifications to be introduced, a discussion of which is beyond our scope here, but which have a quite minor influence on our final balance. We shall split the energy (E) associated with mass, mass either in the system or transported across the system boundaries, into three categories: internal energy (U), kinetic energy (K), and potential energy (P). In addition to the energy transported across the

---

[22] A. W. Porter, *Thermodynamics*, 4th ed., Methuen, London, 1951.

[23] D. Roller, *The Early Development of the Concepts of Temperature and Heat*, Harvard University Press, Cambridge, Mass., 1950.

[24] T. M. Brown, *Am. J. Physics*, v. 33, no. 10, p. 1 (1965).

[25] J. Zernike, *Chem. Weekblad.*, v. 61, pp. 270–274, 277–279 (1965).

[26] L. K. Nash, *J. Chem. Educ.*, v. 42, p. 64 (1965) (a resource paper).

system boundaries associated with mass flow into and out of the system, energy can be transferred by heat (Q) and work (W). For most of the purposes of this text, energy generation or loss will be zero. However, energy transferred to the system by an external electric or magnetic field, or by the slowing down of neutrons, energy transfer not conveniently included in Q or W, may be included conceptually in the energy balance as a generation term. Similarly, radioactive decay can be viewed as a generation term.

Figure 4.9 shows the various types of energy to be accounted for in Eq. (4.22), while Table 4.8 lists the specific individual terms which are to be employed in Eq. (4.22). As to the notation, the subscripts t_1 and t_2 refer to the initial and final time periods over which the accumulation is to be evaluated, with $t_2 > t_1$. The superscript caret ($\hat{}$) means that the symbol stands for energy per unit mass; without the caret, the symbol means energy of the total mass present. Other notation is evident from Figs. 4.9 and 4.10 and can be found in the notation list at the end of the book. Fig. 4.11 shows the energy balance for the earth.

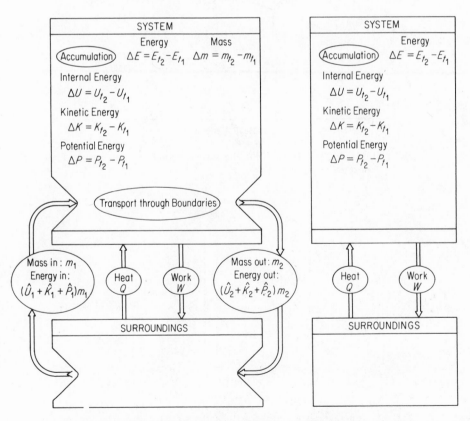

Fig. 4.9. Energy balance over a time period for flow and nonflow systems.

With the aid of the symbols from Table 4.8, the general energy balance, Eq. (4.22), can be written as follows. (An alternative formulation of the balance in terms of differentials is given in Chap. 6, where emphasis is placed on the instantaneous rate of change of energy of a system rather than just the initial and final states of the system; the formulation in this chapter can be considered to be the result of integrating the balance in Chap. 6.):

$$m_{t_2}(\hat{U} + \hat{K} + \hat{P})_{t_2} - m_{t_1}(\hat{U} + \hat{K} + \hat{P})_{t_1} = (\hat{U}_1 + \hat{K}_1 + \hat{P}_1)m_1$$
$$- (\hat{U}_2 + \hat{K}_2 + \hat{P})m_2 + Q - W + p_1\hat{V}_1m_1 - p_2\hat{V}_2m_2 \qquad (4.23)$$

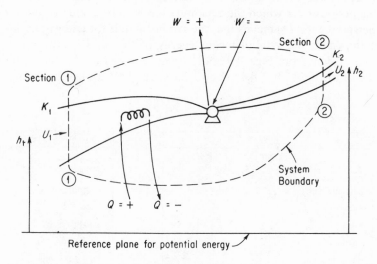

Fig. 4.10. General process showing the system boundary and energy transport across the boundary.

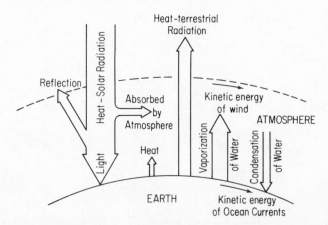

Fig. 4.11. Energy balance on the earth and its atmosphere. The thickness of the arrows are roughly proportional to the magnitude of the energy transfers.

TABLE 4.8 LIST OF SYMBOLS USED IN THE GENERAL ENERGY BALANCE

Accumulation term		
Type of energy	*At time t_1*	*At time t_2*
Internal	U_{t_1}	U_{t_2}
Kinetic	K_{t_1} $\Big\}$ E_{t_1}	K_{t_2} $\Big\}$ E_{t_2}
Potential	P_t	P_{t_2}
Mass	m_{t_1}	m_{t_2}
Energy accompanying mass transport		
Type of energy	*Transport in*	*Transport out*
Internal	U_1	U_2
Kinetic	K_1	K_2
Potential	P_1	P_2
Mass	m_1	m_2
Net heat input to system	Q	
Net work done by system on surroundings		
Mechanical work or work by moving parts:	W	
Work to introduce material into system, less work recovered on removing material from system:	$(p_2 \hat{V}_2)m_2 - (p_1 \hat{V}_1)m_1$	

or in simpler form, which is easier to memorize,

$$\Delta E = E_{t_2} - E_{t_1} = -\Delta[(\hat{H} + \hat{K} + \hat{P})m] + Q - W \qquad (4.24)$$

where Δ = difference operator signifying *out* minus *in*, *exit* minus *entrance*, or *final* minus *initial* in time

$\hat{E} = \hat{U} + \hat{K} + \hat{P}$

$\hat{H} = \hat{U} + p\hat{V}$

Q = heat absorbed *by* the system *from* the surroundings (Q is positive for heat entering the system)

W = mechanical work done *by* the system *on* the surroundings (W is positive for work going from the system to the surroundings)

Note that the expression $\Delta\hat{U} + \Delta p\hat{V} = \Delta\hat{H}$ has been employed in consolidating Eq. (4.23) into (4.24), so that the significance of the enthalpy term in the energy balance manifests itself. A more rigorous derivation of Eq. (4.24) from a microscopic balance may be found in a paper by Bird.[27]

Keep in mind that a system may do work, or have work done on it, without some obvious mechanical device such as a turbine, pump, shaft, etc., being present. Often the nature of the work is implied rather than explicitly stated. For example, a cylinder filled with gas enclosed by a movable piston implies that the surrounding atmosphere can do work on the piston or the reverse; a batch fuel cell does no mechanical work, unless it produces bubbles, but does deliver a current at a potential difference; and so forth.

[27]R. B. Bird, Reprint no. 293, *Univ. Wis. Eng. Expt. Sta. Rept.*, 1957.

Now for a word of warning: Be certain you use *consistent units* for all terms; the use of foot-pound, for example, and Btu in different places in Eq. (4.24) is a common error for the beginner.

The terms $p_1\hat{V}_1$ and $p_2\hat{V}_2$ in Eq. (4.23) and Table 4.7 represent the so-called "pV work," or "pressure energy" or "flow work," or "flow energy" i.e., the work done by the surroundings to put a unit mass of matter into the system at 1 in Fig. 4.10 and the work done by the system on the surroundings as a unit mass leaves the system at 2. If work is defined as

$$W = \int_0^l F \, dl$$

then, since the pressures at the entrance and exit to the system are constant for differential displacements of mass, the work done by the surroundings on the system adds energy to the system at 1:

$$W_1 = \int_0^l F_1 \, dl = \int_0^{\hat{V}_1} p_1 \, d\hat{V} = p_1(\hat{V}_1 - 0) = p_1\hat{V}_1$$

where $\hat{V}$ is the volume per unit mass. Similarly, the work done by the fluid on the surroundings as the fluid leaves the system is $W_2 = p_2\hat{V}_2$, a term that has to be subtracted from the right hand side of Eq. (4.23).

In most problems you do not have to use all of the terms of the general energy balance equation because certain terms may be zero or may be so small that they can be neglected in comparison with the other terms. Several special cases can be deduced from the general energy balance of considerable industrial importance by introducing certain simplifying assumptions:

(a) No mass transfer (*closed* or *batch* system) ($m_1 = m_2 = 0$)

$$\Delta E = Q - W \tag{4.25}$$

Equation (4.25) is known as the *first law of thermodynamics* for a closed system.

(b) No accumulation ($\Delta E = 0$), no mass transfer ($m_1 = m_2 = 0$):

$$Q = W \tag{4.26}$$

(c) No accumulation ($\Delta E = 0$), but with mass flow:

$$Q - W = \Delta[(\hat{H} + \hat{K} + \hat{P})m] \tag{4.27}$$

(d) No accumulation, $Q = 0$, $W = 0$, $\hat{K} = 0$, $\hat{P} = 0$,

$$\Delta\hat{H} = 0 \tag{4.28}$$

Equation (4.28) is the so-called "enthalpy balance."

(e) No accumulation ($\Delta E = 0$), no mass transfer ($m_1 = m_2 = 0$):

$$Q = W \tag{4.29}$$

Take, for example, the flow system shown in Fig. 4.12. Overall, between sections 1 and 5, we would find that $\Delta P = 0$. In fact, the only portion of the

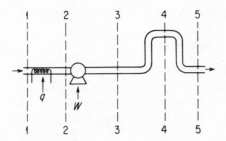

Fig. 4.12. Flow system.

system where ΔP would be of concern would be between section 4 and some other section. Between 3 and 4, ΔP may be consequential, but between 2 and 4 it may be negligible in comparison with the work introduced by the pump. Between section 3 and any further downstream point both Q and W are zero. After reading the problem statements in the examples below but before continuing to read to solution, you should try to apply Eq. (4.24) yourself to test your ability to simplify the energy balance for particular cases.

Some special process names associated with energy balance problems are worth remembering:

   (a) *Isothermal* $(dT = 0)$—constant-temperature process.
   (b) *Adiabatic* $(Q = 0)$—no heat interchange, i.e., an insulated system. If we inquire as to the circumstances under which a process can be called adiabatic, one of the following is most likely:
     (1) The system is insulated.
     (2) Q is very small in relation to the other terms in the energy equation and may be neglected.
     (3) The process takes place so fast that there is no time for heat to be transferred.
   (c) *Isobaric* $(dp = 0)$—constant-pressure process.
   (d) *Isometric* or *isochoric* $(dV = 0)$—constant-volume process.

Occasionally in a problem you will encounter a special term called *sensible heat*. Sensible heat is the enthalpy difference (normally for a gas) between some reference temperature and the temperature of the material under consideration, excluding any enthalpy differences for phase changes that we have previously termed latent heats.

One further remark that we should make is that the energy balance we have presented has included only the most commonly used energy terms. If a change in surface energy, rotational energy, or some other form of energy is important, these more obscure energy terms can be incorporated in an appropriate energy expression by separating the energy of special interest from the term in which it presently is incorporated in Eq. (4.23). As an example, kinetic energy might be split into linear kinetic energy (translation) and angular kinetic energy (rotation).

We turn now to some examples of the application of the general energy balance.

EXAMPLE 4.11   Application of the energy balance

Changes in the heat input that spacecraft encounter constitute a threat both to the occupants and to the payload of instruments. If a satellite passes into the earth's shadow the heat flux that it receives may be as little as 10 percent of that in direct sunlight. Consider the case in which a satellite leaves the earth's shadow and heats up. For simplicity we shall select just the air in the spacecraft as the system. It holds 4.00 kg of air at 20°C (the air has an internal energy of 8.00×10^5 J/kg with reference to fixed datum conditions). Energy from the sun's radiation enters the air as heat until the internal energy is 10.04×10^5 J/kg. (a) How much heat has been transferred to the air? (b) If by some machinery in the spacecraft 0.110×10^5 J of work were done *on* the air over the same interval, how would your answer to part (a) change?

Solution:

(a) The air is chosen as the system, and the process is clearly a closed or batch system. Everything outside the air is the surroundings. See Fig. E4.11. Using Eq. (4.24)

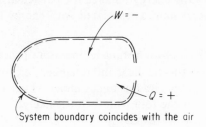

System boundary coincides with the air

Fig. E4.11.

with the absence of K and P in the accumulation term inside the tank, and with no mass flow from the system to the surroundings or the reverse, we have (for $W = 0$)

$$\Delta U = U_{t_1} - U_{t_2} = Q - W$$

Basis: 1 kg air

$$\hat{U}_{t_1} - \hat{U}_{t_2} = (10.04 \times 10^5) - (8.0 \times 10^5) = 2.04 \times 10^5 \text{ J/kg}$$

so that $\hat{Q} = 2.04 \times 10^5$ J/kg. Note that the sign of Q is positive, indicating that heat has been added to the system.

Basis: 4.00 kg of air

$$Q = \frac{2.04 \times 10^5 \text{ J}}{\text{kg}} \Bigg| \frac{4.00 \text{ kg}}{} = 8.16 \times 10^5 \text{ J}$$

(b) If work is done on the air (by compression or otherwise), then the work term is not zero but

$$W = -0.110 \times 10^5 \text{ J}$$

Note that the sign on the work is negative since work is done *on* the gas. In this second case,

$$Q = \Delta U + W = (8.16 \times 10^5) - (0.11 \times 10^5) = 8.05 \times 10^4 \text{ J}$$

Less heat is required than before to reach the same value for the internal energy because some work has been done on the gas.

EXAMPLE 4.12 Application of the energy balance

Air is being compressed from 1 atm and 460°R (where it has an enthalpy of 210.5 Btu/lb) to 10 atm and 500°R (where it has an enthalpy of 219.0 Btu/lb). The exit velocity of the air from the compressor is 200 ft/sec. What is the horsepower required for the compressor if the load is 200 lb/hr of air?

Solution:

This is clearly a flow process or open system as shown in Fig. E4.12 Equation (4.24) can be used with the assumptions that there is no accumulation of air in the com-

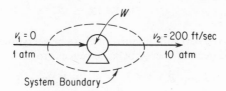

Fig. E4.12.

pressor ($\Delta E = 0$ and $m_1 = m_2$), $Q = 0$ (no heat exchange), and no potential energy terms are of importance. Then

Basis: 1 lb_m air

$$\hat{W} = -\Delta\left(\hat{H} + \frac{v^2}{2}\right)$$

$$\Delta\hat{H} = 219 - 210.5 = 8.5 \text{ Btu/lb}_m$$

Next, we shall assume that the entering velocity of the air is zero, so that

$$\frac{\Delta v^2}{2} = \frac{(200 \text{ ft})^2}{\text{sec}^2} \left|\frac{}{32.2 \frac{(\text{ft})(\text{lb}_m)}{(\text{sec}^2)(\text{lb}_f)}}\right| \frac{1 \text{ Btu}}{778 \text{ (ft)(lb}_f)} \left|\frac{1}{2}\right.$$

$$= 0.80 \text{ Btu/lb}_m$$

$$\hat{W} = -(8.5 + 0.8) = -9.3 \text{ Btu/lb}_m$$

(*Note:* The minus sign indicates work is done on the air.) To convert to power (work/time),

Basis: 200 lb air/hr

$$\text{hp} = \frac{9.3 \text{ Btu}}{\text{lb}} \left|\frac{200 \text{ lb}}{\text{hr}}\right| \frac{1 \text{ hp}}{2545\frac{\text{Btu}}{\text{hr}}} = 0.73 \text{ hp}$$

EXAMPLE 4.13 Application of the energy balance

A steel ball weighing 100 lb is dropped 100 ft to the ground. What are the values of Q and W and the change in U, K, and P for this process? What is the change in the total energy of the universe (system plus surroundings)?

Solution:

See Fig. E4.13.

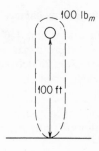

Fig. E4.13.

Let the system be the ball and the surroundings be everything else. This is a batch (nonflow) process. The general energy balance reduces to

$$[(\hat{U}_{t_2} + \hat{K}_{t_2} + \hat{P}_{t_2}) - (\hat{U}_{t_1} + \hat{K}_{t_1} + \hat{P}_{t_1})]m = Q - W$$

After the ball hits the ground and is at rest, we know that its velocity is zero, so that $\Delta K = 0$. We can let the ground be the reference plane for potential energy. Consequently,

$$m(\hat{P}_{t_2} - \hat{P}_{t_1}) = m(g/g_c)(h_2 - h_1)$$
$$= (100 \text{ lb}_m)(g/g_c)(-100 \text{ ft}) = -10,000 \text{ (ft)(lb}_f)$$

If Q and W are zero, then $U_{t_2} - U_{t_1}$ may be calculated. If Q and W are not zero, then $U_{t_2} - U_{t_1}$ is dependent on the values of Q and W, which are not specifically stated in the problem. If $U_{t_2} - U_{t_1}$ were known, then $Q - W$ could be determined. Without more information or without making a series of assumptions concerning Q, W, and $U_{t_2} - U_{t_1}$, it is not possible to complete the requested calculations.

The change in the total energy of the universe is zero.

EXAMPLE 4.14   Application of the energy balance

Water is being pumped from the bottom of a well 150 ft deep at the rate of 200 gal/hr into a vented storage tank 30 ft above the ground. To prevent freezing in the winter a small heater puts 30,000 Btu/hr into the water during its transfer from the well to the storage tank. Heat is lost from the whole system at the constant rate of 25,000 Btu/hr. What is the rise or fall in the temperature of the water as it enters the storage tank, assuming the well water is at 35°F? A 2-hp pump is being used to pump the water. About 55 percent of the rated horsepower goes into the work of pumping and the rest is dissipated as heat to the atmosphere.

Solution:

Let the system consist of the well inlet, the piping, and the outlet at the storage tank. It is a flow process since material is continually entering and leaving the system. See Fig. E4.14.

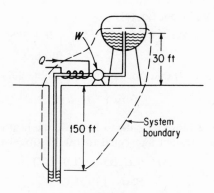

Fig. E4.14.

The general energy equation reduces to

$$Q - W = \Delta[(\hat{H} + \hat{K} + \hat{P})m] \tag{a}$$

The term ΔE is zero because no accumulation of energy takes place in the pipe.

Basis: 1 hr operation

The total amount of water pumped is

$$\frac{200\ \text{gal}}{\text{hr}} \left| \frac{1\ \text{hr}}{} \right| \frac{8.33\ \text{lb}}{1\ \text{gal}} = 1666\ \text{lb/hr}$$

The water initially has a velocity of zero at the well inlet. The problem does not specify the size of pipe through which the water is being pumped so it is impossible to calculate the increase in kinetic energy experienced by the water as it leaves the pipe at the storage tank. Unless the pipe is exceptionally small, however, the kinetic energy term may be considered negligible in comparison with the other energy terms in the equation. Thus

$$\hat{K}_1 = \hat{K}_2 \cong 0$$

The potential energy change is

$$m\ \Delta \hat{P} = mg\ \Delta h = \frac{1666\ \text{lb}_m}{\text{hr}} \left| \frac{32.2\ \text{ft}}{\text{sec}^2} \right| \frac{180\ \text{ft}}{\left| \frac{32.2\ (\text{ft})(\text{lb}_m)}{(\text{sec}^2)(\text{lb}_f)} \right.} = 300,000\ (\text{ft})(\text{lb}_f)/\text{hr}$$

or

$$\frac{300,000\ (\text{ft})(\text{lb}_f)/\text{hr}}{778\ (\text{ft})(\text{lb}_f)/\text{Btu}} = 386\ \text{Btu/hr}$$

The heat lost by the system is 25,000 Btu/hr while the heater puts 30,000 Btu/hr into the system. The net heat exchange is

$$Q = 30,000 - 25,000 = 5000\ \text{Btu/hr}$$

The rate of work being done on the water by the pump is

$$W = -\frac{2\ \text{hp}}{} \left| \frac{0.55}{} \right| \frac{33,000\ (\text{ft}_f)(\text{lb})}{(\text{min})(\text{hp})} \left| \frac{60\ \text{min}}{\text{hr}} \right| \frac{\text{Btu}}{778\ (\text{ft})(\text{lb}_f)}$$

$$= -2800\ \text{Btu/hr}$$

ΔH can be calculated from Eq. (a):

$$5000 - (-2800) = \Delta H + 0 + 386$$

$$\Delta H = 7414 \text{ Btu/hr}$$

We know from Eq. (4.7) that, for an incompressible fluid such as water,

$$\Delta \hat{H} = \int_{T_1}^{T_2} C_p \, dT$$

Because the temperature range considered is small, the heat capacity of liquid water may be assumed to be constant and equal to 1.0 Btu/(lb)(°F) for the problem. Thus

$$\Delta \hat{H} = C_p \Delta T = (1.0)(\Delta T)$$

Basis: 1 lb H_2O

Now

$$\Delta \hat{H} = \frac{7414 \text{ Btu}}{\text{hr}} \bigg| \frac{1666 \text{ lb}}{\text{hr}} = 6.36 \text{ Btu/lb}$$

$$= (1.0)(\Delta T)$$

$$\Delta T \simeq 6.4°\text{F temperature rise}$$

EXAMPLE 4.15 Energy balance

Steam that is used to heat a batch reaction vessel enters the steam chest, which is segregated from the reactants, at 250°C saturated and is completely condensed. The reaction absorbs 1000 Btu/lb of material in the reactor. Heat loss from the steam chest to the surroundings is 5000 Btu/hr. The reactants are placed in the vessel at 70°F and at the end of the reaction the material is at 212°F. If the charge consists of 325 lb of material and both products and reactants have an average heat capacity of C_p = 0.78 Btu/(lb)(°F), how many pounds of steam are needed per pound of charge? The charge remains in the reaction vessel for 1 hr.

Solution:

Analysis of the process reveals that no potential energy, kinetic energy, or work terms are significant in the energy balance. The steam flows in and out, while the reacting material stays within the system. Thus there is an accumulation of energy for the material being processed but no accumulation of mass or energy associated with the steam. See Fig. E4.15(a).

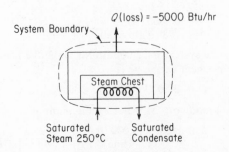

Fig. E4.15(a).

Basis: 1 hr of operation

System: Reaction vessel, including steam chest

The energy balance reduces to

$$(U_{t_2} - U_t)_{\text{material}} = -\Delta H_{\text{steam}} + Q \tag{a}$$

However, since enthalpy is more convenient to work with than internal energy, we let $\hat{U} = \hat{H} - p\hat{V}$,

$$[(\hat{H}_{t_2} - \hat{H}_{t_1}) - (p_{t_2}\hat{V}_{t_2} - p_{t_1}\hat{V}_{t_1})]m = Q - \Delta H_{\text{steam}}$$

Since no information is available concerning the $p\hat{V}$ difference, and it would no doubt be negligible if we could calculate it, it will be neglected, and then

$$(\hat{H}_{t_2} - \hat{H}_{t_1})m = -\Delta H_{\text{steam}} + Q \tag{b}$$

(a) The heat loss is given as $Q = -5000$ Btu/hr.

(b) The enthalpy change for the steam can be determined from the steam tables. The ΔH_{vap} of saturated steam at 250°C is 732.2 Btu/lb, so that

$$\Delta H_{\text{steam}} = -732.2 \text{ Btu/lb}$$

(c) The enthalpy of the initial reactants, with reference to zero enthalpy at 70°F, is

$$\Delta H_{t_1} = m(\Delta \hat{H}) = m \int_{70}^{70} C_p \, dT = 0$$

Note that the selection of zero enthalpy at 70°F makes the calculation of the enthalpy of the reactants at the start quite easy. The enthalpy of the final products relative to zero enthalpy at 70°F is

$$\Delta H_{t_2} = m \int_{70}^{212} C_p \, dT = mC_p(212 - 70)$$

$$= \frac{325 \text{ lb}}{\text{hr}} \left| \frac{0.78 \text{ Btu}}{(\text{lb})(°F)} \right| \frac{(212 - 70)°F}{} = 36,000 \text{ Btu/hr}$$

$$\Delta H_{\text{material}} = 36,000 \text{ Btu/hr}$$

(d) In addition to the changes in enthalpy occurring in the material entering and leaving the system, the reaction absorbs 1000 Btu/lb. This quantity of energy can be conviently thought of as an energy loss term, or alternately as an adjustment to the enthalpy of the reactants.

$$\Delta H = \frac{-1000 \text{ Btu}}{\text{lb}} \left| \frac{325 \text{ lb}}{\text{hr}} \right. = -325,000 \text{ Btu/hr}$$

Introducing these numbers into Eq. (b), we find that

$$36,000 \frac{\text{Btu}}{\text{hr}} = \left(737.2 \frac{\text{Btu}}{\text{lb steam}}\right)\left(S \frac{\text{lb steam}}{\text{hr}}\right)$$

$$- 5000 \frac{\text{Btu}}{\text{hr}} - 325,000 \frac{\text{Btu}}{\text{hr}} \tag{c}$$

from which the pounds of steam per hour, S, can be calculated as

$$S = \frac{366,000 \text{ Btu}}{\text{hr}} \left| \frac{1 \text{ lb steam}}{737.2 \text{ Btu}} \right. = 496 \frac{\text{lb steam}}{\text{hr}}$$

or

$$\frac{366,000 \text{ Btu}}{\text{hr}} \left| \frac{1 \text{ lb steam}}{737.2 \text{ Btu}} \right| \frac{1 \text{ hr}}{325 \text{ lb charge}} = 1.53 \frac{\text{lb steam}}{\text{lb charge}}$$

If the system is chosen to be everything but the steam coils and lines, then we would have a situation as shown in Fig. E4.15(b). Under these circumstances we could talk about the heat transferred into the reactor from the steam chest. From a balance on the steam chest (no accumulation),

$$Q = \Delta H_{\text{steam}}$$

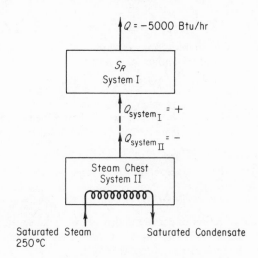

Fig. E4.15(b)

Both have negative values, but remember that

$$Q_{\text{system I}} = -Q_{\text{system II}}$$

so that the value of $Q_{\text{system I}}$ is plus.

The energy balance on system I reduces to

$$(\hat{H}_{t_2} - \hat{H}_{t_1})m = Q \tag{d}$$

where $Q = _{\text{system chest}} + (-5000 \text{ Btu})$. The steam used can be calculated from Eq. (d); the numbers will be identical to the calculation made for the original system.

EXAMPLE 4.16 Energy balance

Ten pounds of water at 35°F, 4.00 lb ice at 32°F, and 6.00 lb steam at 250°F and 20 psia are mixed together. What is the final temperature of the mixture? How much steam condenses?

Solution:

We shall assume that the overall batch process takes place adiabatically. See Fig. E4.16. The system is 20 lb of H_2O in various phases. The energy balance reduces to (superscripts are s = steam, w = water, i = ice)

$$20\hat{U}_{t_2} - (60\hat{U}_{t_1}^s + 10\hat{U}_{t_1}^w + 4\hat{U}_{t_1}^i) = 0 \tag{a}$$

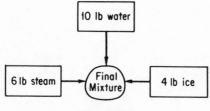

Fig. E4.16.

If U is replaced with $\hat{H} - p\hat{V}$ for convenience in the calculations, we get

$$20\hat{H}_{t_2} - (6\hat{H}^s_{t_1} + 10\hat{H}^w_{t_1} + 4\hat{H}^i_{t_1})$$
$$= 20(p\hat{V})_{t_2} - 6(p\hat{V})^s_{t_1} - 10(p\hat{V})^w_t - 4(p\hat{V})^i_{t_1} \tag{b}$$

The last two terms on the right-hand side of Eq. (b) cannot be more than 1 Btu at the very most and can safely be neglected.

Because of the phase changes that take place as well as the nonlinearity of the heat capacities as a function of temperature, it is not possible to replace the enthalpies in Eq. (b) with functions of temperature and get a linear algebraic equation that is easy to solve. The simplest way to solve the problem is to assume the final conditions of temperature and pressure and check the assumption by Eq. (c)

$$20\hat{H}_{t_2} - 20(p\hat{V})_{t_2} = (6\hat{H}^s_{t_1} + 10\hat{H}^w_{t_1} + 4\hat{H}^i_{t_1}) - 6(p\hat{V})^s_{t_1} \tag{c}$$

If the assumption proves wrong, a new assumption can be made, leading to a solution to the problem by a series of iterative calculations:

$$\text{Basis:} \begin{cases} \text{4 lb ice at } 32°\text{F} \\ \text{10 lb } H_2O \text{ at } 35°\text{F} \\ \text{6 lb steam at } 250°\text{F and 20 psia} \end{cases}$$

Since the enthalpy change corresponding to the heat of condensation of the steam is quite large, we might expect that the final temperature of the mixture would be near or at the saturation temperature of the 20 psia steam. Let us assume that $T_{\text{final}} = 228°\text{F}$ (the saturated temperature of 20 psia of steam). We shall also assume that all the steam condenses as a first guess.

Using the steam tables (ref.: zero enthalpy at 32°F, saturated liquid), we can calculate the following quantities to test our assumption:

Btu

$$6\,\Delta\hat{H}^s_{t_1} = \frac{6 \text{ lb} \,|\, 1168.0 \text{ Btu}}{|\quad \text{lb}} \qquad\qquad = 7008.0$$

$$10\,\Delta\hat{H}^w_{t_1} = \frac{10 \text{ lb} \,|\, 3.02 \text{ Btu}}{|\quad \text{lb}} \qquad\qquad = 30.2$$

The heat of fusion of ice is 143.6 Btu/lb:

$$4\,\Delta\hat{H}^i_{t_1} = -\frac{4 \text{ lb} |\, 143.6 \text{ Btu}}{|\quad \text{lb}} \qquad\qquad = -574.4$$

$$-6(p\hat{V})^s_{t_1} = -\frac{6 \text{ lb}_m \,|\, 20 \text{ lb}_f \,|\, \left(\frac{12 \text{ in.}}{1 \text{ ft}}\right)^2 \,|\, 20.81 \text{ ft}^3 \,|\, 1 \text{ Btu}}{|\quad \text{in.}^2 \,|\qquad\qquad|\quad \text{lb}_m \,|\, 778 \text{ (ft)}(\text{lb}_f)} \quad = -453.2$$

total right-hand side of Eq. (c) $\qquad\qquad\qquad = 6010.6$

The left-hand side of Eq. (c) is

$$20\,\Delta\hat{H}_{t_2} = \frac{20\text{ lb}}{}\bigg|\frac{196.2\text{ Btu}}{\text{lb}} \qquad\qquad\qquad = 3924.0$$

liquid

$$-20(p\hat{V})_{t_2} = -\frac{20\text{ lb}_m}{}\bigg|\frac{19.7\text{ lb}_f}{\text{in.}^2}\bigg|\left(\frac{12\text{ in.}}{1\text{ ft}}\right)^2\bigg|\frac{0.0168\text{ ft}^3}{}\bigg|\frac{1\text{ Btu}}{778\text{ (ft)(lb}_f)} = -1.2$$

total left-hand side of Eq. (c) $= 3922.8$

Evidently our assumption that all the steam condenses is wrong, since the value of the left-hand side of Eq. (c) is too small. With this initial calculation we can foresee that the final temperature of the mixture will be 228°F and that a saturated steam-liquid water mixture will finally exist in the vessel. The next question is, How much steam condenses?

Let S = the amount of steam at 228°F:

$$S[\hat{H}_{t_2} - p\hat{V}]_{\text{vapor}} + (20 - S)[\hat{H}_{t_2} - p\hat{V}]_{\text{liquid}} = 6010.6$$

$$(\Delta\hat{H}_{t_1})_{\text{vapor}} = 1156.1\text{ Btu/lb}$$

$$(\Delta\hat{H}_{t_2})_{\text{liquid}} = 196.2\text{ Btu/lb}$$

$$(p\hat{V})_{\text{vapor}} = (19.7)(12)^2(20.08)/(778) = 73.2\text{ Btu/lb}$$

$$(p\hat{V})_{\text{liquid}} \cong 0.0$$

$$S(1156.1 - 73.2) + (20 - S)(196.2) = 6010.6$$

$$S = 2.35\text{ lb}$$

In effect, $6.00 - 2.35 = 3.65$ lb of steam can be said to condense.

If there had been only 1 lb of steam instead of 6 lb to start with, then on the first calculation the right-hand side of Eq. (c) would have been

$$1168.0 + 30.2 - 574.4 - (1/6)(453.2) = 548.3\text{ Btu}$$

Hence lower temperatures would have been chosen until the left-hand side of Eq. (c) was small enough. Only liquid water would be present.

EXAMPLE 4.17 Application of the energy balance

The peaks of Mount Everest, Mount Fujiyama, Mount Washington, and other mountains are usually covered with clouds even when the surrounding area has nice weather. It is clear that water condenses because the temperature is lower, but why doesn't water condense elsewhere at the same altitude? Is the cloud cap stuck by some force? See Fig. E4.17.

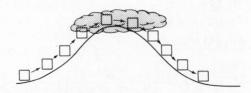

Fig. E4.17.

Solution:

The formation of cloud caps can be predicted with the aid of the general energy balance. As a unit mass of warm moist air, our system, moves up the side of a mountain with only modest breezes, it cannot exchange energy by way of heat with the surrounding air ($Q = 0$), nor does it exchange a significant amount of water vapor ($m_1 = m_2 = 0$). Both kinetic and potential energy effects can be neglected, so that Eq. (4.24) reduces to

$$\Delta \hat{U} = -\hat{W}$$

Now

$$\Delta \hat{U} = \int_{T_1}^{T_2} \hat{C}_v \, dT \cong \hat{C}_v (T_2 - T_1)$$

and

$$\hat{W} = \int_{\hat{V}_1}^{\hat{V}_2} p \, d\hat{V}$$

Assume that an ideal gas expands so that $p\hat{V} = \hat{R}T$. Then, from information as to the pressure change with altitude plus the fact that $C_v = 0.25$ cal/(g mole)(°K), by trial and error the change in temperature can be determined as the air rises. The cooling is roughly 3° to 4°F/1000 ft of elevation. During the descent of the air on the other side of the mountain, the cloud will evaporate. You can also calculate how high a peak has to be for a cloud to form for a given relative humidity and temperature of air at the bottom of the peak or, alternatively, given the peak height, what the highest relative humidity can be for a cloud not to form. As an example of the latter, for Mt. Washington (altitude 6388 ft), the maximum percent relative humidity of 70°F air at sea level is 40 percent.

A more complicated but related problem that we do not have the space to go into is the formation of contrails by supersonic aircraft.

4.6   Reversible processes and the mechanical energy balance

The energy balance of the previous section is concerned with various classes of energy without inquiring into how "useful" each form of energy is to man. Our experience with machines and thermal processes indicates that all types of energy transformation are not equally possible. For example, it is well known that heat cannot be transferred from a low temperature to a high temperature, but only in the reverse direction, nor can internal energy be completely converted into mechanical work. To account for these limitations on energy utilization, the second law of thermodynamics was eventually developed as a general principle.

As applied to fluid dynamic problems, one of the consequences of the second law of thermodynamics is that two categories of energy of different "quality" can be envisioned:

(a) The so-called *mechanical* forms of energy, such as kinetic energy, potential energy, and work, which are *completely* convertible by an *ideal* (frictionless, reversible) engine from one form to another within the class.

(b) Other forms of energy, such as internal energy and heat, which are not so freely convertible.

Of course, in any real process with friction, viscous effects, mixing of components, and other dissipative phenomena taking place which prevent the complete conversion of one form of mechanical energy to another, allowances will have to be made in making a balance on mechanical energy for these "losses" in quality.

To describe the loss or transfer of energy of the freely convertible kind to energy of lesser quality, a concept termed *irreversiblity* has been coined. A process which operates without the degradation of the convertible types of energy is termed a *reversible* process; one which does not is an irreversible process. A reversible process proceeds under conditions of differentially balanced forces. Only an infinitesimal change is required to reverse the process, a concept which leads to the name reversible.

Take, for example, the piston shown in Fig. 4.13. During the expansion process the piston moves the distance x and the volume of gas confined in the

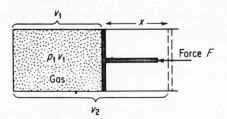

Fig. 4.13. Gas expansion.

piston increases from V_1 to V_2. Two forces act on the reversible (frictionless) piston; one is the force exerted by the gas, equal to the pressure times the area of the piston, and the other is the force on the piston shaft. If the force exerted by the gas equals the force F, nothing happens. If F is greater than the force of the gas, the gas will be compressed, while if F is less than the force of the gas, the gas will expand.

In the latter case, the work done by the expanding gas and the piston will be $W = \int F \, dx$. The work of the gas would be $W = \int p \, dV$ if the process were reversible, i.e., if the force F divided by the piston area were differentially less at all times than the pressure of the gas, and if the piston were frictionless; but in a real process some of the work done by the gas is dissipated by viscous effects, and the piston will not be frictionless so that the work as measured by $\int F \, dx$ will be less than $\int p \, dV$. For these two integrals to be equal, none of the available energy of the system can be degregated to heat or internal energy. To achieve such a situation we shall have to ensure that the movement of the piston is frictionless and that the motion of the piston proceeds under only a differential imbalance of forces so no shock or turbulence is present. Naturally such a process would take a long time to complete.

No real process which involves friction, shock, finite temperature differences, unrestricted expansion, viscous dissipation, or mixing can be reversible, and so we see that the dropping of a weight from a tower, the operation of a heat exchanger, and the combusion of gases in a furnace are all irreversible processes. Since practically all real processes are irreversible you may wonder why we bother paying attention to the theoretical reversible process. The reason is that it represents the best that can be done, the ideal case, and gives us a measure of our maximum accomplishment. In a real process we cannot do as well and so are less effective. A goal is provided by which to measure our effectiveness. And then, many processes are not too irreversible, so that there is not a big discrepancy between the practical and the ideal process.

A balance on mechanical energy can be written on a microscopic basis for an elemental volume by taking the scalar product of the local velocity and the equation of motion.[28] After integration over the entire volume of the system the *steady-state mechanical energy balance* (for a system with mass interchange with the surroundings) becomes, on a per unit mass basis,

$$\Delta(\hat{K} + \hat{P}) + \int_{p_1}^{p_2} \hat{V} \, dp + \hat{W} + \hat{E}_v = 0 \qquad (4.30)$$

where E_v represents the loss of mechanical energy, i.e., the irreversible conversion *by the flowing fluid* of mechanical energy to internal energy, a term which must in each individual process be evaluated by experiment (or, as occurs in practice, by use of already existing experimental results for a similar process). Equation (4.30) is sometimes called the Bernoulli equation, especially for the reversible process for which $\hat{E}_v = 0$. The mechanical energy balance is best applied in fluid-flow calculations when the kinetic and potential energy terms and the work are of importance, and the friction losses can be evaluated from handbooks with the aid of *friction factors* or *orifice coefficients*.

EXAMPLE 4.18   Application of the mechanical energy balance

Calculate the work per minute required to pump 1 lb of water per minute from 100 psia and 80°F to 1000 psia and 100°F.

Solution:

If the process is assumed to be reversible so that $E_v = 0$, if the pump is 100 percent efficient, and if no kinetic or potential energy terms are involved in the process, Eq. (4.30) reduces to

$$\hat{W} = -\int_{p_1}^{p_2} \hat{V} \, dp$$

[28]R. B. Bird, *op. cit.*

From the steam tables, the specific volume of liquid water is 0.01608 ft³/lb$_m$ at 80°F and 0.01613 ft³/lb$_m$ at 100°F. For all practical purposes the water is incompressible, and the specific volume can be taken to be 0.0161 ft³/lb$_m$. Then,

Basis: 1 min

$$W = -\frac{1\ lb_m}{min} \int_{100}^{1000} 0.0161\ dp$$

$$= -\frac{1\ lb_m}{min} \left| \frac{0.0161\ ft^3}{lb_m} \right| \frac{(1000 - 100)\ lb_f}{in.^2} \left| \left(\frac{12\ in.}{1\ ft}\right)^2 \right| \frac{1\ Btu}{778\ (ft)(lb_f)}$$

$$= -2.68 \frac{Btu}{min}$$

About the same value can be calculated using Eq. (4.24) if $\hat{Q} = \hat{K} = \hat{P} = 0$, because the enthalpy change[29] for a reversible process for 1 lb of water going from 100 psia and 100°F to 1000 psia is 2.70 Btu. However, usually the enthalpy data for liquids other than water are missing, or not of sufficient accuracy to be of much value, which forces the engineer to turn to the mechanical energy balance.

One might now well inquire for the purpose of purchasing a pump-motor, say, as to what the work would be for a real process, instead of the fictitious reversible process assumed above. First, you would need to know the efficiency of the combined pump and motor so that the actual input from the surroundings (the electric connection) to the system would be known. Second, the friction losses in the pipe, valves, and fittings must be estimated so that the term $\hat{E}_v$ could be reintroduced into Eq. (4.30). Suppose, for the purposes of illustration, that $\hat{E}_v$ was estimated to be, from an appropriate handbook, 320 (ft)(lb$_f$)/lb$_m$ and the motor-pump efficiency was 60 percent (based on 100 percent efficiency for a reversible pump-motor). Then,

$$E_v = \frac{320\ (ft)(lb_f)}{1\ lb_m} \left| \frac{1\ Btu}{778\ (ft)(lb_f)} \right| \frac{1\ lb_m}{min} = 0.41\ Btu/min$$

$$W = -(2.68 + 0.41) = -3.09\ Btu/min$$

Remember that the minus sign indicates work is done on the system. The pump motor must have the capacity

$$\frac{3.09\ Btu}{1\ min} \left| \frac{1}{0.60} \right| \frac{1\ min}{60\ sec} \left| \frac{1.415\ hp}{1\ Btu/sec} \right. = 0.122\ hp$$

EXAMPLE 4.19 Calculation of work for a batch process

One pound mole of N_2 is in a horizontal cylinder at 100 psia and 70°F. A 1-in.² piston of 5-lb weight seals the cylinder and is fixed by a pin. The pin is released and the N_2 volume is doubled, at which time the piston is stopped again. What is the work done by the gas in this process?

Solution:

(a) Draw a picture. See Fig. E4.19.

(b) Is the process a flow or nonflow process? Since no material leaves or enters the cylinder, let us analyze the process as a nonflow system.

[29]From J. H. Keenan and F. G. Keyes, *Thermodynamic Properties of Steam*, Table 3, Wiley, New York, 1936.

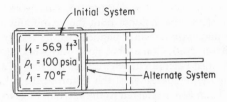

Fig. E4.19.

(c) This process is definitely irreversible because of friction between the piston (and the gas) and the walls of the cylinder, turbulence in the the gas, the large pressure drops, etc.

(d) In selecting a system to work with, we can choose either the gas itself or the gas plus the piston. Let us initially select the N_2 gas as the system, with the surroundings being everything else including the piston and the cylinder.

(e) Since the gas expansion is a nonflow process, Eq. (4.30) does not apply. Even if you had an expression for the unsteady-state mechanical energy balance (which we have not discussed since it involves the concept of entropy), it would not be possible to calculate the work done by the gas because the process is irreversible and the E_v term would not be known. As might be expected, a direct application of $W = \int F\,dl$ fails for this problem because the pressure of the gas does not necessarily equal the force per unit area applied by the piston to the gas, i.e., the work cannot be obtained from $W = \int p\,dV$.

(f) Let us change our system to include the gas, the piston, and the cylinder. With this choice, the system expansion is still irreversible, but the pressure in the surroundings can be assumed to be atmospheric pressure and is constant. The work done in pushing back the atmosphere probably is done almost reversibly from the viewpoint of the surroundings and can be closely estimated by calculating the work done on the surroundings:

$$W = \int_{l_1}^{l_2} F\,dl = \int_{V_1}^{V_2} p\,dV = p\Delta V$$

$$p = \frac{14.7\;lb_f}{in.}\left|\frac{144\;in.^2}{1\;ft^2}\right. = \frac{2120\;lb_f}{ft^2}$$

Assuming that $z = 0.995$ and is constant, the initial volume of gas is

$$V_1 = \frac{359\;ft^3}{1\;lb\;mole}\left|\frac{530°R}{492°R}\right|\frac{14.7}{100}\left|0.995\right. = 56.9\;ft^3/lb\;mole$$

and ΔV of the gas is $2V_1 - V_1 = V_1 = 56.9\;ft^3/lb$ mole so that the volume change of the surroundings is $-56.9\;ft^3$. Then the work done on the surroundings is

$$W = \frac{2120\;lb_f}{ft^2}\left|\frac{-56.9\;ft^3}{lb\;mole}\right|\frac{1\;lb\;mole}{}$$

$$= -120,000\;(ft)(lb_f)$$

The minus sign indicates that work is done on the surroundings. From this we know that the work done by the alternative system is

$$W_{system} = -W_{surroundings} = -(-120,000) = 120,000\;(ft)(lb_f)$$

(g) This still does not answer the question of what work was done by the gas, because our system now includes both the piston and the cylinder as well as the gas. You may wonder what happens to the energy transferred from the gas to the piston itself. Some of it goes into work against the atmosphere, which we have just calculated; where does the rest of it go? This energy can go into raising the temperature of the piston or the cylinder or the gas. With time, part of it can be transferred to the surroundings to raise the temperature of the surroundings. From our macroscopic viewpoint we cannot say specifically what happens to this "lost" energy.

(h) You should also note that if the surroundings were a vacuum instead of air, then *no work* would be done by the system, consisting of the piston plus the cylinder plus the gas, although the gas itself probably still would do some work.

4.7 Heat of reaction

So far we have not discussed how to treat energy changes caused by chemical reactions, a topic that is frequently called *thermochemistry*. The energy change that appears directly as a result of a reaction is termed the *heat of reaction*, ΔH_{rxn}, a term which is a legacy of the days of the caloric theory. ("Heat of reaction" in terms of modern concepts is best thought of as the energy released or absorbed because of the reaction that may be transferred as heat in certain types of experiments which are delineated below but may also end up as internal energy or as another form of energy.) The energy released or absorbed during a reaction comes from the rearrangement of the bonds holding together the atoms of the reacting molecules. For an exothermic reaction, the energy required to hold the products of the reaction together is less than that required to hold the reactants together and the surplus energy is released.

To take account of possible energy changes caused by a reaction, in the energy balance we incorporate in the enthalpy of each individual constituent an additional quantity termed the standard heat (really enthalpy) of formation, $\Delta \hat{H}_f^\circ$, a quantity that is discussed in detail below. Thus, for the case of a single species A without any pressure effect on the enthalpy and omitting phase changes,

$$\Delta \hat{H}_A = \Delta \hat{H}_{fA}^\circ + \int_{T_{ref}}^{T} C_{pA}\, dT \qquad (4.31)$$

For several species in which there is no energy change on mixing we would have

$$\Delta H_{mixture} = \sum_{i=1}^{s} n_i \Delta \hat{H}_{fi}^\circ + \sum_{i=1}^{s} \int_{T_{ref}}^{T} n_i C_{pi}\, dT \qquad (4.32)$$

where i designates each species, n_i is the number of moles of species i, and s is the total number of species.

If a mixture enters and leaves a system without a reaction taking place, we would find that the same species entered and left so that the enthalpy change in either the flow or the accumulation terms in Eq. (4.24) would not be any different with the modification described above than what we have used before, because the terms that account for the heat of reaction in the energy balance would cancel. For example, for the case of two species in a flow system, the output enthalpy would be

$$\Delta H_{\text{output}} = n_1 \Delta \hat{H}^{\circ}_{f1} + n_2 \Delta \hat{H}^{\circ}_{f2} + \int_{T_{\text{ref}}}^{T_{\text{out}}} (n_1 C_{p1} + n_2 C_{p2})\, dT$$

and the input enthalpy would be

$$\Delta H_{\text{input}} = n_1 \Delta \hat{H}^{\circ}_{f1} + n_2 \Delta \hat{H}^{\circ}_{f2} + \int_{T_{\text{ref}}}^{T_{\text{in}}} (n_1 C_{p1} + n_2 C_{p2})\, dT$$

We observe that $\Delta H_{\text{output}} - \Delta H_{\text{input}}$ would only involve the "sensible heat" terms that we have described before.

However, if a reaction takes place, the species that enter and leave will differ, and the terms involving the heat of formation will not cancel. For example, suppose that species 1 and 2 enter the system, react, and species 3 and 4 leave. Then

$$\begin{aligned}
\Delta H_{\text{out}} - \Delta H_{\text{in}} &= (n_3 \Delta \hat{H}^{\circ}_{f3} + n_4 \Delta \hat{H}^{\circ}_{f4}) - (n_1 \Delta \hat{H}^{\circ}_{f1} + n_2 \Delta \hat{H}^{\circ}_{f2}) \\
&+ \int_{T_{\text{ref}}}^{T_{\text{out}}} (n_3 C_{p3} + n_4 C_{p4})\, dT - \int_{T_{\text{ref}}}^{T_{\text{in}}} (n_1 C_{p1} + n_2 C_{p2})\, dT
\end{aligned} \qquad (4.33)$$

To determine the values of the standard heats (enthalpies) of formation, the experimenter usually selects either a simple flow process without kinetic energy, potential energy, or work effects (a flow calorimeter), or a simple batch process (a bomb calorimeter), in which to conduct the reaction. For the flow process in an experiment in which the sensible heat terms on the right hand side of Eq. (4.33) are zero and no work is done, the steadystate (no accumulation term) version of Eq. (4.24) reduces to

$$Q = \Delta[\hat{H}m] = \Delta H_{\text{rxn}}$$

and

$$\Delta H^{\circ}_{\text{rxn}} = \left(\sum_{\text{products}} n_i \Delta \hat{H}^{\circ}_{fi} - \sum_{\text{reactants}} n_i \Delta \hat{H}^{\circ}_{fi} \right) \qquad (4.34)$$

In other words, the energy change caused by the reaction in such an experiment appears as heat transferred to or from the system, and the value of Q that is observed is termed the heat of reaction. A flow calorimeter in which $Q = 0$, $W = 0$, and the temperature change is measured between the input and output streams, can also be used to compute the heat of reaction, because the left hand side of Eq. (4.33) is zero. Refer to Sec. 4.7-3 for an analysis of the batch bomb calorimeter of constant volume.

The calculations that we shall make here will all be for low pressures, and, although the effect of pressure upon heats of reaction is relatively negligible under most conditions, if exceedingly high pressures are encountered, you should make the necessary corrections as explained in most texts on thermochemistry.

There are certain conventions and symbols which you should always keep in mind concerning thermochemical calculations if you are to avoid difficulty later on. These conventions can be summarized as follows:

(a) The reactants are shown on the left-hand side of the equation, and the products are shown on the right—for example,

$$CH_4(g) + H_2O(l) \longrightarrow CO(g) + 3H_2(g)$$

(b) The conditions of phase, temperature, and pressure must be specified unless the last two are the standard conditions, as presumed in the example above, when only the phase is required. This is particularly important for compounds such as H_2O, which can exist as more than one phase under common conditions. If the reaction takes place at other than standard conditions, you might write

$$CH_4(g, 1.5 \text{ atm}) + H_2O(l) \xrightarrow{50°C} CO(g, 3 \text{ atm}) + H_2(g, 3 \text{ atm})$$

(c) Unless otherwise specified, the heats of reaction, the enthalpy changes, and all the constituents are at the standard state of 25°C (77°F) and 1 atm total pressure. The heat of reaction under these conditions is called the *standard heat of reaction*, and is distinguished by the superscript ° symbol.

(d) Unless the amounts of material reacting are stated, it is assumed that the quantities reacting are the stoichiometric amounts shown in the chemical equation.

Data to compute the standard heat of reaction are reported and tabulated in two different but essentially equivalent forms:

(a) Standard heats of formation
(b) Standard heats of combustion

We shall first describe the standard heat of formation ($\Delta \hat{H}_f^°$) and afterward the standard heat of combustion ($\Delta \hat{H}_c^°$). (The superscript ° denotes *standard state.*) The units of both quantities are usually tabulated as energy per mole, such as kilocalories per gram mole. The "per mole" refers to the specified reference substance in the related stoichiometric equation.

4.7.1 Standard Heat of Formation.

This is a special type of heat of reaction, one for the formation of a compound from the elements. The initial reactants and final products must be stable and at 25°C and 1 atm. The reaction does not necessarily represent a real reaction that would proceed at constant temperature but can be a fictitious process for the formation of a compound from the elements. By defining the heat of formation as zero in the standard state for each *element*, it is possible to design a system to express the heats of formation for all *compounds* at 25°C and 1 atm. If you use the conventions discussed above, then the thermochemical calculations will all be consistent, and you should not experience any confusion as to signs. Consequently, with the use of well-defined standard heats of formation, it is not necessary to record the experimental heats of reaction for every reaction that can take place.

A formation reaction is understood by convention to be a reaction which forms 1 mole of compound from the elements which make it up. Standard heats of reaction for any reaction can be calculated from the tabulated (or experimental) standard heats of formation values by using Eq. (4.34), because the standard heat of formation is a state (point) function.

EXAMPLE 4.20 Heats of formation

What is the standard heat of formation of HCl(g)?

Solution:
In the reaction at 25°C and 1 atm,

$$\tfrac{1}{2}H_2(g) + \tfrac{1}{2}Cl_2(g) \longrightarrow HCl(g)$$

Tabulated data: $\Delta \hat{H}_f^\circ\left(\dfrac{cal}{g\ mole}\right)$ 0   0 $-22{,}063$

both $H_2(g)$ and $Cl_2(g)$ would be assigned $\Delta \hat{H}_f^\circ$ values of 0, and the value shown in Appendix F for HCl(g) of $-22{,}063$ cal/g mole is the standard heat of formation for this compound (as well as the standard heat of reaction for the reaction as written above). The value tabulated in Appendix F might actually be determined by carrying out the reaction shown for HCl(g) and measuring the energy liberated in a calorimeter, or by some other more convenient method.

EXAMPLE 4.21 Indirect determination of heats of formation

Suppose that you want to find the standard heat of formation of CO from experimental data. Can you prepare pure CO from C and O_2 and measure the energy liberated? This would be far too difficult. It would be easier experimentally to find the energy liberated for the two reactions shown below and add them as follows:

Basis: 1 g mole CO

$$\Delta \hat{H}_{rxn}^\circ (experimental)$$

$A:$ $C(\beta) + O_2(g) \longrightarrow CO_2(g)$ -94.052 kcal/g mole

$B:$ $CO(g) + \tfrac{1}{2}O_2(g) \longrightarrow CO_2(g)$ -67.636 kcal/g mole

---

$A - B:$ $C(\beta) + \tfrac{1}{2}O_2(g) \longrightarrow CO(g)$

$$\Delta \hat{H}_{rxn\ A-B}^\circ = (-94.052) - (-67.636) = \Delta \hat{H}_f^\circ = -26.416 \text{ kcal/g mole}$$

The energy change for the overall reaction scheme is the desired heat of formation per mole of CO(g). See Fig. E4.21.

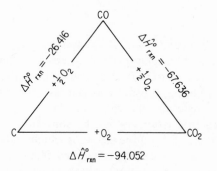

Fig. E4.21.

EXAMPLE 4.22 Calculation of the heat of reaction

Calculate $\Delta \hat{H}_{rxn}^{\circ}$ for the following reaction of 5 moles of NH_3:

$$4NH_3(g) + 5O_2(g) \longrightarrow 4NO(g) + 6H_2O(g)$$

Solution:

Basis: 4 g mole NH_3

tabulated data	$NH_3(g)$	$O_2(g)$	$NO(g)$	$HO_2(g)$
$\Delta \hat{H}_f^{\circ}$ per mole at 25°C and 1 atm (kcal/g mole)	-11.04	0	$+21.60$	-57.80

We shall use Eq. (4.34) to calculate ΔH_{rxn}° for 5 g moles of NH_3:

$$\Delta H_{rxn}^{\circ} = [4(21.60) + 6(-57.80)] - [5(0) + 4(-1104)]$$
$$= -216.24 \text{ kcal/4 g mole} \quad \text{or} \quad \Delta H_{rxn}^{\circ} = -54.0 \text{ kcal/g mole } NH_3$$

$$\Delta H_{rxn}^{\circ} = \frac{(-54.0) \text{ kcal}}{\text{g mole } NH_3} \left| \frac{5 \text{ g mole } NH_3}{} \right. = -270.0 \text{ kcal}$$

EXAMPLE 4.23 Heat of formation with a phase change

If the standard heat of formation for $H_2O(l)$ is $-68,317$ cal/g mole and the heat of evaporation is $+10,519$ cal/g mole at 25°C and 1 atm, what is the standard heat of formation of $H_2O(g)$?

Solution:

Basis: 1 g mole H_2O

We shall proceed as in Example 4.21 to add the known chemical equations to yield the desired chemical equation and carry out the same operations on the heats of reaction. For reaction A, $\Delta \hat{H}_{rxn}^{\circ} = \sum \Delta \hat{H}_f^{\circ} \text{ products} - \sum \hat{H}_f^{\circ} \text{ reactants}$:

A: $\quad H_2(g) + \frac{1}{2}O_2(g) \longrightarrow H_2O(l) \qquad \Delta \hat{H}_{rxn}^{\circ} = -68.317$ kcal/g mole

B: $\quad H_2O(l) \qquad \qquad \longrightarrow H_2O(g) \qquad \Delta \hat{H}_{vap}^{\circ} = +10.519$ kcal/g mole

---

$A + B$: $\quad H_2(g) + \frac{1}{2}O_2(g) \longrightarrow H_2O(g)$

$$\Delta \hat{H}_{rxnA}^{\circ} + \Delta \hat{H}_{vap}^{\circ} = \Delta \hat{H}_{rxnH_2O(g)}^{\circ} = \Delta \hat{H}_{fH_2O(g)}^{\circ} = -57.798 \text{ kcal/g mole}$$

You can see that any number of chemical equations can be treated by algebraic methods, and the corresponding heats of reaction can be added or subtracted in the same fashion as are the equations. By carefully following these rules of procedure, you will avoid most of the common errors in thermochemical calculations.

To simplify matters, the value cited for ΔH_{vap} of water in Example 4.23 was at 25°C and 1 atm. To calculate this value, if the final state for water is specified as $H_2O(g)$ at 25°C and 1 atm, the following enthalpy changes should be taken into account if you start with $H_2O(l)$ at 25°C and 1 atm:

$$\Delta H_{vap} \text{ at } 25°C \text{ and } 1 \text{ atm} \begin{cases} H_2O(l) \text{ 25°C, 1 atm} \\ \qquad \downarrow \Delta H_1 \\ H_2O(l) \text{ 25°C, vapor pressure at 25°C} \\ \qquad \downarrow \Delta H_2 = \Delta H_{vap} \text{ at the vapor pressure of water} \\ H_2O(g) \text{ 25°C, vapor pressure at 25°C} \\ \qquad \downarrow \Delta H_3 \\ H_2O(g) \text{ 25°C, 1 atm} \end{cases}$$

Practically, the value of ΔH_{vap} at 25°C and the vapor pressure of water of 10,495 cal/g mole will be adequate for engineering calculations.

There are many sources of tabulated values for the standard heats of formation. A good source of extensive data is the *National Bureau of Standards Bulletin 500*, and its supplements, by F. K. Rossini. A condensed set of values for the heats of formation may be found in Appendix F. If you cannot find a standard heat of formation for a particular compound in reference books or in the chemical literature, $\Delta \hat{H}_f^\circ$, may be estimated by the methods described in Verma and Doraiswamy[30] or by some of the authors listed as references in their article. Remember that the values for the standard heats of formation are negative for exothermic reactions.

4.7.2 Standard Heat of Combustion. Standard heats of combustion are the second method of expressing thermochemical data useful for thermochemical calculations. The standard heats of combustion do not have the same standard states as the standard heats of formation. The conventions used with the standard heats of combustion are

(a) The compound is oxidized with oxygen or some other substance to the products $CO_2(g)$, $H_2O(l)$, $HCl(aq)$,[31] etc.

(b) The reference conditions are still 25°C and 1 atm.

(c) Zero values of $\Delta \hat{H}_c^\circ$ are assigned to certain of the oxidation products as, for example, $CO_2(g)$, $H_2O(l)$, and $HCl(aq)$.

---

[30] K. K. Verma and L. K. Doraiswamy, *Ind. Eng. Chem. Fundamentals*, v. 4, p. 389 (1965).

[31] HCl(aq) represents an infinitely dilute solution of HCl (see Sec. 4.8).

(d) If other oxidizing substances are present, such as S or N_2, or if Cl_2 is present, it is necessary to make sure that states of the products are carefully specified and are identical to (or can be transformed into) the final conditions which determine the standard state as shown in Appendix F.

The standard heat of combustion can never be positive but is always negative. A positive value would mean that the substance would not burn or oxidize.

In the search for better high-energy fuels, the heat of combustion of each element in terms of energy per unit mass is of interest. You can see from Fig. 4.14 that the best fuels will use H, Be or B as building blocks. With everything

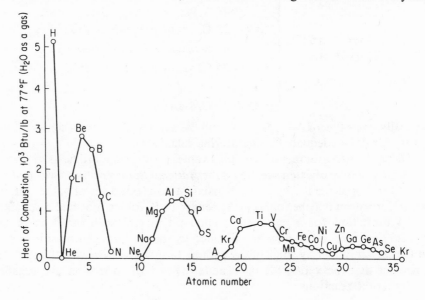

Fig. 4.14.

else held constant, the greater the heat of combustion per unit mass, the longer the range of an airplane or missile.

For a fuel such as coal or oil, the standard heat of combustion is known as the *heating value* of the fuel. To determine the heating value, a weighed sample is burned in oxygen in a calorimeter bomb, and the energy given off is detected by measuring the temperature increase of the bomb and surrounding apparatus.

Because the water produced in the calorimeter is in the vicinity of room temperature (although the gas within the bomb is saturated with water vapor at this temperature), essentially all the water formed in the combustion process is condensed into liquid water. This is not like a real combustion process in a furnace where the water remains as a vapor and passes up the stack. Consequently we have two heating values for fuels containing hydrogen: (a) the gross, or higher, heating value, in which all the water formed is condensed into the liquid state, and (b) the net, or lower, heating value, in which all the water formed remains in the vapor state. The value you determine in the calorimeter is the gross heating value; this is the one reported along with the fuel analysis and should be presumed unless the data state specifically that the net heating value is being reported.

You can estimate the heating value of a coal within about 3 percent from the Dulong formula[32]:

The higher heating value (HHV) in British thermal units per pound

$$= 14{,}544C + 62{,}028\left(H - \frac{O}{8}\right) + 4050\,S$$

where   C = weight fraction carbon
         S = weight fraction net sulfur

$H - \dfrac{O}{8}$ = weight fraction net hydrogen

         = total weight fraction hydrogen $- \frac{1}{8}$ weight fraction oxygen

The values of C, H, S, and O can be taken from the fuel or flue-gas analysis. If the heating value of a fuel is known and the C and S are known, the approximate value of the net H can be determined from the Dulong formula.

EXAMPLE 4.24   Heating value of coal

Coal gasification consists of the chemical transformation of solid coal into gas. The heating values of coal differ, but the higher the heating value, the higher the value of the gas produced (essentially methane, carbon monoxide, hydrogen, etc.). The following coal has a reported heating value of 12,810 Btu/lb as received. Assuming that this is the gross heating value, calculate the net heating value.

comp.	%
C	71.0
H_2	5.6
N_2	1.6
Net S	2.7
Ash	6.1
O_2	13.0
	100.0

Solution:

The corrected ultimate analysis shows 5.6 percent hydrogen on the as-received basis.

Basis: 100 lb coal as received

The water formed on combustion is

$$\frac{5.6\ \text{lb}\ H_2}{}\left|\frac{1\ \text{lb mole}\ H_2}{2.02\ \text{lb}\ H_2}\right|\frac{1\ \text{lb mole}\ H_2O}{1\ \text{lb mole}\ H_2}\left|\frac{18\ \text{lb}\ H_2O}{1\ \text{lb mole}\ H_2O}\right. = 50\ \text{lb}\ H_2O$$

[32]H. H. Lowry, ed., *Chemistry of Coal Utilization*, Wiley, New York, 1945, Chap. 4.

The energy required to evaporate the water is

$$\frac{50 \text{ lb H}_2\text{O}}{100 \text{ lb coal}} \bigg| \frac{1020 \text{ Btu}}{\text{lb H}_2\text{O}} = \frac{510 \text{ Btu}}{\text{lb coal}}$$

The net heating value is

$$12,810 - 510 = 12,300 \text{ Btu/lb}$$

The value 1020 Btu/lb is not the latent heat of vaporization of water at 77°F (1050 Btu/lb) but includes the effect of a change from a heating value at constant volume to one at constant pressure (−30 Btu/lb) as described in a later section of this chapter.

A general relation between the gross heating value and the net heating value is

net Btu/lb coal = gross Btu/lb coal − 91(% total H by weight)

Standard heats of reaction can be calculated from standard heats of combustion by an equation analogous to Eq. (4.34):

$$\Delta H^\circ_{\text{rxn}} = -[\sum \Delta H^\circ_{c \text{ products}} - \sum H^\circ_{c \text{ reactants}}]$$
$$= -[\sum n_{\text{prod}} \Delta \hat{H}^\circ_{c \text{ prod}} - \sum n_{\text{react}} \Delta \hat{H}^\circ_{c \text{ react}}] \tag{4.35}$$

Note: The minus sign in front of the summation expression occurs because the choice of reference states is zero for the right-hand side of the standard equations.

EXAMPLE 4.25 Calculation of heat of reaction from heat of combustion data

What is the heat of reaction of the following reaction:

$$\text{C}_2\text{H}_5\text{OH(l)} + \text{CH}_3\text{COOH(l)} \longrightarrow \text{C}_2\text{H}_5\text{OOCCH}_3\text{(l)} + \text{H}_2\text{O(l)}$$
ethyl alcohol   acetic acid   ethyl acetate

Solution:

Basis: 1 g mole $\text{C}_2\text{H}_5\text{OH}$

Tabulated data	$\text{C}_2\text{H}_5\text{OH(l)}$	$\text{CH}_2\text{COOH(l)}$	$\text{C}_2\text{H}_5\text{OOCCH}_3\text{(l)}$	$\text{H}_2\text{O(l)}$
$\Delta \hat{H}^\circ_c$ per mole at 25°C and 1 atm (kcal/g mole)	−326.70	−208.34	−538.75	0

We use Eq. (4.35):

$$\Delta H^\circ_{\text{rxn}} = -[-538.75 - (-326.70 - 208.34)] = +3.72 \text{ kcal/g mole}$$

EXAMPLE 4.26 Calculation of heat of formation from heats of combustion

Calculate the standard heat of formation of $\text{C}_2\text{H}_2\text{(g)}$ (acetylene) from the standard heat of combustion data.

Solution:

The standard heat of formation for $\text{C}_2\text{H}_2\text{(g)}$ would be expressed in the following manner (basis: 1 g mole C_2H_2):

$$2\text{C}(\beta) + \text{H}_2\text{(g)} \longrightarrow \text{C}_2\text{H}_2\text{(g)} \qquad \Delta H^\circ_f = ?$$

How can we obtain this equation using known standard heat of combustion data? The procedure is to take the chemical equation which gives the standard heat of combustion of $C_2H_2(g)$ (Eq. *A* below) and add or subtract other known combustion equations (*B* and *C*) so that the desired equation is finally obtained algebraically. The details are shown below:

$$\Delta \hat{H}_c^\circ$$
$$\text{(kcal/g mole)}$$

$$
\begin{aligned}
A:&\quad C_2H_2(g) + 2\tfrac{1}{2}O_2(g) \longrightarrow 2CO_2(g) + H_2O(l) &\quad -310.615\\
B:&\quad C(\beta) \;+ O_2(g) \;\longrightarrow CO_2(g) &\quad -94.052\\
C:&\quad H_2(g) \;+ \tfrac{1}{2}O_2(g) \;\longrightarrow H_2O(l) &\quad -68.312\\
\end{aligned}
$$

$-A + 2B + C:$
$$2C(\beta) \;+ H_2(g) \;\longrightarrow C_2H_2(g)$$

$$\Delta H_{net}^\circ = \Delta H_f^\circ = -(-310.615) + 2(-94.052) + (-68.317) = +54.194\,\frac{\text{kcal}}{\text{g mole}}$$

EXAMPLE 4.27   Combination of heats of reaction at 25°C

The following heats of reaction are known from experiments for the reactions below at 25°C in the standard thermochemical state:

	$\Delta \hat{H}_{rxn}^\circ$
rxn	*kcal/g mole*
1. $C_3H_6(g) + H_2(g) \longrightarrow C_3H_8(g)$	-29.6
2. $C_3H_8(g) + 5O_2(g) \longrightarrow 3CO_2(g) + 4H_2O(l)$	-530.6
3. $H_2(g) + \tfrac{1}{2}O_2(g) \longrightarrow H_2O(l)$	-68.3
4. $H_2O(l) \longrightarrow H_2O(g)$	$+10.5\;(\Delta \hat{H}_{vap}^\circ)$
5. $C\text{ (diamond)} + O_2(g) \longrightarrow CO_2(g)$	-94.50
6. $C\text{ (graphite)} + O_2(g) \longrightarrow CO_2(g)$	-94.05

Calculate the following:

(a) The standard heat of formation of propylene (C_3H_6 gas).

(b) The standard heat of combustion of propylene (C_3H_6 gas).

(c) The net heating value of propylene in Btu/ft³ measured at 60°F and 30 in. Hg saturated with water vapor.

Solution:

(a) Reference state: 25°C, 1 atm; $C(\beta)$ (graphite) and $H_2(g)$ are assigned zero values.

$$\text{Desired:}\quad 3C(\beta) + 3H_2(g) \longrightarrow C_3H_6(g)$$
$$\Delta H_f^\circ = ?$$

kcal/g mole

(1) $C_3H_6(g) + H_2(g) \longrightarrow C_3H_8(g)$	$\Delta H_1 = -29.6$
(2) $C_3H_8(g) + 5O_2(g) \longrightarrow 3CO_2(g) + 4H_2O(l)$	$\Delta H_2 = -530.6$
(3) $-4[H_2(g) + \tfrac{1}{2}O_2(g) \longrightarrow H_2O(l)]$	$-4\Delta H_3 = +273.2$
(6) $-3[C(\beta) + O_2(g) \longrightarrow CO_2(g)]$	$-3\Delta H_6 = +282.2$

$(7) = (1) + (2) + (3) + (6):$

$$C_3H_6(g) \longrightarrow 3C(\beta) + 3H_2(g) \qquad\qquad \Delta H_{rxn}^\circ = -4.8$$

$$\Delta H_{f\,C_3H_6(g)}^\circ = +4.8\ \text{kcal/g mole}$$

(b) Reference state: 25°C, 1 atm; $CO_2(g)$ and $H_2O(l)$ have zero values.

Desired: $C_3H_6(g) + \frac{9}{2}O_2(g) \longrightarrow 3CO_2(g) + 3H_2O(l)$

		kcal/g mole

(1) $C_3H_6(g) + H_2(g) \longrightarrow C_3H_8(g)$ $\Delta H_1 = -29.6$
(2) $C_3H_8(g) + 5O_2(g) \longrightarrow 3CO_2(g) + 4H_2O(l)$ $\Delta H_2 = -530.6$
(3) $-[H_2(g) + \frac{1}{2}O_2(g) \longrightarrow H_2O(l)]$ $-\Delta H_3 = +68.3$

(8) = (1) + (2) + (3):
$C_3H_6(g) + \frac{9}{2}O_2(g) \longrightarrow 3CO_2(g) + 3H_2O(l)$ $\Delta H_{rxn}^{\circ} = -491.9$

$$\Delta H_{c\ C_3H_6(g)}^{\circ} = -491.9 \text{ kcal/g mole}$$

(c) Net heating value.

Desired: $C_3H_6(g) + \frac{9}{2}O_2(g) \longrightarrow 3CO_2(g) + 3H_2O(g)$

		kcal/g mole

(8) $C_3H_6(g) + \frac{9}{2}O_2(g) \longrightarrow 3CO_2(g) + 3H_2O(l)$ $\Delta H_8 = -491.9$
(4) $3H_2O(l) \longrightarrow 3H_2O(g)$ $3\Delta H_4 = +31.5$

(9) = (8) + (4):
$C_3H_6(g) + \frac{9}{2}O_2(g) \longrightarrow 3\ CO_2(g) + 3H_2O(g)$ $\Delta H_{rxn} = -460.4$

$$\Delta H_c' = -460,400 \text{ cal/g mole}$$

$$\Delta H_c = (-460,400)(1.8) = -829,000 \text{ Btu/lb mole}$$

At 60°F, 30 in. Hg, saturated with water vapor,

vapor pressure of H_2O at 60°F = 13.3 mm Hg

$$\text{pressure on gas} = (760)\left(\frac{30.00}{29.92}\right) - 13.3$$

$$= 763 - 13.3 = 749.7 \simeq 750 \text{ mm Hg}$$

$$\text{molal volume} = \frac{359}{} \left|\frac{760}{750}\right| \frac{(460+60)}{(460+32)} = 384 \text{ ft}^3/\text{lb mole}$$

$$\text{heating value} = \frac{829,000}{384} = 2160 \text{ Btu/ft}^3 \text{ measured at } 60°F \text{ and } 30 \text{ in. Hg satd.}$$

At 60°F, 30 in. Hg, dry,

$$\text{molal volume} = 379 \text{ ft}^3/\text{lb mole}$$

$$\text{net heating value} = \frac{829,000}{379} = 2185 \text{ Btu/ft}^3$$

One of the common errors made in these thermochemical calculations is to forget that the standard state for heat of combustion calculations is liquid water, and that if gaseous water is present, a phase change (the heat of vaporization or heat of condensation) must be included in the calculations. If the final product in Example 4.25 had been $H_2O(g)$ rather than $H_2O(l)$, we would have incorporated the phase transition for water as follows:

A: $C_2H_5OH(l) + CH_3COOH(l) \longrightarrow C_2H_5OOCCH_3(l) + H_2O(l)$
$$\Delta H^\circ_{rxn} = +3.72$$

B: $H_2O(l) \longrightarrow H_2O(g)$ $\Delta H_{vap} = +10.52$

$A + B$: $C_2H_2OH(l) + CH_3COOH(l) \longrightarrow C_2H_5OOCCH_3(l) + H_2O(g)$
$$\Delta H^\circ_{rxn} = (+3.72) + (10.52) = +14.24 \text{ kcal/g mole}$$

With adequate data available you will find it simpler to use only the heats of combustion or only the heats of formation in the same algebraic calculation for a heat of reaction. Mixing these two sources of enthalpy change data, unless you take care, will only lead to error and confusion. Of course, when called upon, you should be able to calculate the heat of formation from the heat of combustion, or the reverse, being careful to remember and take into account the fact that the standard states for these two types of calculations are different. Since the heat of combustion values are so large, calculations made by subtracting these large values from each other can be a source of considerable error. To avoid other major errors, carefully check all signs and make sure all equations are not only balanced but are written according to the proper convention.

4.7.3 Heats of Reaction at Constant Pressure or at Constant Volume.
Let us first consider the heat of reaction of a substance obtained in a bomb calorimeter, such as in a bomb in which the volume is constant but not the pressure. For such a process (the system is the material in the bomb), the general energy balance, Eq. (4.24), reduces to (with no work, mass flow, nor kinetic or potential energy effects)

$$\Delta U = U_{t_2} - U_{t_1} = Q_v \qquad (4.36)$$

Here Q_v designates the heat transferred from the bomb. Next, let us assume that the heat of reaction is determined in a steady-state flow calorimeter with $W = 0$. Then if the process takes place at constant pressure the general energy balance reduces to:

$$\Delta H = Q_p$$

Here Q_p designates the heat transferred from the flow calorimeter.

If we subtract Q_p from Q_v and introduce $H = U + pV$, we find that

$$Q_p - Q_v = \Delta H - (U_{t_2} - U_{t_1}) = (U_2 - U_1) - (U_{t_2} - U_{t_1}) \qquad (4.37)$$
$$+ (m_2 p_2 \hat{V}_2 - m_1 p_1 \hat{V}_1)$$

Suppose, furthermore, that the internal energy change per unit mass for the batch, constant volume process is made identical to the internal energy difference between the outlet and inlet in the flow process by a suitable adjustment of the temperature of the water bath surrounding the calorimeter. Then

$$(U_{4_2} - U_{t_1}) = (U_2 - U_1)$$

and

$$Q_p - Q_v = (m_2 p_2 \hat{V}_2 - m_1 p_1 \hat{V}_1) \qquad (4.38)$$

To evaluate the terms on the right-hand side of Eq. (4.38), we can assume for solids and liquids that the $\Delta mp\hat{V}$ change is negligible and can be ignored. Therefore, the only change which must be taken into account is for gases present as products and/or reactants. If, for simplicity, the gases are assumed to be ideal, then at constant temperature

$$mp\hat{V} = nRT$$
$$\Delta m\,p\hat{V} = \Delta n\,RT$$

and

$$Q_p - Q_v = \Delta n\,(RT) \tag{4.39}$$

Equation (4.39) gives the difference between the heat of reaction for the constant pressure experiment and the constant volume experiment.

EXAMPLE 4.28   Difference between heat of reaction at constant pressure and at constant volume

Find the difference between the heat of reaction at constant pressure and at constant volume for the following reaction at 25°C (assuming that it could take place):

$$C(s) + \tfrac{1}{2}O_2(g) \longrightarrow CO(g)$$

Solution:

Basis: 1 mole of C(s)

Examine Fig. E4.28. The system is the bomb; $Q = +$ when heat is absorbed by the bomb.

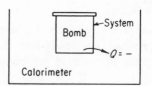

Fig. E4.28.

From Eq. (4.39), $Q_p - Q_v = \Delta n(RT)$.

$$\Delta n = 1 - \tfrac{1}{2} = +\tfrac{1}{2}$$

since C is a solid.

$$Q_p - Q_v = \tfrac{1}{2}RT = 0.5(1.987)(298) = 296 \text{ cal}$$

If the measured heat evolved from the bomb was 26,711 cal,

$$Q_p = Q_v + 296 = -26,711 + 296 = -26,416 \text{ cal}$$

The size of the this correction is relatively insignificant compared to either the quantity Q_p or Q_v. In any case, the heat of reaction calculated from the bomb experiment is

$$\Delta \hat{H}_{\text{rxn}_v} = \frac{Q_v}{n} = \frac{-26,711}{1} = -\frac{26,711 \text{ cal}}{\text{g mole}}$$

Since reported heats of reaction are normally for constant pressure processes, the value of $-26,711$ would not be reported, but instead you would report that

$$\Delta \hat{H}_{rxn_p} = \frac{Q_p}{n} = \frac{-26,416}{1} = -\frac{26,416 \text{ cal}}{\text{g mole}}$$

4.7.4 Incomplete Reactions. If an incomplete reaction occurs, you should calculate the standard heat of reaction only for the products which are actually formed from the reactants which actually react. In other words, only the portion of the reactants which actually undergo some change and liberate or absorb some energy are to be considered in calculating the overall standard heat of the reaction. If some material passes through unchanged, then you should not include it in the *standard heat of reaction* calculations (however, when the reactants or products are at conditions *other* than 25°C and 1 atm, whether they react or not, you must include them in the enthalpy calculations as explained in Sec. 4.7.5).

EXAMPLE 4.29 Incomplete reactions

An iron pyrites ore containing 85.0 percent FeS_2 and 15.0 percent gangue (inert dirt, rock, etc.) is roasted with an amount of air equal to 200 percent excess air according to the reaction

$$4FeS_2 + 11O_2 \longrightarrow 2Fe_2O_3 + 8SO_2$$

in order to produce SO_2. All the gangue plus the Fe_2O_3 end up in the solid waste product (cinder) which analyzes 4.0 percent FeS_2. Determine the standard heat of reaction per kilogram of ore.

Solution:
The material balance for the problem must be worked out prior to determining the standard heat of reaction.

Basis: 100 kg ore

Ore			
Comp.	*kg* = %	*mol. wt*	*kg moles*
FeS_2	85.0	120	0.708
Gangue	15.0		
	100.0		

Required O_2: $\dfrac{0.708 \text{ kg mole FeS}_2}{} \left| \dfrac{11 \text{ kg moles O}_2}{4 \text{ kg moles FeS}_2} \right. = 1.94 \text{ kg moles O}_2$

Entering O_2: $(1 + 2)(1.94) = 5.82 \text{ kg moles O}_2$

Entering N_2: $5.82 \left(\dfrac{79}{21}\right) = 21.9 \text{ kg moles N}_2$

The solid waste comprises Fe_2O_3, gangue, and unburned FeS_2. We need to compute the amount of unburned FeS_2 on the basis of 100 kg of ore. Let $x = \text{kg FeS}_2$

that does not burn. Then the cinder comprises

component	kg
FeS$_2$	x
Gangue	15

Fe$_2$O$_3$ $\quad\dfrac{(85-x)\text{ kg FeS}_2\text{ burned}}{120\dfrac{\text{kg FeS}_2}{\text{kg mole FeS}_2}}\Bigg|\dfrac{2\text{ moles Fe}_2\text{O}_3}{4\text{ moles FeS}_2}\Bigg|\dfrac{160\text{ kg Fe}_2\text{O}_3}{\text{kg mole Fe}_2\text{O}_3}=\left(\dfrac{2}{3}\right)(85-x)$

Thus

$$0.040 = \frac{x}{x + 15 + (2/3)(85 - x)}$$

$$x = 2.90 \text{ kg FeS}_2$$

$$(2/3)(85 - x) = 54.7 \text{ kg Fe}_2\text{O}_3$$

The composition of the solid waste is

Solid waste comp.	kg	mol. wt	kg moles
FeS$_2$	2.90	120	0.0242
Fe$_2$O$_3$	54.7	160	0.342
Gangue	15		

Finally, the FeS$_2$ that is oxidized is

$$\frac{(85 - 2.90)\text{ kg FeS}_2}{}\Bigg|\frac{1\text{ kg mole FeS}_2}{120\text{ kg FeS}_2} = 0.685 \text{ kg mole FeS}_2$$

Corresponding to 0.685 kg mole of FeS$_2$ oxidized:

$$\frac{0.685\text{ kg mole FeS}_2}{}\Bigg|\frac{8\text{ kg moles SO}_2}{4\text{ kg moles FeS}_2} = 1.37 \text{ kg moles SO}_2 \text{ produced}$$

$$\frac{0.685\text{ kg mole FeS}_2}{}\Bigg|\frac{11\text{ kg moles O}_2}{4\text{ kg moles FeS}_2} = 1.88 \text{ kg moles O}_2 \text{ used}$$

Tabulated heats of formation in kilocalories per gram mole are

Comp.:	FeS$_2$(c)	O$_2$(g)	Fe$_2$O$_3$(c)	SO$_2$(g)
$\Delta \hat{H}_f^{\circ}$:	-42.520	0	-196.500	-70.960

$\Delta H_{\text{rxn}}^{\circ} = (1.37)(-70.960) + (0.342)(-196.500) - (0.685)(-42.520)$

$\qquad\quad = -135.292 \text{ kcal}$

$\Delta \hat{H}_{\text{rxn}}^{\circ} = 1.353 \text{ kcal/kg ore}$

Note that only the moles reacting and produced are used in computing the heat of reaction. The unoxidized FeS$_2$ is not included, nor is the N$_2$. The O$_2$ used contributes a value of zero to the $\Delta H_{\text{rxn}}^{\circ}$ because the ΔH_f° of the O$_2$ is zero.

4.7.5 The Energy Balance when the Products and Reactants Are Not at 25°C.

You no doubt realize that the standard state of 25°C for the heats of formation is only by accident the temperature at which the products enter and the reactants leave a process. In most instances the temperatures of the materials entering and leaving will be higher or lower than 25°C. However, since enthalpies (and hence heats of reaction) are point functions, you can use Eq. (4.24) together with Eq. (4.34), or Eq. (4.35), to answer questions about the process being analyzed. Typical questions might be

(a) What is the heat of reaction at a temperature other than 25°C, but still at 1 atm?
(b) What is the temperature of an incoming or exit stream?
(c) What is the temperature of the reaction?
(d) How much material must be introduced to provide a specified amount of heat transfer?

Consider the process illustrated in Fig. 4.15, for which the reaction is

$$aA + bB \longrightarrow cC + dD$$

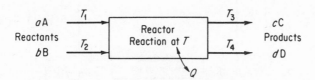

Fig 4.15. Process with reaction.

In employing Eqs. (4.24) and (4.34), you should always first choose a reference state at which the heat of reaction is known. This usually turns out to be 25°C and 1 atm. Then the next step is to calculate the enthalpy changes for each stream entering and leaving *relative to this reference state*. Finally, the enthalpy changes are summed up, including the heat of reaction calculated in the standard state by Eq. (4.34), and the other terms in Eq. (4.24) introduced if pertinent. Usually it proves convenient in calculating the enthalpy change to compute the "sensible heat" changes and the heat of reaction separately. Methods you can use to determine the "sensible heat" ΔH values for the individual streams are

(a) Obtain the enthalpy values from a set of published tables, e.g., the steam tables or tables such as in Appendix D,
(b) Use $\Delta H = C_{p_{m2}}(T_2 - T_{\text{ref}}) - C_{p_{m1}}(T_1 - T_{\text{ref}})$, as previously discussed in Sec. 4.2,
(c) Analytically, graphically, or numerically, find

$$\Delta H = \int_{T_1}^{T_2} C_p \, dT$$

for each component individually, using the respective heat capacity equations.

Any phase changes taking place among the reactants or products not accounted for in the chemical equation must also be taken into account in the calculations.

Let us first demonstrate how to calculate the heat of reaction at a temperature other than 25°C. By this we mean that the reactants enter and the products leave at the same temperature—a temperature different from the standard state of 25°C. Figure 4.16 illustrates the information flow for the calculations assuming a steady-state process ($\Delta E = 0$), no kinetic or potential energy changes, and $W = 0$. The general energy balance, Eq. (4.24), reduces to

$$Q = \Delta H$$

or

$$Q = (\Delta H_{\mathrm{P}} - \Delta H_{\mathrm{R}}) + (\Delta \hat{H}_{\mathrm{rxn}_{T_{\mathrm{ref}}}}) \tag{4.40}$$

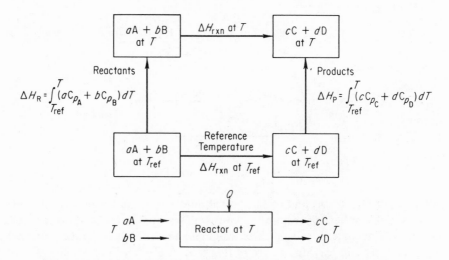

Fig. 4.16. Heat of reaction at a temperature other than standard conditions.

in terms of the notation in Fig. 4.16. By definition, Q, as calculated by Eq. (4.40), is equal to the "heat of reaction at the temperature T" so that

$$\Delta H_{\mathrm{rxn}_T} = \Delta H_{\mathrm{rxn}_{T_{\mathrm{ref}}}} + \Delta H_{\mathrm{P}} - \Delta H_{\mathrm{R}} \tag{4.41}$$

To indicate a simplified way to calculate $\Delta H_{\mathrm{P}} - \Delta H_{\mathrm{R}}$, suppose that the heat capacity equations are expressed as

$$C_p = \alpha + \beta T + \gamma T^2 \tag{4.42}$$

Then, to obtain $\Delta H_{\mathrm{P}} - \Delta H_{\mathrm{R}}$, we add up the enthalpy changes for the products and subtract those for the reactants. Rather than integrate separately, let us consolidate like terms as follows. Each heat capacity equation is multiplied by the proper number of moles:

$$aC_{p_A} = a[\alpha_A + \beta_A T + \gamma_A T^2] \tag{4.43a}$$

$$bC_{p_B} = b[\alpha_B + \beta_B T + \gamma_B T^2] \tag{4.43b}$$

$$cC_{p_C} = c[\alpha_C + \beta_C T + \gamma_C T^2] \tag{4.43c}$$

$$dC_{p_D} = d[\alpha_D + \beta_D T + \gamma_D T^2] \tag{4.43d}$$

Then we define a new term ΔC_p which is equal to

original expression   **equivalent**
new term

$$cC_{pc} + dC_{pD} - (aC_{pA} + bC_{pB}) = \Delta C_p$$

and

$$[(c\alpha_C + d\alpha_D) - (a\alpha_A + b\alpha_B)] = \Delta\alpha \tag{4.44}$$

$$T[c\beta_C + d\beta_D) - (a\beta_A + b\beta_B)] = T\Delta\beta$$

$$T^2[c\gamma_C + d\gamma_D) - (a\gamma_A + b\gamma_B)] = T^2\Delta\gamma$$

Simplified, ΔC_p can be expressed as

$$\Delta C_p = \Delta\alpha + \Delta\beta T + \Delta\gamma T^2 \tag{4.45}$$

Furthermore,

$$(\Delta H_P - \Delta H_R) = \int_{T_R}^{T} \Delta C_p \, dT = \int_{T_R}^{T} (\Delta\alpha + \Delta\beta T + \Delta\gamma T^2) \, dT$$

$$= \Delta\alpha(T - T_R) + \frac{\Delta\beta}{2}(T^2 - T_R^2) + \frac{\Delta\gamma}{3}(T^3 - T_T^3) \tag{4.46}$$

where we let $T_R = T_{\text{ref}}$ for simplicity.

If the integration is carried out without definite limits,

$$(\Delta H_P - \Delta H_R) = \int (\Delta C_p) \, dT = \Delta\alpha T + \frac{\Delta\beta}{2}T^2 + \frac{\Delta\gamma}{3}T^3 + C \tag{4.47}$$

where C is the integration constant.

Finally ΔH_{rxn} at the new temperature T is

$$\Delta H_{\text{rxn}_T} = \Delta H_{\text{rxn}_{T_R}} + \Delta\alpha(T - T_R) + \frac{\Delta\beta}{2}(T^2 - T_R^2) + \frac{\Delta\gamma}{3}(T^3 - T_R^3) \tag{4.48a}$$

or

$$\Delta H_{\text{rxn}_T} = \Delta H_{\text{rxn}_{T_R}} + \Delta\alpha T + \frac{\Delta\beta}{2}T^2 + \frac{\Delta\gamma}{3}T^3 + C \tag{4.48b}$$

as the case may be. Using Eq. (4.48a) and knowing ΔH_{rxn} at the reference temperature T_R, you can easily calculate ΔH_{rxn} at any other temperature.

Equation (4.48b) can be consolidated into

$$\Delta H_{\text{rxn}_T} = \Delta H_0 + \Delta\alpha T + \frac{\Delta\beta}{2}T^2 + \frac{\Delta\gamma}{3}T^3 \tag{4.49}$$

where

$$\Delta H_0 = (\Delta H_{\text{rxn}_{T_R}} + C)$$

Now, if ΔH_{rxn} is known at any temperature T, you can calculate ΔH_0 as follows:

$$\Delta H_0 = \Delta H_{\text{rxn}_T} - \Delta\alpha(T) - \frac{\Delta\beta}{2}(T^2) - \frac{\Delta\gamma}{3}T^3 \tag{4.50}$$

Then you can use this value of ΔH_0 to assist in the calculation of ΔH_{rxn} at any other temperature. Probably the easiest way to compute the necessary enthalpy changes is to use enthalpy data obtained directly from published tables. Do not forget to take into account phase changes, if they take place, in the enthalpy calculations.

EXAMPLE 4.30 Calculation of heat of reaction at a temperature different from standard conditions

An inventor thinks he has developed a new catalyst which can make the gas phase reaction

$$CO_2 + 4H_2 \longrightarrow 2H_2O + CH_4$$

proceed with 100 percent conversion. Estimate the heat which must be provided or removed if the gases enter and leave at 500°C.

Solution:

In effect we need to calculate heat of reaction at 500°C from Eq. (4.41) or Q in Eq. (4.40). For illustrative purposes we first use the technique based on Eqs. (4.49) and (4.50).

Basis: 1 g mole $CO_2(g)$

$$CO_2(g) + 4H_2(g) \longrightarrow 2H_2O(g) + CH_4(g)$$

$$-\Delta \hat{H}_f^\circ \left(\frac{\text{kcal}}{\text{kg mole}} \right) \quad 94,052 \qquad 0 \qquad 57,798 \qquad 17,889$$

$$\Delta \hat{H}_{\text{rxn}298°K} = [-17,889 - (2)(57,798)] - [(4)(0) - 94,052]$$
$$= -39,433 \text{ cal/g mole } CO_2$$

First we shall calculate ΔC_p:

$$C_{p\,CO_2} = 6.393 + 10.100 \times 10^{-3}T - 3.405 \times 10^{-6}T^2 \qquad T \text{ in } °K$$

$$C_{p\,H_2} = 6.424 + 1.039 \times 10^{-3}T - 0.078 \times 10^{-6}T^2 \qquad T \text{ in } °K$$

$$C_{p\,H_2O} = 6.970 + 3.464 \times 10^{-3}T - 0.483 \times 10^{-6}T^2 \qquad T \text{ in } °K$$

$$C_{p\,CH_4} = 3.204 + 18.41 \times 10^{-3}T - 4.48 \times 10^{-6}T^2 \qquad T \text{ in } °K$$

$$\Delta\alpha = [(1)(3.204) + (2)(6.970)] - [(1)(6.393) + (4)(6.424)]$$
$$= -14.945$$

$$\Delta\beta = [(1)(18.41) + (2)(3.464)](10^{-3}) - [(1)(10.100) + (4)(1.039)](10^{-3})$$
$$= 11.087 \times 10^{-3}$$

$$\Delta\gamma = [(1)(-4.48) + (2)(-0.483)](10^{-6}) - [(1)(-3.405) + (4)(-0.078)](10^{-6})$$
$$= -1.729 \times 10^{-6}$$

$$\Delta C_p = -14.945 + 11.082 \times 10^{-3}T - 1.729 \times 10^{-6}T^2$$

Next we find ΔH_0, using as a reference temperature 298°K.

$$\Delta H_0 = \Delta H_{\text{rxn}298} - \Delta\alpha T - \frac{\Delta\beta}{2}T^2 - \frac{\Delta\gamma}{3}T^3$$

$$= -39,443 - (-14.945)(298) - \frac{11.082 \times 10^{-3}}{2}(298)^2$$

$$- \frac{(-1.729 \times 10^{-6})}{3}(298)^3$$

$$= -39,443 + 4450 - 493 + 15 = -35,431 \text{ cal}$$

Then, with ΔH_0 known, the ΔH_{rxn} at 773°K can be determined.

$$\Delta H_{rxn_{773}°K} = \Delta H_0 + \Delta \alpha T + \frac{\Delta \beta}{2} T^2 + \frac{\Delta \gamma}{3} T^3$$

$$= -35,431 - 14.945(773) + \frac{11.082 \times 10^{-3}}{2}(773)^2$$

$$- \frac{1.729 \times 10^{-6}}{3}(773)^3$$

$$= -43,925 \text{ cal}$$

or 43,925 kcal/kg mole CO_2 must be removed.

EXAMPLE 4.31 Calculation of heat of reaction at a temperature different from standard conditions

Repeat the calculation of the previous example using enthalpy values from Table 4.5b and Appendix D.

Solution:
The heat of reaction at 25°C and 1 atm from the previous example is

$$\Delta \hat{H}_{rxn_{298}°K} = -39,433 \frac{\text{kcal}}{\text{kg mole}}$$

$\Delta \hat{H}$ (kcal/kg mole); reference is 0°C

temperature	CO_2	H_2	H_2O	CH_4
25°C	218	172	200	210
500°C	5340	3499	4254	5730

From Eq. (4.41),

$$\Delta H_{rxn_{500}°C} = \Delta H_{rxn_{25}°C} + \Delta H_{products} - \Delta H_{reactants}$$

$$= -39,433 + [(1)(5730 - 210) + 2(4254 - 200)]$$

$$- [(1)(5340 - 218) + 4(3499 - 172)]$$

$$= 44,235 \frac{\text{kcal}}{\text{kg mole } CO_2}$$

Note that the enthalpies of the products and of the reactants are both based on the reference temperature of 25°C. The answer is not quite the same as in the previous example, because the heat capacity data used in Example 4.30 were not quite the same as those used in calculating the ΔH values in the tables.

EXAMPLE 4.32 Application of the energy balance to a reaction

Carbon monoxide at 50°F is completely burned at 2 atm pressure with 50 percent excess air which is at 1000°F. The products of combustion leave the combustion chamber at 800°F. Calculate the heat evolved from the combustion chamber expressed as British thermal units per pound of CO entering.

Solution:

Basis: 1 lb mole of CO

$$CO(g) + \tfrac{1}{2}O_2(g) \longrightarrow CO_2(g)$$

Material balance:

Amount of air entering:

$$\frac{1 \text{ lb mole CO}}{} \Bigg| \frac{0.5 \text{ lb mole O}_2}{\text{lb mole CO}} \Bigg| \frac{1.5 \text{ lb mole O}_2 \text{ used}}{1.0 \text{ lb O}_2 \text{ mole needed}} \Bigg| \frac{1 \text{ lb mole air}}{0.21 \text{ lb mole O}_2}$$

$= 3.57$ lb mole air

Amount of O_2 leaving, unreacted:

$(0.21)(3.57) - 0.5 = 0.25$ lb mole O_2

Amount of N_2 leaving:

$(0.79)(3.57) = 2.82$ lb mole N_2

Enthalpy data have been taken from Table 4.5a and Appendix D. The heat of reaction at 25°C (77°F) and 1 atm from Example 4.21 is $-67,636$ cal/g mole of CO or $-121,745$ Btu/lb mole of CO. We can assume that the slightly higher pressure of 2 atm has no effect on the heat of reaction or the enthalpy values. See Fig. E4.32.

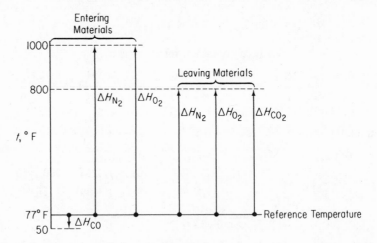

Fig. E4.32.

ΔH (Btu/lb mole); reference is 32°F

temperature	CO	air	O_2	N_2	CO_2
50°F	125.2	—	—	—	—
77°F	313.3	312.7	315.1	312.2	392.2
800°F	—	—	5690	5443	8026
1000°F	—	6984	—	—	—

$$Q = \Delta H_{\text{rxn}_{77°F}} + \Delta H_{\text{products}} - \Delta H_{\text{reactants}}$$

(a)

$$\Delta \hat{H}_{\text{rxn}_{77°F}} = -121,745 \text{ Btu/lb mole}$$

(b)
$$\Delta H_{products} = \Delta H_{800°F} - \Delta H_{77°F}$$
$$= \overset{CO_2}{(1)(8026 - 392.2)} + \overset{N_2}{(2.82)(5443 - 312.2)}$$
$$\overset{O_2}{+ (0.25)(5690 - 315.1)}$$
$$= 7633.8 + 14,480 + 1343$$
$$= 23,457 \text{ Btu/lb mole}$$

(c)
$$\Delta H_{reactants} = \overset{Air}{\Delta H_{1000°F} - \Delta H_{77°F}} + \overset{CO}{\Delta H_{50°F} - \Delta H_{77°F}}$$
$$= (3.57)(6984 - 312.7) + (1)(125.2 - 313.3)$$
$$= 23,800 - 188.1 = 23,612 \text{ Btu/lb mole}$$

(d)
$$Q = -121,745 + 23,457 - 23,612$$
$$= -121,900 \text{ Btu/lb mole}$$

$$\frac{-121,900 \text{ Btu}}{\text{lb mole CO}} \left| \frac{1 \text{ lb mole CO}}{28 \text{ lb CO}} \right. = -4360 \text{ Btu/lb CO}$$

4.7. 6 Temperature of a Reaction. We are now equipped to determine what is called the *adiabatic reaction temperature*. This is the temperature obtained inside the process when (a) the reaction is carried out under adiabatic conditions, i.e., there is no heat interchange between the container in which the reaction is taking place and the surroundings, and (b) when there are no other effects present, such as electrical effects, work, ionization, free radical formation, etc. In calculations of flame temperatures for combustion reactions, the adiabatic reaction temperature assumes complete combustion. Equilibrium considerations may dictate less than complete combustion for an actual case. For example, the adiabatic flame temperature for the combustion of CH_4 with theoretical air has been calculated to be 2010°C; allowing for incomplete combustion, it would be 1920°C. The actual temperature when measured is 1885°C.

The adiabatic reaction temperature tells us the temperature ceiling of a process. We can do no better, but of course the actual temperature may be less. The adiabatic reaction temperature helps us select the types of materials that must be specified for the container in which the reaction is taking place. Chemical combustion with air produces gases at a maximum temperature of 2500°K which can be increased to 3000°K with the use of oxygen and more exotic oxidants, and even this value can be exceeded although handling and safety problems are severe. Applications of such hot gases lie in the preparation of new materials, micromachining, welding using laser beams, and the direct generation of electricity using ionized gases as the driving fluid.

To calculate the adiabatic reaction temperature, you assume that all the energy liberated from the reaction at some base temperature plus that brought in by the entering stream (relative to the same base temperature) is available to raise the temperature of the products. We assume that the products leave at the temperature of the reaction, and thus if you know the temperature of the products, you automatically know the temperature of the reaction. In effect, for this adiabatic process we can apply Eq. (4.40).

Since no heat escapes and the energy liberated is allowed only to increase the enthalpy of the products of the reaction, we have

$$\Delta H_{\text{products}} = \boxed{\Delta H_{\text{reactants}} - \Delta H^{\circ}_{\text{rxn}}} \qquad (4.51)$$

"energy pool"

Any phase changes that take place and are not accounted for by the chemical equation must be incorporated into the "energy pool." Because of the character of the information available, the determination of the adiabatic reaction temperature or flame temperature may involve a trial-and-error solution; hence the iterative solution of adiabatic flame temperature problems is often carried out on the computer.

EXAMPLE 4.33 Adiabatic flame temperature

Calculate the theoretical flame temperature for CO burned at constant pressure with 100 percent excess air, when the reactants enter at 200°F.

Solution:

$$CO(g) + \tfrac{1}{2}O_2(g) \longrightarrow CO_2(g)$$

Basis: 1 g mole CO(g); ref. temp. 25°C (77°F)

$$(200°F \approx 93.3°C)$$

Material balance:

entering reactants		exit products	
comp.	moles	comp.	moles
CO(g)	1.00	CO$_2$(g)	1.00
O$_2$(req.)	0.50	O$_2$(g)	0.50
O$_2$(xs)	0.50	N$_2$(g)	3.76
O$_2$(total)	1.00		
N$_2$	3.76		
Air	4.76		

$$\Delta H_{\text{rxn}} \text{ at } 25°C = -67,636 \text{ cal}$$

Note: For the purposes of illustration, in this example the mean heat capacities have a reference temperature of 25°C rather than 0°C.

Then, for Eq. (4.51), $\Delta H_{\text{reactants}}$:

		reactants		
comp.	moles	ΔT	C_{p_m}	ΔH
CO(g)	1.00	68.3	6.981	476
Air	4.76	68.3	6.993	2270
			$\sum \Delta H_R =$	2746 cal

$$\Delta H_{\text{products}} = 2746 + 67,636 = 70,382 \text{ cal}$$

To find the temperature which yields a $\Delta H_{\text{products}}$ of 70,382 cal, the simplest procedure is to assume various values of the exit temperature of the products until the $\sum \Delta H_P = 70{,}382$.

Assume TFT (theoretical flame temperature) = 1800°C:

$$\Delta T = 1800 - 25 = 1775°C$$

comp.	moles	C_{p_m}	ΔH
CO_2	1.00	12.94	23,000
O_2	0.50	8.349	7,400
N_2	3.76	7.924	52,900

$$\sum \Delta H_P = 83{,}300 \text{ cal}$$

Assume TFT = 1500°C:

$$\Delta T = 1500 - 25 = 1475°C$$

comp.	moles	C_{p_m}	ΔH
CO_2	1.00	12.70	18,740
O_2	0.50	8.305	6,120
N_2	3.76	7.879	43,600

$$\sum \Delta H_P = 68{,}460 \text{ cal}$$

Make a linear interpolation:

$$\text{TFT} = 1500 + \frac{70{,}382 - 68{,}460}{83{,}300 - 68{,}460}(300) = 1500 + 39$$

$$\text{TFT} = 1539°C \backsimeq 2798°F$$

4.8 Heats of solution and mixing

So far in our energy calculations, we have been considering each substance to be a completely pure and separate material. The physical properties of an ideal solution or mixture may be computed from the sum of the properties in question for the individual components. For gases, mole fractions can be used as weighting values, or, alternatively, each component can be considered to be independent of the others. In most instances so far in this book we have used the latter procedure. Using the former technique, as we did in a few instances, we could write down, for the heat capacity of an ideal mixture,

$$C_{p \text{ mixture}} = x_A C_{p_A} + x_B C_{p_B} + x_C C_{p_C} + \cdots \tag{4.52}$$

or, for the enthalpy,

$$\Delta \hat{H}_{\text{mixture}} = x_A \Delta \hat{H}_A + x_B \Delta \hat{H}_B + x_C \Delta \hat{H}_C + \cdots \tag{4.53}$$

These equations are applicable to ideal mixtures only.

When two or more pure substances are mixed to form a gas or liquid solution, we frequently find energy is absorbed or evolved upon mixing. Such a solution would be called a "real" solution. The total heat of mixing (ΔH_{mixing}) has to be determined experimentally, but can be retrieved from tabulated experimental (smoothed) results, once such data are available. This type of

energy change has been given the formal name *heat of solution* when one substance dissoves in another; and there is also the negative of the heat of solution,
the *heat of dissolution*, for a substance which separates from a solution. Tabulated
data for heats of solution appear in Table 4.9 in terms of energy per mole for
consecutively added quantities of solvent to solute; the gram mole refers to the
gram mole of solute. Heats of solution are somewhat similar to heats of reaction
in that an energy change takes place because of differences in the forces of
attraction of the solvent and solute molecules. Of course, these energy changes
are much smaller than those we find accompanying the breaking and combining
of chemical bonds. Heats of solution are conveniently accommodated in exactly
the same way as are the heats of reaction in the energy balance.

TABLE 4.9 HEAT OF SOLUTION OF HCl
(at 25°C and 1 atm)

Composition	Total Moles H_2O Added to 1 mole HCl	$-\Delta \hat{H}°$ for Each Incremental Step (cal/g mole)	Integral Heat of Solution: The Cumulative $-\Delta \hat{H}°$ (cal/g mole)
HCl(g)	0		
HCl·1H$_2$O	1	6268	6,268
HCl·2H$_2$O	2	5400	11,668
HCl·3H$_2$O	3	1920	13,588
HCl·4H$_2$O	4	1040	14,628
HCl·5H$_2$O	5	680	15,308
HCl·8H$_2$O	8	1000	16,308
HCl·10H$_2$O	10	300	16,608
HCl·15H$_2$O	15	359	16,967
HCl·25H$_2$O	25	305	17,272
HCl·50H$_2$O	50	242	17,514
HCl·100H$_2$O	100	136	17,650
HCl·200H$_2$O	200	85	17,735
HCl·500H$_2$O	500	76	17,811
HCl·1000H$_2$O	1,000	39	17,850
HCl·5000H$_2$O	5,000	59	17,909
HCl·50,000H$_2$O	50,000	35	17,944
HCl·∞H$_2$O		16	17,960

SOURCE: *National Bureau of Standards Circular 500* (Reference 15 in Table 4.6).

The solution process can be represented by an equation such as the following:

$$HCl(g) + 5H_2O \longrightarrow HCl:5H_2O$$

or

$$HCl(g) + 5H_2O \longrightarrow HCl(5H_2O)$$
$$\Delta \hat{H}°_{soln} = -15,308 \text{ cal/g mole HCl(g)}$$

The expression $HCl(5H_2O)$ means that 1 mole of HCl has been dissolved in 5 moles of water, and the enthalpy change for the process is $-15,308$ cal/g mole of HCl. Table 4.9 shows the heat of solution for various cumulative numbers of moles of water added to 1 mole of HCl.

The *standard integral heat of solution* is the cumulative $\Delta \hat{H}^\circ_{soln}$ as shown in the last column for the indicated number of molecules of water. As successive increments of water are added to the mole of HCl, the cumulative heat of solution (the integral heat of solution) increases, but the incremental enthalpy change decreases as shown in Table 4.9. Note that both the reactants and products have to be at standard conditions. The heat of dissolution would be just the negative of these values. The integral heat of the solution is plotted in Fig. 4.17, and you can see that an asymptotic value is approached as the solution

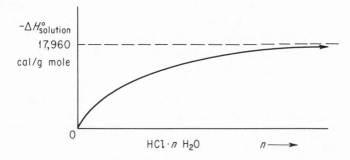

Fig. 4.17. Integral heat of solution of HCl in water.

becomes more and more dilute. At infinite dilution this value is called the *standard integral heat of solution at infinite dilution* and is $-17,960$ cal/g mole of HCl. In the Appendix are other tables presenting standard integral heat of solution data. Since the energy changes for heats of solution are point functions, you can easily look up any two concentrations of HCl and find the energy change caused by adding or subtracting water.

EXAMPLE 4.34   Heat of solution

Calculate the standard heat of formation of HCl in 5 g moles of water.

Solution:
We treat the solution process in an identical fashion to a chemical reaction:

$$\text{kcal/g mole}$$

A:	$\Delta \hat{H}^\circ_f = -22.063$	$\frac{1}{2}H_2(g) + \frac{1}{2}Cl_2(g)$	$= HCl(g)$
B:	$\Delta \hat{H}^\circ_{soln} = -15.308$	$HCl(g) + 5H_2O$	$= HCl(5H_2O)$
$A + B$:	$\Delta \hat{H}^\circ_f = -37.371$	$\frac{1}{2}H_2(g) + \frac{1}{2}Cl_2(g) + 5H_2O$	$= HCl(5H_2O)$

It is important to remember that the heat of formation of H_2O itself does not enter into the calculation. The heat of formation of HCl in an infinitely dilute solution is

$$\Delta \hat{H}_f^\circ = -22.063 - 17.960 = -40.023 \text{ kcal/g mole}$$

Another type of heat of solution which is occasionally encountered is the partial molal heat of solution. Information about this thermodynamic property can be found in most standard thermodynamic texts or in books on thermochemistry, but we do not have the space to discuss it here.

One point of special importance concerns the formation of water in a chemical reaction. When water participates in a chemical reaction in solution as a reactant or product of the reaction, you must include the heat of formation of the water as well as the heat of solution in the energy balance. Thus, if HCl reacts with sodium hydroxide to form water and the reaction is carried out in only 2 moles of water to start with, it is apparent that you will have 3 moles of water at the end of the process. Not only do you have to take into account the heat of reaction when the water is formed, but there is also a heat of solution contribution. If gaseous HCl reacts with crystalline sodium hydroxide and the product is 1 mole of gaseous water vapor, then you could employ the energy balance without worrying about the heat of solution effect.

If an enthalpy-concentration diagram is available for the system in which you are interested, then the enthalpy changes can be obtained directly from the diagram by convenient graphical methods. See Chap. 5 for illustrations of this technique.

EXAMPLE 4.35   Application of heat of solution data

An ammonium hydroxide solution is to be prepared at 77°F by dissolving gaseous NH_3 in water. Prepare charts showing

    (a) The amount of cooling needed in Btu to prepare a solution containing 1 lb mole of NH_3 at any concentration desired.

    (b) The amount of cooling needed in Btu to prepare 100 gal of a solution of any concentration up to 35 percent NH_3.

    (c) If a 10.5 percent NH_3 solution is made up without cooling, at what temperature will the solution end up?

Solution:
Heat of solution data have been taken from *NBS Circular* 500.

    (a)   Basis: 17 lb NH_3 = 1 lb mole NH_3

Reference temperature = 77°F ≎ 25°C

To convert from kilocalories per gram mole to British thermal units per pound mole, multiply by 1800.

description	state	$-\Delta\hat{H}_f^\circ$ kcal/g mole	$-\Delta\hat{H}_f^\circ$ Btu/lb mole	$-\Delta\hat{H}_{soln}^\circ$ Btu/lb mole	weight %NH$_3$
1H$_2$O	g	11.04	19,900	0	100
1H$_2$O	aq	18.1	32,600	12,700	48.5
2H$_2$O	aq	18.7	33,600	13,700	32.0
3H$_2$O	aq	18.87	34,000	14,100	23.9
4H$_2$O	aq	18.99	34,200	14,300	19.1
5H$_2$O	aq	19.07	34,350	14,450	15.9
10H$_2$O	aq	19.23	34,600	14,700	8.63
20H$_2$O	aq	19.27	34,700	14,800	4.51
30H$_2$O	aq	19.28	34,700	14,800	3.05
40H$_2$O	aq	19.28	34,700	14,800	2.30
50H$_2$O	aq	19.29	34,750	14,850	1.85
100H$_2$O	aq	19.30	34,750	14,850	0.94
200H$_2$O	aq	19.32	34,800	14,900	0.47
∞H$_2$O	aq	19.32	34,800	14,900	0.0

Standard heats of solution have been calculated from the cumulative data as follows:

$$\Delta\hat{H}_{soln}^\circ = \Delta\hat{H}_f^\circ - \Delta\hat{H}_{f_{gas}}^\circ$$

For example,

$$\Delta\hat{H}_{soln.}^\circ = [-32,600 - (-19,900)]$$
$$= -12,700 \text{ Btu/lb mole NH}_3$$

Weight percents have been computed as follows:

$$\text{wt \%NH}_3 = \frac{\text{lb NH}_3(100)}{\text{lb H}_2\text{O} + \text{lb NH}_3}$$

$$\text{\%NH}_3 \text{ for 1 H}_2\text{O}: = \frac{17(100)}{18(1) + 17} = 48.5\%$$

The heat of solution values shown in Fig. E4.35(a) are equivalent to the cooling duty required.

(b) Part (b) of the problem requires that a new basis be selected, 100 gal of solution. Additional data concerning densities of NH$_4$OH are shown in Table E4.35. Sample calculations are as follows:

$$\text{density, lb/100 gal} = \frac{(\text{sp gr soln})(62.4 \text{ lb/ft}^3)(1.003)(100)}{(7.48 \text{ gal/ft}^3)}$$

Note: 1.003 is the specific gravity of water at 77°F.

$$\text{density @ 32.0\%NH}_3 = \frac{(0.889)(62.4)(1.003)(100)}{7.48} = 741 \text{ lb/100 gal}$$

$$\left.\begin{array}{c}\text{cooling req'd} \\ \text{Btu/100 gal}\end{array}\right\} = \left(\frac{\text{lb moles NH}_3}{100 \text{ gal}}\right)\left(-\Delta H_{soln}^\circ \frac{\text{Btu}}{\text{lb mole NH}_3}\right)$$

$$\left.\begin{array}{c}\text{cooling req'd} \\ \text{for 100 gal} \\ 32.0\% \text{ NH}_3 \text{ soln}\end{array}\right\} = (13.94)(13,700) = 191,000 \text{ Btu/100 gal soln}$$

These data are portrayed in Fig. E4.35(b).

Basis: 100 gal of solution at concentrations and densities shown

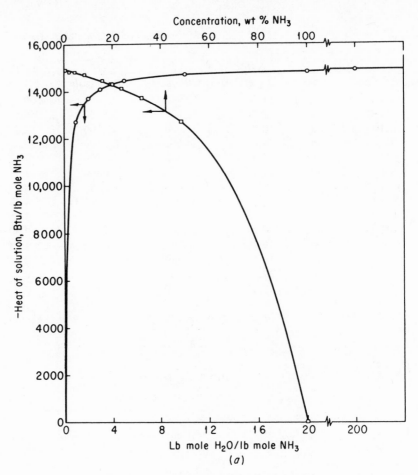

Fig. E4.35(a).

TABLE E4.35

%NH₃	sp gr* @4°C	density, lb/100 gal	lb NH₃ / 100 gal	lb moles NH₃ / 100 gal	−ΔĤ°soln Btu / lb mole NH₃	cooling req'd per 100 gal, Btu
32.0	0.889	741	237	13.94	13,700	191,000
23.9	0.914	761	182	10.70	14,100	151,000
19.1	0.929	774	148	8.70	14,300	124,000
15.9	0.940	784	124.7	7.32	14,450	106,000
8.63	0.965	805	69.4	4.08	14,700	60,000
4.51	0.981	819	37.0	2.18	14,800	32,200
3.05	0.986					
2.30	0.990	826	19.0	1.12	14,800	16,600
1.85	0.992					
0.94	0.995	830	7.8	0.46	14,850	6,800
0.47	0.998					
0.0	1.000	834	0	0	14,900	0

*SOURCE: N. A. Lange, *Handbook of Chemistry*, 8th ed., Handbook Publishers Inc., Sandusky, Ohio, 1958.

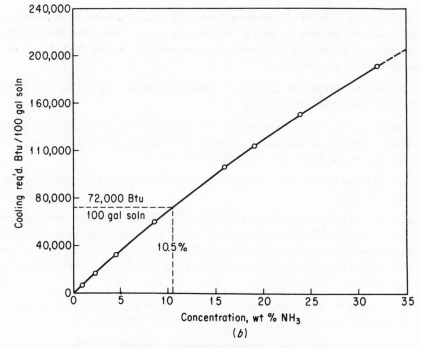

Fig. E4.35(b).

(c)   Basis: 100 gal soln at 10.5% NH_3

$$\Delta E \cong \Delta H = mC_p\Delta T = 72{,}000 \text{ Btu}/100 \text{ gal} \quad [\text{cf. Fig. E4.35(b)}]$$

$$\text{sp gr soln} = 0.955$$

$$m = \frac{(0.955)(62.4)(1.003)(100)}{7.48} = 800 \text{ lb}$$

$$C_p \ 10.5\% \ NH_3 \text{ soln} = 4.261 \text{ J/(g)(°C)}$$
$$= 1.02 \text{ Btu/(lb)(°F)}$$

$$\Delta T = \frac{\Delta H}{mC_p} = \frac{72{,}000}{(800)(1.02)} = 88°F$$

$$T_{\text{final}} = 77 + 88 = 165°F$$

WHAT YOU SHOULD HAVE LEARNED FROM THIS CHAPTER

1. You should know what the energy balance is in words and as a mathematical equation. You should be able to explain what each term in the equation means and be able to apply the equation to practical or simulated problems, i.e., be able to state what each term is in light of the conditions in the problem, know what assumptions can and cannot be made, and know what terms can be omitted.

2. In analyzing problems by means of energy balances, you should be able to set up a suitable system, be able to distinguish between flow and nonflow processes, and be able to judge whether the process can be considered to be reversible or irreversible.

3. You should know the difference between thermal energy and heat and between heat and work.

4. In applying the energy balance, you should be able to locate enthalpy or heat capacity data for a wide range of compounds and, if such data are not available, be able to make suitable approximations.

5. You should know what a state (point) function is and how it differs from a path function.

6. You should be familiar with the principles of thermochemistry, and especially how to use a reference temperature, how to calculate heats of reaction with the products and reactants entering and leaving at various temperatures, and how to calculate standard heats of formation and heats of combustion from experimental data.

7. When applicable, you should be able to compute the heats of solution or mixing and know where to find information of this type.

8. You should be able to calculate the adiabatic reaction temperature for simple or complex reactions, whether they go to completion or not and whether or not the participants are in the proper stoichiometric proportions.

SUPPLEMENTARY REFERENCES

1. Edmister, W. C., *Applied Hydrocarbon Thermodynamics*, Gulf Publishing Co., Houston, Texas, 1961.

2. Henley, E. J., and H. Bieber, *Chemical Engineering Calculations*, McGraw-Hill, New York, 1959.

3. Hougen, O. A., K. M. Watson, and R. A. Ragatz, *Chemical Process Principles*, Part I, 2nd ed., Wiley, New York, 1956.

4. Hougen, O. A., K. M. Watson, and R. A. Ragatz, *Chemical Process Principles*, Part II, 2nd ed., Wiley, New York, 1959.

5. Leland, T. W., and P. S. Chappelear, "The Corresponding States Principle," *Ind. Eng. Chem.*, v. 60, no. 7, p. 15 (1968).

6. Mott-Smith, M., *The Concept of Energy Simply Explained*, Dover, New York, 1964.

7. Smith, J. M., and H. C. Van Ness, *Introduction to Chemical Engineering Thermodynamics*, 2nd ed., McGraw-Hill, New York, 1959.

8. Yesavage, V. F., et al., "Enthalpies of Fluids at Elevated Pressures and Low Temperatures," *Ind. Eng. Chem.*, v. 59, no. 11, p. 35 (1967).

9. Zermansky, M. W., and H. C. Van Ness, *Basic Engineering Thermodynamics*, McGraw-Hill, New York, 1966.

PROBLEMS[32]

4.1. Explain specifically what the system is for each of the following processes; indicate the portions of the energy transfer which are heat and work by the symbols Q and W, respectively:
   (a) A liquid inside a metal can, well insulated on the outside of the can, is shaken very rapidly in a vibrating shaker.
   (b) Hydrogen is exploded in a calometric bomb and the water layer outside the bomb rises in temperature by 1°C.
   (c) A motor boat is driven by an outboard-motor propeller.
   (d) Water flows through a pipe at 10 ft/min, and the temperature of the water and the air surrounding the pipe are the same.

4.2. Draw a simple sketch of each of the following processes, and, in each, label the system boundary, the system, the surroundings, and the streams of material and energy which cross the system boundary:
   (a) Water enters a boiler, is vaporized, and leaves as steam. The energy for vaporization is obtained by combustion of a fuel gas with air outside the boiler surface.
   (b) The steam enters a rotary steam turbine and turns a shaft connected to an electric generator. The steam is exhausted at a low pressure from the turbine.
   (c) A battery is charged by connecting it to a source of current.
   (d) A tree obtains water and minerals through its roots and gives off carbon dioxide from its leaves. The tree, of course, gives off oxygen at night.

4.3. Draw a simple sketch of the following processes; indicate the system boundary; and classify the system as open or closed:
   (a) Automobile engine.   (e) A river.
   (b) Water wheel.   (f) The earth and its atmosphere.
   (c) Pressure cooker.   (g) An air compressor.
   (d) Man himself.   (h) A coffee pot.

4.4. Are the following variables intensive or extensive variables? Explain for each.
   (a) Pressure.
   (b) Volume.
   (c) Specific volume.
   (d) Refractive index.
   (e) Surface tension.

---

[32]An asterisk designates problems appropriate for computer solution. Also refer to the computer problems after Problem 4.111.

4.5. Classify the following measurable physical characteristics of a gaseous mixture of two components as (1) an intensive property, (2) an extensive property, (3) both, or (4) neither:
(a) Temperature.
(b) Composition.
(c) Pressure.
(d) Mass.

4.6. Water can exist in various states depending on its temperature, pressure, etc. From what you have learned in Sec. 3.7, what is the state (gaseous, liquid, solid, or combinations thereof) for water at the following conditions:
(a) 250°F and 1 atm.
(b) −10°C and 720 mm Hg.
(c) 32°F and 100 psia.

4.7. Convert 45.0 Btu/lb_m to the following:
(a) cal/kg
(b) joules/kg
(c) kwhr/kg
(d) (ft)(lb_f)/lb_m

4.8. Suppose that a constant force of 40.0 N is exerted to move an object for 6.00 m. What is the work accomplished (on an ideal system) expressed in the following:
(a) joules   (c) cal
(b) (ft)(lb_f)   (d) Btu

4.9. A sedan and a semi-trailer truck collide head-on on a freeway. The truck weighs 28,000 lb and the sedan weighs 5000 lb. At the moment of impact, both the truck and the sedan are going at 60.0 mi/hr.
(a) What was the car's kinetic energy in (ft)(lb_f)? In joules?
(b) What was the truck's kinetic energy in (ft)(lb_f)? In joules?
(c) How much kinetic energy is changed into other forms of energy in the collision by the time the two vehicles come to rest?

4.10. A missile is fired vertically from the earth's surface so that it reaches a height of 150,000 ft above the ground. Calculate the potential and kinetic energy per pound of the missile at 150,000 ft.

4.11. If a pail weighing $\frac{1}{2}$ lb is dropped into a well 50 ft deep, what is the kinetic and potential energy of the pail (a) just before it hits the water, and (b) after it hits the water surface (at 50 ft)?

4.12. Before it lands, a vehicle returning from space must convert its enormous kinetic energy to heat. To get some idea of what is involved, a vehicle returning from the moon at 25,000 mi/hr can, in converting its kinetic energy, increase the internal energy of the vehicle sufficiently to vaporize it. Obviously a large part of the total kinetic energy must be transferred from the vehicle. How much kinetic energy does the vehicle have in Btu? How much energy must be transferred by heat if the vehicle is to heat up only 20°F/lb?

4.13. A chart for carbon dioxide (see Appendix J) shows that the enthalpy of saturated CO_2 liquid is zero at −40°F. Can this be true?

4.14. Assume that the heat capacity for liquid water is 1 Btu/(lb_m)(°F). Calculate the enthalpy of liquid water at 150°F and 1 atm relative to (a) 32°F and 1 atm and (b) 100°F and 1 atm.

4.15. Given that the heat capacity C_p for mercury at $-20°C$ is 0.0335 cal/g, while at 100°C the value of C_p is 0.0327, compute the change of enthalpy of 10 g of mercury from $-20°C$ to 100°C. Given that the heat capacity at constant volume, C_v, is 0.0294 at $-20°C$ and 0.0276 cal/g at 100°C, compute the change in internal energy for the same temperature range. Then compute ΔpV for mercury for the same temperature range.

4.16. The differential change in the volume of a gas is given by the relation

$$dV = nR\left[\left(\frac{1}{p} + a\right)dT - \frac{T}{p^2}dp\right]$$

Show that the volume V is a point or state function.

4.17. State which of the following variables are point, or state, variables, and which are not; explain your decision in one sentence for each one:

(a) Pressure.   (d) Heat capacity.
(b) Density.   (e) Internal energy.
(c) Molecular weight.   (f) Ionization constant.

4.18. You have calculated that the specific enthalpy of 1 kg mole of an ideal gas at 300 kN/m² and 100°C is 6.05×10^5 J/kg mole (with reference to 0°C and 100 kN/m²). What is the specific internal energy of the gas?

4.19. (a) Ten pound moles of an ideal gas are originally in a tank at 100 atm and 40°F. The gas is heated to 440°F. The specific molar enthalpy, ΔH, of the ideal gas is given by the equation

$$\Delta\hat{H} = 300 + 8.00T$$

where $\Delta\hat{H} =$ Btu/lb mole
$T =$ temperature, °F

(1) Compute the volume of the container (ft³).
(2) Compute the final pressure of the gas (atm).
(3) Compute the enthalpy change of the gas.

(b) Use the equation above to develop an equation giving the molar internal energy, $\Delta\hat{U}$, in cal/g mole as a function of temperature T in °C.
(c) What is the temperature for the reference enthalpy ($\Delta\hat{H} = 0$)?

4.20.* Experimental values of the heat capacity C_p have been determined in the laboratory as follows; fit a second-order polynomial in temperature to the data ($C_p = a + bT + cT^2$):

$T, °C$	$C_p,\ cal/(g\ mole)(°C)$
100	9.69
200	10.47
300	11.23
400	11.79
500	12.25
600	12.63
700	12.94

(The data are for carbon dioxide; if a computer is used in the data-fitting process, also calculate the confidence limits for the predicted value of C_p.)

4.21. One pound mole of NO is heated at constant pressure from 50° to 300°F. What is the enthalpy change? Compare with that for NO as an *ideal* gas.

4.22.* The heat capacity of chlorine has been determined experimentally as follows:

T (°C)	C_p [cal/(g mole)(°K)]
0	8.00
18	8.08
25	8.11
100	8.36
200	8.58
300	8.71
400	8.81
500	8.88
600	8.92
700	8.95
800	8.98
900	9.01
1000	9.03
1100	9.05
1200	9.07

(a) Calculate the enthalpy change required to raise 1 g mole of chlorine from 0° to 1200°C, and show the method of calculation.

(b) From the experimental data, how would you determine the coefficients in an equation of the form $C_p = a + bT + cT^2$?

4.23. The molar heat capacity of CO_2 is

$$C_p = 9.085 + 0.0048T - 0.83 \times 10^{-6}T^2$$

where $T = °C$
$C_p = $ cal/(g mole)(°C)

Calculate the mean molar heat capacity of 1 lb mole of CO_2 between 0° and 1000°C.

4.24. The molar heat capacity of carbon monoxide is

$$C_p = 6.935 + 6.77 \times 10^{-4}T + 1.3 \times 10^{-7}T^2$$

where $C_p = $ cal/(g mole)(°C)
$T = °C$

Calculate the mean molal heat capacity between 500°C and 1000°C.

4.25.* The molar heat capacity of CO_2 is tabulated below:

T (°C)	C_p [cal/(g mole)(°C)]
0	8.61
100	9.69
200	10.47
300	11.23
400	11.79
500	12.25
600	12.63
700	12.94
800	13.20
900	13.41
1000	13.60

(a) Calculate the energy required to raise 1 g mole of CO_2 from 500° to 1000°C.
(b) Calculate the Btu required to raise 1 lb of CO_2 from 932°F (500°C) to 1864°C (1000°C).

4.26. Your assistant has developed the following equation to represent the heat capacity of air (with C_p in cal/(g mole) (°K) and T in °K):

$$C_p = 6.39 + 1.76 \times 10^{-3}T - 0.26 \times 10^{-6}T^2$$

(a) Derive an equation giving C_p but with the temperature expressed in °C.
(b) Derive an equation giving C_p in terms of Btu per pound per degree Fahrenheit with the temperature being expressed in degrees Fahrenheit.

4.27. To evaluate the suitability of Kopp's rule for the heat capacities of solids, compute the heat capacities of sodium sulfate (Na_2SO_4), dextrose ($C_6H_{12}O_6$), and copper ammonium sulfate $[CuSO_4(NH_4)_2 \ SO_4 \cdot 6H_2O]$ and compare your values with the experimental ones at 25°C.

4.28. Repeat Problem 4.27 but for the following liquids at 25°C: acetonitrile (CH_3CN), ethyl ether $[(C_2H_5)_2O]$, and arsenic trichloride ($AsCl_3$).

4.29. Use the Kuloor method to estimate the heat capacity of (a) liquid ethyl formate and (b) liquid benzene at 20°C. Compare your results with the experimental values.

4.30. Estimate the heat capacity of gaseous isobutane at 1000°K and 200 mm Hg by using the Kothari-Doraiswamy relation

$$C_p = A + B \log_{10} T_r$$

from the following experimental data at 200 mm Hg:

C_p [cal/(°K)(g mole)]	23.25	35.62
Temp. (°K)	300	500

The experimental value is 54.40; what is the percentage error in the estimate?

4.31.* Use the tables for the mean heat capacity of the combustion gases to compute the enthalpy change (in Btu) that takes place when a mixture of 6.00 lb moles of H_2O and 4.00 lb moles of CO_2 is heated from 60 to 600°F.

4.32. How many kilojoules are required to convert 5 kg moles of water from the reference state to water vapor at 1000°C? Use the mean heat capacity tables.

4.33. The specific impulse of a rocket engine depends not only on the temperature to which the working fluid is heated but also is roughly proportional to the square root of the reciprocal of the mean molecular weight of the gas as it enters the nozzle. The chemical rocket engine depends on the reaction of an oxidizer and fuel for energy production, and one must perforce accept the combustion products produced, such as water, carbon monoxide, carbon dioxide, and hydrogen. Assume that the combustion gases (83.4 percent CO_2, 12.2 percent CO, and 4.4 percent H_2) are to be cooled from 2700° to 400°C by exchange of heat with the fuel; compute the enthalpy change that takes place in Btu/lb mole of gas and in cal/g mole of gas. Use the tables for the enthalpies of the combustion gases in your computations.

4.34. Calculate the enthalpy change in cooling 10,000 ft³ (measured at the initial conditions) of the following mixture from 500 to 300°F: gas (at 1 atm)— 50 percent N_2, 50 percent C_2H_2; solid—C(graphite), 1 lb/100 ft³ of gas at entering conditions.

4.35. One gram mole of air is heated from 400° to 1000°C. Calculate ΔH by integrating the heat capacity equation. Calculate ΔU also.

4.36. Repeat Problem 4.31 but this time use the heat capacity equations for the gases.

4.37. Repeat Problem 4.33 but this time use the heat capacity equations for the gases.

4.38. What is the change in enthalpy of 10 lb of CO_2 when heated from 140° to 300°F at 1 atm. Use the heat capacity equation for CO_2 from the Appendix.

4.39.* (a) Estimate the heat of vaporization of *n*-butane at its normal boiling point of 31.1°F using the Kistyakowsky equation. The experimental value is 165.8 Btu/lb.

    (b) Make an Othmer plot for *n*-butane, and calculate the latent heat of vaporization at 280°F; compare this value with the one calculated from the following experimental data:

$T°F$	$p^*(atm)$	$V_{(1)}(ft^3/lb)$	$V_{(g)}(ft^3/lb)$
260	24.662	0.0393	0.222
270	27.134	0.0408	0.192
280	29.785	0.0429	0.165
290	32.624	0.0458	0.138
305.56(T_c)	37.47	0.0712	0.0712

Data are from H. W. Prengle, L. R. Greenhaus, and R. York, *Chem. Eng. Progress*, v. 44, p. 863 (1948). The reported value of ΔH is 67.3 Btu/lb.

4.40.* (a) Using the following data on the vapor pressure of Cl_2, plot $\log p^*$ against $1/T$, and calculate the latent heat of vaporization of Cl_2 as a function of temperature:

$T(°F)$	$p^*(psia)$
−22	17.8
32	53.5
86	126.4
140	258.5
194	463.5
248	771
284	1050

How satisfactory are these data?

    (b) The vapor pressure of Cl_2 is given by Lange[33] by the equation

$$\log p^* = A - \frac{B}{(C + T)}$$

where p^* = vapor pressure in mm Hg

T = °C; and for chlorine

$A = 6.86773$

$B = 821.107$

$C = 240$

[33] N. A. Lange, *Handbook of Chemistry*, 9th ed., Handbook Publishers Inc., Sandusky, Ohio, 1956, p. 1426.

Calculate the latent heat of vaporization of chlorine at the normal boiling point in Btu/lb. Compare with experimental value.

4.41. Normal hexane that boils at 69°C is to be used in a heat exchanger to take up energy from a hot hydrocarbon stream. Estimate the heat of vaporization in Btu/lb and joules/kg using Trouton's rule, and compare your answer with the experimental heat of vaporization (obtained from a handbook).

4.42. The vapor pressure of phenylhydrazine has been found to be (from 25° to 240°C)

$$\log_{10} p^* = 7.9046 - \frac{2366.4}{T + 230}$$

where p^* is in atm and T is in °C. What is the heat of vaporization of phenylhydrazine in Btu/lb at 200°F?

4.43. Compute the specific enthalpy (Btu/lb) of carbon tetrachloride (mw = 153.84) as a vapor at 250°F relative to the solid at its melting point. From Kobe and Long,[34] the molar heat capacity of carbon tetrachloride vapor is

$$C_p = 12.24 + 3.40 \times 10^{-2}T - 3.00 \times 10^{-5}T^2$$

where T is the temperature in °K.

4.44. Use of the steam tables and chart:
  - (a) What is the enthalpy change needed to change 3 lb of liquid water at 32°F to steam at 1 atm and 300°F?
  - (b) What is the enthalpy change needed to heat 3 lb of water from 60 psia and 32°F to steam at 1 atm and 300°F?
  - (c) What is the enthalpy change needed to heat 1 lb of water at 60 psia and 40°F to steam at 300°F and 60 psia?
  - (d) What is the enthalpy change needed to change 1 lb of a water-steam mixture of 60 percent quality to one of 80 percent quality if the mixture is at 300°F?
  - (e) Calculate the ΔH value for an isobaric (constant pressure) change of steam from 120 psia and 500°F to saturated liquid.
  - (f) Do the same for an isothermal change to saturated liquid.
  - (g) Does an enthalpy change from saturated vapor at 450° to 210°F and 7 psia represent an enthalpy increase or decrease? A volume increase or decrease?
  - (h) In what state is water at 40 psia and 267.24°F? At 70 psia and 302°F? At 70 psia and 304°F?
  - (i) A 2.5-ft³ tank of water at 160 psia and 363.5°F has how many ft³ of liquid water in it? Assume that you start with 1 lb of H_2O. Could it contain 5 lb of H_2O under these conditions?
  - (j) What is the volume change when 2 lb of H_2O at 1000 psia and 200°F expand to 245 psia and 460°F?
  - (k) Ten pounds of wet steam at 100 psia have an enthalpy of 9000 Btu. Find the quality of the wet steam.

4.45. Pure sodium hydroxide produced as a molten liquid is poured into drums for shipment. If the NaOH placed in a drum is just above its fusion temperature, estimate the enthalpy change in a drum (i.e., the heat lost) that contains 500 lb of NaOH if the drum is to be shipped at 80°F. If your value for the heat of

[34]K. A. Kobe and E. G. Long, *Petrol. Refiner*, v. 29, no. 3, p. 159 (1950).

fusion is off by ±5 percent and your value for the heat capacity is in error by ±20 percent, what is the approximate percentage error in the enthalpy change?

4.46. For an orbital vehicle composed of three basic elements—spacecraft, propellant tanks, and rocket engines—it is generally agreed that the spacecraft and rocket engines should be recovered for reuse. The main point of controversey concerns what portion of the propellant tanks should be made reusable. Propellant containers have been the most weighty and bulky elements of large rocket vehicles. On a cost-per-pound basis, these containers cost the least among major vehicle elements, but still, they are not cheap. The ideal solution would be to eliminate the tanks entirely. But how would the propellants be contained and supported? One answer is to freeze them. Most substances can be frozen, and propellants are no exception. Estimate the sum of the heat of melting of solid oxygen and the heat of vaporization of liquid oxygen per pound at low pressure to determine the amount of energy that will be removed from the rocket exhaust in order to vaporize the solid oxygen. What fraction of the total enthalpy change from $-362°F$ (melting point) to $-297°F$ (boiling point) would you estimate is provided by the latent heats (enthalpy change of the phase transitions)?

4.47. Use the steam tables to determine the answers to the following questions:
   (a) Liquid water exists at a pressure such that its boiling point is 575°F. How many Btu are required to change a pound of liquid water at 575°F to a pound of dry saturated water vapor at 575°F? What is the pressure in psia at which water boils at 575°F?
   (b) Water vapor exists at 240.0 psia and a temperature of 423.86°F. Is this vapor superheated? If so, how many degrees of superheat does it contain?
   (c) Water at 50°F is heated to 495°F, at which temperature it boils. What is the mean heat capacity in Btu/(lb)(°F) for liquid water over this range?
   (d) Wet steam at 315°F has an enthalpy of 1150.2 Btu/lb. What is the quality of this steam?

4.48. One pound mole of an *ideal* gas whose C_p is 7 Btu/(lb mole)(°R) is confined in a reservoir with a floating top such that the pressure on the gas is 60 psig no matter what its volume is. In the morning the gas is at 50°F, but late in the afternoon its temperature rises to 90°F.
   (a) Determine how much heat has been transferred into the tank, the work done by the gas, and the internal energy change for the gas.
   (b) If the system returns to its original state in the evening, again find Q, W, and the internal energy and enthalpy changes for the cooling process.
   (c) What are Q, W, and the internal energy and enthalpy changes for the overall process of heating and cooling?

4.49. Write the simplified energy balances for the following changes:
   (a) A fluid flows steadily through a poorly designed coil in which it is heated from 70° to 250°F. The pressure at the coil inlet is 120 psia, and at the coil outlet is 70 psia. The coil is of uniform cross section, and the fluid enters with a velocity of 2 ft/sec.
   (b) A fluid is expanded through a well-designed adiabatic nozzle from a pressure of 200 psia and a temperature of 650°F to a pressure of 40 psia and

a temperature of 350°F. The fluid enters the nozzle with a velocity of 25 ft/sec.

(c) A turbine directly connected to an electric generator operates adiabatically. The working fluid enters the turbine at 200 psia and 640°F. It leaves the turbine at 40 psia and at a temperature of 350°F. Entrance and exit velocities are negligible.

(d) The fluid leaving the nozzle of part (b) is brought to rest by passing through the blades of an adiabatic turbine rotor and leaves the blades at 40 psia and at 400°F.

(e) A fluid is allowed to flow through a cracked (slightly opened) valve from a region where its pressure is 200 psia and 670°F to a region where its pressure is 40 psia, the whole operation being adiabatic.

4.50. Write the appropriate simplified energy balances for the following changes; in each case the amount of material to be used as a basis of calculation is 1 lb and the initial condition is 100 psia and 370°F:

(a) The substance, enclosed in a cylinder fitted with a movable frictionless piston, is allowed to expand at constant pressure until its temperature has risen to 550°F.

(b) The substance, enclosed in a cylinder fitted with a movable frictionless piston, is kept at constant volume until the temperature has fallen to 250°F.

(c) The substance, enclosed in a cylinder fitted with a movable frictionless piston, is compressed adiabatically until its temperature has risen to 550°F.

(d) The substance, enclosed in a cylinder fitted with a movable frictionless piston, is compressed at constant temperature until the pressure has risen to 200 psia.

(e) The substance is enclosed in a container which is connected to a second evacuated container of the same volume as the first, there being a closed valve between the two containers. The final condition is reached by opening the valve and allowing the pressures and temperatures to equalize adiabatically.

4.51. Dairy scientists at Louisiana State University have learned that cows whose heads are air conditioned produce 20 percent more milk in the summer months than they would normally produce. They had found earlier that cows who live in an air-conditioned chamber *in toto* produce some 30 percent more milk than normal in the summertime, but the cost of the electricity used to operate the air conditioning grows onerous. With a cow as a system, including the air-conditioned head, draw a diagram representing the energy balance, and show how each term in the energy balance corresponds to one of the physical entities within the system. Be sure to include the energy obtained from food, lost by breathing, and so forth.

4.52. In a letter to the editor of a journal the following was written:

In his discussions of the conservation of energy, the author said that teachers should refrain from any suggestions that energy and mass are interconvertible. He states (1) that Einstein's relation $E = mc^2$ is an equality and not a description of a transformation and (2) that it is in no way proper to infer that, *for any change*, either mass or energy is not each separately conserved. From the viewpoint of relativity theory, the total mass of the products of an exothermic chemical reaction will be less than the mass of the reactants by $\Delta m = \Delta E/c^2$ where ΔE is the energy evolved. For example, if $\Delta E = 100$ kcal/mole $= 4.2 \times 10^5$ J/mole, then $\Delta m = 4.7 \times 10^{-9}$ g/mole. One would be hard pressed to measure such a change but yet it is of importance *conceptually*. There may be some semantic problems, but I would say that mass is not conserved—separately from the energy changes—in such reactions.

Comment on this letter. Is mass conserved or not?

4.53. A household freezer is placed inside an insulated sealed room. If the freezer door is left open with the freezer operating, will the temperature of the room increase or decrease? Explain your answer.

4.54. An office building requires water at two different floors. A large pipe brings the city water supply into the building in the basement level where a booster pump is located. The water leaving the pump is transported by smaller insulated pipes to the second and fourth floors where the water is needed. Use the general energy equation to calculate the *minimum* amount of work per unit time (in horsepower) that the pump must do in order to deliver the necessary water, as indicated in Fig. P4.54. (*Minimum* refers to the fact that you should neglect the friction and pump energy losses in your calculations.) The water does not change temperature.

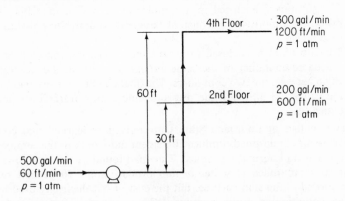

Fig. P4.54

4.55. Water is being pumped from a pond through a long fire hose. A 50-hp motor drives the pump, and the overall efficiency of the motor pump is 75 percent (i.e., 75 percent of the energy supplied by the motor is transferred to the water). The nozzle of the hose is 25 ft above and 250 ft from the pond; 100 gpm are being pumped, and the water velocity at the nozzle exit is 50 ft/sec. Estimate the temperature change in the water between the pond and the nozzle exit.

4.56. The tortuous path of automotive pollution has led Congress to write into law specific emission standards to be effective with cars produced in 1975. These standards have led to the development of several types of steam engines for

automobiles. In a test of the boiler for such an engine for a truck, a closed boiler initially contained 1.20 gal of water at 75°F and at a total pressure just equal to the vapor pressure of water at 75°F. Because the recycle system was not operating properly, 12.0 gal of hot water at 150°F was metered into the boiler, after which the power to an electrical heater inside the boiler was turned on. This heater consumes electricity at a constant rate of 25.0 kW. After turning the power on, the operator left to attend to other duties. One-half hour later he returned to find that the pressure-relief valve (set at 25 psig) had ruptured and that steam was escaping. He turned the power off and closed the relief valve. Determine the following:
(a) How much water remained in the boiler?
(b) At what time after the power was turned on did the pressure relief valve rupture?
Assume that the boiler was well insulated and that the amount of vapor contained in the boiler at any time was very small compared to the liquid water.

4.57. Your company produces small power plants that generate electricity by expanding waste process steam in a turbine. You are asked to study the turbine to determine if it is operating as efficiently as possible. One way to ensure good efficiency is to have the turbine operate adiabatically. Measurements show that for steam at 500°F and 250 psia
(a) The work output of the turbine is 86.5 hp.
(b) The rate of steam usage is 1000 lb/hr.
(c) The steam leaves the turbine at 14.7 psia and consists of 15 percent moisture (i.e., liquid H_2O).
Is the turbine operating adiabatically? Support your answer with calculations.

4.58. A reaction process requires 4000 lb/hr of superheated water vapor at 600°F and 14.7 psia. The following scheme is to be used to provide the superheated vapor. A turbine discharges 1000 lb/hr of saturated steam at 14.7 psi (no moisture), while another process generates 3000 lb/hr of superheated steam at 50 psia and 400°F. These two streams are to be mixed and then heated to 600°F; however, the pressure of the second steam is first reduced to 14.7 psia by expanding it adiabatically through an expansion valve. The mixing is done at a constant pressure of 14.7 psia. Determine the amount of heat that must be supplied after mixing to bring the temperature up to 600°F.

4.59. Solvent emissions are likely to be specifically regulated in areas of the country with present or putative smog problems. These regulations may limit total hydrocarbon emissions. To reduce solvent vapor losses and prevent explosions, a mixture of CO_2 (60.0 percent) and benzene vapor (40.0 percent) at 200°F and 1 atm is cooled to 40°F and 1 atm. Compute the heat removal required per pound of the original gas, and the fraction of the benzene recovered as liquid.

4.60. When a flammable vapor coexists with oxygen under appropriate conditions, combustion can occur. If the combustion takes place in a restricted space, an explosion results. Under some circumstances, the explosion may become reinforced into a detonation, which generates pressures up to 700 times the initial pressure. Even very small quantities of some materials have exploded or detonated, causing loss of life and serious destruction. To improve the safety factor in a hot gas stream leaving a reactor, it is desired to reduce the temperature of the gases from 1000°F as rapidly as possible to 400°F by spraying liquid water into the gas stream. For preliminary calculations, assume that the gas stream has the same properties as air. How many pounds of water at 70°F are required per 1000 ft^3 of total gases (measured at 400°F and 6 in. Hg pressure gauge because the meter cannot be used at higher temperatures). The barometer is 29.75 in. Hg.

4.61. A system consists of 25 lb of water vapor at the dew point. The system is compressed isothermally at 400°F, and 988 Btu of work are done on the system by the surroundings. What volume of liquid was present in the system before and after compression?

4.62. Steam fills tank 1 at 1000 psia and 700°F. The volume of tank 1 equals the volume of tank 2. A vacuum exists in tank 2 initially. See Fig. P4.62. The valve connecting the two tanks is opened, and isothermal expansion of the steam occurs from tank 1 to tank 2. Find the enthalpy change per pound of steam for the entire process.

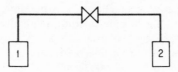

Fig. P4.62

4.63. The process of throttling, i.e., expansion of a gas through a small orifice, is an important part of refrigeration. If wet steam (fraction vapor is 0.975) at 100 psia is throttled to 1 atm, find the following for the process: heat transferred to or from steam, work done, enthalpy change, internal energy change, and exit temperature.

4.64. (a) If 2 lb of $H_2O(l)$ are flashed isothermally at 300°F into steam at 1 atm pressure, what is the change of enthalpy? Assume that $H_2O(l)$ is incompressible and that steam [$H_2O(g)$] is an ideal gas. Assume that the specific heat of $H_2O(l)$ is 1.0 and that of steam is 0.5. The latent heat of vaporization at 212°F and 1 atm is 970.3 Btu/lb.

(b) If the $H_2O(l)$ in part (a) is flashed to 0.20 atm, will this change your answer to part (a)? If so, how much?

(c) Estimate the change in internal energy for the process part (a).

(d) Can you calculate a mean heat capacity for the process? If so, calculate it.

(e) If the water in part (a), after flashing, is compressed back to its original conditions, what is the overall enthalpy change for the 2 lb of water?

(f) Check your calculations using the steam tables.

4.65. A desalted crude oil (40°API) is being heated in a parallel-flow heat exchanger from 70° to 200°F with a straight-run gasoline vapor available at 280°F. If 400 gal/hr of gasoline are available (measured at 60°F, 63°API), how much crude can be heated per hour? Would countercurrent flow of the gasoline to the crude change your answer to this problem? Assume that $K = 11$ for both streams.

4.66. Solid CO_2 is being produced by adiabatically expanding high-pressure CO_2 at 70°F through a valve to a pressure of 20.0 psia. If 30 percent conversion of the initial CO_2 to the solid is desired, at what pressure must the CO_2 be before expansion? Is the high-pressure CO_2 gas, liquid, or a mixture of each? What is the approximate volume change per pound of CO_2 during the expansion?

4.67. In the production of dry ice, liquid CO_2 at 0°F and 400 psia is passed through a valve and expanded adiabatically to a pressure of 20.0 psia. Determine the percentage of CO_2 converted to the solid state in this simple expansion process.

4.68. You have 0.37 lb of CO_2 at 1 atm abs and 100°F contained in a cylinder closed by a piston. The CO_2 is compressed to 70°F and 1000 psia. During the process, 40.0 Btu of heat is removed. Compute the work required for the compression.

4.69. The large increase in the emission of SO_2 has been caused by the large increase in fuel consumption for electricity and other industrial plants. Both the fuel oil and coal used in these plants contain sulfur that forms SO_2 (and a little SO_3) in the exhaust gases. Recovery of the sulfur in the exhaust gases competes with desulfurization of the fuel as a means of preventing air pollution. In a plant, 10,000 ft³/hr of a combustion gas at 1000°F and 1 atm is produced from a high-sulfur fuel. The gas which has the following analysis,

SO_2	1.1%
CO_2	14.5
O_2	3.4
N_2	81.0
	100.0

is to be cooled from 100° to 400°F by mixing with *dry* air whose initial temperature is 65°F and pressure 760 mm Hg. See Fig. P4.69. What volume of air per hour measured at 65°F and 760 mm Hg pressure is required to produce an air-gas mixture having a temperature of 400°F?

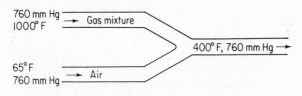

Fig. P4.69

4.70. In one stage of a process for the manufacture of liquid air, air as a gas at 4 atm abs and 250°K is passed through a long, insulated 3-in. ID pipe in which the pressure drops 3 psi because of frictional resistance to flow. Near the end of the line, the air is expanded through a valve to 2 atm abs. State all assumptions.
(a) Compute the temperature of the air just downstream of the valve.
(b) If the air enters the pipe at the rate of 100 lb/hr, compute the velocity just downstream of the valve.

4.71. In a typical refrigeration cycle in a frozen-food storage plant, liquid ammonia at 130 psia and 70°F is expanded adiabatically through a needle valve to 40 psia and enters the cooling coils where it is vaporized by absorbing 500 Btu/lb.

The vaporized ammonia at 40 psia and 10°F is then compressed adiabatically to 130 psia, where it is condensed at temperature of 70°F by giving up 430 Btu/lb of NH$_3$ to cooling water at 60°F. The liquid ammonia is then to be recycled. How much power must be supplied from the compressor if 10.0 lb of NH$_3$ per minute is circulated through the system and the compressor is 100 percent efficient.

4.72. Expansion nozzles are used to produce very high velocities. Steam at 400 psia and 1100°F enters such a nozzle at a negligible velocity and leaves at 20.0 psia with a velocity of 200 ft/sec. Instruments indicate that 5.00 Btu of heat are lost from each pound of steam during passage through the nozzle. If 120 lb of steam per minute are passed through the nozzle, what must the cross-sectional area of the nozzle exit be?

4.73. Frozen-food quality is taking a big step forward with the introduction of ultrafast freezing as a commercial reality. The key to this process is the use of −320°F liquid nitrogen to do the freezing. For example, in the freezing of fish fillets, after precooling by vaporized nitrogen, the fillets enter a freezing section where they pass through a liquid nitrogen spray, which vaporizes into −320°F gas. The resulting cold gas is used to remove sensible heat from the entering fish and exhausts to the atmosphere at −10°F or warmer. A well-insulated tunnel, 6 ft wide by 32 ft long, has a capacity of 1000 lb/hr of fish fillets. Estimate the gallons of liquid N$_2$ used per hour. (The actual consumption was reported to be 130–150 gal/hr.)

4.74. Helium, a monatomic ideal gas, is to be compressed from 65 psia and 95°F to 300 psia. One method of doing this is to use a single-stage reciprocating compressor operating isothermally and then pass the compressed helium through a heat exchanger where enough heat is transferred to produce the desired final temperature. Per 50 lb of gas, measured at entering conditions, compute the following:
(a) The work required if the compressor is assumed to operate reversibly.
(b) The amount of heat which must be transferred in the heat exchanger. State whether heat is added or removed.

4.75. Solid carbon dioxide (dry ice) has innumerable uses in industry and in research. Because it is easy to manufacture, the competition is severe, and it is necessary to make dry ice very cheaply to be successful in selling it. In a proposed plant to make dry ice, the gaseous CO$_2$ is compressed isothermally and essentially reversibly from 6 psia and 40°F to a specific volume of 0.05 ft^3/lb$_m$.
(a) What is the final state of the compressed CO$_2$?
(b) Compute the work of compression.
(c) What is the heat removed?
(d) If the actual efficiency of the compressor (relative to a reversible compressor) is 85 percent and the electricity to run the compressor motor costs $0.02/kWh, what is the cost of compression of the solid CO$_2$ in dollars per pound of dry ice?

4.76.* A thousand pounds of ethane per hour are continuously compressed in a reciprocating compressor from 15 to 100 psia. The compressor is jacketed and sufficient cooling water supplied so that the *p-V* data tabulated below represent the conditions of the gas within the cylinder. Find the compression work done on the gas per hour. Assume the friction to be negligible.

pressure-volume data

pressure, psia	specific volume, ft³/lb
15	13.23
20	9.89
25	7.90
30	6.57
40	4.90
50	3.90
60	3.24
80	2.40
100	1.90

If enthalpy of ethane at 15 psia, 13.23 ft³/lb, is 470 Btu/lb and the enthalpy at 100 psia, 1.90 ft³/lb, is 460 Btu/lb, how much heat is removed in the cooling water per 1000 pounds of ethane compressed? What is the internal energy change of the ethane?

4.77. A cylinder, closed at one end, is fitted with a movable piston. Originally the cylinder contains 1.2 ft³ of gas at 7.3 atm pressure. If the gas pressure is slowly reduced to 1 atm by withdrawing the piston, calculate the work done by the gas on the piston face, assuming that the following relationships hold:
(a) $pV = $ constant.
(b) $pV^{1.3} = $ constant.
(c) Suppose that the pressure is reduced while V is constant.
State all assumptions you must make in order to calculate the work.

4.78. An ideal gas with a heat capacity of $C_p = 6.5 + 1.5T \times 10^{-3}$ [where T is in °R and C_p is in Btu/(lb mole)(°F)] passes through a compressor at the rate of 100 lb moles/min. The entering conditions of the gas are 1 atm and 70°F. The exit conditions are 15 atm and 1200°F. Ignore potential and kinetic energy effects, and assume that the process is adiabatic. Is it a reversible process? Show how you can prove it is or is not by equations.

4.79. A fireman drives a pumper up to a river and pumps water through his hose to a fire 100 ft higher than the river. The water comes out of the fire hose at 100 ft/sec. Ignoring friction, what horsepower engine is required to deliver 5000 gal of water per minute? The river flows at 2 mi/hr.

4.80. Several automobile and truck manufacturers have been developing turbine engines for passenger cars and trucks. Fitting a turbine into a car has many difficulties, but problems with emissions control with internal combustion engines has prompted considerable study. The mechanical energy balance shows that a reversible engine without any kinetic or potential energy effects produces work given by $\int_{p_1}^{p_2} V \, dp$. On the other hand, a frictionless piston in a cylinder produces work given by the relation $\int_{V_1}^{V_2} p \, dV$.
(a) Would these quantities be the same for a turbine vs. a piston engine, if both are reversible?
(b) Would the efficiencies of the two types of engines be the same?

4.81. Calculate the work done when 1 lb mole of water evaporates completely at 212°F. Express your results in Btu.

4.82. Calculate the heat of reaction at 77°F and 1 atm for the following reactions [these are all methods of manufacturing $H_2(g)$]:
   (a) $CH_4(g) + H_2O(g) \longrightarrow CO(g) + 3H_2(g)$.
   (b) $CO(g) + H_2O(g) \longrightarrow CO_2(g) + H_2(g)$.
   (c) $CH_4(g) \longrightarrow C(s) + 2H_2(g)$.
   (d) $3F(s) + 4H_2O(g) \longrightarrow Fe_3O_4(s) + 4H_2(g)$.
   (e) $2H_2O(l) \longrightarrow 2H_2(g) + O_2(g)$.
   (f) $2NH_3(g) \longrightarrow N_2(g) + 3H_2(g)$.

4.83. Calculate the heat of reaction at the standard reference state for the following reactions:
   (a) $CO_2(g) + H_2(g) \longrightarrow CO(g) + H_2O(l)$.
   (b) $2CaO(s) + 2MgO(s) + 4H_2O(l) \longrightarrow 2Ca(OH)_2(s) + 2Mg(OH)_2(s)$.
   (c) $Na_2SO_4(s) + C(s) \longrightarrow Na_2SO_3(s) + CO(g)$.
   (d) $NaCl(s) + H_2SO_4(l) \longrightarrow NaHSO_4(s) + HCl(g)$.
   (e) $NaCl(s) + 2SO_2(g) + 2H_2O(l) + O_2(g) \longrightarrow 2Na_2SO_4(s) + 4HCl(g)$.
   (f) $SO_2(g) + \frac{1}{2}O_2(g) + H_2O(l) \longrightarrow H_2SO_4(l)$.
   (g) $N_2(g) + O_2(g) \longrightarrow 2NO(g)$.
   (h) $Na_2CO_3(s) + 2Na_2S(s) + 4SO_2(g) \longrightarrow 3Na_2S_2O_3(s) + CO_2(g)$.
   (i) $NaNO_3(s) + Pb(s) \longrightarrow NaNO_2(s) + PbO(s)$.
   (j) $Na_2CO_3(s) + 2NH_4Cl(s) \longrightarrow 2NaCl(s) + 2NH_3(g) + H_2O(l) + CO_2(g)$.
   (k) $Ca_3(PO_4)_2(s) + 3SiO_2(s) + 5C \longrightarrow 3CaSiO_3(s) + 2P(s) + 5CO(g)$.
   (l) $P_2O_5(s) + 3H_2O(l) \longrightarrow 2H_3PO_4(l)$.
   (m) $CaCN_2(s) + 2NaCl(s) + C(s) \longrightarrow CaCl_2(s) + 2NaCN(s)$.
   (n) $NH_3(g) + HNO_3(aq) \longrightarrow NH_4NO_3(aq)$.
   (o) $2NH_3(g) + CO_2(g) + H_2O(l) \longrightarrow (NH_4)_2CO_3(aq)$.
   (p) $(NH_4)_2CO_3(aq) + CaSO_4 \cdot 2H_2O(s) \longrightarrow CaCO_3(s) + 2H_2O(l) + (NH_4)_2SO_4(aq)$.
   (q) $CuSO_4(aq) + Zn(s) \longrightarrow ZnSO_4(aq) + Cu(s)$.
   (r) $\underset{\text{benzene}}{C_6H_6(l)} + Cl_2(g) \longrightarrow \underset{\text{chlorobenzene}}{C_6H_5Cl(l)} + HCl(g)$.
   (s) $CS_2(l) + Cl_2(g) \longrightarrow S_2Cl_2(l) + CCl_4(l)$.
   (t) $C_2H_4(g) + HCl(g) \longrightarrow \underset{\text{ethyl chloride}}{CH_3CH_2Cl(g)}$.
       $\underset{\text{ethylene}}{}$
   (u) $\underset{\text{methyl alcohol}}{CH_3OH(g)} + \frac{1}{2}O_2(g) \longrightarrow \underset{\text{formaldehyde}}{H_2CO(g)} + H_2O(g)$.
   (v) $\underset{\text{acetylene}}{C_2H_2(g)} + H_2O(l) \longrightarrow \underset{\text{acetaldehyde}}{CH_3CHO(l)}$.
   (w) $\underset{n\text{-butane}}{n\text{-}C_4H_{10}(g)} \longrightarrow \underset{\text{ethylene}}{C_2H_4(g)} + \underset{\text{ethane}}{C_2H_6(g)}$.
   (x) $\underset{\text{ethylene}}{C_2H_4(g)} + \underset{\text{benzene}}{C_6H_6(l)} \longrightarrow$ ethyl benzene(l).
   (y) $\underset{\text{ethylene}}{2C_2H_4(g)} \longrightarrow \underset{\text{1-butane}}{C_4H_{10}(l)}$.
   (z) $\underset{\text{propene}}{C_3H_6(g)} + \underset{\text{benzene}}{C_6H_6(l)} \longrightarrow \underset{\text{(isopropyl benzene)}}{\text{cumene(l)}}$.

4.84. A major air pollution problem facing many industrial firms is that of oxidizable gaseous emissions. These types of emissions are a major component of the complex mix of materials loosely defined as smog. The means of coping with these types of vapors are rather limited. Absorption on activated carbon is one method, but the relatively high cost of this form of control makes it useful primarily for recovery of potentially valuable by-product emissions. Lacking this economic incentive, most plants faced with the problem fall back on some type of oxidation—thermal or catalytic, to water, and other innocuous by-products. Given that the carbon to hydrogen ratio in a hydrocarbon pollutant gas is 1 mole of C to 4 moles of hydrogen, calculate the standard heat of reaction for the combustion of 1 mole of carbon gas and 4 moles of hydrogen as a gas with O_2 to give the following products:

product	number of moles
$H_2O(g)$	2.00
$CO_2(g)$	0.75
$CO(g)$	0.25

In the absence of any experimental data, assume that the standard heat of reaction of the pollutant gas is the same as methane (CH_4).

4.85. A mixture of gaseous methane (CH_4) and hydrogen (H_2) was found to have a heat of combustion at 25°C of -150.0 kcal/g mole; however, this value was determined under conditions such that the water produced left as a vapor rather than as a liquid as required for the standard values. Calculate the composition of the mixture.

4.86. Determine
(a) ΔH_f° for FeO(s)
(b) ΔH_{rxn}° for FeO(s) $+ 2H^+ \longrightarrow H_2O(l) + Fe^{2+}$
given the following data:

Reaction	ΔH_{rxn}°, cal
Fe(s) $+ 2H^+ \longrightarrow H_2(g) + Fe^{2+}$	$-20,600$
2Fe(s) $+ \frac{3}{2}O_2(g) \longrightarrow Fe_2O_3(s)$	$-196,500$
2FeO(s) $+ \frac{1}{2}O_2(g) \longrightarrow Fe_2O_3(s)$	$-67,900$

4.87. To take care of peak loads in the winter, the local gas company wishes to supplement its limited supply of natural gas (CH_4) by catalytically hydrogenating a heavy liquid petroleum oil that it can store in tanks. The supplier says that the oil has a U.O.P. Characterization Factor of 12.0 and an API gravity of 30°API. To minimize carbon deposition on the catalyst, twice the required amount of hydrogen is introduced with the oil. The reactants enter at 70°F and after heat exchange the products leave at 70°F. Calculate the following, per 1000 ft³ of CH_4 (SCGI):
(a) The pounds oil required.
(b) The heat input or output to the catalytic reactor.

4.88. Calculate the heat of reaction for the following reactions at 25°C and 1 atm. Express your answers in Btu/lb of the first reactant, and state whether energy is evolved or absorbed.
(a) $NaNO_3(c) + H_2SO_4(l) \longrightarrow NaHSO_4(c) + HNO_3(l)$.
(b) $4NH_3(g) + 5O_2(g) \longrightarrow 4NO(g) + 6H_2O(l)$.
(c) $C_2H_5OH(l) + 3O_2(g) \longrightarrow 2CO_2(g) + 3H_2O(g)$.

4.89. The net heating value of CH_4 at 50°C is 191,870 cal/g mole. What is the gross heating value at 0°C?

	T, °C	p_{mm}^*	ΔH_{vap} *cal/g*
C_p for $CH_4 = 8.52 + 0.0096T$	0	4.6	595
$C_p = $ cal/(g mole)(°C)	18	15.5	586
$T = $ °C	25	23.7	582
	50	92.3	568
	100	760.	539

4.90. What is the gross heating value of H_2 at 0°C? Gross heating value means the heat of reaction with the product H_2O as a liquid at 1 atm.

$$H_2(g) + \tfrac{1}{2}O_2(g) \longrightarrow H_2O(l)$$

$$\Delta H_{rxn} \text{ at } 25°C = -68,310 \text{ cal/g mole } H_2$$

$$C_p \text{ of } O_2 \text{ and } N_2 = \tfrac{2}{3} \text{ cal/g mole}$$

What is the net heating value, i.e., heat of reaction with H_2O as a gas?

4.91. The chemist for a gas company finds a gas analyzes CO_2, 9.2; C_2H_4, 0.4; CO, 20.9; H_2, 15.6; CH_4, 1.9; and N_2, 52.0 percent. What should he report as the heating value of the gas?

4.92. Calculate the net heating value of hydrogen at 200°C.

4.93. If 1 lb mole of Cu and 1 lb mole of H_2SO_4(100 percent) react together completely in a bomb calorimeter, how many Btu are absorbed (or evolved)? Assume that the products are $H_2(g)$ and solid $CuSO_4$. The initial and final temperatures in the bomb are 25°C.

4.94. The reaction $SO_2 + \tfrac{1}{2}O_2 \longrightarrow SO_3$ would seem to offer a simple method of making sulfur trioxide. However, both the reaction rate and equilibrium considerations are unfavorable for this reaction. At 600°C the reaction is only about 70 percent complete. If the products leave the reactor at 600°C and the reactants enter at 25°C, what is the heat evolved or absorbed from the system under these conditions?

4.95. Calculate the heat of reaction of the following reactions at the stated temperature:

(a) $\underset{\text{methyl alcohol}}{CH_3OH(g)} + \tfrac{1}{2}O_2(g) \xrightarrow{\text{200°C}} \underset{\text{formaldehyde}}{H_2CO(g)} + H_2O(g).$

(b) $SO_2(g) + \tfrac{1}{2}O_2(g) \xrightarrow{\text{300°C}} SO_3(g).$

4.96. Pyrites is converted to sulfur dioxide by the reaction

$$4FeS_2 + 11O_2 \longrightarrow 2Fe_2O_3 + 8SO_2$$

at about 400°C. Owing to imperfect burner operation, unburned lumps of FeS_2 remain. Also because of equilibrium and rate considerations, the reaction is not complete. If $2\tfrac{1}{2}$ tons of pyrites is burned with 20 percent excess air (based on the equation above) and 1 ton of Fe_2O_3 is produced, calculate the heat of reaction at the following:

(a) 25°C and 1 atm.

(b) 400°C and 1 atm.

4.97. Sulfur can be recovered from the H_2S in natural gas by the following reaction:

$$3H_2S(g) + 1\tfrac{1}{2}O_2(g) \longrightarrow 3S(s) + 3H_2O(g)$$

For the materials as shown, with H_2S and O_2 entering the process at 300°F and the products leaving the reactor at 100°F, calculate the heat of reaction in British thermal units per pound of S formed. Assume that the reaction is complete. If it is only 75 percent complete, will this change your answer, and if so, how much?

4.98. The Smog Abatement Co. is authorized to market and install conversion kits so that automobiles can use natural gas. It requires about 4 hr to install the conversion equipment in each car and costs about \$400–\$500 per car. Customers can recover the expense of conversion because natural gas can be sold to the public at about 20 cents a therm (100 ft³), as compared with about 40 cents/gal for gasoline, the equivalent, according to the company. To provide some specific numbers to promote the conversion service, you are asked to make the following computations. Assume that the automobile engine will burn ethane, C_2H_6.

(a) For theoretical complete combustion with dry air, what weight of air is used per hour when the motor uses 480 ft³ of fuel per hour measured at 32°F and 760 mm Hg pressure?

(b) Assume that the fuel enters the engine at 150°F and the air at 250°F because of heat exchange with the exhaust gases after they leave the engine. Assume that the exhaust gases leave the heat exchanger at 2000°F and that 15 percent of the net heating value of the fuel is lost by radiation from the engine and heat transfer losses. Calculate the amount of energy that is available to operate the engine.

4.99. Propane, butane, or liquefied petroleum gas (LPG) has seen practical service in passenger automobiles for 30 years or more. Because it is used in the vapor phase, it pollutes less than gasoline but more than natural gas. A number of cars in the 1970 Clean Air Car Race ran on LPG. The table below lists their results and those for natural gas. It must be kept in mind that these vehicles were generally equipped with platinum catalyst reactors and with exhaust-gas recycle. Therefore the gains in emission control did not come entirely from the fuels.

	Natural gas, *avg 6 cars*	*LPG,* *avg 13 cars*	*1975* *Fed. Std.*
HC (g/mile)	1.3	0.49	0.22
CO (g/mile)	3.7	4.55	2.3
NO$_x$ (g/mile)	0.55	1.26	0.6

Suppose that in a test butane gas at 100°F is burned completely with the stoichiometric amount of heated air at 400°F and a dew point of 77°F in an engine. To cool the engine, there are 12.5 lb of steam at 100 psia and 95 percent quality generated from water at 77°F per pound of butane burned. It may be assumed that 7 percent of the gross heating value of the butane is lost as radiation from the engine. Will the exhaust gases leaving the engine exceed the temperature limit of the catalyst of 1500°F?

4.100. Coal, petroleum, and natural gas constitute the major sources of energy, with water power and wood supplying supplementary quantities amounting to less than 10 percent of the total. Over the past 50 years the utilization of energy from these sources has increased almost fivefold. One of the characteristics of a fuel that is important in the conversion of energy to work is the adiabatic flame temperature. Calculate the flame temperature for the three fuels, gas, coke, and oil, whose compositions are given below, when they are burned with the theoretical amount of air. Assume that the air enters at 65°F and that the ash contains no C and leaves the fuel bed at 1000°F. The C_p (mean) of the ash $= 0.28$ Btu/(lb)(°F).

compositions of the fuels

	composition	%
Gas (vol %)	CH_4	96.0
	CO_2	3.0
	N_2	1.0
Oil (wt %)	C_6H_{34}	99.0
	S	1.0
Coke (wt %)	C	95.0
	Ash	5.0

4.101. If CO at constant pressure is burned with excess air and the theoretical flame temperature is 1800°F, what was the percentage of excess air used? The reactants enter at 200°F.

4.102. Which substance will give the higher theoretical flame temperature if the inlet percentage of excess air and temperature conditions are identical: (a) CH_4, (b) C_2H_6, (c) C_4H_8?

4.103. A power plant burns natural gas (90 percent CH_4, 10 percent C_2H_6) at 77°F and 1 atm with 70 percent excess air at the same conditions. Calculate the theoretical maximum temperature in the boiler if all products are in the gaseous state.

4.104. In Problem 4.103, if the air is preheated to 500°F, what will be the maximum temperature?

4.105. *n*-Heptane is dehydrocyclicized in the hydroforming process to toluene by means of catalysts as follows:

$$C_7H_{16} \longrightarrow C_6H_5CH_3 + 4H_2$$

Assuming that a yield of 35 percent of the theoretical is obtained under the conditions shown in Fig. P4.105, how much heat is required in the process per 1000 lb of toluene produced? *Note:*

$$C_{pm}(C_7H_{16})_{65-900°F} = 0.56 \text{ Btu/(lb)(°F)}$$

$$C_{pm}(C_6H_5CH_3)_{65-900°F} = 0.62 \text{ Btu/(lb)(°F)}$$

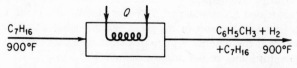

Fig. P4.105

4.106. In the manufacture of benzaldehyde, a mixture of toluene and air is passed through a catalyst bed where it reacts to form benzaldehyde,

$$C_6H_5CH_3 + O_2 \longrightarrow C_6H_5CHO + H_2O$$

Dry air and toluene gas are fed to the converter at a temperature of 350°F and at atmospheric pressure. To maintain a high yield, air is supplied in 100 percent excess over that required for complete conversion of the toluene charged. The degree of completion of the reaction, however, is only 13 percent. Owing to the high temperature in the catalyst bed, 0.5 percent of the toluene charged burns to form CO_2 and H_2O,

$$C_6H_5CH_3 + 9O_2 \longrightarrow 7CO_2 + 4H_2O$$

Cooling water is circulated through a jacket on the converter, entering at 80°F and leaving at 105°F. The hot gases leave the converter at 379°F. During a 4-hr test run, a water layer amounting to 29.3 lb was collected after the exit gases had been cooled. Calculate the following:
(a) The composition of the inlet stream to the converter.
(b) The composition of the exhaust gas from the converter.
(c) The gallons per hour of cooling water required in the converter jacket.
Additional data: The heat capacities of both toluene and benzaldehyde *gas* may be taken as 31 Btu/(lb mole)(°F); C_p for benzaldehyde liquid is 0.43 cal/(g)(°C).

4.107. A pipeline carrying pure ethane at 100°F joins a second line carrying pure methane at 130°F producing a mixture containing 50 mole percent ethane. The pressure of the mixture and the pure gases is 2000 psia. From the data given below calculate the following:
(a) The heat of mixing for a 50 mole percent methane-ethane mixture at 130°F and 2000 psia.
(b) The temperature of the mixture.

Enthalpies, Btu/(lb)(mole), at 2000 psia

Temp (°F)	Pure CH_4	Pure C_2H_6	Mixture, 50% CH_4
70	3247	501	2025
100	3673	1204	2795
130	4028	1958	3408

4.108. (a) In a petrochemical plant 1 ton/hr of 50 percent (by wt) H_2SO_4 is being produced at 250°F by concentrating 30 percent H_2SO_4 supplied at 100°F. From the data provided below, find the heat which has to be supplied per hour:

Btu evolved at 100°F per lb mole H_2SO_4	moles of H_2O per mole of H_2SO_4
18,000	2
21,000	3
23,000	4
24,300	5
27,800	10
31,200	22

C_p $(Btu/(lb)(°F))$	30% H_2SO_4	50% H_2SO_4
100°F	0.80	0.70
250°F	0.75	0.60

The C_p of water vapor is about 0.50, and for water use, 1.0.

(b) Using the same data as in part (a), what is the temperature of a solution of 30 percent (by wt) H_2SO_4 if it is made from 50 percent H_2SO_4 at 100°F and pure water at 200°F if the process is adiabatic?

4.109.* (a) From the data below plot the enthalpy of 1 *mole of solution* at 80°F as a function of the weight percent HNO_3. Use as reference states liquid water at 32°F and liquid HNO_3 at 32°F. You can assume that that C_p for H_2O is 18 cal/(g mole)(°C) and for HNO_3, 30 cal/(g mole)(°C).

$-\Delta H_{soln}$ at 80°F (cal/g mole HNO_3)	moles H_2O added to 1 mole HNO_3
0	0.0
800	0.1
1300	0.2
1650	0.3
2000	0.5
2600	0.67
3400	1.0
4100	1.5
4850	2.0
5750	3.0
6200	4.0
6650	5.0
7300	10.0
7450	20.0

(b) Compute the energy absorbed or evolved at 80°F on making a solution of 4 moles of HNO_3 and 4 moles of water by mixing a solution of $33\frac{1}{3}$ mole percent acid with one of 60 mole percent acid.

4.110.* *National Bureau of Standards Circular* 500 gives the following data for calcium chloride (mol. wt 111) and water:

formula	state	$-\Delta H_f$ at 25°C (kcal/g mole)
H_2O	liq	68.317
	g	57.798
$CaCl_2$	c	190.0
	in 25 moles of H_2O	208.51
	50	208.86
	100	209.06
	200	209.20
	500	209.30
	1000	209.41
	5000	209.60
	∞	209.82
$CaCl_2 \cdot H_2O$	c	265.1
$CaCl_2 \cdot 2H_2O$	c	335.5
$CaCl_2 \cdot 4H_2O$	c	480.2
$CaCl_2 \cdot 6H_2O$	c	623.15

Calculate the following:

(a) The energy evolved when 1 lb mole of $CaCl_2$ is made into a 20 percent solution at 77°F.

(b) The heat of hydration of the dihydrate to the hexahydrate.

(c) The energy evolved when a 20 percent solution containing 1 lb mole of $CaCl_2$ is diluted with water to 5 percent at 77°F.

4.111. A vessel contains 100 lb of an $NH_4OH–H_2O$ liquid mixture at 25°C and 1 atm that is 15.0 wt percent NH_4OH. Just enough aqueous H_2SO_4 is added to the vessel from an H_2SO_4 liquid mixture at 25°C and 1 atm, that is, 25.0 mole percent H_2SO_4, so that the reaction to $(NH_4)_2SO_4$ is complete. After the reaction, the products are again at 25°C and 1 atm. How much heat (Btu) is absorbed or evolved by this process? It may be assumed that the final volume of the products is equal to the sum of the volumes of the two initial mixtures.

PROBLEMS TO PROGRAM ON THE COMPUTER

4.1. The enthalpy of a mixture of constant composition at any constant temperature T is

$$\Delta H_{\text{mixture}}(T, p) = \sum_{i=1}^{n} x_i \sum H_i^0(T, p) - \Delta H_{\text{mixture}}^D(T, p)$$

where the deviation, ΔH^D, is a correction because of high pressure. Use of the Benedict-Webb-Rubin (BWR) equation (see Table 3.1) gives the following for ΔH^D:

$$\Delta H^D = \left(B_0 RT - 2A_0 - 4\frac{C_0}{T^2}\right)\rho + \frac{1}{2}(2bRT - 3a)\rho^2 + \frac{6}{5}(a\alpha\rho^5)$$
$$+ \left(\frac{c\rho}{T^2}\right)\left[\frac{3[1 - \exp(-\gamma\rho^2)]}{\gamma\rho^2} - \left(\frac{1}{2} - \gamma\rho^2\right)\exp(-\gamma\rho^2)\right]$$

Use the density values calculated in computer Problem 3.3 for ρ, and prepare an additional subroutine to give $\Delta H_{\text{mixture}}^D(T, p)$ for selected T and p pairs. Refer to Problem 3.3 for other pertinent information.

4.2. Prepare a computer program to find the adiabatic flame temperature of a gas such as in Problem 4.33. Assume that the principal reactant is completely burned.

4.3. Prepare a more general program than the one in Problem 4.2 immediately above to compute the adiabatic flame temperature. Assume that several hydrocarbons of the form CH_n can be burned with various amounts of excess air. The percentage of excess air is to be read into the program as well as the type of fuel gas or solid. The only combustion products that need to be considered are carbon dioxide, carbon monoxide, water, hydrogen, oxygen, and nitrogen. Make provision for the following additional parameters:

(a) Pressure in the burner.

(b) Humidity of the incoming air.

(c) Number of moles of each hydrocarbon and of nitrogen in the fuel stream.

(d) Standard heats of formation for all the hydrocarbon gases burned and the products. (Assume that gaseous water is formed in the reaction.)

(e) Heat capacities of all components.

(f) Inlet reactant temperatures, which may be different for each reactant.

The gases can be treated as an ideal mixture.

Combined Material and Energy Balances 5

Now that you have accumulated some experience in making energy balances and in the principles of thermochemistry, it is time to apply this knowledge to situations and problems involving both material and energy balances. You have already encountered some simple examples of combined material and energy balances, as, for example, in the calculation of the adiabatic reaction temperature, where a material balance provides the groundwork for the writing of an energy balance. In fact, in all energy balance problems, however trifling they may seem, you must know the amount of material entering and leaving the process if you are to apply successfully the appropriate energy balance equation(s).

In this chapter we shall consider both those problems which require you to make a preliminary material balance prior to making an energy balance and also problems which require you to solve simultaneous material and energy balances. In line with this discussion we shall take up two special types of charts, enthalpy-concentration charts and humidity charts, which are prepared by combining material and energy balances and which are useful in a wide variety of practical problems. Figure 5.0 shows the relationships among the topics to be discussed in this chapter and the relationships with topics in previous chapters.

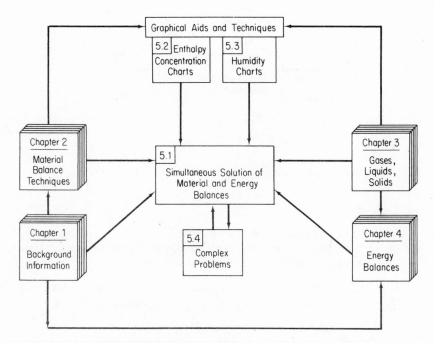

Fig. 5.0. Hierarchy of topics to be studied in this chapter (section numbers are in the upper left-hand corner of the boxes).

5.1 Simultaneous use of material and energy balances for the steady state

For any specified system or piece of equipment you can write

(a) A total material balance.

(b) A material balance for each chemical component (or each atomic species, if preferred).

You can use the energy balance to add one additional independent overall equation to your arsenal, but you cannot make energy balances for each individual chemical component in a mixture or solution. As we mentioned previously, you need one independent equation for each unknown in a given problem. The energy balance often provides the extra piece of information to help you resolve an apparently insuperable calculation composed solely of material balances.

We can illustrate any given system or piece of equipment *in the steady state* by drawing a diagram such as Fig. 5.1. Figure 5.1 could represent a boiler, a distillation column, a drier, a human being, or a city, and takes into account processes involving a chemical reaction.

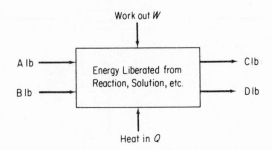

Fig. 5.1. Sketch of generalized flow process with chemical reaction.

Let x_i = weight fraction of any component

$\hat{H}_i$ = enthalpy per unit mass of any component with respect to some reference temperature

If the subscript numbers 1, 2, 3, etc., represent the components in each stream, you can write the following material balances in algebraic form:

balance	in	=	out
Total:	$A + B$	$=$	$C + D$
Component 1:	$Ax_{A_1} + Bx_{B_1}$	$=$	$Cx_{C_1} + Dx_{D_1}$
Component 2:	$Ax_{A_2} + Bx_{B_2}$	$=$	$Cx_{C_2} + Dx_{D_2}$
etc.			

As previously discussed, the number of independent equations is one less than the total number of equations.

In addition, an overall energy balance can be written as (ignoring kinetic and potential energy changes)

$$Q - W = (C\hat{H}_C + D\hat{H}_D) - (A\hat{H}_A + B\hat{H}_B)$$

For a more complex situation, examine the interrelated pieces of equipment in Fig. 5.2.

Again the streams are labeled, and the components are 1, 2, 3, etc. How should you attack a problem of this nature? From an overall viewpoint, the problem reduces to that shown in Fig. 5.1. In addition, a series of material balances and an energy balance can be made around each piece of equipment I, II, and III. Naturally, all these equations are not independent, because, if you add up the total material balances around boxes I, II, and III, you simply have the grand overall material balance around the whole process. Similarly, if you add up the component balances for each box, you have a grand component balance for the entire process, and the sum of the energy balances around boxes I, II, and III gives the overall energy balance. Following Fig. 5.2 (on p. 365) is a list of the material and energy balances that you can write for it.

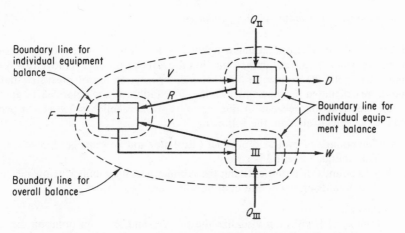

Fig. 5.2. Sketch for interrelated processes and streams.

balance	in	out

Entire process:

Total
$$F = D + W$$

Component
$$Fx_{F_i} = Dx_{D_i} + Wx_{W_i}$$

Energy
$$Q_{II} + Q_{III} + F\,\Delta\hat{H}_F = D\,\Delta\hat{H}_D + W\,\Delta\hat{H}_W$$

Process I:

Total
$$F + R + Y = V + L$$

Component
$$Fx_{F_i} + Rx_{R_i} + Yx_{Y_i} = Vx_{V_i} + Lx_{L_i}$$

Energy
$$F\,\Delta\hat{H}_F + R\,\Delta\hat{H}_R + Y\,\Delta\hat{H}_Y = V\,\Delta\hat{H}_V + L\,\Delta\hat{H}_L$$

Process II:

Total
$$V = R + D$$

Component
$$Vx_{V_i} = Rx_{R_i} + Dx_{D_i}$$

Energy
$$Q_{II} + V\,\Delta\hat{H}_V = R\,\Delta\hat{H}_R + D\,\Delta\hat{H}_D$$

Process III:

Total
$$L = Y + W$$

Component
$$Lx_{L_i} = Yx_{Y_i} + Wx_{W_i}$$

Energy
$$Q_{III} + L\,\Delta\hat{H}_L = Y\,\Delta\hat{H}_Y + W\,\Delta\hat{H}_W$$

Note the symmetry among these equations. The methods of writing and applying these balances are now illustrated by an example.

EXAMPLE 5.1 Simultaneous material and energy balances

A distillation column separates 10,000 lb/hr of a 40 percent benzene, 60 percent chlorobenzene liquid solution which is at 70°F. The liquid product from the top of the column is 99.5 percent benzene, while the bottoms (stream from the reboiler) contains 1 percent benzene. The condenser uses water which enters at 60°F and leaves at 140°F,

365

while the reboiler uses saturated steam at 280°F. The reflux ratio (the ratio of the liquid overhead returned to the column to the liquid overhead product removed) is 6 to 1. Assume that both the condenser and reboiler operate at 1 atm pressure, that the temperature calculated for the condenser is 178°F and for the reboiler 268°F, and that the calculated fraction benzene in the vapor from the reboiler is 3.9 weight percent (5.5 mole percent). Calculate the following:

(a) The pounds of overhead product (distillate) and bottoms per hour.
(b) The pounds of reflux per hour.
(c) The pounds of liquid entering the reboiler and the reboiler vapor per hour.
(d) The pounds of steam and cooling water used per hour.

Solution:

(a) Figure E5.1 will help visualize the process and assist in pointing out what additional data have to be determined.

First we have to get some pertinent enthalpy or heat capacity data. Then we can make our material and energy balances. (The exit streams have the same composition as the solutions in the condenser or reboiler.)

(b) Convert the reboiler analysis into mole fractions.

Basis: 100 lb B

comp.	lb	mol. wt	lb moles	mole fr.
Bz	1	78.1	0.0128	0.014
Cl	99	112.6	0.88	0.986
			0.8928	1.000

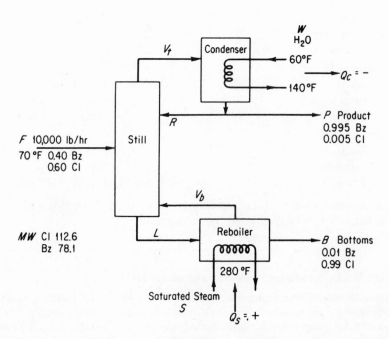

Fig. E5.1.

(c) The heat capacity data for liquid benzene and chlorobenzene will be assumed to be as follows[1] (no enthalpy tables are available):

Temp. (°F)	C_p, Btu/(lb)(°F)		$\Delta \hat{H}_{vaporization}$, Btu/lb	
	Cl	Bz	Cl	Bz
70	0.31	0.405		
90	0.32	0.415		
120	0.335	0.43		
150	0.345	0.45		
180	0.36	0.47	140	170
210	0.375	0.485	135	166
240	0.39	0.50	130	160
270	0.40	0.52	126	154

Basis: 10,000 lb feed/hr

(d) Overall material balances:

Overall total material balance:

$$F = P + B$$
$$10,000 = P + B$$

Overall benzene balance:

$$Fx_F = Px_P + Bx_B$$
$$10,000(0.40) = P(0.995) + B(0.01)$$
$$10,000(0.40) = P(0.995) + (10,000 - P)(0.01)$$
$$P = 3960 \text{ lb/hr} \longleftarrow \text{(a}_1)$$
$$B = 6040 \text{ lb/hr} \longleftarrow \text{(a}_2)$$

(e) Material balances around the condenser:

$$\frac{R}{P} = 6 \quad \text{or} \quad R = 6P = 6(3960) = 23,760 \text{ lb/hr} \longleftarrow \text{(b)}$$
$$V_t = R + P = 23,760 + 3960 = 27,720 \text{ lb/hr}$$

(f) Material balances around the reboiler:

Total:

$$L = B + V_b$$

Benzene:

$$Lx_L = Bx_B + V_b x_{V_b}$$
$$L = 6040 + V_b$$
$$Lx_L = 6040(0.01) + V_b(0.039)$$

We have three unknowns and only two *independent* equations. We can write additional equations around the still, but these will not resolve the problem since we would still be left with one unknown more than the number of *independent* equations. This is the stage at which energy balances can be used effectively.

[1]Data estimated from tabulations in the Appendix of *Process Heat Transfer* by D. Q. Kern, McGraw-Hill, New York, 1950.

(g) Overall energy balance: Let the reference temperature be 70°F; this will eliminate the feed from the enthalpy calculations. Assume that the solutions are ideal so that the thermodynamic properties (enthalpies and heat capacities) are additive. No work or potential or kinetic energy is involved in this problem. Thus

$$Q_{\text{steam}} + Q_{\text{condenser}} = \underbrace{P \int_{70}^{178} C_{P_P}\, dt}_{\Delta H_P} + \underbrace{B \int_{70}^{268} C_{P_B}\, dt}_{\Delta H_B} - \underbrace{F \int_{70}^{70} C_{P_F}\, dt}_{\Delta H_F = 0}$$

We know all the terms except Q_{steam} and $Q_{\text{condenser}}$ and still need more equations.

(h) Energy balance on the condenser: This time we shall let the reference temperature equal 178°F; this choice simplifies the calculation, because then neither the R nor P streams have to be included. Assume that the product leaves at the saturation temperature in the condenser of 178°F. With the condenser as the system and the water as the surroundings we have

system (condenser)   **surroundings (water)**

$$\Delta H_{\text{condenser}} = Q_{\text{condenser}} \qquad \Delta H_{\text{water}} = Q_{\text{water}}$$

and, since $Q_{\text{system}} = -Q_{\text{surroundings}}$, $\Delta H_{\text{condenser}} = -\Delta H_{\text{water}}$

$$V_t(-\Delta \hat{H}_{\text{vaporization}}) = -WC_{P_{\text{H}_2\text{O}}}(t_2 - t_1)$$

$$27{,}720[170(0.995) + 140(0.005)] = W(1)(140 - 60) = Q_{\text{water}} = -Q_{\text{condenser}}$$

Solving, $Q_C = -4.71 \times 10^6$ Btu/hr (heat evolved), and the cooling water used is

$$W = 5.89 \times 10^4 \text{ lb H}_2\text{O/hr} \longleftarrow \boxed{d_1}$$

(i) Amount of steam used: We shall use the overall energy balance from step (g).

$$Q_{\text{steam}} = 3960\frac{\text{lb}}{\text{hr}}\left(46.9\frac{\text{Btu}}{\text{lb}}\right) + 6040\frac{\text{lb}}{\text{hr}}\left(68.3\frac{\text{Btu}}{\text{lb}}\right) + 4.71 \times 10^6\frac{\text{Btu}}{\text{hr}}$$

In the preceding equation the $\int C_p\, dt$ was determined by first graphically integrating $\int C_p\, dt$ to get $\Delta \hat{H}$ for each component and then weighting the $\Delta \hat{H}$ values by their respective weight fractions:

$$\Delta \hat{H}_P = \int_{70}^{178} C_{P_P}\, dt \quad \text{Btu/lb} \qquad \Delta \hat{H}_B = \int_{70}^{268} C_{P_B}\, dt \quad \text{Btu/lb}$$

Bz	Cl	Avg		Bz	Cl	Avg
47.0	36.2	46.9		88.1	68.0	68.3

An assumption that the P stream was pure benzene and the B stream was pure chlorobenzene would be quite satisfactory since the value of Q_C is one order of magnitude larger than the "sensible heat" terms

$$Q_{\text{steam}} = 5.31 \times 10^6 \text{ Btu/hr}$$

From the steam tables, $\Delta \hat{H}_{\text{vap}}$ at 280°F is 923 Btu/lb. Assume that the steam leaves at its saturation temperature and is not subcooled. Then

$$\text{lb steam used/hr} = \frac{5.31 \times 10^6 \text{ Btu/hr}}{923 \text{ Btu/lb}} = 5760 \text{ lb/hr} \longleftarrow \boxed{d_2}$$

(j) Energy balance around the reboiler:

$$Q_{\text{steam}} + L(\Delta \hat{H}_L) + V_b(\Delta \hat{H}_{V_b}) + B(\Delta \hat{H}_B)$$

We know that $Q_{steam} = 5.31 \times 10^6$ Btu/hr. We do not know L or the value of $\Delta \hat{H}_L$. However, even if $\Delta \hat{H}_L$ is not known, the temperature of stream L entering the reboiler would not be more than 20°F lower than 268°F at the very most, and the heat capacity would be about that of chlorobenzene. We shall assume a difference of 20°F, but you can note in the calculations below that the effect of including $\Delta \hat{H}_L$, as opposed to ignoring it, is minor.

The value of V_b is still unknown, but all the other values ($\Delta \hat{H}_{V_s}$, B, $\Delta \hat{H}_B$) are known or can be calculated. Thus, if we combine the energy balance around the reboiler with the overall material balance around the reboiler, we can find both L and V_b.

Reference temperature: 268°F

Energy balance:

$$5.31 \times 10^6 \frac{\text{Btu}}{\text{hr}} + (L \text{ lb})\left(0.39 \frac{\text{Btu}}{(\text{lb})(°\text{F})}\right)(-20°\text{F}) = V_b[(0.99)(126)$$
$$+ (0.01)(154)] + B(0)$$

Material balance:

$$L = 6040 + V_b$$
$$5.31 \times 10^6 - (6040 + V_b)(7.8) = 126.3 V_b$$
$$5.31 \times 10^6 - 0.047 \times 10^6 = 126.3 V_b + 7.8 V_b$$
$$V_b = \frac{5.26 \times 10^6}{134} = 39,300 \text{ lb/hr} \longleftarrow \boxed{c_1}$$
$$L = V_b + B = 39,300 + 6040 = 45,340 \text{ lb/hr} \longleftarrow \boxed{c_2}$$

If we had ignored the enthalpy of the L stream, then V_b would have been

$$V_b = \frac{5.31 \times 10^6}{126.3} = 42,100 \text{ lb/hr}$$

a difference of about 7.1 percent.

In practice, the solution of steady-state material and energy balances is quite tedious, a situation which has led engineers to the use of digital and hybrid computer routines. Certain special problems are encountered in the solution of large-scale sets of equations, problems of sufficient complexity to be beyond our scope here.

An important aspect of combined material and energy balance problems is how to ensure that the process equations are determinate, i.e., have at least one solution. We can use the phase rule (discussed in Sec. 3.7-1) plus some common sense to provide a guide as to the number of variables that must be specified and the number of variables that can be unknown in any problem. The phase rule gives the relationship between the number of degrees of freedom and the *intensive* variables such as temperature, pressure, specific volume, and composition that are associated with each stream. According to Eq. (3.57), the number of degrees of freedom for a single phase at equilibrium is

$$F = C - 1 + 2 = C + 1$$

However, to completely specify a stream, we need to know how much material is flowing, so that in addition to the $C + 1$ intensive variables, one extensive variable must be given, for a total of

$$N_t = C + 2$$

where N_t is the total number of variables to be specified for a single stream. Each *independent* material and energy balance that is written reduces the degrees of freedom in a problem by one. Consequently, in any problem you can add up the total number of variables and subtract the number of equality constraints and specified input conditions to get the difference, which is the number of variables that have to be specified if the problem is to be determinate. If a chemical reaction takes place, the number of independent material balances that can be made is equal to the number of atomic species, not C, the number of components.

We shall now illustrate the application of the above concept to several simple cases.

EXAMPLE 5.2   Determining the degrees of freedom in a process

We shall consider five typical processes, as depicted by the respective figures below, and for each ask the question, How many variables have to be specified, i.e., what are the degrees of freedom (d.f.) to make the problem of solving the combined material and energy balances determinate. All the processes will be steady-state ones.

(a) *Simple stream junction, single phase* [Fig. E5.2(a)]. We assume that $Q = 0$ and $W = 0$, and that the pressure and temperature of P_1 and P_2 are the same as that

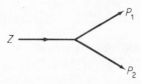

Fig. E5.2(a).

of Z, the input stream. The statement of the total number of variables, constraints, and degrees of freedom can be summarized as follows.

Total number of variables (3 streams):	$3(C + 2)$
Number of independent equality constraints:	
Material balances:	C
Energy balance:	1
$T_{P_1} = T_Z; T_{P_2} = T_Z$:	2
$p_{P_1} = p_Z; p_{P_2} = p_Z$:	2
Input conditions in Z specified:	$C + 2$

$$\text{Total no. d.f.} = 3(C + 2) - 2C - 7 = C - 1$$

Note that although $C + 1$ material balances can be written corresponding to C components and one total balance, only C of the balances are independent and included in the tally above.

We can illustrate the relation between the count for the degrees of freedom above and the constraints by considering a two-component system for which d.f. $= C - 1$ $= 2 - 1 = 1$:

Material balances:

$$Z = P_1 + P_2$$

$$Zx_Z = P_1 x_{P_1} + P_2 x_{P_2}$$

$$Zy_Z = P_1 y_{P_1} + P_2 y_{P_2}$$

Energy balance:

$$P_1 \hat{H}_{P_1} + P_2 \hat{H}_{P_2} - Z\hat{H}_Z = 0$$

Definitions:

$$x_{P_1} + y_{P_1} = 1$$

$$x_{P_2} + y_{P_2} = 1$$

$$x_Z + y_Z = 1$$

If Z, x_Z, T_Z, and p_Z are specified in the input, then $\hat{H}_Z$ is known because H is a function only of T, p, and x. We have as given $T_{P_1} = T_Z = T_{P_2}$ and $p_{P_1} = p_Z = p_{P_2}$, so that $\hat{H}_{P_1}$ and $\hat{H}_{P_2}$ depend only on x_{P_1} and x_{P_2}, respectively. You can see by inspection that if P_1 is specified, representing the 1 degree of freedom available, the total material balance can be solved for P_2, and then the (nonlinear) energy balance and one-component material balance can be solved simultaneously for x_{P_1} and x_{P_2}.

If the initial assumption had been that the compositions of P_1 and P_2 were the same as Z and we reduced the total d.f. by an additional factor of $2(C - 1)$, there being $C - 1$ compositions to be specified, we would be making an error because the number of independent equations would not be $C + 1$ anymore but would simply be 1. The component material balances and energy balance would all be the same. Consequently the number of degrees of freedom would be

$$3(C + 2) - 1 - 2 - 2 - (C + 2) - 2(C - 1) = 1$$

the same as before.

(b) *Mixer* [Fig. E5.2(b)]. For this process we assume that $W = 0$.

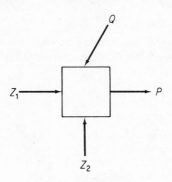

Fig. E5.2(b).

Total number of variables (3 streams $+ Q$): $\quad 3(C + 2) + 1$
Number of independent equality constraints:
    Material balances: $\qquad\qquad\qquad\qquad\qquad\qquad C$
    Energy balance: $\qquad\qquad\qquad\qquad\qquad\qquad\quad 1$

$\qquad$ Total no. d.f. $= 3(C + 2) + 1 - C - 1 = 2C + 6$

(c) *Heat exchanger* [Fig. E5.2(c)]. For this process we assume that $W = 0$.

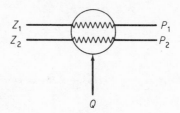

Fig. E5.2(c).

Total number of variables (4 streams $+ Q$): $\quad 4(C + 2) + 1$
Number of independent equality constraints:
    Material balances (2 separate lines): $\qquad\qquad 2C$
    Energy balance: $\qquad\qquad\qquad\qquad\qquad\qquad\quad 1$

$\qquad$ Total no. d.f. $= 4(C + 2) + 1 - 2C - 1 = 2C + 8$

(d) *Pump* [Fig. E5.2(d)]. Here $Q = 0$.

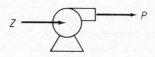

Fig. E5.2(d)

Total number of variables (2 streams $+ W$): $\quad 2(C + 2) + 1$
Number of independent equality constraints:
    Material balances: $\qquad\qquad\qquad\qquad\qquad\qquad C$
    Energy balance: $\qquad\qquad\qquad\qquad\qquad\qquad\quad 1$

$\qquad$ Total no. d.f. $= 2(C + 2) + 1 - C - 1 = C + 4$

(e) *Two-phase well-mixed tank* (*stage*) *at equilibrium* [Fig. E5.2(e)] (*L*, liquid phase; *V*, vapor phase). Here $W = 0$. Although two phases exist at equilibrium inside

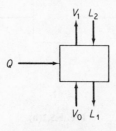

Fig. E5.2(e).

the system, the streams entering and leaving are single-phase streams. By equilibrium we mean that both phases are at the same temperature and pressure and that an equation is known that relates the composition in one phase to that in the other for each component.

Total number of variables (4 streams + Q):	$4(C + 2) + 1$
Number of independent equality constraints:	
Material balances:	C
Energy balance:	1
Composition relations at equilibrium:	C
Temperatures in the two phases equal:	1
Pressures in the two phases equal:	1

$$\text{Total no. d.f.} = 4(C + 2) + 1 - 2C - 3 = 2C + 6$$

In general you might specify the following variables to make the problem determinate:

Input stream L_2:	$C + 2$
Input stream V_0:	$C + 2$
Pressure:	1
Q:	1
Total:	$2C + 6$

Other choices are of course possible.

You can compute the degrees of freedom for combinations of like or different simple processes by proper combination of their individual degrees of freedom. In adding the degrees of freedom for units, you must eliminate any double counting either for variables or constraints and take proper account of interconnecting streams whose characteristics are often fixed only by implication.

EXAMPLE 5.3   Degrees of freedom in combined units

Consider the mixer-separator shown in Fig. E5.3. Superficially it would seem that we could carry out an analysis as follows to combine the units. For the mixer considered as a separate unit, from the previous example, d.f. $= 2C + 6$. For the separator, an equilibrium unit, we find from the previous example that d.f. $= C + 4$. Hence the total d.f. $= 2C + 6 + C + 4 = 3C + 10$. However, we have to eliminate certain redundant counts of variables and constraints for the combination:

Eliminate variable Z:	$C + 2$
Eliminate Q_S:	1
Total:	$C + 3$
Eliminate one energy balance:	1
Net change:	$C + 2$

The net d.f. for the entire unit would be $2C + 8$, but this count is incorrect because you can see from the previous example that the correct number of d.f. $= 2C + 6$.

We can locate the source of this discrepancy if we note for the mixer that stream Z is actually a *two-phase stream*; hence $N_t = C + 1$, not $C + 2$. The proper analysis

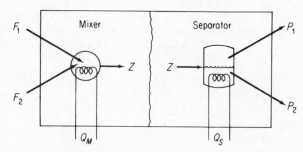

Fig. E5.3. Degrees of freedom in combined units.

is as follows:

For the mixer:

	single-phase streams	two-phase stream	Q_M
Total no. variables:	$2(C + 2) + 1(C + 1) + 1 = 3C + 6$		
Constraints (as before):			$C + 1$
Net:			$2C + 5$

For the separator:

Total no. variables:	$2(C + 2) + 1(C + 1) + 1 = 3C + 6$	
Constraints (as before):		$2C + 3$
Net:		$C + 3$

Eliminate redundant counting:

Drop Q_S:		1	
Eliminate Z:		$C + 1$	
Together:			$C + 2$

Consequently the total degrees of freedom are

$$(2C + 5) + (C + 3) - (C + 2) = 2C + 6$$

as in Example 5.2.

We do not have the space to illustrate additional combinations of simple units to form more complex units, but Kwauk[2] prepared several excellent tables summarizing the variables and degrees of freedom for distillation columns, absorbers, heat exchangers, and the like.

5.2 Enthalpy-concentration charts

An enthalpy-concentration chart is a convenient graphical method of representing enthalpy data for a binary mixture. If available,[3] such charts are useful

[2]M. Kwauk, *A.I.Ch.E. J.*, v. 2, p. 240 (1956).

[3]For a literature survey as of 1957 see Robert Lemlich, Chad Gottschlich, and Ronald Hoke, *Chem. Eng. Data Series*, v. 2, p. 32 (1957). Additional references: for CCl_4, see M. M.

in making combined material and energy balances calculations in distillation, crystallization, and all sorts of mixing and separation problems. You will find a few examples of enthalpy-concentration charts in Appendix I.

At some time in your career you may find you have to make numerous repetitive material and energy balance calculations on a given binary system and would like to use an enthalpy-concentration chart for the system, but you cannot find one in the literature or in your files. How do you go about constructing such a chart?

5.2-1 Construction of an Enthalpy-Concentration Chart. As

usual, the first thing to do is choose a basis—some given amount of the mixture, usually 1 lb or 1 lb mole. Then choose a reference temperature ($H_0 = 0$ at T_0) for the enthalpy calculations. Assuming 1 lb is the basis, you then write an energy balance for solutions of various compositions at various temperatures:

$$\Delta \hat{H}_{\text{mixture}} = x_A \Delta \hat{H}_A + x_B \Delta \hat{H}_B + \Delta \hat{H}_{\text{mixing}} \qquad (5.1)$$

where $\Delta \hat{H}_{\text{mixture}}$ = the enthalpy of 1 lb of the mixture

$\Delta \hat{H}_A, \Delta \hat{H}_B$ = the enthalpies of the pure components per lb relative to the reference temperature (the reference temperature does not have to be the same for A as for B)

$\Delta \hat{H}_{\text{mixing}}$ = the heat of mixing (solution) per lb at the temperature of the given calculation

In the common case where the $\Delta \hat{H}_{\text{mixing}}$ is known only at one temperature (usually 77° F), a modified procedure as discussed below would have to be followed to calculate the enthalpy of the mixture.

The choice of the reference temperatures for A and B locates the zero enthalpy datum on each side of the diagram (see Fig. 5.3). From enthalpy tables such as the steam tables, or by finding

$$(\hat{H}_A - \hat{H}_{A_0}) = \int_{t_0}^{t=77°F} C_{p_A} \, dt$$

you can find $\Delta \hat{H}_A$ at 77°F and similarly obtain $\Delta \hat{H}_B$ at 77°F. Now you have located points A and B. If one of the components is water, it is advisable to choose 32°F as t_0 because then you can use the steam tables as a source of data. If both components have the same reference temperature, the temperature isotherm will intersect on both sides of the diagram at 0, but there is no particular advantage in this.

Krishnaiah et al., *J. Chem. Eng. Data*, v. 10, p. 117 (1965); for EtOH-EtAc, see Robert Lemlich, Chad Gottschlich, and Ronald Hoke, *Brit. Chem. Eng.*, v. 10, p. 703 (1965); for methanol-toluene, see C. A. Plank and D. E. Burke, *Hydrocarbon Processing*, v. 45, no. 8, p. 167 (1966); for acetone-isopropanol, see S. N. Balasubramanian, *Brit. Chem. Eng.*, v. 11, p. 1540 (1966); for alcohol-aromatic systems, see C. C. Reddy and P. S. Murti, *Brit. Chem. Eng.*, v. 12, p. 1231 (1967); for acetonitrile-water-ethanol, see Reddy and Murti, *ibid.*, v. 13, p. 1443 (1968); and for alcohol-aliphatics, see Reddy and Murti, *ibid.*, v. 16, p. 1036 (1971).

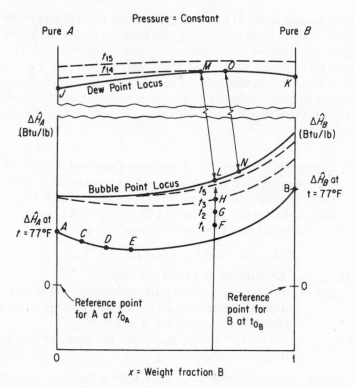

Fig. 5.3. Enthalpy-concentration diagram.

Now that you have $\Delta\hat{H}_A$ and $\Delta\hat{H}_B$ at 77°F (or any other temperature at which $\Delta\hat{H}_{\text{mixing}}$ is known), you can calculate $\Delta\hat{H}_{\text{mixture}}$ at 77°F by Eq. (5.1). You can review the details about how to get $\Delta\hat{H}_{\text{mixing}}$ in Chap. 4. For various mixtures such as 10 percent A and 90 percent B, 20 percent A and 80 percent B, etc., plot the calculated $\Delta\hat{H}_{\text{mixture}}$ as shown by the points C, D, E, etc., in Fig 5.3, and connect these points with a continuous line.

You can calculate $\Delta\hat{H}_{\text{mixture}}$ at any other temperature T, once this 77°F isotherm has been constructed, by again making an enthalpy balance as follows:

$$\Delta\hat{H}_{\text{mixture at any } t} = \Delta\hat{H}_{\text{mixture at } 77°F} + \int_{77°F}^{T} C_{p_x}\, dt \qquad (5.2)$$

where C_{p_x} is the heat capacity of the solution at concentration x. These heat capacities must be determined experimentally, although in a pinch they might be estimated. Points F, G, H, etc., can be determined in this way for a given composition, and then additional like calculations at other fixed concentrations will give you enough points so that all isotherms can be drawn up to the bubble-point line for the mixture (1 atm).

To include the vapor region on the enthalpy concentration chart, you need to know

(a) The heat of vaporization of the pure components (to get points J and K).

(b) The composition of the vapor in equilibrium with a given composition of liquid (to get the tie lines *L-M*, *N-O*, etc.).

(c) The dew-point temperatures of the vapor (to get the isotherms in the vapor region).

(d) The heats of mixing in the vapor region, usually negligible [to fix the points *M*, *O*, etc., by means of Eq. (5.1)]. (Alternatively, the heat of vaporization of a given composition could be used to fix points *M*, *O*, etc.)

The construction and use of enthaply-concentration diagrams for combination material-energy balance problems will now be illustrated.

EXAMPLE 5.4   Preparation of an enthalpy-concentration chart for ideal mixtures

Prepare an enthalpy-concentration chart from the data listed below for the *n*-butane, *n*-heptane system. Assume that heats of mixing (solution) are negligible.

Solution:
The compositions of the vapor and liquid phases as a function of temperature are also essential data and are listed in the first columns of the calculations below.

100 PSIA, *n*-BUTANE–*n*-HEPTANE SYSTEM*

Enthalpy of Saturated Hydrocarbon Liquids at 100 psia
Reference State: 32°F and 14.7 psia

Temperature, °F	$\Delta \hat{H}_{C_4}$ (*n*-butane) Btu/lb mole	$\Delta \hat{H}_{C_7}$ (*n*-heptane) Btu/lb mole
140	3,800	7,100
160	4,500	7,700
200	6,050	9,100
240	7,700	11,400
280	9,400	14,000
320	11,170	16,700
360	13,350	19,400

Enthalpy of Saturated Hydrocarbon Vapors at 100 psia
Reference State: 32°F and 14.7 psia

Temperature, °F	$\Delta \hat{H}_{C_4}$ (*n*-butane) Btu/lb mole	$\Delta \hat{H}_{C_7}$ (*n*-heptane) Btu/lb mole
140	11,300	19,300
160	11,800	20,100
200	12,820	21,600
240	13,900	23,250
280	15,030	25,000
320	16,200	26,900
360	17,550	29,300

*Calculated from data of E. G. Scheibel, *Petroleum Refiner*, v. 26, p. 116 (1947).

Intermediate values of enthalpy were interpolated from a plot of enthalpy vs. temperature prepared from the data given on p. 377.

Basis: 1 lb mole mixture

Reference conditions: 32°F and 14.7 psia

Step 1: Plot the boiling points of pure C_4 (146°F) and pure C_7 (358°F). Fix the enthalpy scale from 0 to 30,000 Btu/lb mole mixture. By plotting the temperature-enthalpy data on separate graphs you can find that the enthalpy of the pure C_4 at 146°F is 3970 Btu/lb mole and that of C_7 at 358°F is 19,250 Btu/lb mole. These are marked in Fig. E5.4 as points *A* and *B*, respectively.

Step 2: Now choose various other compositions of liquid between *A* and *B* (the corresponding temperatures determined from experimental measurements are shown in the table), and multiply the enthalpy of the pure component per mole by the mole fraction present:

$$\Delta \hat{H}_{\text{liquid mixture}} = \Delta \hat{H}_{C_4} x_{C_4} + \Delta \hat{H}_{C_7} x_{C_7}$$

No heat of mixing is present since ideality was assumed. In this way the bubble-point line between *A* and *B* can be established.

H-x Calculations for C_4-C_7 Liquid (Saturated) at 100 psia

Enthalpy, Btu/lb mole

Temp., °F	Mole fr. x_{C_4}	x_{C_7}	$\Delta \hat{H}_{C_4}$	$\Delta \hat{H}_{C_7}$	$\Delta \hat{H}_{C_4} x_{C_7}$	$\Delta \hat{H}_{C_7} x_{C_7}$	$\Delta \hat{H}_L = \Delta \hat{H}_{C_4} x_{C_4} + \Delta \hat{H}_{C_7} x_{C_7}$
358	0	1.0	13,200	19,250	0	19,250	19,250
349	0.02	0.98	12,700	18,650	254	18,250	18,504
340.5	0.04	0.96	12,220	18,000	490	17,280	17,770
331.5	0.06	0.94	11,750	17,380	710	16,300	17,010
315	0.10	0.90	11,000	16,200	1,100	14,600	15,700
296	0.15	0.85	10,100	14,900	1,515	12,650	14,165
278.5	0.20	0.80	9,350	13,750	1,870	11,000	12,870
248	0.30	0.70	8,050	11,850	2,420	8,300	10,720
223.5	0.40	0.60	7,000	10,400	2,800	6,240	9,040
204	0.50	0.50	6,200	9,400	3,100	4,700	7,800
188	0.60	0.40	5,600	8,700	3,360	3,480	6,840
173.8	0.70	0.30	5,000	8,120	3,500	2,440	5,940
160.5	0.80	0.20	4,500	7,700	3,600	1,540	5,140
151.0	0.90	0.10	4,150	7,400	3,740	740	4,480
147.8	0.95	0.05	4,000	7,310	3,800	366	4,166
146.0	1.0	0	3,970	7,280	3,970	0	3,970

Step 3: Plot the enthalpy values for the saturated pure vapor (dew point), points *C* and *D*. A supplementary graph again is required to assist in interpolating enthalpy values.

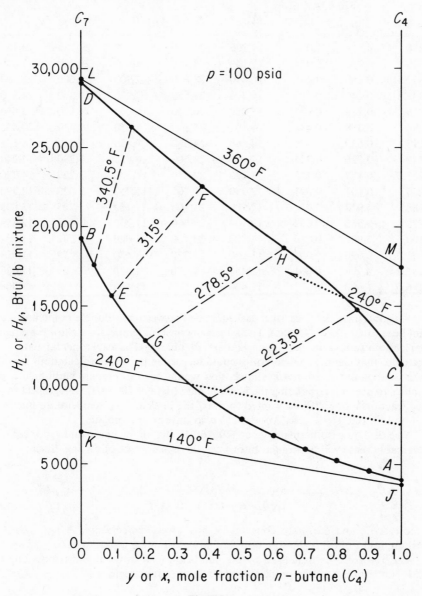

Fig. E5.4.

Step 4: Fill in the saturated-vapor curve between *C* and *D* by the following calculations:

$$\Delta \hat{H}_V = \Delta \hat{H}_{C_4} y_{C_4} + \Delta \hat{H}_{C_7} y_{C_7}$$

Enthalpy, Btu/lb mole

Temp., °F	Mole fr.		$\Delta \hat{H}_{C_4}$	$\Delta \hat{H}_{C_7}$	$\Delta \hat{H}_{C_4} y_{C_4}$	$\Delta \hat{H}_{C_7} y_{C_7}$	$\Delta \hat{H}_V =$ $\Delta \hat{H}_{C_4} y_{C_4}$ $+ \Delta \hat{H}_{C_7} y_{C_7}$
	y_{C_4}	y_{C_7}					
358	0	1.0	17,500	29,250	0	29,250	29,250
349	0.084	0.914	17,150	28,700	1,440	26,200	27,640
340.5	0.160	0.84	16,850	28,100	2,700	23,600	26,300
331.5	0.241	0.759	16,500	27,550	3,970	20,860	24,830
315	0.378	0.622	16,050	26,700	6,060	16,600	22,660
296	0.518	0.482	15,800	25,800	8,050	12,420	20,470
278.5	0.630	0.370	14,980	24,900	9,425	9,210	18,635
248	0.776	0.224	14,100	23,600	10,950	5,290	16,240
223.5	0.857	0.143	13,420	22,550	11,500	3,220	14,720
204	0.907	0.093	12,950	21,700	11,750	2,020	13,770
188	0.942	0.058	12,500	21,100	11,780	1,225	13,005
173.8	0.968	0.032	12,100	20,600	11,700	659	12,359
160.5	0.986	0.014	11,800	26,100	11,620	282	11,902
151.0	0.996	0.004	11,500	19,720	11,450	79	11,529
147.8	0.999	0.001	11,450	19,600	11,430	20	11,450
146.0	1.000	0	11,400	19,540	11,400	0	11,400

Step 5: Draw tie lines (the dashed lines) between the bubble-point and dew-point lines (lines *E-F*, *G-H*, etc.). These lines of constant temperature show the equilibrium concentrations in the vapor and liquid phases. The experimental data were selected so that identical temperatures could be used in the above calculations for each phase; if such data were not available, you would have to prepare additional gaphs or tables to use in interpolating the isobaric data (data at 100 psia) with respect to the mole fraction y or x. If you wanted to draw the lines at even temperature intervals, such as 140, 160, 180°F, etc., you would also have to interpolate.

Step 6: Draw isothermal lines at 140, 160°F, etc., in the liquid and vapor regions. Presumably these will be straight lines since there is no heat of mixing. Since:

$$x_{C_7} + x_{C_4} = 1 \tag{a}$$

$$\Delta \hat{H}_L = \Delta \hat{H}_{\text{mixture}} = x_{C_4} \Delta \hat{H}_{C_4} + x_{C_7} \Delta \hat{H}_{C_7} = x_{C_4} \Delta \hat{H}_{C_4} + (1 - x_{C_4}) \Delta \hat{H}_{C_7}$$
$$= (\Delta \hat{H}_{C_4} - \Delta \hat{H}_{C_7}) x_{C_4} + \Delta \hat{H}_{C_7} \tag{b}$$

Because $\Delta \hat{H}_{C_4}$ and $\Delta \hat{H}_{C_7}$ are constant at any given temperature, $\Delta \hat{H}_{C_4} - \Delta \hat{H}_{C_7}$ is constant, and Eq. (b) is the equation of a straight line on a plot of $\Delta \hat{H}_{\text{mixture}}$ vs. x_{C_4}.

The 140°F isotherm is fixed by the points *J* and *K* in the liquid region, and the 360°F isotherm is fixed by points *L* and *M* in the vapor region:

	140°F, *liquid* $\Delta \hat{H}_L$ (*Btu/mole*)		360°F, *vapor* $\Delta \hat{H}_V$ (*Btu/mole*)
$J(C_4)$	3800	(C_4)	17,550
$K(C_7)$	7100	(C_7)	29,300

The 240°F isotherm (and others higher than 140°F) in the liquid region is real only up to the saturated-liquid line—the dotted portion is fictitious. Portions of isotherms in the vapor region lower than 360°F would also be fictitious, as, for example, at 240°F.

EXAMPLE 5.5 Application of the enthalpy-concentration chart

One hundred pounds of a 73 percent NaOH solution at 350°F are to be diluted to give a 10 percent solution at 80°F. How many pounds of water at 80°F and ice at 32°F are required if there is no external source of cooling available? See Fig. E5.5.

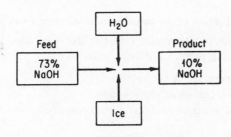

Fig. E5.5.

Use the steam tables and the NaOH-H_2O enthalpy-concentration chart in Appendix I as your source of data. (The reference conditions for the latter chart are $H = 0$ at 32°F for liquid water and $H = 0$ for an infinitely dilute solution of NaOH, with pure caustic having an enthalpy at 68°F of 455 Btu/lb above this datum.)

Solution:
The data required to make a material and an energy balance are as follows:

From Appendix I			*From the steam tables*
NaOH			$\Delta \hat{H}$ of liquid H_2O
conc.	*temp.*, °F	$\Delta \hat{H}$, *Btu/lb*	= 48 Btu/lb
73	350	468	$\Delta \hat{H}$ of ice = -143 Btu/lb
10	80	42	(or minus the heat of fusion)

We can make a material balance first.

Basis: 100 lb 73 percent at 350°F

The tie element in the material balance is the NaOH, and using it we can find the total H_2O added (ice plus H_2O):

$$\frac{100 \text{ lb product}}{10 \text{ lb NaOH}} \Bigg| \frac{73 \text{ lb NaOH}}{100 \text{ lb feed}} = 730 \text{ lb product/100 lb feed}$$

$$\text{less } 100 \text{ lb feed}$$

$$\text{gives } \underline{\underline{630}} \text{ lb } H_2O \text{ added/100 lb feed}$$

Next we make an energy (enthalpy) balance,

$$\Delta H_{\text{overall}} = 0 \quad \text{or} \quad \Delta H_{\text{in}} = \Delta H_{\text{out}}$$

To differentiate between the ice and the liquid water added, let us designate the ice as x lb and then the H_2O becomes $(630 - x)$ lb:

in			out
73% NaOH solution	H_2O	ice	10% NaOH solution

$$\frac{100\,\text{lb} \mid 468\,\text{Btu}}{\mid \text{lb}} + \frac{(630-x)\,\text{lb} \mid 48\,\text{Btu}}{\mid \text{lb}} + \frac{x\,\text{lb} \mid -143\,\text{Btu}}{\mid \text{lb}} = \frac{730\,\text{lb} \mid 42\,\text{Btu}}{\mid \text{lb}}$$

$$46{,}800 + 30{,}240 - 191x = 30{,}660$$

$$x = 243 \text{ lb ice at } 32°F$$

$$\text{liquid } H_2O \text{ at } 80°F = 387 \text{ lb}$$

EXAMPLE 5.6   Application of the enthalpy-concentration chart

Six hundred pounds of 10 percent NaOH at 200°F are added to 400 lb of 50 percent NaOH at the boiling point. Calculate the following:

(a) The final temperature of the solution.
(b) The final concentration of the solution.
(c) The pounds of water evaporated during the mixing process.

Solution:

Basis: 1000 lb final solution

Use the same NaOH-H_2O enthalpy-concentration chart as in the previous problem to obtain the enthalpy data. We can write the following material balance:

comp.	10% solution	+ 50% solution	= final solution	wt %
NaOH	60	200	260	26
H_2O	540	200	740	74
total	600	400	1000	100

Next, the energy (enthalpy) balance (in Btu) is

$$
\begin{array}{ccccc}
10\% \text{ solution} & & 50\% \text{ solution} & & \text{final solution} \\
600(152) & + & 400(290) & = & \Delta H \\
91{,}200 & + & 116{,}000 & = & 207{,}200
\end{array}
$$

Note that the enthalpy of the 50 percent solution at its boiling point is taken from the bubble-point curve at $x = 0.50$. The enthalpy per pound is

$$\frac{207{,}200 \text{ Btu}}{1000 \text{ lb}} = 207 \text{ Btu/lb}$$

On the enthalpy-concentration chart for NaOH-H_2O, for a 26 percent NaOH solution with an enthalpy of 207 Btu/lb, you would find that only a two-phase mixture of (a) saturated H_2O vapor and (b) NaOH-H_2O solution at the boiling point could exist. To get the fraction H_2O vapor, we have to make an additional energy (enthalpy) balance. By interpolation, draw the tie line through the point $x = 0.26$, $H = 207$ (make it parallel to the 220° and 250°F tie lines). The final temperature appears from Fig. E5.6 to be 232°F; the enthalpy of the liquid at the bubble point is about 175 Btu/lb. The enthalpy of the saturated water vapor (no NaOH is in the vapor phase) from the steam tables at 232°F is 1158 Btu/lb. Let $x = $ lb H_2O evaporated.

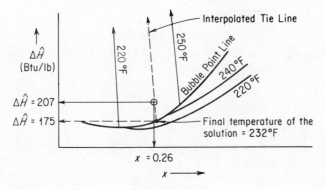

Fig. E5.6.

Basis: 1000 lb final solution

$$x(1158) + (1000 - x)\,175 = 1000(207)$$
$$983x = 32{,}000$$
$$x = 32.6 \text{ lb } H_2O \text{ evaporated}$$

5.2-2 Graphical Solutions on an Enthalpy-Concentration Chart.

One of the major advantages of an enthalpy-concentration chart is that the same problems we have just used as examples, and a wide variety of other problems, can be solved graphically directly on the chart. We shall just indicate the scope of this technique, which in its complete form is usually called the Ponchon-Savaritt method; you can refer to some of the references for enthalpy-concentration charts at the end of this chapter for additional details.

Any steady-state process with three streams and a net heat interchange Q with the surroundings can be represented by a simple diagram as in Fig. 5.4. A, B, and C are in pounds or moles. An overall material balance gives us

$$A + B = C \tag{5.3}$$

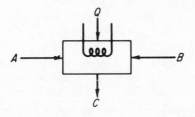

Fig. 5.4. A typical process with mass and heat interchange.

and a component material balance gives

$$Ax_A + Bx_B = Cx_C \qquad (5.4)$$

An energy balance for a flow process (neglecting the work and the kinetic and potential energy effects) gives us[4]

$$Q = \Delta H$$

or

$$Q + AH_A + BH_B = CH_C \qquad (5.5)$$

First we must choose some basis for carrying out the calculations, and then Q is dependent on this basis. Thus, if A is chosen as the basis, Q_A will be Q per unit amount of A, or

$$Q_A = \frac{Q}{A} \qquad (5.6)$$

Similarly, if B is the basis,

$$Q_B = \frac{Q}{B} \qquad (5.7)$$

Thus

$$Q = AQ_A = BQ_B = CQ_C \qquad (5.8)$$

Let us choose A as the basis for working the problem, and then Eq. (5.5) becomes

$$AQ_A + AH_A + BH_B = CH_C \qquad (5.9)$$

or

$$A(H_A + Q_A) + BH_B = CH_C \qquad (5.10)$$

Similar equations can be written if B or C is chosen as the basis.

(a) *Adiabatic processes.* As a special case, consider the adiabatic process $(Q = 0)$. Then (with A as the basis) the energy balance becomes

$$AH_A + BH_B = CH_C = (A + B)H_C = AH_C + BH_C$$

or

$$A(H_A - H_C) = B(H_C - H_B)$$

or

$$\frac{A}{B} = \frac{H_C - H_B}{H_A - H_C} \qquad (5.11)$$

Combining material balances,

$$Ax_A + Bx_B = Cx_C = (A + B)x_C = Ax_C + Bx_C$$

or

$$A(x_A - x_C) = B(x_C - x_B)$$

or

$$\frac{A}{B} = \frac{x_C - x_B}{x_A - x_C} \qquad (5.12)$$

---

[4]In what follows H is always the enthalpy relative to some reference value, and we suppress the caret ($\wedge$) and the symbol Δ to make the notation more compact.

Obviously then,

$$\frac{A}{B} = \frac{x_C - x_B}{x_A - x_C} = \frac{H_C - H_B}{H_A - H_C} \tag{5.13}$$

Similarly, solving the energy and material balances for $\acute{A}$ and C instead of A and B, we can get

$$\frac{A}{C} = \frac{x_C - x_B}{x_A - x_B} = \frac{H_C - H_B}{H_A - H_B} \tag{5.14}$$

and in terms of B and C we could get a like relation. Rearranging the two right-hand members of Eq. (5.13), we can find that

$$\frac{H_C - H_B}{x_C - x_B} = \frac{H_A - H_C}{x_A - x_C} \tag{5.13a}$$

Also, from Eq. (5.14),

$$\frac{H_C - H_B}{x_C - x_B} = \frac{H_A - H_B}{x_A - x_B} \tag{5.14a}$$

On an enthalpy-concentration diagram (an H-x plot) the distances H_C-H_B, H_A-H_B, and H_A-H_C are projections on the H axis of a curve, while the distances x_C-x_B, x_A-x_B, x_A-x_C are the corresponding projections on the x axis (see Fig. 5.5) The form of Eqs. (5.13a) and (5.14a) is such that they tell us that they are

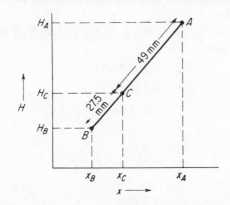

Fig. 5.5. Projections of the line BCA on the H and x axes.

each equations of a straight line on an H-x plot and that these lines pass through the points A, B, and C which are represented on an H-x diagram by the coordinates $A(H_A, x_A)$, $B(H_B, x_B)$, and $C(H_C, x_C)$, respectively.

Since Eqs. (5.13a) and (5.14a) both pass through point C, a common point, and have the same slope $(H_C - H_B)/(x_C - x_B)$, mathematical logic tells us they must be equations of the same line, i.e., the lines are actually a single line passing through B,C, and A, as shown in Fig. 5.5. The symmetry of this type of arrangement leads to use of the *lever arm* rule.

If $A + B = C$, then point C lies on a line *between* A and B, and *nearer* to the quantity present in the *largest* amount. If more B is added, C will lie nearer B. When two streams are subtracted, the resulting stream will lie *outside* the region of a line between the two subtracted streams, and *nearer* the one present in the larger amount. Thus if $C - B = A$, and C is larger than A, then A will lie on a straight line through C and B nearer to C than B. Look at Fig. 5.5 and imagine that B is removed from C. A represents what is left after B is taken from C.

The *inverse* lever arm principle tells us how far to go along the line through A and B to find C if $A + B = C$, or how far to go along the line through C and B if $C - B = A$. In the case of $A + B = C$, we saw that the weight (or mole) ratio of A to B was given by Eq. (5.13). From the symmetry of Fig. 5.5, you can see that if the ratio of A to B is equal to the ratio of the projections of line segments BC and AC on the H or x axes, then the amount of $A + B$ should be proportional to the sum of the line segments $(x_C - x_B) + (x_A - x_C)$, or $(H_C - H_B) + (H_A - H_C)$. Furthermore, the amount of A should be proportional to the measured distance $\overline{BC}$ and the amount of B proportional to the distance $\overline{AC}$. Also the ratio of A to B should be

$$\frac{A}{B} = \frac{\text{distance } \overline{BC}}{\text{distance } \overline{AC}} \tag{5.15}$$

This relation is the *inverse lever arm rule*. The distances $\overline{BC}$, $\overline{AC}$, and $\overline{AB}$ can be measured with a ruler.

As an example of the application of Eq. (5.15), if A were 100 lb and the measured distances were

$$\overline{BC} = 27.5 \text{ mm}$$

$$\overline{AC} = 49.0 \text{ mm}$$

$$\overline{AB} = 76.5 \text{ mm}$$

then

$$\frac{A}{B} = \frac{100}{B} = \frac{\overline{BC}}{\overline{AC}} = \frac{27.5 \text{ mm}}{49.0 \text{ mm}} = 0.561$$

and

$$B = \frac{100}{0.561} = 178 \text{ lb}$$

The sum of $A + B = C$ is $100 + 178 = 278$ lb. Alternatively,

$$\frac{A}{C} = \frac{BC}{AB} = \frac{27.5 \text{ mm}}{76.5 \text{ mm}} = 0.360$$

$$C = \frac{A}{0.360} = \frac{100}{0.360} = 278 \text{ lb}$$

(b) *Nonadiabatic processes.* Let us now consider the circumstances under which Q is not equal to zero in Eq. (5.5). This time, for variety, let us take C as a basis so that $Q = CQ_C$. Then we have two pairs of equations to work with again,

$$AH_A + BH_B = C(H_C - Q_C) \tag{5.16}$$

$$Ax_A + Bx_B = Cx_C \tag{5.4}$$

Briefly, Eqs. (5.16) and (5.4) tell us that on an *H-x* diagram points *A*, *B*, and *C* lie on a straight line if the coordinates of the points are $A(H_A, x_A)$, $B(H_B, x_B)$, and $C[(H_C - Q_C), x_C]$, respectively,[5] as illustrated in Fig. 5.6 Since the coordi-

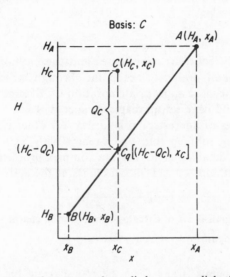

Fig. 5.6. Inverse lever arm rule applied to a nonadiabatic process.

---

[5] Proof of these relations is analogous to the previous development.

Material balances:

$$Ax_A + Bx_B = Cx_C = (A + B)x_C = Ax_C + Bx_C \tag{a}$$

$$A(x_A - x_C) = B(x_C - x_B) \tag{b}$$

Energy balance:

$$AH_A + BH_B = C(H_C - Q_C) = (A + B)(H_C - Q_C)$$
$$= A(H_C - Q_C) + B(H_C - Q_C) \tag{c}$$

$$A[H_A - (H_C - Q_C)] = B(H_C - Q_C) - H_B \tag{d}$$

Thus

$$\frac{A}{B} = \frac{x_C - x_B}{x_A - x_C} = \frac{(H_C - Q_C) - H_B}{H_A - (H_C - Q_C)} \tag{e}$$

and, rearranging,

$$\frac{H_A - (H_C - Q_C)}{x_A - x_C} = \frac{(H_C - Q_C) - H_B}{x_C - x_B} \tag{f}$$

A similar equation can be obtained for the ratio A/C.

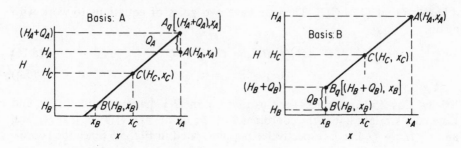

Figs. 5.7 and 5.8. Inverse lever arm principle for different bases.

nates of C are really fictitious coordinates and the enthalpy projection of C is $(H_C - Q_C)$, we shall designate the point through which the line passes $[(H_C - Q_C), x_C]$ by the more appropriate notation C_Q. If A or B had been chosen as the basis, we would have set up diagrams such as Figs. 5.7 and 5.8.

A different series of figures could be drawn for other typical problems, as, for example, C being split up into two streams with heat being absorbed or evolved. Additional details related to more complex problems will be found in the references on enthalpy-concentration charts at the end of the chapter.

EXAMPLE 5.7 Graphical use of enthalpy-concentration charts

Do Example 5.5 again by graphical means.

Solution:

Plot the known points A and C on the H-x diagram for NaOH. The known points are as follows (with the given coordinates or data):

solution		x	H, Btu/lb	T, °F
Initial	(A)	0.73	468	350
Final	(C)	0.10		80
Added	(D)	0.00		80
	(E)	0.00		32 (ice)

What we shall do is remove solution A from the final solution C to obtain solution B, i.e., subtract $C - A$ to get B as in Fig. E5.7a. This is the reverse of adding A to B to get C.

Draw a line through A and C until it intersects the $x = 0$ axis (for pure water). The measured (with a ruler) ratio of the lines $\overline{BC}$ to $\overline{AC}$ is the ratio of 73 percent NaOH solution to pure water.

Basis: 100 lb 73 percent NaOH solution at 350°F (A)

$$\frac{100 \text{ lb } (A)}{\text{lb } H_2O \ (B)} = \frac{\overline{BC}}{\overline{AC}} = \frac{2.85 \text{ units}}{17.95 \text{ units}} = 0.159$$

$$\frac{100}{0.159} = \text{lb } H_2O = 630 \text{ lb ice plus liquid water } (B)$$

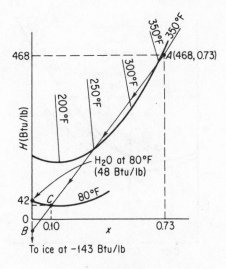

Fig. E5.7(a).

To obtain the proportions of ice and water, we want to subtract D from B to get E as in Fig. E5.7(b).

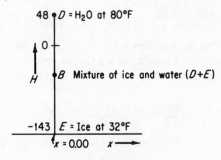

Fig. E5.7(b).

We measure the distances $\overline{DE}$ and $\overline{BE}$. Then,

$$\frac{\text{lb H}_2\text{O }(D)}{\text{lb total }(B)} = \frac{\overline{BE}}{\overline{DE}} = \frac{3.65 \text{ units}}{5.95 \text{ units}} = 0.613$$

$$\text{lb H}_2\text{O }(D) = 630(0.613) = 386 \text{ lb}$$

$$\text{lb ice }(E) = 630 - 386 = 244 \text{ lb}$$

EXAMPLE 5.8 Graphical use of enthalpy-concentration charts

Do Example 5.6 again by graphical means.

Solution:
See Fig. E5.8.

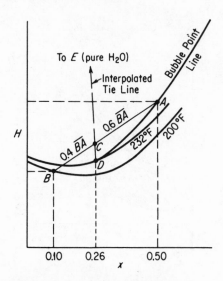

Fig. E5.8. Graphical use of enthalpy charts.

Basis: 1000 lb of total solution

Again plot the known data on an *H-x* chart. Point *C*, the sum of solutions *A* and *B*, is found graphically by noting by the inverse lever arm rule that

$$\frac{\text{distance } \overline{BC}}{\text{distance } \overline{BA}} = \frac{\text{lb } A}{\text{lb } A + B} = \frac{400}{1000} = 0.4$$

Now measure $\overline{BA}$ on Fig. E5.8 and plot $\overline{BC}$ as $0.4(\overline{BA})$. This fixes point *C*.

Since *C* is in the two-phase region, draw a tie line through *C* (parallel to the nearby tie lines) and measure the distances $\overline{CD}$ and $\overline{ED}$. They will give the H_2O evaporated as follows:

$$\frac{\overline{CD}}{\overline{ED}} = \frac{\text{lb } H_2O \text{ evaporated}}{1000 \text{ lb mixture}}$$

$$\text{lb } H_2O \text{ evaporated} \cong 33 \text{ lb}$$

The final concentration of the solution and the temperature of the solution can be read off the *H-x* chart as before as

$$T = 232°F, \qquad x = 26\%$$

EXAMPLE 5.9   Graphical use of enthalpy-concentration charts with processes involving heat transfer

An acetic acid-water mixture is being concentrated as shown in Fig. E5.9(a). How much heat is added or removed?

Solution:

First we have to prepare an enthalpy concentration chart by plotting the data for the acetic acid-water system from Appendix I.

The coordinates of all the points, *A*, *B*, and *C*, are known. These points can be plotted on the *H-x* chart, and then some basis chosen, *A*, *B*, or *C*.

Basis: 1 lb mole A

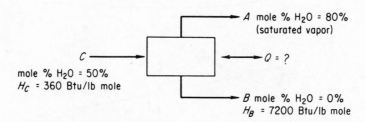

Fig. E5.9(a). Graphical use of enthalpy-concentration charts with processes involving heat transfer.

The governing material and energy balances in this case are (assuming that Q is added and is plus)

$$Cx_C = Ax_A + Bx_B$$
$$Q + CH_C = AH_A + BH_B = AQ_A + CH_C$$

or

$$A(H_A - Q_A) + BH_B = CH_C$$

Connect points B and C with a line, and extend the line toward the right-hand axis (see Fig. E5.9(b)). What we are looking for is the point A_Q located at the coordinates

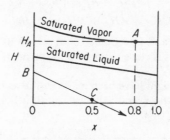

Fig. E5.9(b)

$H = (H_A - Q_A)$ and $x = 0.80$ on the extension of the line $\overline{BC}$ through B and C. To reach A_Q from A, we must subtract Q_A; by measuring on Fig. E5.9(c), we find a plus value has to be subtracted from H_A, or

$$Q_A = +24,000 \text{ Btu/lb mole } A \qquad \text{(heat added)}$$

If point A_Q had fallen above A, then heat would have been removed from the system and Q_A would have been negative.

If we convert our answer to the basis of 1 lb mole of C, we find that

$$\frac{A}{C} = \frac{\overline{BC}}{\overline{AB}} = \frac{0.5 - 0}{0.8 - 0} = 0.625$$

or

$$C = \frac{1}{0.625} = 1.60 \text{ lb mole}$$

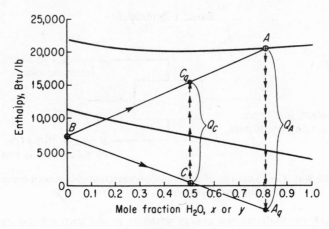

Fig. E5.9(c).

Basis: 1 lb mole C

$$Q_C = \frac{24{,}000 \text{ Btu}}{\text{lb mole } A} \Big| \frac{1.00 \text{ lb mole } A}{1.60 \text{ lb mole } C} = 15{,}000 \text{ Btu/lb mole } C$$

As an alternative solution, we could have selected C as the basis to start with and used the equations

$$C x_C = A x_A + B x_B$$

$$Q + C H_C = A H_A + B H_B = C(H_C + Q_C)$$

In this case, join A and B and find point C_Q at $x = 0.50$ on the line $\overline{AB}$. Measure $\overline{C_Q C}$; it is 14,900 Btu, which is close enough to the value of 15,000 Btu/lb mole C calculated above.

A similar calculation might have been made with B as the basis, by joining points A and C and extending to the $x = 0$ axis in order to get B_Q. Q_B would be measured from B_Q to B on the $x = 0$ axis.

5.3 Humidity charts and their use

In Chap. 3 we discussed humidity, condensation, and vaporization. In this section we are going to apply simultaneous material and energy balances to humidification, air conditioning, water cooling towers, and the like. Before proceeding further, you should review briefly the sections in Chap. 3 dealing with vapor pressure.

Recall that the *humidity* $\mathcal{H}$ is the pounds of water vapor per pound of bone-dry air (some texts use moles of water vapor per mole of dry air as the humidity) or, as indicated by Eq. (3.52),

$$\mathcal{H} = \frac{18 p_{H_2O}}{29(p_T - p_{H_2O})} = \frac{18 n_{H_2O}}{29(n_T - n_{H_2O})} \tag{5.17}$$

You will also find it indispensable to learn the following special definitions and relations.

(a) The *humid heat* is the heat capacity of an air-water vapor mixture expressed on the *basis of 1 lb of bone-dry air*. Thus the humid heat C_S is

$$C_S = C_{p_{\text{air}}} + (C_{p_{\text{H}_2\text{O vapor}}})(\mathcal{3C}) \tag{5.18}$$

where the heat capacities are all per pound and not per mole. Assuming that the heat capacities of air and water vapor are constant under the narrow range of conditions found for air conditioning and humidification calculations, we can write

$$C_S = 0.240 + 0.45(\mathcal{3C}) \tag{5.19}$$

where C_S is in Btu/(°F)(lb dry air).

(b) The *humid volume* is the volume of 1 lb of dry air plus the water vapor in the air,

$$\hat{V} = \frac{359 \text{ ft}^3}{1 \text{ lb mole}} \left| \frac{1 \text{ lb mole air}}{29 \text{ lb air}} \right| \frac{T_{\circ\text{F}} + 460}{32 + 460}$$

$$+ \frac{359 \text{ ft}^3}{1 \text{ lb mole}} \left| \frac{1 \text{ lb mole H}_2\text{O}}{18 \text{ lb H}_2\text{O}} \right| \frac{T_{\circ\text{F}} + 460}{32 + 460} \left| \frac{\mathcal{3C} \text{ lb H}_2\text{O}}{\text{lb air}} \right.$$

$$= (0.730 T_{\circ\text{F}} + 336)\left(\frac{1}{29} + \frac{\mathcal{3C}}{18}\right) \tag{5.20}$$

where $\hat{V}$ is in ft³/lb dry air.

(c) The *dry-bulb temperature* (T_{DB}) is the ordinary temperature you always have been using for a gas in °F.

(d) The *wet-bulb temperature* (T_{WB}) you may guess, even though you may never have heard of this term before, has something to do with water (or other liquid, if we are concerned not with humidity but with saturation) evaporating from around an ordinary mercury thermometer bulb. Suppose that you put a wick, or porous cotton cloth, on the mercury bulb of a thermometer and wet the wick. Next you either (1) whirl the thermometer in the air as in Fig. 5.9 (this apparatus is called a sling psychrometer when the wet-bulb and dry-bulb thermometers are mounted together), or (2) set up a fan to blow rapidly on the bulb at 1000 ft³/min or more. What happens to the temperature recorded by the wet-bulb thermometer?

As the water from the wick evaporates, the wick cools down and continues to cool until the rate of energy transferred to the wick by the air blowing on it equals the rate of loss of energy caused by the water evaporating from the wick. We say that the temperature at equilibrium with the wet wick is the wet-bulb temperature. (Of course, if water continues to evaporate, it eventually will all disappear, and the wick temperature will rise.) The final temperature for the process described above will lie on the 100 percent relative humidity curve (saturated-air curve), while the so-called wet-bulb line showing how the wet-bulb temperature changes on approaching equilibrium is approximately a straight line and has a negative slope, as illustrated in Fig. 5.10.

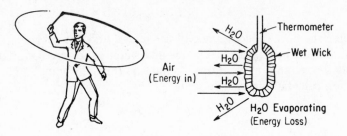

Fig. 5.9. The wet-bulb temperature obtained with a sling psychrometer.

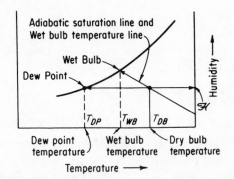

Fig. 5.10. General layout of the humidity chart showing the location of the wet-bulb and dry-bulb temperatures, the dew point and dew-point temperature, and the adiabatic saturation line and wet-bulb line.

Now that we have an idea of what the various features portrayed on the *humidity chart* (psychrometric chart) are, let us look at the chart itself, Fig. 5.11 (a) and 5.11 (b) (inside back cover). It is nothing more than a graphical means of presenting the relationships for and between the material and energy balances in water vapor-air mixtures. Its skeleton consists of a humidity ($\mathcal{H}$) — temperature (T_{DB}) set of coordinates together with the additional parameters (lines) of

(a) Constant relative humidity (10–90 percent).

(b) Constant moist volume (humid volume).

(c) Adiabatic cooling lines which are the same (for water vapor only[6]) as the wet-bulb or psychrometric lines.

(d) The 100 percent relative humidity (identical to the 100 percent absolute humidity) curve or saturated-air curve.

With any two values known, you can pinpoint the air-moisture condition on the chart and determine all the other required values.

Off to the left of the 100 percent relative humidity line you will observe scales showing the enthalpy per pound of dry air of a saturated air-water vapor mixture. Enthalpy corrections for air less than saturated are shown on the chart itself. The enthalpy of the wet air, in Btu/lb dry air, is

$$\Delta \hat{H} = \Delta \hat{H}_{air} + \Delta \hat{H}_{H_2O \; vapor}(\mathcal{H}) \tag{5.21}$$

We should mention at this point that the reference conditions for the humidity chart are liquid water at 32°F and 1 atm (not the vapor pressure of H_2O) for water, and 0°F and 1 atm for air. The chart is suitable for use only at normal atmospheric conditions and must be modified[7] if the pressure is significantly different than 1 atm. If you wanted to, you could calculate the enthalpy values shown on the chart directly from tables listing the enthalpies of air and water vapor by the methods described in Chap. 4, or you could, by making use of Eq. (5.21), compute the enthalpies with reasonable accuracy from the following equation for 1 lb of air:

$$\Delta \hat{H} = \underbrace{0.240(T_{°F} - 0)}_{C_p(\Delta T) \text{ for air}} + \underbrace{\mathcal{H}[1075}_{\substack{\text{heat of vaporization} \\ \text{of water at 32}}} + \underbrace{0.45(T_{°F} - 32)]}_{\substack{C_p(\Delta T) \text{ for} \\ \text{water vapor}}} \tag{5.22}$$

Consolidating terms:

$$\Delta \hat{H} = 0.240 T_{°F} + \mathcal{H}(1061 + 0.45 T_{°F}) \tag{5.23}$$

You will recall that the idea of the wet-bulb temperature is based on the equilibrium between the *rates* of energy transfer to the bulb and evaporation of water. Rates of processes are a topic that we have not discussed. The fundamental idea is that a large amount of air is brought into contact with a little bit of water and that presumably the evaporation of the water leaves the temperature and humidity of the air unchanged. Only the temperature of the water changes. The equation of the wet-bulb line is

$$h_c(T - T_{WB}) = k'_g \Delta \hat{H}_{vap}(\mathcal{H}_{WB} - \mathcal{H}) \tag{5.24}$$

---

[6] For a detailed discussion of the uniqueness of this coincidence, consult any of the references at the end of this chapter.

[7] See G. E. McElroy, *U.S. Bur. Mines Rept. Invest.*, No. 4165, Dec. 1947.

where h_c = heat transfer coefficient for convection to the bulb
$\quad k'_g$ = mass transfer coefficient
$\quad \Delta \hat{H}_{vap}$ = latent heat of vaporization
$\quad \mathfrak{K}$ = humidity of moist air
$\quad T$ = temperature of moist air in °F

The equation for the wet-bulb lines is based on a number of assumptions, a discussion of which is beyond the scope of this book. However, we can form the ratio

$$\frac{(\mathfrak{K}_{WB} - \mathfrak{K})}{(T_{WB} - T)} = -\frac{h_c}{(k'_g)\Delta\hat{H}_{vap}} \tag{5.25}$$

For water only, it so happens that $h_c/k'_g \cong C_S$, i.e., the numerical value is about 0.25, which gives the wet-bulb lines the slope of

$$\frac{(\mathfrak{K}_{WB} - \mathfrak{K})}{(T_{WB} - T)} = -\frac{C_S}{\Delta\hat{H}_{vap}} \tag{5.26}$$

For other substances, the value of h_c/k'_g can be as much as twice that of water.

Another type of process of some importance occurs when an adiabatic cooling or humidification takes place between air and water that is recycled as in Fig. 5.12. In this process the air is both cooled and humidified (its water

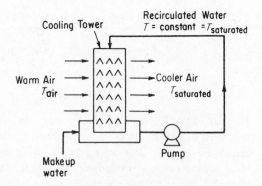

Fig. 5.12. Adiabatic humidification with recycle of water.

content rises) while a little bit of the recirculated water is evaporated. At equilibrium, in the steady state, the temperature of the air is the same as the temperature of the water, and the exit air is saturated at this temperature. By making an overall energy balance around the process ($Q = 0$), we can obtain the equation for the adiabatic cooling of the air. The equation, when plotted on the humidity chart, yields what is known as an adiabatic cooling line. Employing a version of Eq. (5.22) with the equilibrium temperature of the water, T_s, taken as a reference temperature rather than 0 or 32°F, we get, ignoring the small amount of makeup water or assuming that it enters at T_S,

$$\overbrace{0.240(T_{air} - T_S)}^{\substack{\textbf{enthalpy of air} \\ \textbf{entering}}} + \mathfrak{IC}_{air}[\overbrace{\Delta\hat{H}_{\text{vap H}_2\text{O at } T_S} + 0.45(T_{air} - T_S)]}^{\substack{\textbf{enthalpy of water vapor} \\ \textbf{in air entering}}}$$

$$= \underbrace{0.240(T_S - T_S)}_{\substack{\textbf{enthalpy of air} \\ \textbf{leaving}}} + \mathfrak{IC}_S[\underbrace{\Delta\hat{H}_{\text{vap H}_2\text{O at } T_S} + 0.45(T_S - T_S)]}_{\substack{\textbf{enthalpy of water vapor} \\ \textbf{in air leaving}}} \quad (5.27)$$

This can be reduced to

$$T_{air} = \frac{\Delta\hat{H}_{\text{vap H}_2\text{O at } T_S}(\mathfrak{IC}_S - \mathfrak{IC}_{air})}{0.240 + 0.45\mathfrak{IC}_{air}} + T_S \quad (5.28)$$

which is the equation for adiabatic cooling.

Notice that this equation can be written as

$$\frac{(\mathfrak{IC}_S - \mathfrak{IC})}{(T_S - T_{air})} = -\frac{C_S}{\Delta\hat{H}_{\text{vap at } T_S}} \quad (5.29)$$

because $C_S = 0.240 + 0.45\mathfrak{IC}$ and $T_{WB} = T_S$. Thus the wet-bulb process equation, for water only, is essentially the same as the adiabatic cooling equation. For other materials these two equations have different slopes.

Only two of the quantities in Eq. (5.28) are variables, if T_S is known, because $\mathfrak{IC}_S$ is the humidity of saturated air at T_S and $\Delta\hat{H}_{\text{vap H}_2\text{O at } T_S}$ is fixed by T_S. Thus, for any value of T_S, you can make a plot of Eqs. (5.25) and/or (5.29) on the humidity chart in the form of $\mathfrak{IC}$ vs. T_{air}. These curves, which are essentially linear, will intersect the 100 percent relative humidity curve at $\mathfrak{IC}_S$ and T_S, as described earlier.

The adiabatic cooling lines are lines of almost constant enthalpy for the entering air-water mixture, and you can use them as such without much error (1 or 2 percent). However, if you want to correct a saturated enthalpy value for the deviation which exists for a less-than-saturated air-water vapor mixture, you can employ the enthalpy deviation lines which appear on the chart and which can be used as illustrated in the examples below. Any process which is not a wet-bulb process or an adiabatic process with recirculated water can be treated by the usual material and energy balances, taking the basic data for the calculation from the humidity charts. If there is any increase or decrease in the moisture content of the air in a psychrometric process, the small enthalpy effect of the moisture added to the air or lost by the air may be included in the energy balance for the process to make it more exact as illustrated in Examples 5.11 and 5.13.

You can find further details regarding the construction of humidity charts in the references at the end of this chapter. Tables are also available listing all the thermodynamic properties (p, $\hat{V}$, $\mathfrak{IC}$, $\Delta\hat{H}$, and S) in great detail.[8] Although

---

[8]Microfilms of Psychrometric Tables, 1953, by Byron Engelbach are available from University Microfilms, Ann Arbor, Mich.

we shall be discussing humidity charts exclusively, charts can be prepared for mixtures of any two substances in the vapor phase, such as CCl_4 and air or acetone and nitrogen, by use of Eqs. (5.17)–(5.28) if all the values of the physical constants for water and air are replaced by those of the desired gas and vapor. The equations themselves can be used for humidity problems if charts are too inaccurate or are not available.

EXAMPLE 5.10 Properties of moist air from the humidity chart

List all the properties you can find on the humidity chart for moist air at a dry-bulb temperature of 90°F and a wet-bulb temperature of 70°F.

Solution:

A diagram will help explain the various properties obtained from the humidity chart. You can find the location of point *A* for 90°F DB (dry bulb) and 70°F WB (wet bulb) by following a vertical line at $T_{DB} = 90°F$ until it crosses the wet-bulb line for 70°F. This wet-bulb line can be located by searching along the 100 percent humidity line until the saturation temperature of 70°F is reached, or, alternatively, by proceeding up a vertical line at 70°F until it intersects the 100 percent humidity line. From the wet-bulb temperature of 70°F, follow the adiabatic cooling line (which is the same as the wet-bulb temperature line on the humidity chart) to the right until it intersects the 90°F DB line. Now that point *A* has been fixed, you can read the other properties of the moist air from the chart. See Fig. E5.10.

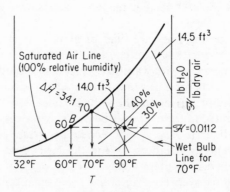

Fig. E5.10.

(a) *Dew point.* When the air at *A* is cooled at constant pressure (and in effect at *constant humidity*), as described in Chap. 3, it eventually reaches a temperature at which the moisture begins to condense. This is represented by a horizontal line, a constant-humidity line, on the humidity chart, and the dew point is located at *B*, or about 60°F.

(b) *Relative humidity.* By interpolating between the 40 percent ℛℋ and 30 percent ℛℋ lines with a ruler, you can find that point *A* is at about 37 percent ℛℋ.

(c) *Humidity* ($\mathcal{3C}$). You can read the humidity from the right-hand ordinate as 0.0112 lb H_2O/lb air.

(d) *Humid volume.* By interpolation again between the 14.0- and the 14.5-ft^3 lines, you can find the humid volume to be 14.097 ft^3/lb dry air.

(e) *Enthalpy.* The enthalpy value for saturated air with a wet-bulb temperature of 70°F is $\Delta \hat{H} = 34.1$ Btu/lb dry air (a more accurate value can be obtained from psychrometric tables if needed). The enthalpy deviation for less- than-saturated air is about -0.2 Btu/lb dry air; consequently the actual enthalpy of air at 37 percent $\mathcal{R3C}$ is $34.1 - 0.2 = 33.9$ Btu/lb dry air.

EXAMPLE 5.11   Heating at constant humidity

Moist air at 50°F and 50 percent $\mathcal{R3C}$ is heated in your furnace to 100°F. How much heat has to be added per cubic foot of initial moist air, and what is the final dew point of the air?

Solution:

As shown in Fig. E5.11, the process goes from point A to point B on a horizontal line of constant humidity. The initial conditions are fixed at $T_{DB} = 50°F$ and 50 percent

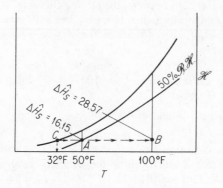

Fig. E5.11.

$\mathcal{R3C}$. Point B is fixed by the intersection of the horizontal line from A and the vertical line at 100°F. The dew point is unchanged in this process and is located at C at 32.5°F.

The enthalpy values are as follows (all in Btu/lb dry air):

point	$\Delta \hat{H}_{satd}$	δH	$\Delta \hat{H}_{actual}$
A	16.15	−0.02	16.13
B	28.57	−0.26	28.31

Also at A the volume of the moist air is 12.92 ft^3/lb dry air. Consequently the heat added is ($Q = \Delta \hat{H}$) $28.31 - 16.13 = 12.18$ Btu/lb dry air.

$$\frac{12.18 \text{ Btu}}{\text{lb dry air}} \left| \frac{1 \text{ lb dry air}}{12.92 \text{ ft}^3} \right. = 0.942 \text{ Btu/ft}^3 \text{ initial moist air}$$

EXAMPLE 5.12 Cooling and humidification

One way of adding moisture to air is by passing it through water sprays or air washers. See Fig. E5.12(a). Normally, the water used is recirculated rather than

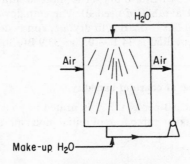

Fig. E5.12(a).

wasted. Then, in the steady state, the water is at the adiabatic saturation temperature which is the same as the wet-bulb temperature. The air passing through the washer is cooled, and if the contact time between the air and the water is long enough, the air will be at the wet-bulb temperature also. However, we shall assume that the washer is small enough so that the air does not reach the wet-bulb temperature; instead the following conditions prevail:

	T_{DB}, °F	T_{WB}, °F
Entering air	100	70
Exit air	80	

Find the moisture added per pound of dry air.

Solution:

The whole process is assumed to be *adiabatic*, and, as shown in Fig. E5.12(b), takes place between points A and B along the adiabatic cooling line. The wet-bulb temperature remains constant at 70°F.

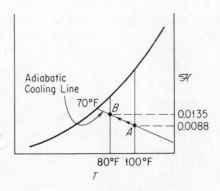

Fig. E5.12(b).

Humidity values are

$$\mathfrak{K}, \frac{\text{lb } H_2O}{\text{lb air}}$$

B	0.0135
A	0.0088

Difference:

$$0.0047 \frac{\text{lb } H_2O}{\text{bl dry air}} \quad \text{added.}$$

EXAMPLE 5.13   Cooling and dehumidification

A process which takes moisture out of the air by passing the air through water sprays sounds peculiar but is perfectly practical as long as the water temperature is below the dew point of the air. Equipment such as shown in Fig. E5.13(a) would do the

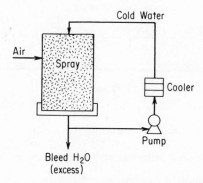

Fig. E5.13(a).

trick. If the entering air has a dew point of 70°F and is at 40 percent $\mathfrak{RK}$, how much heat has to be removed by the cooler, and how much water vapor is removed, if the exit air is at 56°F with a dew point of 54°F?

Solution:
From Fig. 5.11(b) the initial and final values of the enthalpies and humidities are

	A	B
$\mathfrak{K}\left(\frac{\text{grains } H_2O}{\text{lb dry air}}\right)$	111	62
$\Delta\hat{H}\left(\frac{\text{Btu}}{\text{lb dry air}}\right)$	$41.3 - 0.2 = 41.1$	$23.0 - 0 = 23.0$

Look at Fig. E5.13(b) to locate A and B. The grains of H_2O removed are

$$111 - 62 = 49 \text{ grains/lb dry air}$$

The cooling duty is approximately

$$41.1 - 23.0 = 18.1 \text{ Btu/lb dry air}$$

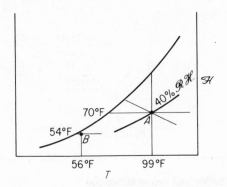

Fig. E5.13(b).

In the upper left of the humidity chart is a little insert which gives the value of the small correction factor for the water condensed from the air which leaves the system. Assuming that the water leaves at the dew point of 54°F, read for 49 grains a correction of −0.15 Btu/lb dry air. You could calculate the same value by taking the enthalpy of liquid water from the steam tables and saying,

$$\frac{22 \text{ Btu}}{\text{lb H}_2\text{O}} \left| \frac{1 \text{ lb H}_2\text{O}}{7000 \text{ grains}} \right| \frac{49 \text{ grains rejected}}{1 \text{ lb dry air}} = 0.154 \text{ Btu/lb dry air}$$

The energy (enthalpy) balance will then give us the cooling load:

$$\text{air in   air out H}_2\text{O out}$$
$$41.1 - 23.0 = 0.15 = 17.9 \text{ Btu/lb dry air}$$

EXAMPLE 5.14   Combined material and energy balances for a cooling tower

You have been requested to redesign a water-cooling tower which has a blower with a capacity of 8.30×10^6 ft³/hr of moist air (at 80°F and a wet-bulb temperature of 65°F). The exit air leaves at 95°F and 90°F wet bulb. How much water can be cooled in pounds per hour if the water to be cooled is not recycled, enters the tower at 120°F, and leaves the tower at 90°F?

Solution:
Enthalpy, humidity, and humid volume data taken from the humidity chart are as follows (see Fig. E5.14)

	A	B
$\mathcal{H}\left(\dfrac{\text{lb H}_2\text{O}}{\text{lb dry air}}\right)$	0.0098	0.0297
$\mathcal{H}\left(\dfrac{\text{grains H}_2\text{O}}{\text{lb dry air}}\right)$	69	208
$\Delta \hat{H}\left(\dfrac{\text{Btu}}{\text{lb dry air}}\right)$	$30.05 - 0.12 = 29.93$	$55.93 - 0.10 = 55.83$
$\hat{V}\left(\dfrac{\text{ft}^3}{\text{lb dry air}}\right)$	13.82	14.65

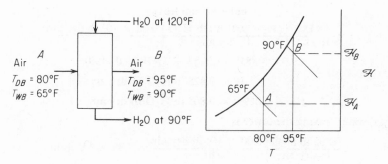

Fig. E5.14.

The cooling water exit temperature can be obtained from an energy balance around the process.

$$\text{Basis: } 8.30 \times 10^6 \text{ ft}^3/\text{hr moist air}$$

$$\frac{8.30 \times 10^6 \text{ ft}^3 \;|\; \text{lb dry air}}{|\; 13.82 \text{ ft}^3} = 6.00 \times 10^5 \text{ lb dry air/hr}$$

The enthalpy of the entering water stream is (reference temperature is 32°F and 1 atm)

$$\Delta \hat{H} = C_{p_{H_2O}} \Delta T = 1(120 - 32) = 88 \text{ Btu/lb } H_2O$$

while that of the exit stream is 58 Btu/lb H_2O. [The value from the steam tables at 120°F for liquid water of 87.92 Btu/lb H_2O is slightly different since it represents water at its vapor pressure (1.69 psia) based on reference conditions of 32°F and liquid water at its vapor pressure.] Any other datum could be used instead of 32°F for the liquid water. For example, if you chose 90°F, one water stream would not have to be taken into account because its enthalpy would be zero.

The loss of water to the air is

$$0.0297 - 0.0098 = 0.0199 \text{ lb } H_2O/\text{lb dry air}$$

(a) *Material balance for water stream:*

Let W = lb H_2O entering the tower in the water stream per lb dry air

Then $(W - 0.0199)$ = lb H_2O leaving tower in the water stream per lb dry air

(b) *Energy balance (enthalpy balance) around the entire process:*

air and water in air entering

$$\frac{29.93 \text{ Btu} \;|\; 6.00 \times 10^5 \text{ lb dry air}}{\text{lb dry air} \;|}$$

water stream entering

$$+ \frac{88 \text{ Btu} \;|\; W \text{ lb } H_2O \;|\; 6.00 \times 10^5 \text{ lb dry air}}{\text{lb } H_2O \;|\; \text{lb dry air} \;|}$$

air and water in air leaving

$$= \frac{55.83 \text{ Btu} \;|\; 6.00 \times 10^5 \text{ lb dry air}}{\text{lb dry air} \;|}$$

water stream leaving

$$+ \frac{58 \text{ Btu}}{\text{lb H}_2\text{O}} \bigg| \frac{(W - 0.0199) \text{ lb H}_2\text{O}}{\text{lb dry air}} \bigg| \frac{6.00 \times 10^5 \text{ lb dry air}}{}$$

$$29.93 + 88W = 55.83 + 58(W - 0.0199)$$

$$W = 0.825 \text{ lb H}_2\text{O/lb dry air}$$

$$(W - 0.0199) = 0.805 \text{ lb H}_2\text{O/lb dry air}$$

The total water leaving the tower is

$$\frac{0.805 \text{ lb H}_2\text{O}}{\text{lb dry air}} \bigg| \frac{6.00 \times 10^5 \text{ lb dry air}}{\text{hr}} = 4.83 \times 10^5 \text{ lb/hr}$$

5.4 Complex problems

In this book we have treated only small segments of material and energy balance problems. Put a large number of these segments together and you will have a real industrial process. We do not have the space to describe the details of any specific process, but for such information you can consult the references at the end of this chapter. It is always wise to read about a process and gain as much information as you can about the stoichiometry and energy relations involved before undertaking to make any calculations. Do not be unnerved by the complexity of a large-scale detailed plant. With the techniques you have accumulated while studying this text you will find that you will be able to break down the overall scheme into smaller sections involving a manageable number of streams that can be handled in the same way you have handled the material and energy balances in this text. Perhaps you can find a tie element (or make one up) that will permit you to work from one unit of the process to the next. Example 5.15 briefly outlines, without much descriptive explanation, how you should apply your knowledge of material and energy balances to a not-too-complex alcohol plant.

Problems in which simultaneous material and energy balances appear can often be solved using special-purpose digital computer programs intended for the design and operation of chemical plants. Such computer programs include techniques of solving the material and energy balances for an almost endless combination of pumps, compressors, heat exchangers, distillation columns, reactors, and so forth. To be able to solve the balances for even one of these pieces of equipment, you need to have a bank of data for the stream properties. Most programs furnish the necessary physical data and make provision for you to provide any missing information. Assuming that subprograms have been prepared and are available that simulate the behavior of individual pieces of process equipment and produce numerical outputs for varying inputs and process parameters, as the user your main task is to designate the interconnections between the units and the choice of the subprograms to represent each

TABLE 5.1

Name	Where Developed	Reference
APACHE	General Electric Co., Bethesda, Md.	
CHEOPS	Shell Development Co., Emeryville, Calif.	R. R. Hughes et al., *6th World Petrol. Conf. Proc., Frankfurt,* Sec. VII, paper 17, p. 93 (1963)
CHESS	University of Houston	R. L. Motard, H. M. Lee, and R. W. Barkley, *CHESS—Chemical Engineering Simulation System,* Tech Publishing Co., Houston, 1968
CHIPS		
FLOWTRAN	Service Bureau Corp, I.B.M. Monsanto Co., St. Louis, Mo.	R. L. Rorschack and R. E. Harris, "Process Simulation Made by Computer," *Oil and Gas J.,* p. 62 (Aug. 17, 1970)
GEMCS	McMaster University Hamilton, Ont., Canada	A. I. Johnson, *GEMCS Manual,* Dept. Chem. Eng., McMaster Univ., Hamilton, Ontario, 1969
NETWORK 67	I.C.I., Ltd., Runcorn, England	S. M. Andrew, "Computer Flowsheeting Using Network 67; an Example," *Trans. Inst. Chem. Eng.,* v. 46, p. T123 (1967)
PACER	Purdue University and Dartmouth College	C. M. Crowe et al., *Chemical Plant Simulation,* Prentice-Hall, Englewood Cliffs, N.J., 1971
REMUS	University of Pennsylvania	P. T. Shannon et al., "Computer Simulation of a Sulfuric Acid Plant," *Chem. Eng. Progress,* v. 62, No. 6, p. 49 (1966) P. E. Ham, *Users' Manual for Remus (Routine for Executive Multi-Unit Simulation),* School of Chem. Eng., Univ. of Pennsylvania, Philadelphia, 1969
SWAPSCO	Stone and Webster, Boston, Mass.	
UOS	Bonner and Moore Assoc., Houston, Texas	T. Utsumi, "Computer Package Aids Systems Engineering," *Oil and Gas J.,* p. 100 (June 8, 1970)

405

unit. In some instances you can also select the subroutine to be used to solve the large sets of nonlinear equations that must be solved simultaneously. Table 5.1 lists some of the programs available that are user-oriented. Evans et al.[9] give a very lucid state of the art survey of heat and material balance programs, and additional references can be found at the end of this chapter.

EXAMPLE 5.15 A more complex problem

The Blue Ribbon Sour Mash Company plans to make commercial alcohol by a process shown in Fig. E5.15(a). Grain mash is fed through a heat exchanger where it is heated to 170°F. The alcohol is removed as 60 percent (weight) alcohol from the

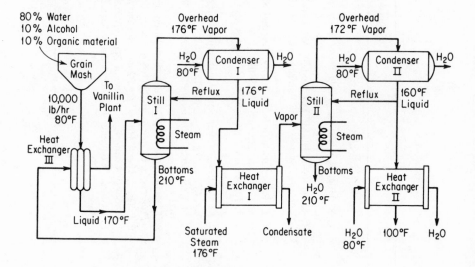

Fig. E5.15(a).

[9]L. B. Evans et al., *Chem. Eng. Progress*, v. 64, no. 4, p. 39 1968).

first fractionating column; the bottoms contain no alcohol. The 60 percent alcohol is further fractionated to 95 percent alcohol and essentially pure water in the second column. Both stills operate at a 3-to-1 reflux ratio and heat is supplied to the bottom of the columns by steam. Condenser water is obtainable at 80°F. The operating data and physical properties of the streams have been accumulated and are listed for convenience:

operating information and data

stream	state	b.p. (°F)	C_p, Btu/(lb)(°F) liquid	vapor	heat of vaporization (Btu/lb)
Feed	Liquid	170	0.96		950
60% alcohol	Liquid or vapor	176	0.85	0.56	675
Bottoms (I)	Liquid	212	1.00	0.50	970
95% alcohol	Liquid or vapor	172	0.72	0.48	650
Bottoms (II)	Liquid	212	1.00	0.50	970

Make a complete material balance on the process and

(a) Determine the weight of the following streams per hour:
   (1) Overhead product, column (I).
   (2) Reflux, column (I).
   (3) Bottoms, column (I).
   (4) Overhead product, column (II).
   (5) Reflux, column (II).
   (6) Bottoms, column (II).
(b) Calculate the temperature of the bottoms leaving heat exchanger III.
(c) Determine the total heat input to the system in Btu/hr.
(d) Calculate the water requirements for each condenser and heat exchanger II in gal/hr if the maximum exit temperature of water from this equipment is 130°F.

Solution:
Let us call the alcohol A for short and the organic matter M.

STEP 1: MATERIAL BALANCES

Basis: 10,000 lb/hr mash

(a) *Material balance around column I and condenser:*

In

Feed $\begin{cases} \text{A} \\ \text{H}_2\text{O} \\ \text{M} \end{cases}$ $\begin{matrix} (0.10)(10,000) = & 1,000 \text{ lb} \\ (0.80)(10,000) = & 8,000 \\ (0.10)(10,000) = & 1,000 \end{matrix}$

$\overline{10,000 \text{ lb}}$

Out

Product $\begin{cases} \text{A} \\ \text{H}_2\text{O} \end{cases}$ $\begin{matrix} & = & 1,000 \text{ lb} \\ \left(\dfrac{1000}{0.60}\right) - 1000 = & 667 \end{matrix}$

Bottoms $\begin{cases} \text{H}_2\text{O} \\ \text{M} \end{cases}$ $\begin{matrix} (8000 - 667) = & 7,333 \\ & = & 1,000 \end{matrix}$

$\overline{10,000 \text{ lb}}$

(b) *Material balance around column I only:*

Feed $\begin{cases} \text{A} \\ \text{H}_2\text{O} \\ \text{M} \end{cases}$ $\begin{matrix} & = & 1,000 \text{ lb} \\ & = & 8,000 \\ & = & 1,000 \end{matrix}$

Reflux $\begin{cases} \text{A} \\ \text{H}_2\text{O} \end{cases}$ $\begin{matrix} (3)(1000) = & 3,000 \\ (3)(667) = & 2,000 \end{matrix}$

$\overline{15,000 \text{ lb}}$

Overhead vapor $\begin{cases} \text{A} \\ \text{H}_2\text{O} \end{cases}$ $\begin{matrix} 4(1000) = & 4,000 \text{ lb} \\ 4(667) = & 2,667 \end{matrix}$

Bottoms $\begin{cases} \text{H}_2\text{O} \\ \text{M} \end{cases}$ $\begin{matrix} = & 7,333 \\ = & 1,000 \end{matrix}$

$\overline{15,000 \text{ lb}}$

(c) *Material balance around column II and condenser:*

Feed $\begin{cases} \text{A} \\ \text{H}_2\text{O} \end{cases}$ $\begin{matrix} = & 1,000 \text{ lb} \\ = & 667 \end{matrix}$

$\overline{1,667 \text{ lb}}$

(d) Overhead product $\begin{cases} \text{A} \\ \text{H}_2\text{O} \end{cases}$ $\begin{matrix} & = & 1,000 \text{ lb} \\ \left(\dfrac{1000}{0.95} - 1000\right) = & 50 \end{matrix}$

Bottoms $\{\text{H}_2\text{O}$ $\quad (667 - 50) = \quad 617$

$\overline{1,667 \text{ lb}}$

(d) *Material balance on column II cutting reflux and overhead vapor:*

In

Feed	{A		= 1,000 lb
	H₂O		= 667
(e)	{A	3(1000)	= 3,000
Reflux	H₂O	(3)(50)	= 150
			4,817 lb

Out

Vapor	{A	(4)(1000)	= 4,000 lb
	H₂O	(4)(50)	= 200
Bottoms	{H₂O	(667 − 50)	= 617
			4,817 lb

STEP 2: ENERGY BALANCES

(a) *Energy balance on heat exchanger III* [Fig. E5.15(b)]:

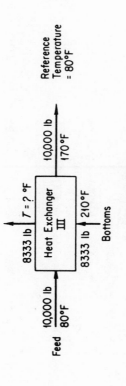

Fig. E5.15(b).

In

Feed	(10,000)(0.96)(80 − 80) =	0 Btu
Bottoms	(8333)(1.0)(210 − 80) =	1,084,000 Btu

Out

Feed	(10,000)(0.96)(170 − 80)	= 864,000 Btu
Bottoms	(833)(1.0)(T − 80)	= (8333T − 666,000) Btu

$$1,084,000 = 8333T - 666,000 + 864,000$$
$$886,000 = 8333T$$
$$T = 106.5°F$$

409

(b) *Energy balance around heat exchanger III inlet, and overhead vapor and reflux line from I:*

Reference temperature = 176°F

In

Feed	(10,000)(0.96)(80 − 176) =	−921,000
Reflux	(5000)(0.85)(176 − 176) =	0
Steam		ΔH_{S_1}

$(\Delta H_{S_1} - 921,000)$

Out

Bottoms to
vanillin plant (8333)(1.0)(106.5 − 176) = −579,000
Vapor (6667)(0.85)(176 − 176) + (6667)(675) = 4,500,000

3,921,000

$\Delta H_{S_1} - 921,000 = 3,921,000$

$\Delta H_{S_1} = 4,842,000$ Btu/hr

(c) *Energy balance around heat exchanger I outlet and overhead vapor and reflux from II:*

Reference temperature = 172°F

In

Feed to II (1667)(0.85)(176 − 172) = 5,650
$\Delta H_{\text{vaporization}=\text{steam}@176°F}$
$+$(1667)(675) = 1,125,000
Reflux (3150)(0.72)(160 − 172) = −27,200
Steam ΔH_{S_2}

$(\Delta H_{S_2} + 1,103,450)$

Out

Vapor (4200)(0.72)(172 − 172) + (4200)(650) = 2,725,000

Bottoms (617)(1.0)(210 − 172) = 23,400

2,748,400

$\Delta H_{S_2} + 1,103,450 = 2,748,400$

$\Delta H_{S_2} = 1,645,000$ Btu/hr

(d) *Total energy input into system:*

$\Delta H_{S_1} + \Delta H_{S_2} +$ steam heater = 4,842,000 + 1,645,000 + 1,125,000 = 7,612,000 Btu/hr ⓒ

(e) *Energy balance on condenser I:*

$W = $ lb H_2O/hr;   Reference temperature = 176°F

In

Vapor	$(6667)(0.85)(176 - 176) + (6667)(675) =$	4,500,000
Cooling H_2O	$(W)(1.0)(80 - 176) =$	$-96W$
		$(-96W + 4,500,000)$

Out

Condensate	$(6667)(0.85)(176 - 176) =$	0
Cooling H_2O	$(W)(1.0)(130 - 176) =$	$-46W$
		$-46W$

$50W = 4,500,000$ or $W = \dfrac{4,500,000}{50} = 90,000$ lb H_2O/hr

$\dfrac{90,000 \text{ lb } H_2O/\text{hr}}{8.345 \text{ lb/gal}} = 10,800$ gal H_2O/hr ⟵ (d₁)

(f) *Energy balance on condenser II:*

$W = $ lb H_2O/hr;   Reference temperature = 172°F

In

Vapor	$(4200)(0.72)(172 - 172) + (4200)(650) =$	2,725,000
Cooling H_2O	$(W)(1.0)(80 - 172) =$	$-92W$
		$(-92W + 2,725,000)$

Out

Condensate	$(4200)(0.72)(160 - 172) =$	$-36,000$
Cooling H_2O	$(W)(1.0)(172 - 130) =$	$-42W$
		$(-42W - 36,000)$

$50W = 2,761,000$ or $W = \dfrac{2,761,000}{50} = 55,200$ lb H_2O/hr

$\dfrac{55,200 \text{ lb } H_2O}{8.345 \text{ lb/gal}} = 6620$ gal H_2O/hr ⟵ (d₂)

(g) *Energy balance on heat exchanger II:*

$W = $ lb H_2O/hr;   Reference temperature = 80°F

In

Condensate	$(1050)(0.72)(160 - 80) =$	60,500
Cooling H_2O	$(W)(1.0)(80 - 80) =$	0
		60,500

Out

Product	$(1050)(0.72)(100 - 80) =$	15,120
Cooling H_2O	$(W)(1.0)(130 - 80) =$	$50W$
		$50W + 15,120$

$50W + 15,120 = 60,500$ or   109 gal H_2O/hr ⟵ (d₃)

$W = 907$ lb H_2O/hr

411

WHAT YOU SHOULD HAVE LEARNED FROM THIS CHAPTER

1. By the end of this chapter you should have perfected your techniques in using material and energy balances separately or in combination to such an extent that you should be able to analyze any type of process and write and solve the appropriate balances.
2. You should know how to use humidity charts and enthalpy-concentration charts as aids in the solution of problems.

SUPPLEMENTARY REFERENCES

Combustion of Solid, Liquid, and Gaseous Fuels

1. Gaydon, *Flames*, 3rd ed., Chapman & Hall, London, 1970.
2. Griswold, John, *Fuels, Combustion and Furnaces*, McGraw-Hill, New York, 1946.
3. Lewis, W. K., A. H. Radasch, and H. C. Lewis, *Industrial Stoichiometry*, 2nd ed., McGraw-Hill, New York, 1954.
4. Popovich, M., and C. Hering, *Fuels and Lubricants*, Wiley, New York, 1959.

Gas Producers and Synthetic Gas

1. Arne, Francis, "Manufactured Gas," *Chem. Eng.*, pp. 121–123 (Mar. 24, 1958).
2. Griswold, John, *Fuels, Combustion and Furnaces*, McGraw-Hill, New York, 1946.
3. Gumz, Wilhelm, *Gas Producers and Blast Furnaces*, Wiley, New York, 1958.
4. Lewis, W. K., A. H. Radasch, and H. C. Lewis, *Industrial Stoichiometry*, 2nd ed., McGraw-Hill, New York, 1954.
5. Mills, G. A., *Environmental Sci. & Technology*, v. 12, no. 5, p. 1178 (1971).

Enthalpy-Concentration Charts

*1. Badger, W. L., and J. T. Banchero, *Introduction to Chemical Engineering*, McGraw-Hill, New York, 1955.
*2. Brown, G. G., et al., *Unit Operations*, Wiley, New York, 1950.
*3. Ellis, S. R. M., *Chem. Eng. Sci.*, v. 3, p. 287 (1954).
**4. Lemlich, Robert, Chad Gottschlich, and Ronald Hoke, *Chem. Eng. Data Series*, v. 2, p. 32 (1957).
*5. McCabe, W. L., *Am. Inst. Chem. Eng. Trans.*, v. 31, p. 129 (1935).
**6. Othmer, D. F., et al., *Ind. Eng. Chem.*, v. 51, p. 89 (1959).
*7. Robinson, C. S., and E. R. Gilliland, *Elements of Fractional Distillation*, 3rd ed., McGraw-Hill, New York, 1939.
*8. White, R. R., *Petroleum Refiner*, v. 24, no. 8, p. 101; v. 24, no. 9, p. 127 (1945).

Humidity Charts and Calculations

1. Badger, W. L., and J. T. Banchero, *Introduction to Chemical Engineering*, McGraw-Hill, New York, 1955.

*Concerned with *H-x* chart calculations.
**Concerned with *H-x* chart construction.

2. McCabe, W. L., and J. C. Smith, *Unit Operations of Chemical Engineering*, McGraw-Hill, New York, 1956.

3. Treybal, R. E., *Mass Transfer Operations*, McGraw-Hill, New York, 1955.

4. Woolrich, W. R., and W. R. Woolrich, Jr., *Air Conditioning*, Ronald, New York, 1957.

Industrial Process Calculations

1. Hougen, O. A., K. M. Watson, and R. A. Ragatz, *Chemical Process Principles*, Part I, 2nd ed., Wiley, New York, 1954.

2. Lewis, W. K., A. H. Radasch, and H. C. Lewis, *Industrial Stoichiometry*, 2nd ed., McGraw-Hill, New York, 1954.

3. Nelson, W. L., *Petroleum Refinery Engineering*, 4th ed., McGraw-Hill, New York, 1958.

4. Shreve, R. N., *The Chemical Process Industries*, 2nd ed., McGraw-Hill, New York, 1956.

5. Williams, F. A., *Combustion Theory*, Addison-Wesley, Reading, Mass., 1965.

Solution of Material and Energy Balances Via Digital Computers

1. Andrew, S. M., *British Chem. Eng.*, v. 14, p. 1057 (1969).

2. Beckett, K. A., "Computer-aided Process Design," *Chem. Eng.*, no. 228, p. 163 (1969).

3. Hutchison, H. P., and G. F. Forder, *Chemical Plant Network Specification Using a Visual Display*, Symposium Series No. 23, The Institution of Chemical Engineers, London, 1967, p. 169.

4. Kenny, L. N., and J. W. Prados, *A Generalized Digital Computer Program for Performing Process Material and Energy Balances*, University of Tennessee, Knoxville, 1966.

5. Lee, W., and D. F. Rudd, *A.I.Ch.E. J.*, v. 12, p. 1184 (1966).

6. Sargent, R. W. H., *Chem. Eng. Progress*, v. 63, no. 9, p. 71 (1967).

PROBLEMS

Combined Material and Energy Balance Problems

5.1. Limestone ($CaCO_3$) is converted into CaO in a continuous vertical kiln (see Fig. P5.1). Heat is supplied by combustion of natural gas (CH_4) in direct contact

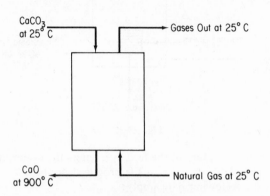

Fig. P5.1.

with the limestone using 50 percent excess air. Determine the pounds of $CaCO_3$ that can be processed per pound of natural gas. Assume that the following mean heat capacities (relative to 25°C) apply:

$$C_{p_m} \text{ of } CaCO_3 = 56.0 \text{ Btu/(lb mole)(°F)}$$

$$C_{p_m} \text{ of } CaO = 26.7 \text{ Btu/(lb mole)(°F)}$$

5.2. A vertical lime kiln is charged with pure limestone ($CaCO_3$) and pure coke (carbon), both at 25°C. Dry air at 25°C is blown in at the bottom to burn the coke to CO_3, which provides the necessary heat for decomposition of the carbonate. The lime (CaO) leaves the bottom of the kiln at 950°F and contains no carbon or $CaCO_3$. The kiln gases leave at 600°F and contain no free oxygen. The molal ratio of $CaCO_3 : C$ in the charge is 1.5 to 1. Calculate the following:
(a) The analysis of the kiln gases.
(b) The heat loss from the kiln per pound mole of coke fed to the kiln.
You may use the following heat capacities as constants:

Comp.	*Btu* (lb mole)(°F)
$CaCO_3$	$C_p = 24.2$
CaO	$C_p = 12.3$
C	$C_p = 3.72$

5.3. A feed stream of 16,000 lb/hr of 7 percent (by weight) NaCl solution is concentrated to a 40 percent (by weight) solution in an evaporator. The feed enters the evaporator, where it is heated to 180°F. The water vapor from the solution and the concentrated solution leave at 180°F. Steam at the rate of 15,000 lb/hr enters at 230°F and leaves as condensate at 230°F. See Fig. P5.3.

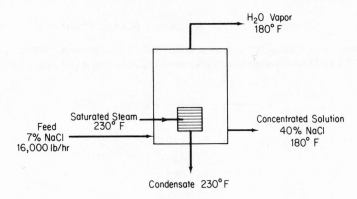

Fig. P5.3.

(a) What is the temperature of the feed as it enters the evaporator?
(b) What weight of 40 percent NaCl is produced per hour?
Assume that the following data apply:

$$\text{Mean } C_p \text{ 7\% NaCl soln:} \qquad 0.92 \text{ Btu/(lb)(°F)}$$
$$\text{Mean } C_p \text{ 40\% NaCl soln:} \qquad 0.85 \text{ Btu/(lb)(°F)}$$
$$\Delta \hat{H}_{\text{vap}} \text{ H}_2\text{O at 180°F} = 990 \text{ Btu/lb}$$
$$\Delta \hat{H}_{\text{vap}} \text{ H}_2\text{O at 230°F} = 959 \text{ Btu/lb}$$

5.4. One way to get pure N_2 is to remove the CO_2 from a combustion gas that contains no O_2. A gas composed of 20 percent CO_2 (by volume) and 80 percent N_2 is passed into a tower where it contacts a 20 percent (by weight) NaOH solution. The CO_2 forms Na_2CO_3 and the N_2 is sent to be dried. See Fig. P5.4.

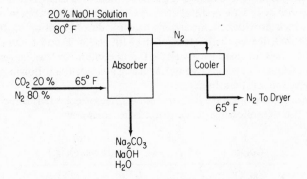

Fig. P5.4.

(a) Assuming that 10,000 ft³/hr of the gas mixture measured at 65°F, 760 mm Hg is passed through the tower, what weight of 20 percent NaOH solution is required if 20 percent excess is desired in the entering liquid?
(b) If the temperature of the N_2 and the Na_2CO_3 solution leaving the tower

is the same, what is this temperature if the NaOH solution enters at 80°F and the gas enters at 65°F? [C_p liquid NaOH soln = 0.9 Btu/(lb)(°F); C_p liquid NaOH-Na$_2$CO$_3$ soln = 0.92 Btu/(lb)(°F).]

(c) How many Btu must be removed in the cooler per hour if the N$_2$ is to be cooled to 65°F?

Data: Assume that the $\Delta \hat{H}_f^\circ$ of NaOH is -101.96 kcal/g mole.

5.5. One of the ways in which steel can be welded is to take a mixture of powdered aluminum and finely divided iron oxide and ignite it. A vigorous reaction takes place, and aluminum oxide and molten iron are formed at a temperature up to 2300°C:

$$2Al + Fe_2O_3 \longrightarrow Al_2O_3 + 2Fe$$

A mixture of this type is called thermite; it has been used to weld together the ends of steel rails, to repair machinery, etc. A manufacturer is considering making a new type of thermite which combines powdered magnesium with iron oxide. Which stoichiometric mixture weighs less, the 2Al + Fe$_2$O$_3$ or the 3Mg + Fe$_2$O$_3$, to get the desired temperature of 3000°F if the heat lost by convection and radiation is 20 percent of the heat of reaction? Assume that the initial thermite mixture is at 65°F. The heat capacities of Al$_2$O$_3$ and MgO can be considered constant at 0.20 Btu/(lb)(°F). For iron and steel use the following data:

$$C_p(s) = 0.12 \text{ Btu/(lb)(°F)}$$
$$C_p(l) = 0.25 \text{ Btu/(lb)(°F)}$$
$$\Delta H_{\text{fusion}} = 86.5 \text{ Btu/lb}, \quad \text{melting point} = 2800°F$$

5.6. C. C. Hunicke and C. L. Wagner in U.S. Patent 2,005,422 (June 18, 1935) concentrate sulfite waste liquor (s.w.l.) from the pulping of wood by blowing hot stack gas through sparger pipes immersed in the liquid. The foam that forms is knocked down by a spray of liquid s.w.l. feed. Because most of the substances in s.w.l. are colloidal, its vapor pressure and thermal properties may be considered to be the same as those of water. The flue gas analyzes CO$_2$, 12.0 percent; O$_2$, 7.0 percent; N$_2$, 81.0 percent, and has a dew point of 90°F. It enters the apparatus at 800°F. The s.w.l. from the blowpits is fed into the apparatus at 120°F. The weather is partly cloudy, temperature 80°F, barometer 29.84 in. Hg, and the wind is 10 mi/hr NNE. Calculate the following:

(a) The boiling or equilibrium temperature of the s.w.l. in the evaporator.

(b) The cubic feet of stack gas entering per pound of water evaporated.

5.7. Determine the number of degrees of freedom for the condenser shown in Fig. P5.7.

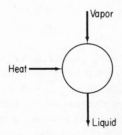

Fig. P5.7.

5.8. Determine the number of degrees of freedom for the reboiler shown .in Fig. P5.8. What variables should be specified to make the solution of the material and energy balances determinate?

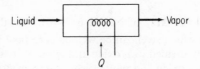

Fig. P5.8.

5.9. If to the equilibrium stage shown in Example 5.2 you add a feed stream, determine the number of degrees of freedom. See Fig. P5.9.

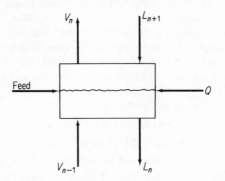

Fig. P5.9.

5.10. How many variables must be specified for the furnace shown in Fig. P5.10?

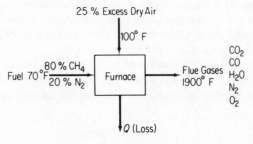

Fig. P5.10.

5.11. Read Problem 4.79. How many additional variables have to be specified, i.e., assumed, in order to make the problem determinate?

5.12. Figure P5.12 shows a simple absorber or extraction unit. S is the absorber oil (or fresh solvent), and F is the feed from which material is to be recovered. Each stage has a Q (not shown); the total number of equilibrium stages is N. What is the number of degrees of freedom for the column? What variables should be specified?

Fig. P5.12.

Enthalpy-Concentration Chart Problems

5.13. Draw the saturated-liquid line (1 atm isobar) on an enthalpy concentration chart for caustic soda solutions. Show the 200, 250, 300, 350, and 400°F isotherms in the liquid region, and draw tie lines for these temperatures to the vapor region. Using the data from *N.B.S. Circular* 500, show where molten NaOH would be on this chart. Take the remaining data from the following table and the steam tables.

temp. (°F)	x satd. liquid conc., wt fr.	$\Delta\hat{H}$ enthalpy, satd. liquid, Btu/lb
200	0	168
250	0.34	202
300	0.53	314
350	0.68	435
400	0.78	535

C_p for NaOH solutions, Btu/(lb)(°F)

wt % NaOH	temperature, °F				
	32	60	100	140	180
10	0.882	0.897	0.911	0.918	0.922
20	0.842	0.859	0.875	0.884	0.886
30		0.837	0.855	0.866	0.869
40		0.815	0.826	0.831	0.832
50			0.769	0.767	0.765

SOURCE: J. W. Bertetti and W. L. McCabe, *Ind. Eng. Chem.*, v. 28, p. 375 (1936).
NOTE: All H_2O tie lines extend to pure water vapor.

5.14. Refer to Fig. I.1 in Appendix I, the enthalpy-composition chart for the system *A-B* at 1 atm, where *A* = ethanol and *B* = water.
  (a) What are the reference states?
  (b) What is the normal boiling point (°F) of pure *A*?
  (c) What is the normal boiling point (°F) of pure *B*?
  (d) What is the heat of vaporization (Btu/lb) of pure *B* at the normal boiling point?
  (e) What is the bubble point (°F) of a mixture of *A* and *B* which is 45 weight percent *A*?
  (f) Estimate the heat capacity in Btu/(lb)(°F) of a vapor mixture of *A* and *B* which is 45 weight percent at 185°F.
  (g) One hundred pounds of a liquid *A-B* mixture at 150°F and 90 weight percent *A* is mixed with 38.5 lb of pure *B* vapor at 215°F in a steady-flow process.

The process is adiabatic. Determine the temperature (°F), the compositions (wt % *A*), and the masses (lb) of the vapor and liquid streams leaving
the process.

5.15. A 50 percent (by weight) sulfuric acid solution is to be made by mixing the following:
(a) Ice at 32°F.
(b) 80 percent H_2SO_4 at 100°F.
(c) 20 percent H_2SO_4 at 100°F.
How much of each must be added to make 1000 lb of the 50 percent solution with
a final temperature of 100°F if the mixing is adiabatic.

5.16. A 10 percent (by weight) sulfuric acid solution at 70°F is placed in an open kettle
evaporator and evaporated at atmospheric pressure to a 60 percent solution. If
the steam leaving the kettle contains no sulfuric acid and is considered as leaving
at the average temperature of the boiling range (arithmetic mean temperature),
(a) What are the initial and final boiling temperatures?
(b) How much heat must be supplied per pound of 60 percent acid produced?

5.17. One thousand pounds of a 70 percent H_2SO_4 solution at 70°F is to be concentrated to an 80 percent solution by *flash* evaporation in which the solution is
heated under pressure while flowing through a pipe and then passes into a chamber maintained at 1 atm pressure where part of the water flashes off as steam at
1 atm. The liquid and vapor phases formed are essentially in equilibrium with
each other at the boiling point of the liquid.
(a) What is the temperature in the flash chamber?
(b) How much heat was added to the solution in the heating section?

5.18. A dilute sulfuric acid (62 percent by weight) is to be made by diluting concentrated acid (95 percent) with water; however, due to the highly exothermic nature
of the mixing, some provision is necessary to remove the energy generated on
mixing. The scheme shown in Fig. P5.18 is to be used. The concentrated acid is

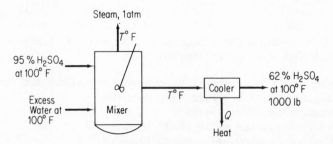

Fig. P5.18.

mixed with an excess of water and the mixture boils, driving off steam (no H_2SO_4
evaporates). Hot acid of the desired concentration leaves the mixer (both the
steam and hot acid leave at the same temperature) and is finally cooled to
100°F. Determine the following:
(a) The amount of concentrated acid required per 1000 lb of dilute acid produced.
(b) The amount of steam that leaves.

(c) The amount of water added to mixer.

(d) The temperature of the contents of the mixer.

(e) The amount of heat removed from the acid by the cooler.

5.19. To a closed tank containing 100 lb of water at 90°F is added 200 lb of 80 percent H_2SO_4 solution at 294°F and 100 lb of 40 percent H_2SO_4 solution at 140°F. To cool the solution, a 30-lb cake of ice is dropped into the tank. What is the final composition and temperature of the mixture in the tank? Use the chart provided for enthalpies of H_2SO_4 solutions in Appendix I (note that the abscissa is wt % of H_2SO_4). Latent heat of fusion of ice is 79.7 cal/g. Heat capacity of H_2O is about 1 Btu/lb. Heat of formation of $H_2O(g)$ is -57.80 kcal/g mole.

5.20. One thousand pounds of 10 percent NaOH solution at 100°F is to be fortified to 30 percent NaOH by adding 73 percent NaOH at 200°F. How much 73 percent solution must be used? How much cooling must be provided so that the final temperature will be 70°F?

5.21. For the ammonia-water system at . . . psia, calculate the unknown quantities for each of the three cases below (see Fig. P5.21):

	stream	wt % NH_3	enthalpy Btu/lb	amount lb
(a)	A	10	Satd. liquid	150
	B	70	Satd. vapor	300
	C	Unknown	Unknown	Unknown
	$Q = -400,000$ Btu			
(b)	A	80	Satd. vapor	Unknown
	B	10	1700	Unknown
	C	50	100	100
	$Q = $ unknown			
(c)	A	90	1200	100
	B	Unknown	1500	Unknown
	C	35	800	400
	$Q = $ unknown			

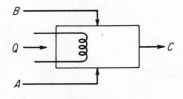

Fig. P5.21.

5.22. A mixture of ammonia and water in the vapor phase, saturated at 250 psia and containing 80 weight percent ammonia, is passed through a condenser at a rate of 10,000 lb/hr. Heat is removed from the mixture at the rate of 5,800,000 Btu/hr while the mixture passes through a cooler. The mixture is then expanded to a pressure of 100 psia and passes into a separator. A flow sheet of the process is given in Fig. P5.22. If the heat loss from the equipment to the surroundings is

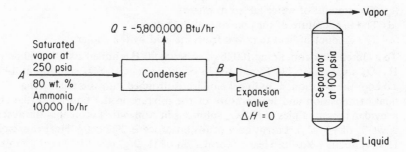

Q = -5,800,000 Btu/hr

Saturated
vapor at
250 psia

A

80 wt. %
Ammonia
10,000 lb/hr

Condenser

B

Expansion
valve
Δ*H* = 0

Separator
at 100 psia

Vapor

Liquid

Fig. P5.22.

neglected, determine the composition of the liquid leaving the separator, by

(a) A material and energy balance set of equations.

(b) Using the enthalpy-concentration diagram method.

5.23. (a) Plot an enthalpy-concentration diagram for the ethyl alcohol-water system
at 760 mm Hg total pressure (enthalpy of the mixture in Btu/lb vs. weight
fraction ethanol). Plot the diagram on 11- by 17-in. graph paper. Be sure
to include the following: (1) saturated-vapor line, (2) saturated-liquid line,
and (3) lines of constant temperature in the subcooled-liquid region from 0
to 200°F. Notice that even though the EtOH-water system is a nonideal
system, the saturated-vapor and liquid lines are almost straight lines and
could have been drawn from the enthalpies of the pure components without
serious error.

(b)

In each of the four cases diagramed in Fig. P5.23 the following data apply:

stream	wt %EtOH	condition	amount
A	80	Unknown	Unknown
B	10	70°F	Unknown
C	60	Superheated vapor at 800 Btu/lb	100 lb

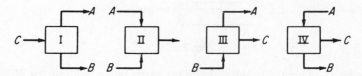

Fig. P5.23.

(1) For diagram I:
(a) By a material and energy balance, find the weights of streams *A*
and *B* and the enthalpy of stream *A*.

  (b) Prove by analytical geometry that *A*, *B*, and *C* will lie on a straight line when plotted on an *H-x* diagram.
  (c) Repeat part (a), but this time use the method of Ponchon instead of a system of material and energy balance equations.
 (2) Repeat 1(c) for diagram II.
 (3) Repeat 1(c) for diagram III.
 (4) Repeat 1(c) for diagram IV.

 Note: For part (a) you must plot the actual data and show your points clearly. For part (b), use $8\frac{1}{2}$-by-11 graph paper with the abscissa marked from 0.1 to 0.9 wt fraction. If you use more than one graph for the solution of part (b), you may plot the saturated-vapor and liquid lines from the enthalpies of the pure components and connect them with a straight line. See Appendix I.

 References for data: L. W. Cornell and R. E. Montonna, *Ind. Eng. Chem.*, v. 25, pp. 1331–1335 (1933); J. H. Perry, *Chemical Engineers' Handbook*, 2nd ed., McGraw-Hill, New York, 1941, p. 1364; W. A. Noyes and R. R. Warfle, *J. Am. Chem. Soc.*, v. 23, p. 463 (1901).

5.24. Ethanol-water are separated as shown in the information flow diagram (Fig. P5.24). In the diagram, the vessels labeled *A*, *B*, and *C* are vapor-liquid separators

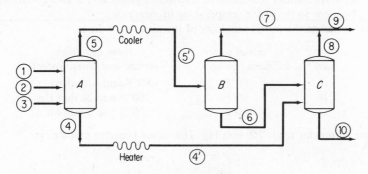

Fig. P5.24.

in which an equilibrium is reached between the liquid and vapor (i.e., streams 8 and 10 are in equilibrium, 6 and 7 are in equilibrium, and 4 and 5 are in equilibrium). The total pressure is 1 atm and the following data are available:

Stream No.	Quantity lb/hr	Composition wt fr. EtOH	Enthalpy Btu/lb
1	1090	0.20	50.0
2	233	0.15	1400.0
3	128	0.05	700.0
9	—	0.395	—
10	—	0.04	—

Compute the heat loads on the cooler and heater in Btu/hr.

Humidity and Saturation Chart Problems

5.25. Construct a "saturation" chart for the system ethyl alcohol-air at normal atmospheric pressure. Follow the style of Fig. 5.11 inside the back cover. Cover the temperature range from 50 to 200°F. Include the following:

(a) Percent absolute saturation lines from 100 to 25 percent.

(b) Saturated ethyl alcohol vs. temperature.

(c) Adiabatic saturation lines.

(d) Constant wet-bulb temperature lines.

Obtain the required data from any handbook or from the literature. [An older reference is *Chem. Metal. Eng.*, v. 47, p. 287 (1940).]

5.26. Moist air at 1 atm, a dry-bulb temperature of 195°F, and a wet-bulb temperature of 115°F is enclosed in a rigid container. The container and its contents are cooled to 110°F.

(a) What is the molar humidity of the cooled moist air?

(b) What is the final total pressure in atm in the container?

(c) What is the dew point in °F of the cooled moist air?

(d) What is the final wet-bulb temperature in °F?

5.27. Your nephew has seen his teacher demonstrate a sling psychrometer in high school and is slightly puzzled as to how it works.

(a) Tell precisely what happens in each of the cases given below (i.e., direction of mass transfer, direction of transfer of latent and sensible heat, and what happens to the water temperature in each case).

(b) Calculate the initial humidity of the air in each case:

case	initial water temp. on wick	room condition (temp. and press. water)
A	60°F	80°F and saturated
B	60°F	100°F and 26 mm Hg
C	80°F	100°F and 5 mm Hg

The barometer reads 760 mm Hg. The vapor pressure of water is

°F	mm Hg
40	6.4
50	13.6
80	26
100	49

5.28. Derive the expressions for humid heat, saturated volume, and adiabatic cooling lines.

5.29. A rotary dryer operating at atmospheric pressure dries 10 tons/day of wet grain at 70°F, from a moisture content of 10 to 1 percent moisture. The air flow is countercurrent to the flow of grain, enters at 225°F dry-bulb and 110°F wet-bulb temperature, and leaves at 125°F dry-bulb. See Fig. P5.29. Determine the humidity of the entering and leaving air, the water removal in pounds per hour, the daily product output in pounds per day, and the heat input to the dryer. (Assume that there is no heat loss from the dryer, that the grain is discharged at 110°F, and that the specific heat is 0.18.)

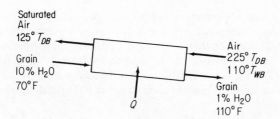

Fig. P5.29.

5.30. A dryer produces 400 lb/hour of a product containing 8 percent water from a feed stream that contains 1.25 g of water per gram of dry material. The air enters the dryer at 212°F dry-bulb and 100°F dew point; the exit air leaves at 128°F dry-bulb and 60 percent relative humidity. Part of the exit air is mixed with the fresh air supplied at 70°F, 52 percent relative humidity, as shown in Fig. P5.30.

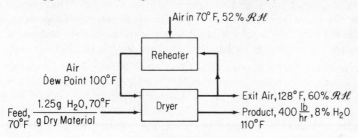

Fig. P5.30.

Calculate the air and heat supplied to the dryer, neglecting any heat lost by radiation, used in heating the solid, the conveyor trays, and so forth. The specific heat of the product is 0.18.

5.31. Clean and air-conditioned air must be furnished to two rooms in which solid-state devices for digital computers are produced. Low-humidity air is supplied to the first room and flows from it to the second room. Additional low-humidity air is supplied to the second room, as shown in Fig. P5.31. The humidity of the

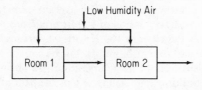

Fig. P5.31.

low-humidity air is 0.0040 lb water/lb dry air, and the humidity of the air leaving the first room is 0.0055. The humidity of the air leaving the second room is 0.0065. Water is evaporated in the first room at the rate of 2.03 lb/hr and in the second room at the rate of 2.53 lb/hr. Calculate the low-humidity air flow rate to the two rooms and the total air flow rate, both in terms of pounds of dry air per hour.

5.32. A humidifier is conditioning air to 120°F dry-bulb and 90°F wet-bulb by heating outside air and then passing it through a spray chamber in which it reaches 90 percent humidity and then reheating it to the desired temperature. The outside air is foggy at 40°F, carrying as liquid water 0.0004 lb water/ft^3 wet air. What temperature (of the air) must be reached in each heating operation, and how many Btu are required in each heating stage per 100 lb of dry air entering from the outside?

5.33. Leaf tobacco cannot be handled without breakage unless it contains over 14 percent water, and it cannot be stored without molding if the water exceeds 13 percent. A variation of $\frac{1}{2}$ percent from the optimum will seriously impair the cigarette machines. Cigarette manufacture, therefore, must be a cycle of drying and humidifying regardless of what else may be involved. You receive a leaf tobacco from the storage warehouse containing 12 percent water. Ten thousand pounds of this tobacco is stored in a *sweat room* for 1 week. The air in the room is maintained at 100°F and 85 percent relative humidity. The tobacco removed from the sweat room at the end of the week contains 14 percent moisture. Assuming that only 5 percent of the water in the air is absorbed by the tobacco, calculate the volume of air handled per week for the 10,000 lb of leaf.

5.34. In one of the hotter regions of the country a home owner decides to keep his home at an average temperature of 80°F and 40 percent humidity. On a typical day the outside conditions are as follows: dry-bulb temperature = 95°F and wet-bulb = 85°F. The city water supply is at 70°F and scarce. He therefore decides to use an electric refrigeration unit to cool the air entering the ventilating ducts. Summary of conditions:

Size of home: 50,000 ft^3.
Recirculation rate: 1 complete change every 3 min.
Average outlet temperature of air: 82°F.
Increase in humidity in house may be considered as zero.
Makeup air may be considered at 15 percent of inlet air.

Give a complete flow sheet of the process and determine the amount of refrigeration and reheating necessary to maintain the above conditions.

5.35. Your boss wants to air-condition a service building 100 ft long by 60 ft wide by 16 ft average height by cooling and dehumidifying the necessary fresh air with cold water in a spray chamber. The average occupancy of the building is 100 persons/hr with a total emission of 800 Btu/(person)(hr). The severest atmospheric conditions for the city are 100°F with 95 percent humidity. It is believed that air at 70°F and 60 percent humidity will be satisfactory, provided the total circulation is sufficient to hold the temperature rise of the air to 2°F. Neglecting radiation from the building and using the sensible heat of the recirculated air for reheating, calculate the following:
(a) The volume of recirculated air at the inlet conditions.
(b) The volume of fresh air under the worst conditions.
(c) The tons of refrigeration (1 ton = 12,000 Btu/hr) required.
(d) The volume of the dehumidifier spray chamber, assuming that the air approaches within 2°F of the water temperature.

5.36. Processes involving evaporation, drying, combustion, and gas scrubbing inherently omit water vapor that condenses in the cooler air surrounding the exhaust

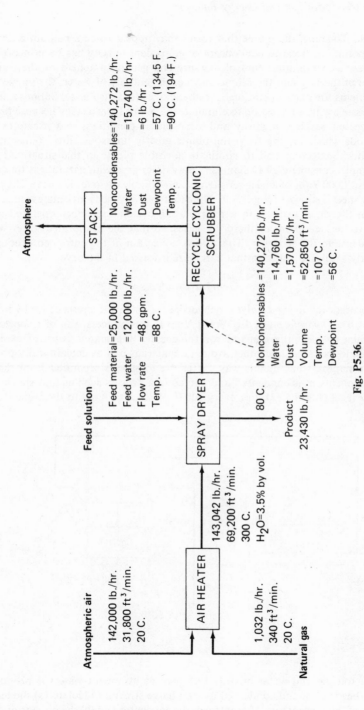

Fig. P5.36.

stack. To some, the plume that forms may appear to be a flagrant example of air pollution. Because the concept of equivalent opacity has been incorporated by law as a pollution control measure and generally upheld by the courts, in spite of the fact that the plume may consist entirely of water, the plume causes problems for plant operations. Aesthetic questions also arise. Suppose that it is necessary without question to eliminate the plume from a stack because the plant is located beside a highway and atmospheric conditions may create extended periods where the fog from the plume greatly limits visibility. What methods would you recommend to eliminate a visible plume in this situation? As an example, consider a 20-ft-diameter spray dryer fired with natural gas that evaporates 12,000 lb/hr of a heat-sensitive, water-soluble organic chemical. The process flow sheet is shown in Fig. P5.36. Note that the entrained particulates are removed from the dryer exhaust in a recirculating cyclonic scrubber operated slightly above the dew point of the dryer gases, so that the evaporated water will not condense in the scrubber. This allows collection of the entrained product as a nearly saturated solution that can be reprocessed in the dryer.

General Problems

5.37. A process involving catalytic dehydrogenation in the presence of hydrogen is known as *hydroforming*. In World War II this process was of importance in helping to meet the demand for toluene for the manufacture of explosives. Toluene, benzene, and other aromatic materials can be economically produced from naptha feeds in this way. After the toluene is separated from the other components, it is condensed and cooled in a process such as that shown in the flow sheet (Fig. P5.37). For every 100 lb of stock charged to the system, 27.5 lb

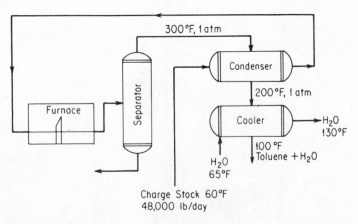

Fig. P5.37.

of a toluene and water mixture (9.1 percent by weight water) is produced as overhead vapor and condensed by the charge stream. Calculate (a) the temperature of the charge stock after it leaves the condenser and (b) the pounds of cooling water required per hour. (Continued on the next page.)

Additional data:

stream	C_p Btu/(lb)(°F)	b.p. (°F)	latent heat vap (Btu/lb)
$H_2O(l)$	1.0	212	970
$H_2O(g)$	0.5	—	—
Toluene(l)	0.4	260	100
Toluene(g)	0.3	—	—
Charge stock	0.5	—	—

5.38. As shown in Fig. P5.38, a mixture of 10 weight percent C_2H_6, 20 weight percent C_3H_8, and 70 weight percent C_4H_{10} is charged as liquid from storage to a dehydrogenation furnace at the rate of 5000 lb/hr. In the furnace the C_2H_6 goes through unchanged, 10 percent of the C_3H_8 passing through the furnace is dehydrogenated ($C_3H_8 \rightarrow C_3H_6 + H_2$), and 70 percent of the n-butane is dehydrogenated ($C_4H_{10} \rightarrow C_4H_8 + H_2$). In the separator all the H_2 and C_2H_6 are separated and withdrawn from the system. The C_3H_6, C_3H_8, C_4H_8, and n-C_4H_{10} are sent to fractionator no. 1, where the C_3H_8 is separated. Fractionator no. 2 separates the C_3H_6, C_4H_{10}, and C_4H_8. All C_4H_{10} and 20 lb/hr of C_3H_8 are recycled to the furnace. The distillation columns (fractionators) operate at a 5 to 1 reflux ratio, i.e., $R/D = 5/1$.
  (a) Determine the volume of gas in cubic feet from the separator per hour measured at 200°F and 200 psia.
  (b) What weight of C_3H_8 is being refluxed into the distillation tower per hour?
  (c) What weight of C_3H_8 has to be removed from the separator per hour?
  (d) Determine the ratio of the recycle n-C_4H_{10} to the fresh feed of C_4H_{10} from storage.
  (e) What weights of C_4H_8 and C_3H_6 are being recovered per hour?
  (f) Determine the heat duty (cooling or heating) required for the furnace, assuming no heat losses to the surroundings.
  (g) If cooling water is available at 70°F, how many gallons per hour are required for all three condensers if the maximum allowable discharge temperature of the water is 100°F?
      Note: In computing enthalpies, preferably use actual data from *p-H* charts or tables. If ideal gas values at 1 atm are used, assume that ΔH from 1 atm to 200 psia is as follows:

PRESSURE CORRECTION

	$\Delta H^*(Btu/lb)$	
	200°F	100°F
$C_2H_6(g)$	-8	-12
$C_3H_6(g)$	-14	-20
$C_3H_8(g)$	-12	-18
$C_4H_{10}(g)$	-18	-7†
$H_2(g)$	0	0

$^*\Delta H_{200\ psia} - \Delta H_{1\ atm}$.
†From 2 atm to 52 psia where C_4H_{10} is satd. vapor.
SOURCE: W. C. Edmister, *Applied Hydrocarbon Thermodynamics*, Gulf Publishing Co., Houston, 1961.

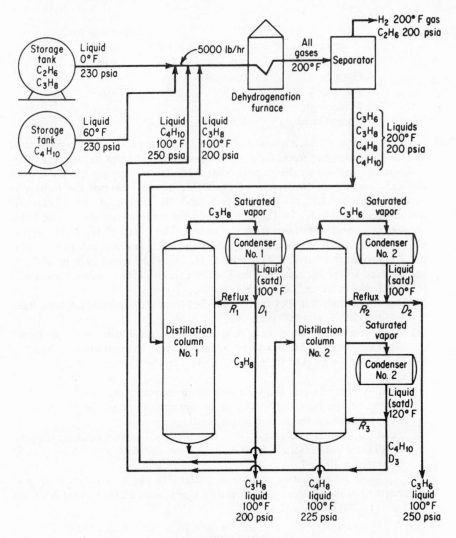

Fig. P5.38.

5.39. The steam flows for a plant are shown in Fig. P5.39. Write the material and energy balances for the system and calculate the unknown quantities in the diagram (*A* to *F*). There are two main levels of steam flow: 600 psig and 50 psig. Use the steam tables for the enthalpies.

5.40. Sulfur dioxide emission from coal-burning power plants causes serious atmospheric pollution in the eastern and midwestern portions of the United States. Unfortunately, the supply of low-sulfur coal is insufficient to meet the demand.

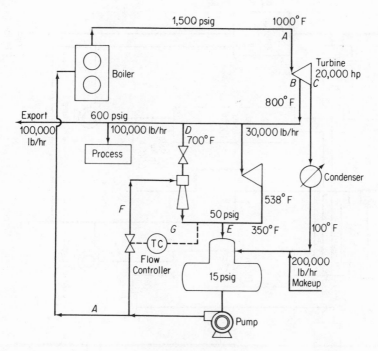

Fig. P5.39.

Processes presently under consideration to alleviate the situation include coal gasification followed by desulfurization and stack-gas cleaning. One of the more promising stack-gas-cleaning processes involves reacting SO_2 and O_2 in the stack gas with a solid metal oxide sorbent to give the metal sulfate and then thermally regenerating the sorbent and absorbing the resulting SO_3 to produce sulfuric acid. Recent laboratory experiments indicate that sorption and regeneration can be carried out with several metal oxides, but no pilot or full-scale processes have yet been put into operation.

You are asked to provide a preliminary design for a process which will remove 95 percent of the SO_2 from the stack gas of a 1000-MW power plant. Some data are given below and in the flow diagram of the process (Fig. P5.40). The sorbent consists of fine particles of a dispersion of 30 weight percent CuO in a matrix of inert porous Al_2O_3. This solid reacts in the fluidized bed sorber at 600°F. Exit solid is sent to the regenerator where SO_3 is evolved at 1300°F, converting all the $CuSO_4$ present back to CuO. The fractional conversion of CuO to $CuSO_4$ which occurs in the sorber is called α and is an important design variable. You are asked to carry out your calculations for $\alpha = 0.2, 0.5,$ and 0.8. The SO_3 produced in the regenerator is swept out by recirculating air. The SO_3 laden air is sent to the acid tower, where the SO_3 is absorbed in recirculating sulfuric acid and oleum, part of which is withdrawn as salable by-products. You will notice that the sorber, regenerator, and perhaps the acid tower are adiabatic; their temperatures are adjusted by heat exchange with incoming streams. Some of the heat exchangers (nos. 1 and 3) recover heat by countercurrent exchange between the feed and exit streams. Additional heat is provided by withdrawing

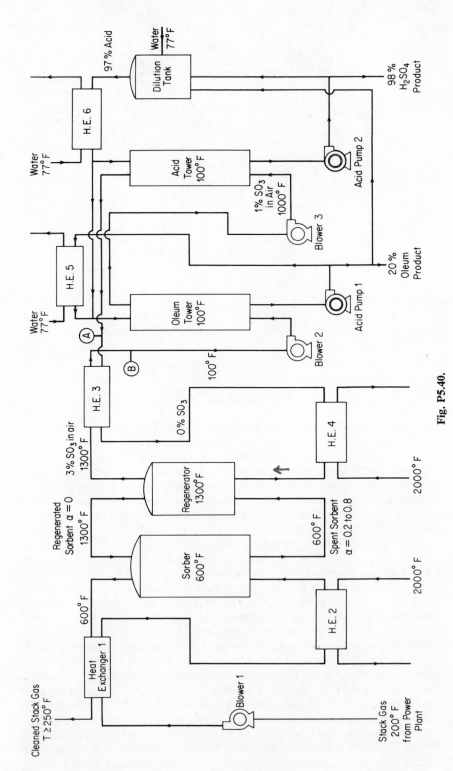

Fig. P5.40.

432

flue gas from the power plant at any desired high temperature up to 2000°F and then returning it at a lower temperature. Cooling is provided by water at 77°F. As a general rule, the temperature difference across heat-exchanger walls separating the two streams should average about 50°F. The nominal operating pressure of the whole process is 1 atm. The three blowers provide 25 in. of water additional head for the pressure losses in the equipment, and the acid pumps have a discharge pressure of 30 ft of water gauge. You are asked to write the material and energy balances and some equipment specifications as follows:

(a) Sorber, regenerator, and acid tower. Determine the flow rate, composition, and temperature of all streams entering and leaving.
(b) Heat exchangers. Determine the heat load, and flow rate, temperature, and enthalpy of all streams.
(c) Blowers. Determine the flow rate and theoretical horsepower.
(d) Acid pump. Determine the flow rate and theoretical horsepower.

Use units of Btu, °F, lb, lb mole, ft³/hr, Btu/hr, gal/hr, etc. Also, use a basis of 100 lb of coal burned for all of your calculations; then convert to the operating basis at the end of the calculations.

Supplementary data:

Power plant operation. The power plant burns 309 tons/hr of coal having the analysis given below. The coal is burned with 18 percent excess air, based on complete combustion to CO_2, H_2O, and SO_2. In the combustion only the ash and nitrogen is left unburned; all the ash has been removed from the stack gas.

Element	wt %
C	76.6
H	5.2
O	6.2
S	2.3
N	1.6
Ash	8.1

DATA ON SOLIDS (SOURCE: JANAF Thermochemical Tables)

	Al_2O_3		CuO		$CuSO_4$	
$T(°K)$	C_p	H_T-H_{298}	C_p	H_T-H_{298}	C_p	H_T-H_{298}
298	18.89	0.00	10.07	0.00	23.63	0.00
400	22.99	2.15	11.24	1.09	27.47	2.61
500	25.35	4.58	11.96	2.25	30.40	5.51
600	26.89	7.19	12.50	3.48	32.58	8.66
700	27.97	9.94	12.98	4.75	34.16	12.01
800	28.76	12.78	13.43	6.07	35.30	15.48
900	29.35	15.69	13.87	7.44	36.10	19.05
1000	29.81	18.64	14.31	8.85	36.75	22.70

Units of C_p are cal/(g mole)(°K); units of H are kcal/g mole.

5.41. The initial process in most refineries is a simple distillation in which the crude oil is separated into various fractions. The flow sheet for one such process is illustrated in Fig. P5.41. Make a complete material and energy balance around the entire distillation system and for each unit including the heat exchangers and condensers. Also,

(a) Calculate the heat load that has to be supplied by the furnace in Btu/hr.
(b) Determine the additional heat that would have to be supplied by the furnace if the charge oil were not preheated to 200°F before it entered the furnace.

Do the calculated temperatures of the streams going into storage from the heat exchangers seem reasonable?

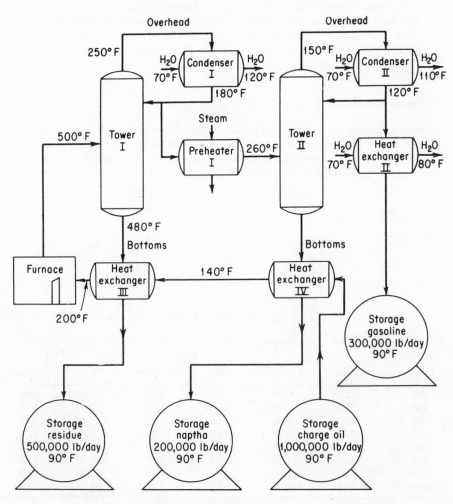

Fig. P5.41.

Additional data:

	specific heat of liquid Btu/(lb)(°F)	latent heat of vaporization Btu/lb	specific heat of vapor Btu/(lb)(°F)	conden- sation temp., °F
Charge oil	0.53	100	0.45	480
Overhead, tower I	0.59	111	0.51	250
Bottoms, tower I	0.51	92	0.42	500
Overhead, tower II	0.63	118	0.58	150
Bottoms, tower II	0.58	107	0.53	260

The reflux ratio of tower I is 3 recycle to 1 product.
The reflux ratio of tower II is 2 recycle to 1 product.

5.42. Sulfuric acid is a basic raw material used in a wide range of industries. It has been claimed that the state of civilization of a country can be determined by the amount of sulfuric acid consumed per capita. Sulfuric acid was one of the earliest known acids because it could be made by the absorption in H_2O of the SO_3 formed when naturally occurring sulfates or sulfides were roasted. Innumerable man-years of design effort have gone into developing methods of obtaining economic production of sulfuric acid from various starting materials. The source of sulfur in this problem is sulfur itself. From an overall viewpoint the preparation of sulfuric acid is quite simple. The sulfur is burned to SO_2,

$$S(s) + O_2(g) \longrightarrow SO_2(g)$$

which is then oxidized to SO_3 with additional air,

$$SO_2(g) + \tfrac{1}{2}O_2(g) \longrightarrow SO_3(g)$$

and the SO_3 formed is finally absorbed in water to yield H_2SO_4,

$$SO_3(g) + H_2O(l) \longrightarrow H_2SO_4(l)$$

The details of the sulfuric acid plant are shown in the process flow sheet (Fig. P5.42). Fifty percent excess air (based on the oxidation of sulfur to SO_2) is used.
(a) Determine the amount of heat necessary to add to the melting tank per day.
(b) Calculate the quantity of heat lost from the burner per day.
(c) Calculate the heat transferred per day from the converter by the cooling air.
(d) Calculate the amount of 98 percent H_2SO_4 recirculated through the 98 percent absorber per day.
(e) What percentage of the total amount of SO_3 produced per day is absorbed in the oleum absorber and what percentage in the 98 percent H_2SO_4 absorber?

Additional data: Heat of solution data can be assumed to be as follows:

solution	$\Delta H_{solution}$
SO_3 in 20% oleum	5 Btu/lb of 30% oleum formed
SO_3 in 98% H_2SO_4	50 Btu/lb of 10% oleum formed
20% oleum in 98% H_2SO_4	40 Btu/lb of 10% oleum formed
H_2O in 10% oleum	120 Btu/lb of 98% H_2SO_4 formed

Note: % oleum equals the wt % free SO_3 in the H_2SO_4-SO_3 mixture. All H_2O is considered combined with SO_3 to form H_2SO_4.

Use the following values for the heat capacities of the streams:

stream	C_p^*, Btu/(lb)(°F)
S(s)	0.163
S(l)	0.235
98% H$_2$SO$_4$	0.5
5% oleum	0.45
10% oleum	0.43
20% oleum	0.40
35% oleum	0.38

*C_p in Btu/(lb mole)(°F) and T in °R.

Other data can be found in the Appendix.

5.43. When coal is distilled by heating without contact with air, a wide variety of solid, liquid, and gaseous products of commercial importance is produced, as well as some significant air pollutants. The nature and amounts of the products produced depend on the temperature used in the decomposition and the type of coal. At low temperatures (400–750°C) the yield of synthetic gas is small relative to the yield of liquid products, whereas at high temperatures (above 900°C) the reverse is true.

For the typical process flow sheet, shown in Fig. P5.43,

(a) How many tons of the various products are being produced?

(b) Make an energy balance around the primary distillation tower and benzol tower.

(c) How much (in pounds) of 40 percent NaOH solution is used per day for the purification of the phenol?

(d) How much 50 percent H$_2$SO$_4$ is used per day in the pyridine purification?

(e) What weight of Na$_2$SO$_4$ is produced per day by the plant?

(f) How many cubic feet of gas per day is produced? What percent of the gas (volume) is needed for the ovens?

DATA

Products Produced Per Ton of Coal Charged		Mean C_p Liquid, cal/g	Mean C_p Vapor, cal/g	Mean C_p Solid, cal/g	Melting Point, °C	Boiling Point, °C
Synthetic gas—10,000 ft³						
(555 Btu/ft³)						
(NH₄)₂SO₄	22 lb					
Benzol	15 lb	0.50	0.30	—	—	60
Toluol	5 lb	0.53	0.35	—	—	109.6
Pyridine	3 lb	0.41	0.28	—	—	114.1
Phenol	5 lb	0.56	0.45	—	—	182.2
Naphthalene	7 lb	0.40	0.35	0.281	80.2	218
				$+0.00111T_{°F}$		
Cresols	20 lb	0.55	0.50	—	—	202
Pitch	40 lb	0.65	0.60	—	—	400
Coke	1500 lb	—	—	0.35	—	—

	ΔH_{vap} cal/g	ΔH_{fusion} cal/g
Benzol	97.5	—
Toluol	86.53	—
Pyridine	107.36	—
Phenol	90.0	—
Naphthalene	75.5	35.6
Cresols	100.6	—
Pitch	120	—

5.44. A butadiene plant flow sheet is shown in Fig. P5.44. See Shreve [R. N. Shreve, The Chemical Process Industries, McGraw-Hill, N.Y., 1945] for a discussion of the process chemistry and economics. For the plant shown,

(a) Prepare the overall material balance about the entire plant.

(b) Prepare the energy balance(s) about the entire plant.

(c) Determine the water requirements of the plant in gallons per day. Assume that the cooling water is available at 80°F and that the maximum temperature to which it can be heated without fouling is 130°F.

(d) Assuming that C_4H_8 is an ideal gas (which it is not), calculate the total length of tubes (6-in. internal diameter) in the furnace if the butene gas has to remain in the furnace 10 min.

(e) Calculate the overall yield in both sections of the plant.

DATA

60% by weight of the butane is converted to butenes in furnace 1.

30% by weight of the butane charged remains unchanged as butane.

45 Btu/lb is the energy input in compressors 1.

45% by weight of the butenes are converted to butadienes in the catalyst chamber.

55% by weight of the butenes remain unchanged in the exit stream from the catalyst chamber.

32 Btu/lb is the energy input in compressors 2.

The input to the plant is 3×10^5 ft³ (S.C.) per day.

The cooling water streams for the compressors enter at 80°F and exit at 125°F.

The reflux to recycle ratio in Towers 1 and 2 is 4 to 1.

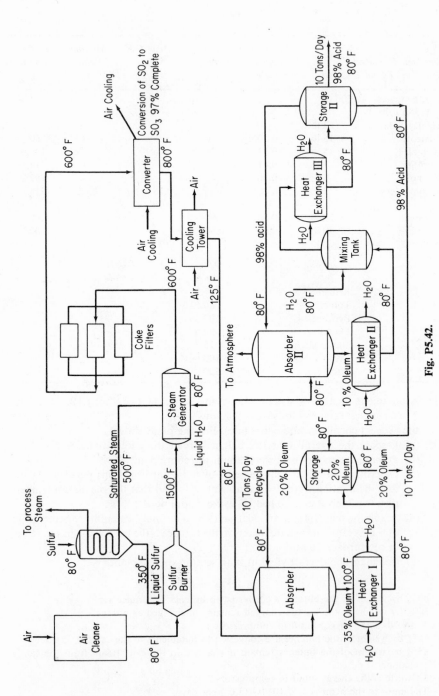

Fig. P5.42.

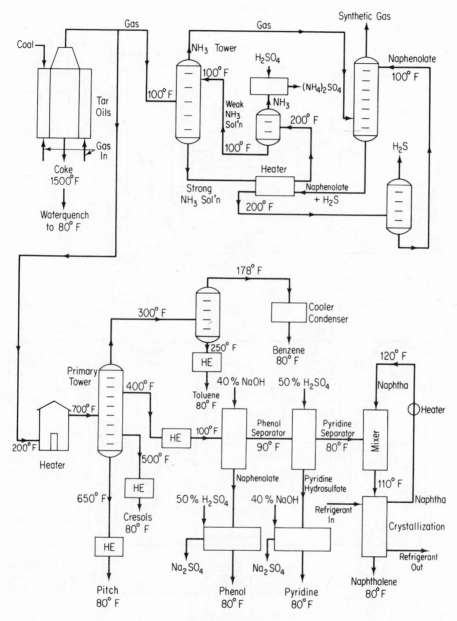

HE = Heat Exchanger

Fig. P5.43.

Butadiene Plant

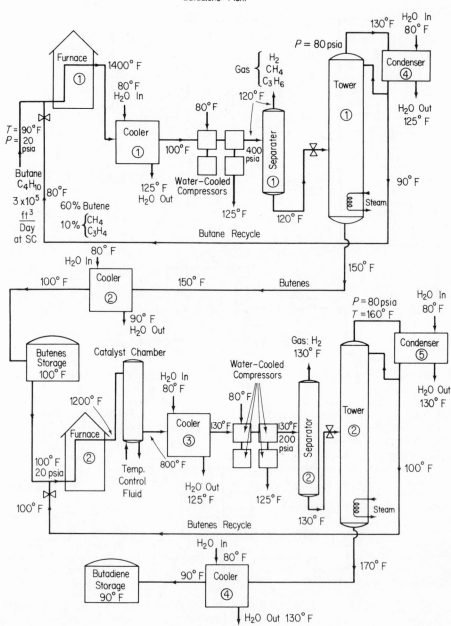

Fig. P5.44.

HEATS OF FORMATION OR COMBUSTION

Compound	$\Delta \hat{H}_f^\circ$, cal/g mole	$\Delta \hat{H}_c^\circ$, cal/g mole
Charge		−688,000
Butenes	+1,080	−649,760
Butadiene	+25,870	−608,400

ADDITIONAL DATA

Material	Composition (Gas: vol % Liq: wt %)	State	Boiling Point (at p)	C_p Liquid, Btu/ (lb)(°F)	C_p Vapor, Btu/ (lb)(°F)	ΔH_{vap}, cal/ g mole
Charge to system	C_4H_{10}: 100	Vapor 50%		—	0.48	534
Gas from separator 1	CH_4: 12.5 C_3H_6: 12.5 H_2: 75.0	Vapor	−200°F	—	0.63 (0.55?)	—
Liquid from separator 1	C_4H_{10}: 34.1 C_4H_8: 65.9	Liquid	At 100 psia, 140°F	0.55	—	5400
Liquid charge to tower 1	C_4H_{10}: 34.1 C_4H_8: 65.9	Liquid	At 100 psia, 140°F	0.54	—	5400
Overhead tower 1	C_4H_{10}: 100	Vapor	At 100 psia, 130°F	0.54	0.48	5340
Residue tower 1	C_4H_8: 100	Liquid	At 100 psia, 158°F	0.56	0.366	5510
Charge to furnace 2	C_4H_8: 100	Vapor	At 100 psia, 158°F	0.56	0.366	5510
Gas from separator 2	H_2: 100	Vapor	—	—	0.69	—
Liquid from separator 2	C_4H_8 C_4H_6	Liquid	At 100 psia, 165°F	0.58	0.345	5660
Charge to tower 2	C_4H_8 C_4H_6	Liquid	At 100 psia, 165°F	0.58	0.345	5660
Overhead tower 2	C_4H_8: 100	Vapor	At 100 psia, 160°F	0.56	0.366	5510
Residue tower 2	C_4H_6: 100	Liquid	At 100 psia, 175°F	0.61	0.335	5840
Hydrogen	H_2: 100	—	—	—	0.69	—
Methane	CH_4	—	—	—	0.52	—
Propene	C_3H_6	—	0 psia, −47.6°C	0.50	0.371	4570
Butane	C_4H_{10}	—	0 psia, 0.5°C	0.54	0.48	5340
Butenes	C_4H_8	—	0 psia, −6.24°C	0.56	0.366	5510
Butadiene	C_4H_6	—	0 psia, −4.5°C	0.61	0.335	5840

PROBLEM TO PROGRAM ON THE COMPUTER

5.1. Determine the values of the unknown quantities in Fig. CP5.1 by solving the following set of linear material and energy balances that represent the steam balance:

(1) $181.60 - x_3 - 132.57 - x_4 - x_5 = -y_1 - y_2 + y_3 + y_4 = 5.1$

(2) $1.17x_3 - x_6 = 0$

(3) $132.57 - 0.745x_7 = 61.2$

(4) $x_5 + x_7 - x_8 - x_9 - x_{10} + x_{15} = y_7 + y_8 - y_3 = 99.1$

(5) $x_8 + x_9 + x_{10} + x_{11} - x_{12} - x_{13} = -y_7 = -8.4$

(6) $x_6 - x_{15} = y_6 - y_5 = 24.2$

(7) $-1.15(181.60) + x_3 - x_6 + x_{12} + x_{16} = 1.15y_1 - y_9 + 0.4 = 19.7$

(8) $181.60 - 4.594x_{12} - 0.11x_{16} = -y_1 + 1.0235y_9 + 2.45 = 35.05$

(9) $-0.0423(181.60) + x_{11} = 0.0423y_1 = 2.88$

(10) $-0.016(181.60) + x_4 = 0$

(11) $x_8 - 0.0147x_{16} = 0$

(12) $x_5 - 0.07x_{14} = 0$

(13) $-0.0805(181.60) + x_9 = 0$

(14) $x_{12} - x_{14} + x_{16} = 0.4 - y_9 = -97.9$

There are four levels of steam: 680, 215, 170, and 37 psia. The 14 x_i, $i = 3, \ldots,$ 16, are the unknowns and the y_i are given parameters for the system. Both x_i and y_i have the units of 10^3 lb/hr.

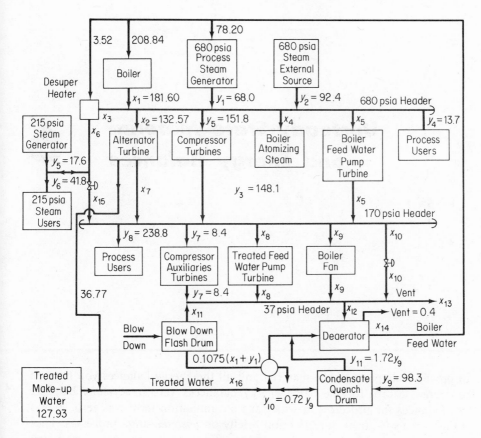

Fig. CP5.1.

Unsteady-State Material and Energy Balances

6

In previous chapters all the material and all the energy balances you have encountered, except for the batch energy balances, were *steady-state* balances, i.e., balances for processes in which the accumulation term was zero. Now it is time for us to focus our attention briefly on *unsteady-state* processes. These are processes in which quantities or operating conditions *change with time*. Sometimes you will hear the word *transient state* applied to such processes. The unsteady state is somewhat more complicated than the steady state and in general problems involving unsteady-state processes are more difficult to handle mathematically than those involving steady-state processes. However, a wide variety of important industrial problems falls into this category, such as the startup of equipment, batch heating or reactions, the change from one set of operating conditions to another, and the perturbations that develop as process conditions fluctuate. In this chapter we consider only one category of unsteady-state processes, but it is the one that is the most widely used.

The basic expression in words, Eqs. (2.1) or (4.22), for either the material or the energy balance should by now be well known to you and is repeated as Eq. (6.1) below. However, you should also realize that this equation can be applied at various levels, or strata, of description. In other words, the engineer can portray the operation of a real process by writing balances on a number of physical scales. A typical illustration of this concept might be in meterology, where the following different degrees of detail can be used on a descending scale of magnitude in the real world:

Global weather pattern
Local weather pattern
Individual clouds
Convective flow in clouds
Molecular transport
The molecules themselves

Similarly, in chemical engineering we write material and energy balances from the viewpoint of various scales of information:

(a) Molecular and atomic balances,
(b) Microscopic balances,
(c) Multiple gradient balances,
(d) Maximum gradient balances, and
(e) Macroscopic balances (overall balances)

in decreasing order of degree of detail about a process.[1] In this chapter, the type of balance to be described and applied is the simplest one, namely, the macroscopic balance [(e) in the tabulation above].

The macroscopic balance ignores all the detail within a system and consequently results in a balance about the entire system. Only time remains as an independent variable in the balance. The dependent variables, such as concentration and temperature, are not functions of position but represent overall averages throughout the entire volume of the system. In effect, the system is assumed to be sufficiently well mixed so that the output concentrations and temperatures are equivalent to the concentrations and temperatures inside the system.

To assist in the translation of Eq. (6.1)

$$
\begin{Bmatrix} \text{Accumulation} \\ \text{or depletion} \\ \text{within the} \\ \text{system} \end{Bmatrix} = \begin{Bmatrix} \text{Transport into} \\ \text{system through} \\ \text{system} \\ \text{boundary} \end{Bmatrix} - \begin{Bmatrix} \text{Transport out} \\ \text{of system} \\ \text{through system} \\ \text{boundary} \end{Bmatrix}
$$

$$
+ \begin{Bmatrix} \text{Generation} \\ \text{within} \\ \text{system} \end{Bmatrix} - \begin{Bmatrix} \text{Consumption} \\ \text{within} \\ \text{system} \end{Bmatrix} \tag{6.1}
$$

into mathematical symbols, you should refer to Fig. 6.1. Equation (6.1) can be applied to the mass of a single component or to the total amount of material or energy in the system. Let us write each of the terms in Eq. (6.1) in mathematical symbols for a very small time interval Δt. Let the accumulation be positive

---

[1]Additional information together with applications concerning these various types of balances can be found in D. M. Himmelblau and K. B. Bischoff, *Process Analysis and Simulation*, Wiley, New York, 1968.

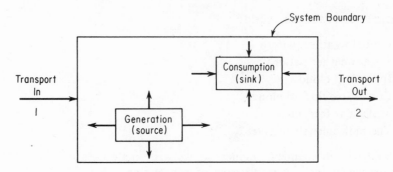

Fig. 6.1. A general unsteady-state process with transport in and out and internal generation and consumption.

in the direction in which time is positive, i.e., as time increases from t to $t + \Delta t$. Then, using the component mass balance as an example, the accumulation will be the mass of A in the system at time $t + \Delta t$ minus the mass of A in the system at time t,

$$\text{Accumulation} = \rho_A V |_{t+\Delta t} - \rho_A V |_t$$

where[2]

$$\rho_A = \text{mass of component } A \text{ per unit volume}$$
$$V = \text{volume of the system}$$

Note that the net dimensions of the accumulation term are the mass of A.

We shall split the mass transport across the system boundary into two parts, transport through defined surfaces S_1 and S_2, whose areas are known, and transport across the system boundary through other (undefined) surfaces. The net transport of A into and out of the system through defined surfaces can be written as

$$\text{Net flow across boundary via } S_1 \text{ and } S_2 = \rho_A v S \, \Delta t |_{S_1} - \rho_A v S \, \Delta t |_{S_2}$$

where

$$v = \text{fluid velocity in a duct of cross section } S$$
$$S = \text{defined cross-sectional area perpendicular to material flow}$$

Again note that the net dimensions of the transport term are the mass of A. Other types of transport across the system boundary can be represented by

$$\text{Net residual flow across boundary} = \tilde{w}_A \, \Delta t$$

where

$$\tilde{w}_A = \text{rate of mass flow of component } A \text{ through the system boundaries}$$
$$\text{other than the defined surfaces } S_1 \text{ and } S_2$$

---

[2]The symbol $|_t$ means that the quantities preceding the vertical line are evaluated at time t, or time $t + \Delta t$, or at surface S_1, or at surface S_2, as the case may be.

Finally, the net generation-consumption term will be assumed to be due to a chemical reaction r_A:

$$\text{Net generation-consumption} = \tilde{r}_A \, \Delta t$$

where $\tilde{r}_A$ = net rate of generation-consumption of component A by chemical reaction

Introduction of all these terms into Eq. (6.1) gives Eq. (6.2). Equations (6.3) and (6.4) can be developed from exactly the same type of analysis.

Component material balances:

$$\underset{\text{accumulation}}{\rho_A V|_{t+\Delta t} - \rho_A V|_t} = \underset{\substack{\text{transport through defined} \\ \text{boundaries}}}{\rho_A vS \, \Delta t|_{S_1} - \rho_A vS \, \Delta t|_{S_2}} + \underset{\substack{\text{transport} \\ \text{through} \\ \text{other} \\ \text{boundaries}}}{\tilde{w}_A \, \Delta t} + \underset{\substack{\text{genera-} \\ \text{tion or} \\ \text{consump-} \\ \text{tion}}}{\tilde{r}_A \, \Delta t}$$

$$(6.2)$$

Total material balance:

$$\underset{\text{accumulation}}{\rho V|_{t+\Delta t} - \rho V|_t} = \underset{\substack{\text{transport through defined} \\ \text{boundaries}}}{\rho vS \, \Delta t|_{S_1} + \rho vS \, \Delta t|_{S_2}} + \underset{\substack{\text{transport} \\ \text{through} \\ \text{other} \\ \text{boundaries}}}{\tilde{w} \, \Delta t} \qquad (6.3)$$

Energy balance:

$$\underset{\text{accumulation}}{E|_{t+\Delta t} - E|_t} = \underset{\text{transport through defined boundaries}}{\left|\left(\hat{H} + \frac{v^2}{2} + gh\right)\tilde{m} \, \Delta t\right|_{S_1} - \left|\left(\hat{H} + \frac{v^2}{2} + gh\right)\tilde{m} \, \Delta t\right|_{S_2}}$$

$$+ \underset{\text{heat}}{\tilde{Q} \, \Delta t} - \underset{\text{work}}{\tilde{W} \, \Delta t} + \underset{\substack{\text{transport} \\ \text{through} \\ \text{other} \\ \text{boundaries}}}{\tilde{B} \, \Delta t} \qquad (6.4)$$

where

$\tilde{B}$ = rate of energy transfer accompanying $\tilde{w}$

$\tilde{m}$ = rate of mass transfer through defined surfaces

$\tilde{Q}$ = rate of heat transfer

$\tilde{W}$ = rate of work done by the system

$\tilde{w}$ = rate of total mass flow through the system boundaries other than through the defined surfaces S_1 and S_2

ρ = total mass, per unit volume

The other notation for the material and energy balances is identical to that of Chaps. 2 and 4; note that the work, heat, generation, and mass transport are now all expressed as rate terms (mass or energy per unit time).

If each side of Eq. (6.2) is divided by Δt, we obtain

$$\frac{\rho_A V|_{t+\Delta t} - \rho_A V|_t}{\Delta t} = \rho_A vS|_{S_1} - \rho_A vS|_{S_2} + \tilde{w}_A + \tilde{r}_A \qquad (6.5)$$

Similar relations can be obtained from Eqs. (6.3) and (6.4). Next, if we take the limit of each side of Eq. (6.5) as $\Delta t \rightarrow 0$, we get

$$\frac{d(\rho_A V)}{dt} = -\Delta(\rho_A vS) + \tilde{w}_A + \tilde{r}_A \qquad (6.6)$$

Similar treatment of the total mass balance and the energy balance yields the following two equations:

$$\frac{d(\rho V)}{dt} = -\Delta(\rho v S) + \tilde{w} \tag{6.7}$$

$$\frac{d(E)}{dt} = -\Delta\left[\left(\hat{H} + \frac{v^2}{2} + gh\right)\tilde{m}\right] + \tilde{Q} - \tilde{W} + \tilde{B} \tag{6.8}$$

The relation between the energy balance given by Eq. (6.8), which has the units of energy per unit time, and the energy balance given by Eq. (4.24), which has the units of energy, should be fairly clear. Equation (4.24) represents the integration of Eq. (6.8) with respect to time, expressed formally as follows:

$$E_{t_2} - E_{t_1} = \int_{t_1}^{t_2} \{-\Delta[(\hat{H} + \hat{K} + \hat{P})\tilde{m}] + \tilde{Q} + \tilde{B} - \tilde{W}\}\, dt$$

The quantities designated in Eq. (4.24) without the tilde ($\sim$) are the respective integrated values from Eq. (6.9).

To solve one or a combination of the very general equations (6.6), (6.7), or (6.8) analytically is usually quite difficult, and in the following examples we shall have to restrict our analyses to simple cases. If we make enough (reasonable) assumptions and work with simple problems, we can consolidate or eliminate enough terms of the equations to be able to integrate them and develop some analytical answers. Numerical solutions are also possible.

In the solution of unsteady-state problems you have to apply the usual procedures of problem solving initially discussed in Chap. 1. Two major tasks exist:

(a) To set up the unsteady-state equation.
(b) To solve it once the equation is established.

After you draw a diagram of the system and set down all the information available, you should try to recognize the important variables and represent them by letters. Next, decide which variable is the independent one, and label it. The independent variable is the one you select to have one or a series of values, and then the other variables are all dependent ones—they are determined by the first variable(s) you selected. Which is chosen as an independent and which as a dependent variable is usually fixed by the problem but may be arbitrary. Although there are no general rules applicable to all cases, the quantities which appear prominently in the statement of the problem are usually the best choices. You will need as many equations as you have dependent variables.

As mentioned before, in the macroscopic balance the independent variable is time. In mathematics, when the quantity x varies with time, we consider dx to be the change in x that occurs during time dt if the process continues through the interval beginning at time t. Let us say that t is the independent variable and x is the dependent one. In solving problems you can use one or a combination of Eqs. (6.6), (6.7), and (6.8) directly, or alternatively you can proceed, as shown in some of the examples, to set up the differential equations from scratch exactly in the same fashion as Eqs. (6.2)–(6.4) were formulated. For either approach the objective is to translate the problem statement in words into one or more simultaneous differential equations having the form

$$\frac{dx}{dt} = f(x, t) \tag{6.10}$$

Then, assuming this differential equation can be solved, x can be found as a function of t. Of course, we have to know some initial condition(s) or at one (or more) given time(s) know the value(s) of x.

If dt and dx are always considered positive when increasing, then you can use Eq. (6.1) without having any difficulties with signs. However, if you for some reason transfer an output to the left-hand side of the equation, or the equation is written in some other form, then you should take great care in the use of signs (see Example 6.2 below). We are now going to examine some very simple unsteady-state problems which are susceptible to reasonably elementary mathematical analysis. You can (and will) find more complicated examples in texts dealing with all phases of mass transfer, heat transfer, and fluid dynamics.

EXAMPLE 6.1   Unsteady-state material balance without generation

A tank holds 100 gal of a water-salt solution in which 4.0 lb of salt are dissolved. Water runs into the tank at the rate of 5 gal/min and salt solution overflows at the same rate. If the mixing in the tank is adequate to keep the concentration of salt in the tank uniform at all times, how much salt is in the tank at the end of 50 min? Assume that the density of the salt solution is essentially the same as that of water.

Solution:
We shall set up the differential equations which describe the process from scratch.
Step 1: Draw a picture, and put down the known data. See Fig. E6.1.

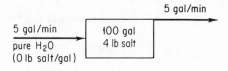

Fig. E6.1.

Step 2: Choose the independent and dependent variables. Time, of course, is the independent variable, and either the salt quantity or concentration in the tank can be the dependent variable. Suppose that we make the mass (quantity) of salt the dependent variable. Let $x =$ lb of salt in the tank at time t.

Step 3: Write the known value of x at a given value of t. This is the initial condition:

$$\text{at } t = 0, \quad x = 4.0 \text{ lb}$$

Step 4: It is easiest to make a total material balance and a component material balance on the salt. (No energy balance is needed because the system can be assumed to be isothermal.)

Total balance:

$$\text{accumulation} \qquad\qquad = \qquad\qquad \text{in}$$

$$[m_{\text{tot}}\ \text{lb}]_{t+\Delta t} - [m_{\text{tot}}\ \text{lb}]_t = \frac{5\ \text{gal}}{\text{min}}\left|\frac{1\ \text{ft}^3}{7.48\ \text{gal}}\right|\frac{\rho_{\text{H}_2\text{O}}\ \text{lb}}{\text{ft}^3}\left|\Delta t\ \text{min}\right.$$

$$-\ \text{out}$$

$$-\ \frac{5\ \text{gal}}{\text{min}}\left|\frac{1\ \text{ft}^3}{7.48\ \text{gal}}\right|\frac{\rho_{\text{soln}}\ \text{lb}}{\text{ft}^3}\left|\Delta\ t\ \text{min}\right. = 0$$

This equation tells us that the flow of water into the tank equals the flow of solution out of the tank if $\rho_{\text{H}_2\text{O}} = \rho_{\text{soln}}$ as assumed. Otherwise there is an accumulation term.

Salt balance:

$$\text{accumulation} \qquad = \text{in} \qquad - \qquad \text{out}$$

$$[x\ \text{lb}]_{t+\Delta t} - [x\ \text{lb}]_t = 0 - \frac{5\ \text{gal}}{\text{min}}\left|\frac{x\ \text{lb}}{100\ \text{gal}}\right|\Delta t\ \text{min}$$

Dividing by Δt and taking the limit as Δt approaches zero,

$$\lim_{\Delta t\to 0}\frac{[x]_{t+\Delta t} - [x]_t}{\Delta t} = -0.05x$$

or

$$\frac{dx}{dt} = -0.05x \qquad\qquad\qquad (a)$$

Notice how we have kept track of the units in the normal fashion in setting up these equations. Because of our assumption of uniform concentration of salt in the tank, the concentration of salt leaving the tank is the same as that in the tank, or x lb/100 gal of solution.

Step 5: Solve the unsteady-state material balance on the salt. By separating the independent and dependent variables, we get

$$\frac{dx}{x} = -0.05\ dt$$

This equation is easily integrated between the definite limits of

$$t = 0, \qquad x = 4.0$$

$$t = 50, \qquad x = \text{the unknown value } X\ \text{lb}$$

$$\int_{4.0}^{X}\frac{dx}{x} = -0.05\int_{0}^{50}dt$$

$$\ln\frac{X}{4.0} = -2.5, \qquad \ln\frac{4.0}{X} = 2.5$$

$$\frac{4.0}{X} = 12.2, \qquad X = \frac{4.0}{12.2} = 0.328\ \text{lb salt}$$

An equivalent differential equation to Eq. (a) can be obtained directly from the component mass balance in the form of Eq. (6.6) if we let $\rho_A = $ concentration of salt in the tank at any time t in terms of lb/gal:

$$\frac{d(\rho_A V)}{dt} = -\left(\frac{5\ \text{gal}}{\text{min}}\left|\frac{\rho_A\ \text{lb}}{\text{gal}}\right. - 0\right)$$

If the tank holds 100 gal of solution at all times, V is a constant and equal to 100, so that

$$\frac{d\rho_A}{dt} = -\frac{5\rho_A}{100} \qquad\qquad\qquad (b)$$

The initial conditions are

$$\text{at } t = 0, \quad p_A = 0.04$$

The solution of Eq. (b) would be carried out exactly as the solution of Eq. (a).

EXAMPLE 6.2   Unsteady-state material balance without generation

A square tank 4 ft on a side and 10 ft high is filled to the brim with water. Find the time required for it to empty through a hole in the bottom 1 in.² in area.

Solution:
Step 1: Draw a diagram of the process, and put down the data. See Fig. E6.2(a).

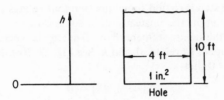

Fig. E6.2(a).

Step 2: Select the independent and dependent variables. Again, time will be the independent variable. We could select the quantity of water in the tank as the dependent variable, but since the cross section of the tank is constant, let us choose h, the height of the water in the tank, as the dependent variable.

Step 3: Write the known value of h at a given value of t:

$$\text{at } t = 0, \quad h = 10 \text{ ft}$$

Step 4: Develop the unsteady-state balance(s) for the process. In an elemental time Δt, the height of the water in the tank drops Δh. The mass of water leaving the tank is in the form of a cylinder 1 in.² in area, and we can calculate the quantity as

$$\frac{1 \text{ in.}^2 \ \left| \ \frac{1 \text{ ft}^2}{144 \text{ in.}^2} \ \right| \ \frac{v^* \text{ ft}}{\text{sec}} \ \left| \ \frac{p \text{ lb}}{\text{ft}^3} \ \right| \ \Delta t \text{ sec}}{} = p\frac{v^* \Delta t}{144} \text{lb}$$

where

$$p = \text{the density of water}$$
$$v^* = \text{the average velocity of the water leaving the tank}$$

The depletion of water inside the tank in terms of the variable h, expressed in lb, is

$$\frac{16 \text{ ft}^2 \ \left| \ \frac{p \text{ lb}}{\text{ft}^3} \ \right| \ h \text{ ft}}{}\bigg|_{t+\Delta t} - \frac{16 \text{ ft}^2 \ \left| \ \frac{p \text{ lb}}{\text{ft}^3} \ \right| \ h \text{ ft}}{}\bigg|_{t} = 16p\,\Delta h$$

An overall material balance indicates that

$$\textbf{accumulation} = \textbf{in} - \textbf{out}$$
$$16p\,\Delta h = 0 - \frac{pv^* \Delta t}{144} \qquad (a)$$

Although Δh is a negative value, our equation takes account of this automatically. The accumulation is positive if the right-hand side is positive, and the accumulation is negative if the right-hand side is negative. Here the accumulation is really a depletion. You can see that the term ρ, the density of water, cancels out, and we could just as well have made our material balance on a volume of water.

Equation (a) becomes

$$\frac{\Delta h}{\Delta t} = -\frac{v^*}{(16)(144)}$$

Taking the limit as Δh and Δt approach zero, we get

$$\frac{dh}{dt} = -\frac{v^*}{(16)(144)} \tag{b}$$

Unfortunately, this is an equation with one independent variable, t, and two dependent variables, h and v^*. We must find another equation to eliminate either h or v^* if we want to obtain a solution. Since we want our final equation to be expressed in terms of h, the next step is to find some function that relates v^* to h and t, and then we can substitute for v^* in the unsteady-state equation.

We shall employ the steady-state mechanical energy balance for an incompressible fluid, discussed in Chap. 4, to relate v^* and h. See Fig. E6.2(b). Recall from Eq. (4.30) with $W = 0$ and $E_v = 0$ that

$$\Delta\left(\frac{v^2}{2} + gh\right) = 0 \tag{c}$$

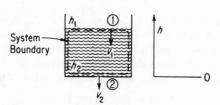

Fig. E6.2(b).

We assume that the pressures are the same at sections ① and ② of the system consisting of the water in the tank. Equation (c) reduces to

$$\frac{v_2^2 - v_1^2}{2} + g(h_2 - h_1) = 0 \tag{d}$$

where

v_2 = the exit velocity through the 1-in.² hole at boundary ②
v_1 = the velocity of the water in the tank at boundary ①

If $v_1 \cong 0$, a reasonable assumption for the water in the large tank at any time, at least compared to v_2, and the reference plane is located at the 1-in.² hole,

$$v_2^2 = -2g(0 - h_1) = 2gh$$
$$v_2 = \sqrt{2gh} \tag{e}$$

Because the exit-stream flow is not frictionless and because of turbulence and orifice effects in the exit hole, we must correct the value of v given by Eq. (e) for frictionless flow by an empirical adjustment factor as follows:

$$v_2 = c\sqrt{2gh} = v^* \tag{f}$$

where c is an orifice correction which we could find (from a text discussing fluid dynamics) has a value of 0.62 for this case. In the American engineering system ($g = 32.2$ ft/sec²), $v^* = 0.62\sqrt{2(32.2)h} = 4.97\sqrt{h}$ ft/sec. Let us substitute this relation into Eq. (b) in place of v^*. Then we obtain

$$\frac{dh}{dt} = -\frac{4.97(h)^{1/2}}{(16)(144)}$$

or

$$-464\frac{dh}{h^{1/2}} = dt \tag{g}$$

Equation (g) can be integrated between

$$h = 10 \text{ ft} \qquad \text{at } t = 0$$

and

$$h = 0 \text{ ft} \qquad \text{at } t = \theta, \qquad \text{the unknown time}$$

$$-464\int_{10}^{0} \frac{dh}{h^{1/2}} = \int_{0}^{\theta} dt$$

to yield θ,

$$\theta = 464\int_{0}^{10} \frac{dh}{h^{1/2}} = 464\left[2\sqrt{h}\right]_{0}^{10} = 2940 \text{ sec}$$

EXAMPLE 6.3   Material balance in batch distillation

A small still is separating propane and butane at 275°F and initially contains 10 lb moles of a mixture whose composition is $x = 0.30$ ($x =$ mole fraction butane). Additional mixture ($x_F = 0.30$) is fed at the rate of 5 lb moles/hr. If the total volume of the liquid in the still is constant, and the concentration of the vapor from the still (x_D) is related to x_S as follows:

$$x_D = \frac{x_S}{1 + x_S}$$

how long will it take for the value of x_S to change from 0.30 to 0.40? What is the steady-state ("equilibrium") value of x_S in the still (i.e., when x_S becomes constant)? See Fig. E6.3.

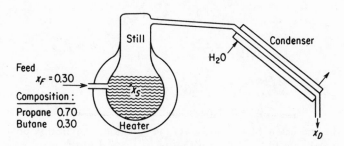

Fig. E6.3.

Solution:

Since butane and propane form ideal solutions, we do not have to worry about volume changes on mixing or separation. Only the material balance is needed to answer the questions posed. If t is the independent variable and x_S the dependent variable, we can say that

Butane balance (C_4):

The input to the still is

$$\frac{5 \text{ moles feed}}{\text{hr}} \bigg| \frac{0.30 \text{ mole } C_4}{\text{mole feed}} \bigg| \Delta t \text{ hr}$$

The output from the still is equal to the amount condensed,

$$\frac{5 \text{ moles cond.}}{\text{hr}} \bigg| \frac{x_D \text{ moles } C_4}{\text{mole cond.}} \bigg| \Delta t \text{ hr}$$

The accumulation is

$$\frac{10 \text{ moles in still}}{} \bigg| \frac{x_S \text{ moles } C_4}{\text{mole in still}} \bigg|_{t+\Delta t} - \frac{10 \text{ moles in still}}{} \bigg| \frac{x_S \text{ moles } CH_4}{\text{mole in still}} \bigg|_{t} = 10\Delta x_S$$

Our unsteady-state material balance is then

$$\textbf{accumulation} = \textbf{in} - \textbf{out}$$
$$10 \, \Delta x_S = 1.5 \, \Delta t - 5x_D \, \Delta t$$

or, dividing by Δt and taking the limit as Δt approaches zero,

$$\frac{dx_S}{dt} = 0.15 - 0.5x_D$$

As in the previous example, it is necessary to reduce the equation to two variables by substituting for x_D:

$$x_D = \frac{x_S}{1 + x_S}$$

Then

$$\frac{dx_S}{dt} = 0.15 - \frac{x_S}{1 + x_S}(0.5)$$

$$\frac{dx_S}{0.15 - \frac{0.5x_S}{1 + x_S}} = dt$$

The integration limits are

$$\text{at } t = 0, \quad x_S = 0.30$$
$$t = \theta, \quad x_S = 0.40$$

$$\int_{0.30}^{0.40} \frac{dx_S}{0.15 - [0.5x_S/(1 + x_S)]} = \int_0^\theta dt = \theta$$

$$\int_{0.30}^{0.40} \frac{(1 + x_S)\, dx_S}{0.15 - 0.35x_S} = \theta = \left[-\frac{x_S}{0.35} - \frac{1}{(0.35)^2} \ln (0.15 - 0.35x_S) \right]_{0.30}^{0.40}$$

$$\theta = 5.85 \text{ hr}$$

If you did not know how to integrate the equation analytically, or if you only had experimental data for x_D as a function of x_S instead of the given equation, you could always integrate the equation graphically as shown in Example 6.6 or integrate numerically on a digital computer.

The steady-state value of x_S is established at infinite time or, alternatively, when the accumulation is zero. At that time,

$$0.15 = \frac{0.5x_S}{1 + x_S} \quad \text{or} \quad x_S = 0.428$$

The value of x_S could never be greater than 0.428 for the given conditions.

EXAMPLE 6.4   Unsteady-state chemical reaction

A compound dissolves in water at a rate proportional to the product of the amount undissolved and the difference between the concentration in a saturated solution and the concentration in the actual solution at any time. A saturated solution of compound contains 40 g/100 g H_2O. In a test run starting with 20 lb of undissolved compound in 100 lb of pure water it is found that 5 lb are dissolved in 3 hr. If the test continues, how many pounds of compound will remain undissolved after 7 hr? Assume that the system is isothermal.

Solution:
Step 1: See Fig. E6.4.

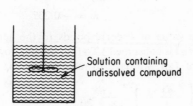

Fig. E6.4.

Step 2: Let the dependent variable $x =$ the pounds of *undissolved* compound at time t (the independent variable).

Step 3: At any time t the concentration of dissolved compound is

$$\frac{(20 - x) \text{ lb compound}}{100 \quad \text{lb } H_2O}$$

The rate of dissolution of compound according to the problem statement is

$$\text{rate } \frac{\text{lb}}{\text{hr}} = k(x \text{ lb})\left(\frac{40 \text{ lb compound}}{100 \text{ lb } H_2O} - \frac{(20 - x) \text{ lb compound}}{100 \text{ lb } H_2O}\right)$$

where k is the proportionality constant. We could have utilized a concentration measured on a volume basis rather than one measured on a weight basis, but this would just change the value and units of the constant k.

Next we make an unsteady-state balance for the compound in the solution (the system):

$$\text{accumulation} = \text{in} - \text{out} + \text{generation}$$

$$\frac{dx}{dt} = 0 - 0 + kx\left(\frac{40}{100} - \frac{20 - x}{100}\right)$$

or

$$\frac{dx}{dt} = \frac{kx}{100}(20 + x)$$

Step 4: Solution of the unsteady-state equation:

$$\frac{dx}{x(20 + x)} = \frac{k}{100} \, dt$$

This equation can be split into two parts and integrated,

$$\frac{dx}{x} - \frac{dx}{x + 20} = \frac{k}{5} \, dt$$

or integrated directly for the conditions

$$t_0 = 0, \qquad x_0 = 20$$
$$t_1 = 3, \qquad x_1 = 15$$
$$t_2 = 7, \qquad x_3 = \, ?$$

Since k is an unknown, one extra pair of conditions is needed in order to evaluate k.

To find k, we integrate between the limits stated in the test run:

$$\int_{20}^{15} \frac{dx}{x} - \int_{20}^{15} \frac{dx}{x + 20} = \frac{k}{5} \int_0^3 dt$$

$$\left[\ln \frac{x}{x + 20} \right]_{20}^{15} = \frac{k}{5}(3 - 0)$$

$$k = -0.257$$

(Note that the negative value of k corresponds to the actual physical situation in which undissolved material is consumed.) To find the unknown amount of undissolved compound at the end of 7 hr, we have to integrate again:

$$\int_{20}^{x_2} \frac{dx}{x} - \int_{20}^{x_2} \frac{dx}{x + 20} = \frac{-0.257}{5} \int_0^7 dt$$

$$\left[\ln \frac{x}{x + 20} \right]_{20}^{x_2} = \frac{-0.0257}{5}(7 - 0)$$

$$x_2 = 10.7 \text{ lb}$$

EXAMPLE 6.5   Overall unsteady-state process

A 15 percent Na_2SO_4 solution is fed at the rate of 12 lb/min into a mixer which initially holds 100 lb of a 50-50 mixture of Na_2SO_4 and water. The exit solution leaves at the rate of 10 lb/min. Assuming uniform mixing, what is the concentration of Na_2SO_4 in the mixer at the end of 10 min? Ignore any volume changes on mixing.

Solution:

In contrast to Example 6.1, both the water and the salt concentrations, and the total material in the vessel, change with time in this problem.

Step 1: See Fig. E6.5.

Step 2:

Let x = fraction Na_2SO_4 in the tank at time t in $\left(\dfrac{\text{lb } Na_2SO_4}{\text{lb total}} \right)$

y = total lb of material in the tank at time t

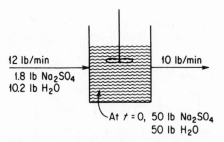

Fig. E6.5.

Step 3: At time $t = 0$, $x = 0.50$, $y = 100$.

Step 4: Material balances between time t and $t + \Delta t$:

(a) *Total balance:*

$$\underset{\text{accumulation}}{[y]_{t+\Delta t} - [y]_t = \Delta y} = \underset{\text{in}}{\frac{12 \text{ lb}}{\text{min}} \bigg| \frac{\Delta t \text{ min}}{}} - \underset{\text{out}}{\frac{10 \text{ lb}}{\text{min}} \bigg| \frac{\Delta t \text{ min}}{}}$$

or

$$\frac{dy}{dt} = 2 \qquad \text{(a)}$$

(b) Na_2SO_4 *balance:*

$$\underset{\text{accumulation}}{\left[\frac{x \text{ lb } Na_2SO_4}{\text{lb total}} \bigg| y \text{ lb total}\right]_{t+\Delta t} - \left[\frac{x \text{ lb } Na_2SO_4}{\text{lb total}} \bigg| y \text{ lb total}\right]_t}$$

$$= \underset{\text{in}}{\frac{1.8 \text{ lb } Na_2SO_4}{\text{min}} \bigg| \Delta t \text{ min}} - \underset{\text{out}}{\frac{10 \text{ lb total}}{\text{min}} \bigg| \frac{x \text{ lb } Na_2SO_4}{\text{lb total}} \bigg| \Delta t \text{ min}}$$

or

$$\frac{d(xy)}{dt} = 1.8 - 10x \qquad \text{(b)}$$

Step 5: There are a number of ways to handle the solution of these two differential equations with two unknowns. Perhaps the easiest is to differentiate the xy product in Eq. (b) and substitute Eq. (a) into the appropriate term containing dy/dt:

$$\frac{d(xy)}{dt} = x\frac{dy}{dt} + y\frac{dx}{dt}$$

To get y, we have to integrate Eq. (a):

$$\int_{100}^{y} dy = \int_{0}^{t} 2 \, dt$$

$$y - 100 = 2t$$

$$y = 100 + 2t$$

Then

$$\frac{d(xy)}{dt} = x(2) + (100 + 2t)\frac{dx}{dt} = 1.8 - 10x$$

$$\frac{dx}{dt}(100 + 2t) = 1.8 - 12x$$

$$\int_{0.50}^{x} \frac{dx}{1.8 - 12x} = \int_{0}^{10} \frac{dt}{100 + 2t}$$

$$-\frac{1}{12}\ln\frac{1.8 - 12x}{1.8 - 6.0} = \frac{1}{2}\ln\frac{100 + 20}{100}$$

$$x = 0.267\frac{\text{lb Na}_2\text{SO}_4}{\text{lb total solution}}$$

An alternative method of setting up the dependent variable would have been to let $x = $ lb Na_2SO_4 in the tank at the time t. Then the Na_2SO_4 balance would be

$$\frac{dx}{dt}\frac{\text{lb Na}_2\text{SO}_4}{\text{min}} = \frac{1.8 \text{ lb Na}_2\text{SO}_4}{\text{min}} - \frac{10 \text{ lb total}}{\text{min}}\left|\frac{x \text{ lb Na}_2\text{SO}_4}{y \text{ lb total}}\right.$$

$$\frac{dx}{dt} = 1.8 - \frac{10x}{100 + 2t}$$

or

$$\frac{dx}{dt} + \frac{10}{100 + 2t}x = 1.8$$

The solution to this linear differential equation gives the pounds of Na_2SO_4 after 10 min; the concentration then can be calculated as x/y. To integrate the linear differential equation, we introduce the integrating factor $e^{\int(dt/10+0.2t)}$:

$$xe^{\int(dt/10+0.2t)} = \int (e^{\int(dt/10+0.2t)})(1.8)\,dt + C$$

$$x(10 + 0.2t)^5 = 1.8\int (10 + 0.2t)^5\,dt + C$$

at $x = 50$, $t = 0$, and $C = 35 + 10^5$;

$$x = 1.5(10 + 0.2t) + \frac{35 \times 10^5}{(10 + 0.2t)^5}$$

at $t = 10$ min;

$$x = 1.5(12) + \frac{35 \times 10^5}{(12)^5} = 32.1$$

The concentration is

$$\frac{x}{y} = \frac{32.1}{100 + 20} = 0.267\frac{\text{lb Na}_2\text{SO}_4}{\text{lb total solution}}$$

EXAMPLE 6.6   Unsteady-state energy balance with graphical integration

Five thousand pounds of oil initially at 60°F are being heated in a stirred (perfectly mixed) tank by saturated steam which is condensing in the steam coils at 40 psia. If the rate of heat transfer is given by Newton's heating law, i.e.,

$$\tilde{Q} = \frac{dQ}{dt} = h(T_{\text{steam}} - T_{\text{oil}})$$

where Q is the heat transferred in Btu and h is the heat transfer coefficient in the proper units, how long does it take for the discharge from the tank to rise from 60 to 90°F? What is the maximum temperature that can be achieved in the tank?

Additional data:

$$\text{entering oil flow rate} = 1018 \text{ lb/hr at a temperature of } 60°F$$

$$\text{discharge oil flow rate} = 1018 \text{ lb/hr at a temperature of } T$$

$$h = 291 \text{ Btu/(hr)(°F)}$$

$$C_{p_{oil}} = 0.5 \text{ Btu/(lb)(°F)}$$

Solution:

The process is shown in Fig. E6.6(a). The independent variable will be t, the time; the dependent variable will be the temperature of the oil in the tank, which is

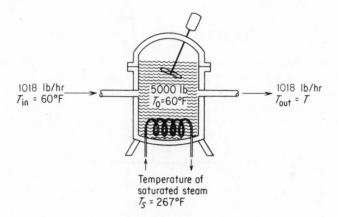

Fig. E6.6(a).

the same as the temperature of the oil discharged. The material balance is not needed because the process, insofar as the quantity of oil is concerned, is assumed to be in the steady state.

Our next step is to set up the unsteady-state energy balance for the interval Δt. Let T_S = the steam temperature and T = the oil temperature:

$$\text{accumulation} = \text{input} - \text{output}$$

A good choice for a reference temperature for the enthalpies is 60°F because this makes the input enthalpy zero.

enthalpy of the input stream	$\dfrac{1018 \text{ lb}}{\text{hr}} \Bigg\| \dfrac{0.5 \text{ Btu}}{\text{(lb)(°F)}} \Bigg\| \dfrac{\overbrace{(T_{in} - T_{ref})}^{60°F - 60°F}}{} \Bigg\| \dfrac{\Delta t \text{ hr}}{} = 0$	
transfer heat	$h(T_S - T) = \dfrac{291 \text{ Btu}}{\text{(hr)(°F)}} \Bigg\| \dfrac{(267 - T)°F}{} \Bigg\| \dfrac{\Delta t \text{ hr}}{}$	**input (Btu)**
enthalpy of the output stream	$\dfrac{1018 \text{ lb}}{\text{hr}} \Bigg\| \dfrac{0.5 \text{ Btu}}{\text{(lb)(°F)}} \Bigg\| \dfrac{(T - 60)°F}{} \Bigg\| \dfrac{\Delta t \text{ hr}}{}$	**output (Btu)**
enthalpy change inside tank	$\left\{ \begin{array}{l} \left[\dfrac{5000 \text{ lb}}{} \Bigg\| \dfrac{0.5 \text{ Btu}}{\text{(lb)(°F)}} \Bigg\| \dfrac{(T - 60)°F}{} \right]_{t+\Delta t} \\[2ex] - \left[\dfrac{5000 \text{ lb}}{} \Bigg\| \dfrac{0.5 \text{ Btu}}{\text{(lb)(°F)}} \Bigg\| \dfrac{(T - 60)°F}{} \right]_{t} \end{array} \right.$	**accumulation (Btu)**

The rate of change of energy inside the tank is nothing more than the rate of change of internal energy, which, in turn, is essentially the same as the rate of change of enthalpy, i.e., $dE/dt = dU/dt = dH/dt$, because $d(pV)/dt \cong 0$.

We use Eq. (6.8) with $\tilde{B} = 0$. We assume that all the energy introduced into the motor enters the tank as $\tilde{W}$:

$$\tilde{W} = \frac{3 \text{ hp}}{4} \left| \frac{0.7608 \text{ Btu}}{(\text{sec})(\text{hp})} \right| \frac{3600 \text{ sec}}{1 \text{ hr}} = \frac{1910 \text{ Btu}}{\text{hr}}$$

$$2500\frac{dT}{dt} = 291(267 - T) - 509(T - 60) + 1910$$

$$\frac{dT}{dt} = 44.1 - 0.32T$$

$$\int_{60}^{90} \frac{dT}{44.1 - 0.32T} = \int_{0}^{\theta} dt = \theta$$

An analytical solution to this equation gives $\theta = 1.52$ hr. For illustrative purposes, so you can see how equations too complex to integrate directly can be handled, let us graphically integrate the left-hand side of the equation. We set up a table and choose values of T. We want to calculate $1/(44.1 - 0.32T)$ and plot this quantity vs. T. The area under the curve from $T = 60°$ to $T = 90°$F should be 1.52 hr.

T	$0.32T$	$44.1 - 0.32T$	$\dfrac{1}{44.1 - 0.32T} = \phi$
60	19.2	24.9	0.0402
70	22.4	21.7	0.0462
80	25.6	18.5	0.0540
90	28.8	15.3	0.0655

These data are plotted in Fig. E6.6(b).

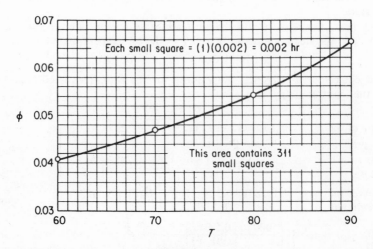

Fig. E6.6(b).

The area below $\phi = 0.030$ is

$$(90 - 60)(0.030 - 0) = 0.90 \text{ hr}$$

The area above $\phi = 0.030$ but under the curve is

$$(311)(0.002) = 0.622 \text{ hr}$$

The total area $= 0.90 + 0.622 = 1.52$ hr.

WHAT YOU SHOULD HAVE LEARNED FROM THIS CHAPTER

This chapter was intended to acquaint you with the basic concepts underlying unsteady-state material and energy balances and to enable you to set up the necessary differential equations for very simple physical systems.

SUPPLEMENTARY REFERENCES

1. Himmelblau, D. M., and K. B. Bischoff, *Process Analysis and Simulation*, Wiley, New York, 1968.
2. Jenson, V. G., and G. V. Jeffreys, *Mathematical Methods in Chemical Engineering*, Academic Press, New York, 1963.
3. Marshall, W. R., and R. L. Pigford, *The Application of Differential Equations to Chemical Engineering Problems*, University of Delaware, Newark, 1942.
4. Sherwood, T. K., and C. E. Reid, *Applied Mathematics in Chemical Engineering*, 2nd ed., McGraw-Hill, New York, 1957.

PROBLEMS

6.1. A defective tank of 1500 ft³ volume containing 100 percent propane gas (at 1 atm) is to be flushed with air (at 1 atm) until the propane concentration is reduced to less than 1 percent. At that concentration of propane the defect can be repaired by welding. If the flow rate of air into the tank is 30 ft³/min, for how many minutes must the tank be flushed out? Assume that the flushing operation is conducted so that the gas in the tank is well mixed.

6.2. The catalyst in a fluidized bed reactor of 200-ft³ volume is to be regenerated by contact with a hydrogen stream. Before the hydrogen can be introduced into the reactor, the O_2 content of the air in the reactor must be reduced to 0.1 percent. If pure N_2 can be fed into the reactor at the rate of 20 ft³/min, for how long should the reactor by purged with N_2? Assume that the catalyst solids occupy 6 percent of the reactor volume and that the gases are well mixed.

6.3. A plant at Canso, Nova Scotia, makes fish-protein concentrate (FPC). It takes 6.6 lb of whole fish to make 1 lb of FPC, and therein is the problem—to make money, the plant must operate most of the year. One of the operating problems is the drying of the FPC. It dries in the fluidized dryer at a rate approximately proportional to its moisture content. If a given batch of FPC loses one-half of its initial moisture in the first 15 min, how long will it take to remove 90 percent of the water in the batch of FPC?

6.4. The growth of yeast and bacteria on hydrocarbons is a well-known phenomenon and is the key to making wine, baker's yeast, and, more recently, fodder yeast. These processes are nearly all carried out in batch or at best in a multiple batch system giving a continuous output. Commercial processes for the production of single-cell proteins have evolved, one based on gas oil as a feedstock and another that feeds high-purity n-paraffins in the C_{10} to C_{20} range. If in the culture of protein the amount of active material doubles in 3 hr, how much growth may be expected at the end of 12 hr at the same rate of growth?

6.5. A mixer containing 100 gal of Na_2SO_4 solution (4 lb of Na_2SO_4/gal) is attached to a reservoir containing a solution of 1 lb of Na_2SO_4/gal. If the solution is pumped from the reservoir at the rate of 5 gal/min and the mixed solution runs out of the mixer at the rate of 5 gal/min, how long will it take to reduce the concentration of salt in the tank to one-half its original value? To one-fourth its original value?

6.6. A sewage disposal plant has a big concrete holding tank of 100,000 gal capacity. It is three-fourths full of liquid to start with and contains 60,000 lb of organic material in suspension. Water runs into the holding tank at the rate of 20,000 gal/hr and the solution leaves at the rate of 15,000 gal/hr. How much organic material is in the tank at the end of 3 hr?

6.7. Suppose that in Problem 6.6 the bottom of the tank is covered with sludge (precipitated organic material) and that the stirring of the tank causes the sludge to go into suspension at a rate proportional to the difference between the concentration of sludge in the tank at any time and 10 lb sludge/gal. If no organic material were present, the sludge would go into suspension at the rate of 0.05 lb/(min) (gal solution) when 75,000 gal of solution are in the tank. How much organic material is in the tank at the end of 3 hr?

6.8. If the average person takes 18 breaths per minute and on each breath exhales 100 in.^{3} containing 4 percent CO_2, find the percent CO_2 in the air of a bomb shelter $\frac{1}{2}$ hr after a group of 50 enters. Assume that the air in the shelter was fresh to start with and that the ventilation system introduces 100 ft^3 of fresh air (CO_2 content is 0.04 percent by volume) per minute. The shelter size is 10^4 ft^3.

6.9. A tank is filled with water. At a given instant two orifices in the side of the tank are opened to discharge the water. The water at the start is 10 ft deep and one orifice is 6 ft below the top while the other one is 8 ft below the top. The coefficient of discharge of each orifice is known to be 0.61. The tank is a vertical right circular cylinder 6 ft in diameter. The upper and lower orifices are 2 and 4 in. in diameter, respectively. How long will be required for the tank to be drained so that the water level is at a depth of 5 ft?

6.10. Two tanks are connected in series. Each contains 100 gal of liquid. The liquid in the first is initially saturated with a salt, the concentration being 2 lb/gal. In addition, 20 lb of solid salt is in the bottom of this tank. The second tank initially contains pure water. If pure water is fed to the first tank at a rate of 2 gal/min, the overflow from this tank is fed to the second tank, and the second tank overflows at the rate of 2 gal/min, write the differential equations for the total weight of salt and the concentration in each tank as functions of time. Assume that the fluids in the tanks are well mixed and that the salt dissolves rapidly.

6.11. A well-mixed tank has a maximum capacity of 100 gal and it is initially half full. The discharge pipe at the bottom is very long and thus it offers resistance to the flow of water through it. The force that causes the water to flow is the height of the water in the tank, and in fact that flow is just proportional to this height. Since the height is proportional to the total volume of water in the tank, the volumetric flow rate of water out, q_o, is

$$q_o = kV$$

The flow rate of water into the tank, q_i, is constant. Use the information given in Fig. P6.11 to decide whether the amount of water in the tank increases, decreases, or remains the same. If it changes, how much time is required to completely empty or fill the tank, as the case may be?

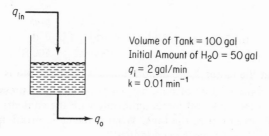

Volume of Tank = 100 gal
Initial Amount of H_2O = 50 gal
q_i = 2 gal/min
k = 0.01 min^{-1}

Fig. P6.11.

6.12. As a chemical engineer in a nuclear reactor design group, you have been asked to design a holding tank for a 10 gal/min effluent stream from a homogeneous reactor (one with the fuel in solution in water). Normally the concentration of a dangerous isotope, having a half-life of 10 min, is 0.010 mg/ft^3 in the reactor. If you design the holding tank in such a way that mixing is complete inside the tank and the solution overflows at the top, how big does the tank have to be to reduce the normal concentration of the isotope at the overflow to 0.001 mg/ft^3? *Note:* If the half-life of the isotope ($t_{1/2}$) is 10 min, its rate of decay is

$$\frac{dy}{dt} = \frac{0.693}{t_{1/2}} y$$

where y is the amount of material at any time t.

6.13. A radioactive waste that contains 1500 ppm of ^{92}Sr is pumped into a holding tank that contains 100 gal at the rate of 5 gal/min. ^{92}Sr decays as follows:

$$^{92}Sr \longrightarrow {}^{92}Y \longrightarrow {}^{92}Zr$$

half-life: 2.7 hr   3.5 hr

If the tank contains clear water initially and the solution runs out at the rate of 5 gal/min, assuming perfect mixing, what is the concentration of Sr, Y, and Zr after 1 day? What is the equilibrium concentration of Sr and Y in the tank? *Additional information:* The rate of decay of such isotopes is $dN/dt = -\lambda N$, where $\lambda = 0.693/t_{1/2}$ and the half-life is $t_{1/2}$. $N =$ moles.

6.14. Suppose that an organic compound decomposes as follows:

$$C_6H_{12} \longrightarrow C_4H_8 + C_2H_4$$

If 1 mole of C_6H_{12} exists at $t = 0$ but no C_4H_8 and C_2H_4, set up equations showing the moles of C_4H_8 and C_2H_4 as a function of time. The rates of formation of C_4H_8 and C_4H_8 are each proportional to the number of moles of C_6H_{12} present.

6.15. A large tank is connected to a smaller tank by means of a valve. The large tank contains N_2 at 100 psia, while the small tank is evacuated. If the valve leaks between the two tanks and the rate of leakage of gas is proportional to the pressure difference between the two tanks $(p_1 - p_2)$, how long does it take for the pressure in the small tank to be one-half its final value? The instantaneous initial flow rate with the small tank evacuated is 0.2 lb mole/hr.
Other data:

	tank 1	*tank* 2
Initial pressure (psia)	100	0
Volume	1000 ft^3	500 ft^3

Assume that the temperature in both tanks is constant and is 70°F.

6.16. A cylinder contains 3 ft^3 of pure oxygen at atmospheric pressure. Air is slowly pumped into the tank and mixes uniformly with the contents, an equal volume of which is forced out of the tank. What is the concentration of oxygen in the tank after 9 ft^3 of air has been admitted?

6.17. Two well-mixed tanks are connected in series as shown in Fig. P6.17. The entering concentration of a substance in tank 1 is c_i. No reactions occur. Tanks 1 and 2

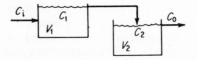

Fig. P6.17.

both overflow. Set up an equation which will predict the concentration c_0 as a function of c_i.

6.18. The following chain reactions take place in a constant-volume batch tank:

$$A \xrightarrow{k_1} B \xrightarrow{k_2} C$$

Each reaction is first order and irreversible. If the initial concentration of A is C_{A_0} and if only A is present initially, find an expression for C_B as a function of time. Under what conditions will the concentration of B be dependent primarily on the rate of reaction of A?

6.19. Consider the following chemical reaction in a constant-volume batch tank:

$$A \underset{k_2}{\overset{k_1}{\rightleftharpoons}} B$$
$$k_3 \downarrow$$
$$C$$

All the indicated reactions are first order. The initial concentration of A is C_{A_0}, and nothing else is present at that time. Determine the concentrations of A, B, and C as functions of time.

6.20. A ground material is to be suspended in water and heated in preparation for a chemical reaction. It is desired to carry out the mixing and heating simultaneously in a tank equipped with an agitator and a steam coil. The cold liquid and solid are to be added continuously and the heated suspension will be withdrawn at the same rate. One method of operation for starting up is to (1) fill the tank initially with water and solid in the proper proportions, (2) start the agitator, (3) introduce fresh water and solid in the proper proportions and simultaneously begin to withdraw the suspension for reaction, and (4) turn on the steam. An estimate is needed of the time required, after the steam is turned on, for the temperature of the effluent suspension to reach a certain elevated temperature.

(a) Using the nomenclature given below, formulate a differential equation for this process. Integrate the equation to obtain n as a function of B and ϕ (see nomenclature).

(b) Calculate the time required for the effluent temperature to reach 180°F if the initial contents of the tank and the inflow are both at 120°F and the steam temperature is 220°F. The surface area for heat transfer is 23.9 ft², and the heat transfer coefficient is 100 Btu/(hr)(ft²)(°F). The tank contains 6000 lb, and the rate of flow of both streams is 1200 lb/hr. In the proportions used, the specific heat of the suspension may be assumed to be 1.00.

If the area available heat transfer is doubled, how will the time required be affected? Why is the time with the larger area less than half that obtained previously? The heat transferred is $Q = UA(T_{\text{tank}} - T_{\text{steam}})$.

Nomenclature

W = weight of tank contents, lb
G = rate of flow of suspension, lb/hr
T_s = temperature of steam, °F
T = temperature in tank at any instant, perfect mixing assumed, °F
T_0 = temperature of suspension introduced into tank; also initial temperature of tank contents, °F
U = heat-transfer coefficient, Btu/(hr)(ft²)(°F)
A = area of heat-transfer surface, ft²
C_p = specific heat of suspension, Btu/(lb)(°F)
t = time elapsed from the instant the steam is turned on, hr
n = dimensionless time, Gt/W
B = dimensionless ratio, UA/GC_p
ϕ = dimensionless temperature (relative approach to the steam temperature) $(T - T_0)/(T_s - T_0)$

6.21. Tanks *A*, *B*, and *C* are each filled with 1000 gal of water. See Fig. P6.21. Workmen have instructions to dissolve 2000 lb of salt in each tank. By mistake, 3000 lb

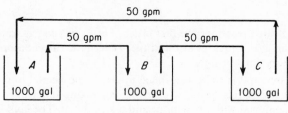

Fig. P6.21.

are dissolved in each of tanks *A* and *C* and none in *B*. You wish to bring all the compositions to within 5 percent of the specified 2 lb/gal. If the units are connected *A-B-C-A* by three 50-gpm pumps,

(a) Express concentrations C_A, C_B, and C_C in terms of *t* (time).

(b) Find the shortest time at which all concentrations are within the specified range.

6.22. Consider a well-agitated cylindrical tank in which the heat-transfer surface is in the form of a coil which is distributed uniformly from the bottom of the tank to the top of the tank. The tank itself is completely insulated. Liquid is introduced into the tank at a uniform rate, starting with no liquid in the tank, and the steam is turned on at the instant that liquid flows into the tank.

(a) Using the nomenclature of Problem 6.20, formulate a differential equation for this process. Integrate this expression to obtain an equation for ϕ as a function of *B* and *f*, where $f =$ fraction filled $= W/W_{\text{filled}}$.

(b) If the heat-transfer surface consists of a coil of 10 turns of 1-in. OD tubing 4 ft in diameter, the feed rate is 1200 lb/hr, the heat capacity of the liquid is 1.0 Btu/(lb)(°F), the heat-transfer coefficient is 100 Btu/(hr)(°F)(ft²) of covered area, the steam temperature is 200°F, and the temperature of the liquid introduced into the tank is 70°F, what is the temperature in the tank when it is completely full? What is the temperature when the tank is half full? The heat transfer is given by $Q = UA(T_{\text{tank}} - T_{\text{steam}})$.

6.23. A cylindrical tank 5 ft in diameter and 5 ft high is full of water at 70°F. The water is to be heated by means of a steam jacket around the sides only. The steam temperature is 230°F, and the overall coefficient of heat transfer is constant at 40 Btu/(hr)(ft²)(°F). Use Newton's law of cooling (heating) to estimate the heat transfer. Neglecting the heat losses from the top and the bottom, calculate the time necessary to raise the temperature of the tank contents to 170°F. Repeat, taking the heat losses from the top and the bottom into account. The air temperature around the tank is 70°F, and the overall coefficient of heat transfer for both the top and the bottom is constant at 10 Btu/(hr)(ft²)(°F).

6.24. A desired chemical B is manufactured according to the reaction

$$A \underset{k_2}{\overset{k_1}{\rightleftharpoons}} B + H_2O$$

where k is in hr^{-1}. The forward reaction is first order, and, under the conditions of a large excess of water present, the reverse reaction may be taken to be first order also, depending only on the B present. The following procedure is usually employed: 100 ft^3 of the initial water-A mixture is added to a well-stirred tank (concentration of $A = C_A$ lb moles/ft^3), and the catalyst to initiate the reaction is then added. During the next 4 hr, further water-A mixture is added to the tank at a rate of 100 ft^3/hr while the reaction proceeds, and, after the complete charge of 500 ft^3 has been added, the reaction is allowed to proceed for 6 hr. At the end of this time, the contents of the reactor are dumped into a chiller (to stop the reaction immediately).

(a) Set up the necessary differential equations, and give the necessary initial conditions needed to calculate the concentration of B when the material was dumped into the chiller. (Assume the volume change on reaction to be negligible.)

(b) On a particular day the operator notices after a period of 2 hr that the valve in the line to the chiller was not closed and that the chiller contains 100 ft^3 of material. The operator closes the valve. He then decides to carry out the rest of the run but to allow more time so as to bring the overall concentration of B (material in the reactor + material already in chiller) up to the usual level. Assuming that the 100 ft^3 of material had leaked into the chiller at a *uniform rate* during the first 2 hr after the catalyst was placed in the tank, set up the necessary differential equations and initial conditions to determine how long he should let the reaction proceed before dumping the remainder of the material into the chiller.

Define all your symbols (and units).

6.25. Consider a flow process in which a precipitation is being formed by mixing two streams A and B to form a third stream C in which the precipitate is carried away. The reaction is extremely rapid, and agitation will be assumed to be so efficient that the material in the tank has substantially the same composition at all points. Streams A and B enter at the rate of a and b cfm, respectively. Change in volume due to reaction may be neglected, so that stream C leaves at the rate of $a + b$ cfm. To ensure proper quality of the precipitate, it is necessary to maintain the acidity of the tank contents at n_0 lb/ft^3 acid, with an allowable variation of $\pm \alpha$ lb/ft^3. The acidity of the bath is maintained by a negligibly small volume of acid carried by stream A. On the assumption that the acid supply fails suddenly at a time when the concentration is n_0, it is desired to develop the law relating acid concentration and time in order to estimate what time would elapse before this concentration falls below the allowable level. During this time, all other flows will be assumed constant.

6.26. Determine the time required to heat a 10,000-lb batch of liquid from 60 to 120°F using an external, counterflow heat exchanger having an area of 300 ft². Water at 180°F is used as the heating medium and flows at a rate of 6000 lb/hr. An overall heat-transfer coefficient of 40 Btu/(hr)(ft²)(°F) may be assumed; use Newton's law of heating. The liquid is circulated at a rate of 6000 lb/hr, and the specific heat of the liquid is the same as that of water (1.0). Assume that the residence time of the liquid in the external heat exchanger is very small and that there is essentially no holdup of liquid in this circuit.

Conversion Factors[*]

TABLE A.1 VOLUME EQUIVALENTS

in.³	ft³	U.S. gal	liters	m³
1	5.787×10^{-4}	4.329×10^{-3}	1.639×10^{-2}	1.639×10^{-5}
1.728×10^3	1	7.481	28.32	2.832×10^{-2}
2.31×10^2	0.1337	1	3.785	3.785×10^{-3}
61.03	3.531×10^{-2}	0.2642	1	1.000×10^{-3}
6.102×10^4	35.31	264.2	1000	1

TABLE A.2 MASS EQUIVALENTS

avoir oz	pounds	grains	grams
1	6.25×10^{-2}	4.375×10^2	28.35
16	1	7×10^3	4.536×10^2
2.286×10^{-3}	1.429×10^{-4}	1	6.48×10^{-2}
3.527×10^{-2}	2.20×10^{-3}	15.432	1

TABLE A.3 LINEAR MEASURE EQUIVALENTS

meter	inch	foot	mile
1	39.37	3.2808	6.214×10^{-4}
2.54×10^{-2}	1	8.333×10^{-2}	1.58×10^{-5}
0.3048	12	1	1.8939×10^{-4}
1.61×10^3	6.336×10^4	5280	1

TABLE A.4 POWER EQUIVALENTS

hp	kw	ft-lb/sec	Btu/sec	J/sec
1	0.7457	550	0.7068	7.457×10^2
1.341	1	737.56	0.9478	1.000×10^3
1.818×10^{-3}	1.356×10^{-3}	1	1.285×10^{-3}	1.356
1.415	1.055	778.16	1	1.055×10^3
1.341×10^{-3}	1.000×10^{-3}	0.7376	9.478×10^{-4}	1

[*]Read across.

TABLE A.5 HEAT, ENERGY, OR WORK EQUIVALENTS

ft-lb	kw-hr	hp-hr	Btu	calorie*	Joule
0.7376	2.773×10^{-7}	3.725×10^{-7}	9.478×10^{-4}	0.2390	1
7.233	2.724×10^{-6}	3.653×10^{-6}	9.296×10^{-3}	2.3438	9.80665
1	3.766×10^{-7}	5.0505×10^{-7}	1.285×10^{-3}	0.3241	1.356
2.655×10^{6}	1	1.341	3.4128×10^{3}	8.6057×10^{5}	3.6×10^{6}
1.98×10^{6}	0.7455	1	2.545×10^{3}	6.4162×10^{5}	2.6845×10^{6}
74.73	2.815×10^{-5}	3.774×10^{-5}	9.604×10^{-2}	24.218	1.0133×10^{2}
3.086×10^{3}	1.162×10^{-3}	1.558×10^{-3}	3.9657	1×10^{3}	4.184×10^{3}
7.7816×10^{2}	2.930×10^{-4}	3.930×10^{-4}	1	2.52×10^{2}	1.055×10^{3}
3.086	1.162×10^{-6}	1.558×10^{-6}	3.97×10^{-3}	1	4.184

*The thermochemical calorie = 4.184 J; the IT calorie = 4.1867 J (see Sec. 4.1).

TABLE A.6 PRESSURE EQUIVALENTS

mm Hg	in. Hg	bar	atm	N/m²
1	3.937×10^{-2}	1.333×10^{-3}	1.316×10^{-3}	1.333×10^{2}
25.40	1	3.387×10^{1}	3.342×10^{-2}	3.387×10^{3}
750.06	29.53	1	0.9869	1.000×10^{5}
760.0	29.92	1.013	1	1.013×10^{5}
7.502×10^{-2}	2.954×10^{-4}	1.000×10^{-5}	9.872×10^{-6}	1

Atomic Weights and Numbers

TABLE B.1 RELATIVE ATOMIC WEIGHTS, 1965

Based on the Atomic Mass of $^{12}C = 12$

The values for atomic weights given in the table apply to elements as they exist in nature, without artificial alteration of their isotopic composition, and, further, to natural mixtures that do not include isotopes of radiogenic origin.

Name	Symbol	Atomic Number	Atomic Weight	Name	Symbol	Atomic Number	Atomic Weight
Actinium	Ac	89	—	Mercury	Hg	80	200.59
Aluminium	Al	13	26.9815	Molybdenum	Mo	42	95.94
Americium	Am	95	—	Neodymium	Nd	60	144.24
Antimony	Sb	51	121.75	Neon	Ne	10	20.183
Argon	Ar	18	39.948	Neptunium	Np	93	—
Arsenic	As	33	74.9216	Nickel	Ni	28	58.71
Astatine	At	85	—	Niobium	Nb	41	92.906
Barium	Ba	56	137.34	Nitrogen	N	7	14.0067
Berkelium	Bk	97	—	Nobelium	No	102	—
Beryllium	Be	4	9.0122	Osmium	Os	75	190.2
Bismuth	Bi	83	208.980	Oxygen	O	8	15.9994
Boron	B	5	10.811	Palladium	Pd	46	106.4
Bromine	Br	35	79.904	Phosphorus	P	15	30.9738
Cadmium	Cd	48	112.40	Platinum	Pt	78	195.09
Caesium	Cs	55	132.905	Plutonium	Pu	94	—
Calcium	Ca	20	40.88	Polonium	Po	84	—
Californium	Cf	98	—	Potassium	K	19	39.102
Carbon	C	6	12.01115	Praseodym	Pr	59	140.907
Cerium	Ce	58	140.12	Promethium	Pm	61	—
Chlorine	Cl	17	35.453[b]	Protactinium	Pa	91	—
Chromium	Cr	24	51.996[b]	Radium	Ra	88	—
Cobalt	Co	27	58.9332	Radon	Rn	86	—
Copper	Cu	29	63.546[b]	Rhenium	Re	75	186.2
Curium	Cm	96	—	Rhodium	Rh	45	102.905
Dysprosium	Dy	66	162.50	Rubidium	Rb	37	84.57
Einsteinium	Es	99	—	Ruthenium	Ru	44	101.07
Erbium	Er	68	167.26	Samarium	Sm	62	150.35
Europium	Eu	63	151.96	Scandium	Sc	21	44.956
Fermium	Fm	100	—	Selenium	Se	34	78.96
Fluorine	F	9	18.9984	Silicon	Si	14	28.086
Francium	Fr	87	—	Silver	Ag	47	107.868
Gadolinium	Gd	64	157.25	Sodium	Na	11	22.9898
Gallium	Ga	31	69.72	Strontium	Sr	38	87.62
Germanium	Ge	32	72.59	Sulfur	S	16	32.064
Gold	Au	79	196.967	Tantalum	Ta	73	180.948
Hafnium	Hf	72	178.49	Technetium	Tc	43	—
Helium	He	2	4.0026	Tellurium	Te	52	127.60
Holmium	Ho	67	164.930	Terbium	Tb	65	158.924
Hydrogen	H	1	1.00797	Thallium	Tl	81	204.37
Indium	In	49	114.82	Thorium	Th	90	232.038
Iodine	I	53	126.9044	Thulium	Tm	59	168.934
Iridium	Ir	77	192.2	Tin	Sn	50	118.69
Iron	Fe	26	55.847	Titanium	Ti	22	47.90
Krypton	Kr	36	83.80	Tungsten	W	74	183.85
Lanthanum	La	57	138.91	Uranium	U	92	238.03
Lawrencium	Lr	103	—	Vanadium	V	23	50.942
Lead	Pb	82	207.19	Xenon	Xe	54	131.30
Lithium	Li	3	6.939	Ytterbium	Yb	70	173.04
Lutetium	Lu	71	174.97	Yttrium	Y	39	88.905
Magnesium	Mg	12	24.312	Zinc	Zn	30	65.37
Manganese	Mn	25	54.9380	Zirconium	Zr	40	91.22
Mendelevium	Md	101	—				

SOURCE: *Comptes Rendus*, 23rd IUPAC Conference, 1965, Butterworth's, London, 1965, pp. 177–178.

Steam Tables

SOURCE: Combustion Engineering, Inc.

Absolute pressure = atmospheric pressure − vacuum.

Barometer and vacuum columns may be corrected to mercury at 32°F by subtracting $0.00009 \times (t - 32) \times$ column height, where t is the column temperature in °F.

One inch of mercury at 32°F = 0.4912 lb/in.^{2}

Example:

Barometer reads 30.17 in. at 70°F. Vacuum column reads 28.26 in. at 80°F. Pounds pressure = $(30.17 - 0.00009 \times 38 \times 30.17) - 28.26 - 0.00009 \times 48 \times 28.26) = 1.93$ in. of mercury at 32°F.

Saturation temperature (from table) = 100°F.

TABLE C.1 SATURATED STEAM: TEMPERATURE TABLE

Temp. Fahr. t	Absolute pressure		Specific volume			Enthalpy		
	Lb/in.^{2} p	In. Hg 32°F	Sat. Liquid v_f	Evap. v_{fg}	Sat. Vapor v_g	Sat. Liquid h_f	Evap. h_{fg}	Sat. Vapor h_g
32	0.0886	0.1806	0.01602	3305.7	3305.7	0	1075.1	1075.1
34	0.0961	0.1957	0.01602	3060.4	3060.4	2.01	1074.9	1076.0
36	0.1041	0.2120	0.01602	2836.6	2836.6	4.03	1072.9	1076.9
38	0.1126	0.2292	0.01602	2632.2	2632.2	6.04	1071.7	1077.7
40	0.1217	0.2478	0.01602	2445.1	2445.1	8.05	1070.5	1078.6
42	0.1315	0.2677	0.01602	2271.8	2271.8	10.06	1069.3	1079.4
44	0.1420	0.2891	0.01602	2112.2	2112.2	12.06	1068.2	1080.3
46	0.1532	0.3119	0.01602	1965.5	1965.5	14.07	1067.1	1081.2
48	0.1652	0.3364	0.01602	1829.9	1829.9	16.07	1065.9	1082.0
50	0.1780	0.3624	0.01602	1704.9	1704.9	18.07	1064.8	1082.9
52	0.1918	0.3905	0.01603	1588.4	1588.4	20.07	1063.6	1083.7
54	0.2063	0.4200	0.01603	1482.4	1482.4	22.07	1062.5	1084.6
56	0.2219	0.4518	0.01603	1383.5	1383.5	24.07	1061.4	1085.5
58	0.2384	0.4854	0.01603	1292.7	1292.7	26.07	1060.2	1086.3
60	0.2561	0.5214	0.01603	1208.1	1208.1	28.07	1059.1	1087.2
62	0.2749	0.5597	0.01604	1129.7	1129.7	30.06	1057.9	1088.0
64	0.2949	0.6004	0.01604	1057.1	1057.1	32.06	1056.8	1088.9
66	0.3162	0.6438	0.01604	989.6	989.6	34.06	1055.7	1089.8
68	0.3388	0.6898	0.01605	927.0	927.0	36.05	1054.5	1090.6
70	0.3628	0.7387	0.01605	868.9	868.9	38.05	1053.4	1091.5
72	0.3883	0.7906	0.01606	814.9	814.9	40.04	1052.3	1092.3
74	0.4153	0.8456	0.01606	764.7	764.7	42.04	1051.2	1093.2
76	0.4440	0.9040	0.01607	718.0	718.0	44.03	1050.1	1094.1
78	0.4744	0.9659	0.01607	674.4	674.4	46.03	1048.9	1094.9
80	0.5067	1.032	0.01607	633.7	633.7	48.02	1047.8	1095.8
82	0.5409	1.101	0.01608	595.8	595.8	50.02	1046.6	1096.6
84	0.5772	1.175	0.01608	560.4	560.4	52.01	1045.5	1097.5
86	0.6153	1.253	0.01609	527.6	527.6	54.01	1044.4	1098.4
88	0.6555	1.335	0.01609	497.0	497.0	56.00	1043.2	1099.2
90	0.6980	1.421	0.01610	468.4	468.4	58.00	1042.1	1100.1
92	0.7429	1.513	0.01611	441.7	441.7	59.99	1040.9	1100.9
94	0.7902	1.609	0.01611	416.7	416.7	61.98	1039.8	1101.8
96	0.8403	1.711	0.01612	393.2	393.2	63.98	1038.7	1102.7
98	0.8930	1.818	0.01613	371.3	371.3	65.98	1037.5	1103.5
100	0.9487	1.932	0.01613	350.8	350.8	67.97	1036.4	1104.4
102	1.0072	2.051	0.01614	331.5	331.5	69.96	1035.2	1105.2
104	1.0689	2.176	0.01614	313.5	313.5	71.96	1034.1	1106.1
106	1.1338	2.308	0.01615	296.5	296.5	73.95	1033.0	1107.0
108	1.2020	2.447	0.01616	280.7	280.7	75.94	1032.0	1107.9

v = specific volume, ft^3/lb. h = enthalpy, Btu/lb.

TABLE C.1 (CONT.)

Temp. Fahr. t	Absolute pressure		Specific volume			Enthalpy		
	Lb/in.² p	In. Hg 32°F	Sat. Liquid v_f	Evap. v_{fg}	Sat. Vapor v_g	Sat. Liquid h_f	Evap. h_{fg}	Sat. Vapor h_g
110	1.274	2.594	0.01617	265.7	265.7	77.94	1030.9	1108.8
112	1.350	2.749	0.01617	251.6	251.6	79.93	1029.7	1109.6
114	1.429	2.909	0.01618	238.5	238.5	81.93	1028.6	1110.5
116	1.512	3.078	0.01619	226.2	226.2	83.92	1027.5	1111.4
118	1.600	3.258	0.01620	214.5	214.5	85.92	1026.4	1112.3
120	1.692	3.445	0.01620	203.45	203.47	87.91	1025.3	1113.2
122	1.788	3.640	0.01621	193.16	193.18	89.91	1024.1	1114.0
124	1.889	3.846	0.01622	183.44	183.46	91.90	1023.0	1114.9
126	1.995	4.062	0.01623	174.26	174.28	93.90	1021.8	1115.7
128	2.105	4.286	0.01624	165.70	165.72	95.90	1020.7	1116.6
130	2.221	4.522	0.01625	157.55	157.57	97.89	1019.5	1117.4
132	2.343	4.770	0.01626	149.83	149.85	99.89	1018.3	1118.2
134	2.470	5.029	0.01626	142.59	142.61	101.89	1017.2	1119.1
136	2.603	5.300	0.01627	135.73	135.75	103.88	1016.0	1119.9
138	2.742	5.583	0.01628	129.26	129.28	105.88	1014.9	1120.8
140	2.887	5.878	0.01629	123.16	123.18	107.88	1013.7	1121.6
142	3.039	6.187	0.01630	117.37	117.39	109.88	1012.5	1122.4
144	3.198	6.511	0.01631	111.88	111.90	111.88	1011.3	1123.2
146	3.363	6.847	0.01632	106.72	106.74	113.88	1010.2	1124.1
148	3.536	7.199	0.01633	101.82	101.84	115.87	1009.0	1124.9
150	3.716	7.566	0.01634	97.18	97.20	117.87	1007.8	1125.7
152	3.904	7.948	0.01635	92.79	92.81	119.87	1006.7	1126.6
154	4.100	8.348	0.01636	88.62	88.64	121.87	1005.5	1127.4
156	4.305	8.765	0.01637	84.66	84.68	123.87	1004.4	1128.3
158	4.518	9.199	0.01638	80.90	80.92	125.87	1003.2	1129.1
160	4.739	9.649	0.01639	77.37	77.39	127.87	1002.0	1129.9
162	4.970	10.12	0.01640	74.00	74.02	129.88	1000.8	1130.7
164	5.210	10.61	0.01642	70.79	70.81	131.88	999.7	1131.6
166	5.460	11.12	0.01643	67.76	67.78	133.88	998.5	1132.4
168	5.720	11.65	0.01644	64.87	64.89	135.88	997.3	1133.2
170	5.990	12.20	0.01645	62.12	62.14	137.89	996.1	1134.0
172	6.272	12.77	0.01646	59.50	59.52	139.89	995.0	1134.9
174	6.565	13.37	0.01647	57.01	57.03	141.89	993.8	1135.7
176	6.869	13.99	0.01648	56.64	54.66	143.90	992.6	1136.5
178	7.184	14.63	0.01650	52.39	52.41	145.90	991.4	1137.3
180	7.510	15.29	0.01651	50.26	50.28	147.91	990.2	1138.1
182	7.849	15.98	0.01652	48.22	48.24	149.92	989.0	1138.9
184	8.201	16.70	0.01653	46.28	46.30	151.92	987.8	1139.7
186	8.566	17.44	0.01654	44.43	44.45	153.93	986.6	1140.5
188	8.944	18.21	0.01656	42.67	42.69	155.94	985.3	1141.3
190	9.336	19.01	0.01657	40.99	41.01	157.95	984.1	1142.1
192	9.744	19.84	0.01658	39.38	39.40	159.95	982.8	1142.8
194	10.168	20.70	0.01659	37.84	37.86	161.96	981.5	1143.5
196	10.605	21.59	0.01661	36.38	36.40	163.97	980.3	1144.3
198	11.057	22.51	0.01662	34.98	35.00	165.98	979.0	1145.0
200	11.525	23.46	0.01663	33.65	33.67	167.99	977.8	1145.8
202	12.010	24.45	0.01665	32.37	32.39	170.01	976.6	1146.6
204	12.512	25.47	0.01666	31.15	31.17	172.02	975.3	1147.3
206	13.031	26.53	0.01667	29.99	30.01	174.03	974.1	1148.1
208	13.568	27.62	0.01669	28.88	28.90	176.04	972.8	1148.8
210	14.123	28.75	0.01670	27.81	27.83	178.06	971.5	1149.6
212	14.696	29.92	0.01672	26.81	26.83	180.07	970.3	1150.4
215	15.591		0.01674	25.35	25.37	186.10	968.3	1151.4
220	17.188		0.01677	23.14	23.16	188.14	965.2	1153.3
225	18.915		0.01681	21.15	21.17	193.18	961.9	1155.1
230	20.78		0.01684	19.371	19.388	198.22	958.7	1156.9
235	22.80		0.01688	17.761	17.778	203.28	955.3	1158.6
240	24.97		0.01692	16.307	16.324	208.34	952.1	1160.4
245	27.31		0.01696	15.010	15.027	213.41	948.7	1162.1
250	29.82		0.01700	13.824	13.841	218.48	945.3	1163.8

v = specific volume, ft³/lb. h = enthalpy, Btu/lb.

Table C.1 (Cont.)

Temp. Fahr. t	Absolute pressure Lb/in.² p	In. Hg 32°F	Specific volume Sat. Liquid v_f	Evap. v_{fg}	Sat. Vapor v_g	Enthalpy Sat. Liquid h_f	Evap. h_{fg}	Sat. Vapor h_g
255	32.53		0.01704	12.735	12.752	223.56	942.0	1165.6
260	35.43		0.01708	11.754	11.771	228.65	938.6	1167.3
265	38.54		0.01713	10.861	10.878	233.74	935.3	1169.0
270	41.85		0.01717	10.053	10.070	238.84	931.8	1170.6
275	45.40		0.01721	9.313	9.330	243.94	928.2	1172.1
280	49.20		0.01726	8.634	8.651	249.06	924.6	1173.7
285	53.25		0.01731	8.015	8.032	254.18	921.0	1175.2
290	57.55		0.01735	7.448	7.465	259.31	917.4	1176.7
295	62.13		0.01740	6.931	6.948	264.45	913.7	1178.2
300	67.01		0.01745	6.454	6.471	269.60	910.1	1179.7
305	72.18		0.01750	6.014	6.032	274.76	906.3	1181.1
310	77.68		0.01755	5.610	5.628	279.92	902.6	1182.5
315	83.50		0.01760	5.239	5.257	285.10	898.8	1183.9
320	89.65		0.01765	4.897	4.915	290.29	895.0	1185.3
325	96.16		0.01771	4.583	4.601	295.49	891.1	1186.6
330	103.03		0.01776	4.292	4.310	300.69	887.1	1187.8
335	110.31		0.01782	4.021	4.039	305.91	883.2	1189.1
340	117.99		0.01788	3.771	3.789	311.14	879.2	1190.3
345	126.10		0.01793	3.539	3.557	316.38	875.1	1191.5
350	134.62		0.01799	3.324	3.342	321.64	871.0	1192.6
355	143.58		0.01805	3.126	3.144	326.91	866.8	1193.7
360	153.01		0.01811	2.940	2.958	332.19	862.5	1194.7
365	162.93		0.01817	2.768	2.786	337.48	858.2	1195.7
370	173.33		0.01823	2.607	2.625	342.79	853.8	1196.6
375	184.23		0.01830	2.458	2.476	348.11	849.4	1197.5
380	195.70		0.01836	2.318	2.336	353.45	844.9	1198.4
385	207.71		0.01843	2.189	2.207	358.80	840.4	1199.2
390	220.29		0.01850	2.064	2.083	364.17	835.7	1199.9
395	233.47		0.01857	1.9512	1.9698	369.56	831.0	1200.6
400	247.25		0.01864	1.8446	1.8632	374.97	826.2	1201.2
405	261.67		0.01871	1.7445	1.7632	380.40	821.4	1201.8
410	276.72		0.01878	1.6508	1.6696	385.83	816.6	1202.4
415	292.44		0.01886	1.5630	1.5819	391.30	811.7	1203.0
420	308.82		0.01894	1.4806	1.4995	396.78	806.7	1203.5
425	325.91		0.01902	1.4031	1.4221	402.28	801.6	1203.9
430	343.71		0.01910	1.3303	1.3494	407.80	796.5	1204.3
435	362.27		0.01918	1.2617	1.2809	413.35	791.2	1204.6
440	381.59		0.01926	1.1973	1.2166	418.91	785.9	1204.8
445	401.70		0.01934	1.1367	1.1560	424.49	780.4	1204.9
450	422.61		0.01943	1.0796	1.0990	430.11	774.9	1205.0
455	444.35		0.0195	1.0256	1.0451	435.74	769.3	1205.0
460	466.97		0.0196	0.9745	0.9941	441.42	763.6	1205.0
465	490.43		0.0197	0.9262	0.9459	447.10	757.8	1204.9
470	514.70		0.0198	0.8808	0.9006	452.84	751.9	1204.7
475	539.90		0.0199	0.8379	0.8578	458.59	745.9	1204.5
480	566.12		0.0200	0.7972	0.8172	464.37	739.8	1204.2
485	593.28		0.0201	0.7585	0.7786	470.18	733.6	1203.8
490	621.44		0.0202	0.7219	0.7421	476.01	727.3	1203.3
495	650.59		0.0203	0.6872	0.7075	481.90	720.8	1202.7
500	680.80		0.0204	0.6544	0.6748	487.80	714.2	1202.0
505	712.19		0.0206	0.6230	0.6436	493.8	707.5	1201.3
510	744.55		0.0207	0.5932	0.6139	499.8	700.6	1200.4
515	777.96		0.0208	0.5651	0.5859	505.8	693.6	1199.4
520	812.68		0.0209	0.5382	0.5591	511.9	686.5	1198.4
525	848.37		0.0210	0.5128	0.5338	518.0	679.2	1197.2
530	885.20		0.0212	0.4885	0.5097	524.2	671.9	1196.1
535	923.45		0.0213	0.4654	0.4867	530.4	664.4	1194.8
540	962.80		0.0214	0.4433	0.4647	536.6	656.7	1193.3

v = specific volume, ft³/lb. h = enthalpy, Btu/lb.

TABLE C.1 (CONT.)

Temp. Fahr. t	Absolute pressure			Specific volume			Enthalpy		
	Lb/in.² p	In. Hg 32°F	Sat. Liquid v_f	Evap. v_{fg}	Sat. Vapor v_g		Sat. Liquid h_f	Evap. h_{fg}	Sat. Vapor h_g
545	1003.6		0.0216	0.4222	0.4438		542.9	648.9	1191.8
550	1045.6		0.0218	0.4021	0.4239		549.3	640.9	1190.2
555	1088.8		0.0219	0.3830	0.4049		555.7	632.6	1188.3
560	1133.4		0.0221	0.3648	0.3869		562.2	624.1	1186.3
565	1179.3		0.0222	0.3472	0.3694		568.8	615.4	1184.2
570	1226.7		0.0224	0.3304	0.3528		575.4	606.5	1181.9
575	1275.7		0.0226	0.3143	0.3369		582.1	597.4	1179.5
580	1326.1		0.0228	0.2989	0.3217		588.9	588.1	1177.0
585	1378.1		0.0230	0.2840	0.3070		595.7	578.6	1174.3
590	1431.5		0.0232	0.2699	0.2931		602.6	568.8	1171.4
595	1486.5		0.0234	0.2563	0.2797		609.7	558.7	1168.4
600	1543.2		0.0236	0.2432	0.2668		616.8	548.4	1165.2
605	1601.5		0.0239	0.2306	0.2545		624.1	537.7	1161.8
610	1661.6		0.0241	0.2185	0.2426		631.5	526.6	1158.1
615	1723.4		0.0244	0.2068	0.2312		638.9	515.3	1154.2
620	1787.0		0.0247	0.1955	0.2202		646.5	503.7	1150.2
625	1852.4		0.0250	0.1845	0.2095		654.3	491.5	1145.8
630	1919.8		0.0253	0.1740	0.1993		662.2	478.8	1141.0
635	1989.0		0.0256	0.1638	0.1894		670.4	465.5	1135.9
640	2060.3		0.0260	0.1539	0.1799		678.7	452.0	1130.7
645	2133.5		0.0264	0.1441	0.1705		687.3	437.6	1124.9
650	2208.8		0.0268	0.1348	0.1616		696.0	422.7	1118.7
655	2286.4		0.0273	0.1256	0.1529		705.2	407.0	1112.2
660	2366.2		0.0278	0.1167	0.1445		714.4	390.5	1104.9
665	2448.0		0.0283	0.1079	0.1362		724.5	372.1	1096.6
670	2532.4		0.0290	0.0991	0.1281		734.6	353.3	1087.9
675	2619.2		0.0297	0.0904	0.1201		745.5	332.8	1078.3
680	2708.4		0.0305	0.0810	0.1115		757.2	310.0	1067.2
685	2800.4		0.0316	0.0716	0.1032		770.1	284.5	1054.6
690	2895.0		0.0328	0.0617	0.0945		784.2	254.9	1039.1
695	2992.7		0.0345	0.0511	0.0856		801.3	219.1	1020.4
700	3094.1		0.0369	0.0389	0.0758		823.9	171.7	995.6
705	3199.1		0.0440	0.0157	0.0597		870.2	77.6	947.8
705.34*	3206.2		0.0541	0	0.0541		910.3	0	910.3

*Critical temperature. v = specific volume, ft³/lb. h = enthalpy, Btu/lb.

TABLE C.2 SUPERHEATED STEAM

Abs. Press. lb/in.² (Sat. Temp.)		Sat. Water	Sat. Steam	200°	250°	300°	350°	400°	450°	500°	600°	700°	800°	900°	1000°	1100°	1200°
								Temperature—Degrees Fahrenheit									
1 (101.76)	Sh	0.0161	333.79	98.24	148.24	198.24	248.24	298.24	348.24	398.24	498.24	598.24	698.24	798.24	898.24	998.24	1098.24
	v	69.72	1105.2	392.5	422.5	452.1	482.1	511.7	541.8	571.3	630.9	690.6	750.2	809.8	869.4	929.1	988.7
	h			1149.2	1171.9	1194.4	1217.3	1240.2	1263.5	1286.7	1333.9	1382.1	1431.0	1480.8	1531.0	1583.0	1635.4
5 (162.25)	Sh	0.0164	73.600	37.75	87.75	137.75	187.75	237.75	287.75	337.75	437.75	537.75	637.75	737.75	837.75	937.75	1037.75
	v	130.13	1130.8	78.17	84.24	90.21	96.26	102.19	108.23	114.16	126.11	138.05	149.99	161.91	173.83	185.80	197.72
	h			1148.3	1171.1	1193.6	1216.6	1239.8	1263.0	1286.1	1333.5	1381.8	1430.8	1480.6	1531.3	1582.9	1635.3
10 (193.21)	Sh	0.0166	38.462	6.79	56.79	106.79	156.79	206.79	256.79	306.79	406.79	506.79	606.79	706.79	806.79	906.79	1006.79
	v	161.17	1143.3	38.88	41.96	44.98	48.02	51.01	54.04	57.02	63.01	68.99	74.96	80.92	86.89	92.88	98.85
	h			1146.7	1170.2	1192.8	1216.0	1239.3	1262.5	1285.8	1333.3	1381.6	1430.6	1480.5	1531.2	1582.8	1635.2
14.696 (212.00)	Sh	0.0167	26.828		38.00	88.00	138.00	188.00	238.00	288.00	388.00	488.00	588.00	688.00	788.00	888.00	988.00
	v	180.07	1150.4		28.44	30.52	32.61	34.65	36.73	38.75	42.83	46.91	50.97	55.03	59.00	63.19	67.25
	h				1169.2	1192.0	1215.0	1238.0	1262.1	1285.4	1333.0	1381.4	1430.5	1480.4	1531.1	1582.7	1635.1
15 (213.03)	Sh	0.0167	26.320		36.97	86.97	136.97	186.97	236.97	286.97	386.97	486.97	586.97	686.97	786.97	886.97	986.97
	v	181.11	1150.7		27.86	29.90	31.94	33.95	35.98	37.97	41.98	45.97	49.95	53.93	57.91	61.91	65.89
	h				1169.2	1192.0	1215.4	1238.9	1262.1	1285.4	1333.0	1381.4	1430.5	1480.5	1531.1	1582.7	1635.1
20 (227.96)	Sh	0.0168	20.110		22.04	72.04	122.04	172.04	222.04	272.04	372.04	472.04	572.04	672.04	772.04	872.04	972.04
	v	196.16	1156.1		20.81	22.36	23.91	25.43	26.95	28.45	31.46	34.46	37.44	40.43	43.42	46.43	49.41
	h				1168.0	1191.1	1214.8	1238.4	1261.6	1285.0	1332.7	1381.2	1430.3	1480.3	1531.0	1582.6	1635.1
25 (240.07)	Sh	0.0169	16.321		9.93	59.93	109.93	159.93	209.93	259.93	359.93	459.93	559.93	659.93	759.93	859.93	959.93
	v	208.41	1160.4		16.58	17.84	19.08	20.30	21.53	22.73	25.15	27.55	29.94	32.33	34.73	37.14	39.52
	h				1166.3	1190.2	1214.1	1237.9	1261.1	1284.6	1332.4	1381.0	1430.1	1480.0	1530.9	1582.5	1635.0
30 (250.34)	Sh	0.0170	13.763			49.66	99.66	149.66	199.66	249.66	349.66	449.66	549.66	649.66	749.66	849.66	949.66
	v	218.83	1164.0			14.82	15.87	16.89	17.91	18.92	20.94	22.94	24.94	26.93	28.93	30.94	32.93
	h					1189.2	1213.4	1237.4	1260.6	1284.2	1332.1	1380.8	1429.9	1479.9	1530.8	1582.4	1634.9
35 (259.28)	Sh	0.0171	11.907			40.72	90.72	140.72	190.72	240.72	340.72	440.72	540.72	640.72	740.72	840.72	940.72
	v	227.92	1167.0			12.66	13.57	14.45	15.33	16.20	17.94	19.66	21.36	23.08	24.79	26.52	28.22
	h					1188.2	1212.7	1236.9	1260.1	1283.8	1331.9	1380.6	1429.8	1479.8	1530.7	1582.3	1634.8
40 (267.24)	Sh	0.0172	10.506			32.76	82.76	132.76	182.76	232.76	332.76	432.76	532.76	632.76	732.76	832.76	932.76
	v	236.02	1169.7			11.04	11.84	12.62	13.40	14.16	15.68	17.19	18.69	20.18	21.68	23.20	24.69
	h					1187.1	1211.9	1236.4	1259.6	1283.4	1331.6	1380.4	1429.6	1479.6	1530.6	1582.2	1634.8
45 (274.45)	Sh	0.0172	9.408			25.55	75.55	125.55	175.55	225.55	325.55	425.55	525.55	625.55	725.55	825.55	925.55
	v	243.38	1172.0			9.785	10.50	11.20	11.89	12.57	13.93	15.27	16.60	17.94	19.27	20.62	21.95
	h					1185.9	1211.1	1235.8	1259.1	1283.0	1331.0	1380.1	1429.4	1479.4	1530.5	1582.1	1634.7

Sh = superheat, °F. v = specific volume, ft³/lb. h = enthalpy, Btu/lb.

Superheated Steam — Pressures 50 to 75 lb/sq in.

Press. (Sat. °F)		Sat. liq.	Sat. vap.	300°	350°	400°	450°	500°	600°	700°	800°	900°	1000°	1100°	1200°
50 (281.01)	Sh			18.99	68.99	118.99	168.99	218.99	318.99	418.99	518.99	618.99	718.99	818.99	918.99
	v	0.0173	8.522	8.777	9.430	10.06	10.69	11.30	12.53	13.74	14.93	16.14	17.34	18.55	19.75
	h	250.09	1174.0	1184.6	1210.3	1235.2	1258.6	1282.6	1331.0	1379.9	1429.3	1479.3	1530.4	1582.0	1634.6
55 (287.07)	Sh			12.93	62.93	112.93	162.93	212.93	312.93	412.93	512.93	612.93	712.93	812.93	912.93
	v	0.0173	7.792	7.950	8.553	9.130	9.703	10.26	11.38	12.48	13.57	14.67	15.76	16.86	17.95
	h	256.30	1175.8	1183.2	1209.4	1234.6	1258.2	1282.2	1330.7	1379.7	1429.1	1479.2	1530.3	1581.9	1634.5
60 (292.71)	Sh			7.29	57.29	107.29	157.29	207.29	307.29	407.29	507.29	607.29	707.29	807.29	907.29
	v	0.0174	7.179	7.260	7.821	8.353	8.882	9.398	10.42	11.44	12.44	13.44	14.44	15.45	16.45
	h	262.10	1177.5	1181.8	1208.5	1234.0	1257.7	1281.8	1330.4	1379.5	1428.9	1479.0	1530.2	1581.8	1634.4
65 (297.97)	Sh			2.03	52.03	102.03	152.03	202.03	302.03	402.03	502.03	602.03	702.03	802.03	902.03
	v	0.0174	6.654	6.674	7.202	7.696	8.187	8.665	9.614	10.55	11.48	12.40	13.33	14.26	15.19
	h	267.51	1179.1	1180.4	1207.6	1233.4	1257.2	1281.4	1330.1	1379.3	1428.8	1478.9	1530.1	1581.6	1634.4
70 (302.92)	Sh				47.08	97.08	147.08	197.08	297.08	397.08	497.08	597.08	697.08	797.08	897.08
	v	0.0175	6.210		6.671	7.132	7.592	8.036	8.920	9.791	10.65	11.51	12.37	13.24	14.10
	h	272.61	1180.5		1206.7	1232.8	1256.7	1281.0	1329.8	1379.0	1428.6	1478.7	1530.0	1581.6	1634.3
75 (307.60)	Sh				42.40	92.40	142.40	192.40	292.40	392.40	492.40	592.40	692.40	792.40	892.40
	v	0.0175	5.820		6.210	6.644	7.076	7.492	8.319	9.133	9.938	10.74	11.54	12.36	13.16
	h	277.44	1181.9		1205.8	1232.2	1256.2	1280.6	1329.6	1378.8	1428.4	1478.6	1529.8	1581.5	1634.2

Superheated Steam — Pressures 80 to 110 lb/sq in.

Press. (Sat. °F)		Sat. liq.	Sat. vap.	340°	360°	380°	400°	420°	450°	500°	600°	700°	800°	900°	1000°	1100°	1200°
80 (312.03)	Sh			27.97	47.97	67.97	87.97	107.97	137.97	187.97	287.97	387.97	487.97	587.97	687.97	787.97	887.97
	v	0.0176	5.476	5.889	6.055	6.217	6.384	6.623	7.015	7.793	8.558	9.313	10.07	10.82	11.51	11.58	12.33
	h	282.02	1183.1	1200.0	1211.0	1221.2	1231.5	1240.5	1255.7	1280.2	1329.3	1378.5	1428.2	1478.2	1529.7	1581.4	1634.1
85 (316.25)	Sh			23.75	43.75	63.75	83.75	103.75	133.75	183.75	283.75	383.75	483.75	583.75	683.75	783.75	883.75
	v	0.0176	5.169	5.368	5.528	5.685	5.839	5.995	6.226	6.594	7.329	8.050	8.762	9.472	10.18	10.90	11.61
	h	286.40	1184.3	1198.5	1210.0	1220.5	1230.7	1239.7	1255.1	1279.7	1329.3	1378.2	1428.0	1478.2	1529.6	1581.3	1634.0
90 (320.27)	Sh			19.73	39.73	59.73	79.73	99.73	129.73	179.73	279.73	379.73	479.73	579.73	679.73	779.73	879.73
	v	0.0177	4.898	5.055	5.208	5.357	5.504	5.653	5.869	6.220	6.916	7.599	8.272	8.943	9.626	10.29	10.96
	h	290.57	1185.4	1197.3	1209.0	1219.8	1230.0	1239.1	1254.5	1279.3	1328.7	1378.1	1427.9	1478.1	1529.5	1581.2	1634.0
95 (324.13)	Sh			15.87	35.87	55.87	75.87	95.87	125.87	175.87	275.87	375.87	475.87	575.87	675.87	775.87	875.87
	v	0.0177	4.653	4.773	4.921	5.063	5.205	5.346	5.552	5.886	6.547	7.195	7.834	8.481	9.117	9.751	10.38
	h	294.58	1186.4	1196.0	1208.0	1219.0	1229.3	1238.6	1254.0	1278.9	1328.4	1377.8	1427.7	1478.0	1529.4	1581.1	1633.9
100 (327.83)	Sh			12.17	32.17	52.17	72.17	92.17	122.17	172.17	272.17	372.17	472.17	572.17	672.17	772.17	872.17
	v	0.0177	4.433	4.520	4.663	4.801	4.936	5.070	5.266	5.589	6.217	6.836	7.448	8.055	8.659	9.262	9.862
	h	298.43	1187.3	1194.9	1207.0	1218.3	1228.4	1238.6	1253.7	1278.6	1327.9	1377.5	1427.5	1478.0	1529.2	1581.0	1633.7
105 (331.38)	Sh			8.62	28.62	48.62	68.62	88.62	118.62	168.62	268.62	368.62	468.62	568.62	668.62	768.62	868.62
	v	0.0178	4.232	4.292	4.429	4.562	4.691	4.820	5.007	5.316	5.916	6.507	7.090	7.670	8.245	8.819	9.391
	h	302.13	1188.2	1193.5	1205.9	1217.2	1227.6	1237.5	1252.9	1278.0	1327.6	1377.4	1427.3	1477.7	1529.2	1580.9	1633.7
110 (334.79)	Sh			5.21	25.21	45.21	65.21	85.21	115.21	165.21	265.21	365.21	465.21	565.21	665.21	765.21	865.21
	v	0.0178	4.050	4.084	4.217	4.345	4.469	4.592	4.773	5.069	5.643	6.208	6.765	7.319	7.869	8.417	8.963
	h	305.69	1189.0	1192.2	1204.9	1216.4	1226.8	1236.9	1252.4	1277.5	1327.1	1377.1	1427.1	1477.5	1529.1	1580.8	1633.6

Sh = superheat, °F.

v = specific volume, ft³/lb.

h = enthalpy, Btu/lb.

TABLE C.2 (CONT.)

Temperature—Degrees Fahrenheit

Abs. Press. lb/in.² (Sat. Temp.)		Sat. Water	Sat. Steam	340°	360°	380°	400°	420°	450°	500°	600°	700°	800°	900°	1000°	1100°	1200°
115 (338.08)	Sh	0.0179	3.882		21.92	41.92	61.92	81.92	111.92	161.92	261.92	361.92	461.92	561.92	661.92	761.92	861.92
	v	309.13	1189.8		4.022	4.146	4.266	4.384	4.558	4.843	5.393	5.935	6.469	6.999	7.525	8.049	8.572
	h				1203.8	1215.6	1226.2	1236.3	1251.9	1277.1	1327.1	1376.9	1427.0	1477.4	1528.9	1580.7	1633.6
120 (341.26)	Sh	0.0179	3.728		18.74	38.74	58.74	78.74	108.74	158.74	258.74	358.74	458.74	558.74	658.74	758.74	858.74
	v	312.46	1190.6		3.845	3.963	4.079	4.194	4.361	4.635	5.165	5.685	6.197	6.705	7.210	7.713	8.215
	h				1202.7	1214.7	1225.4	1235.7	1251.4	1276.7	1326.8	1376.7	1426.8	1477.2	1528.8	1580.6	1633.5
125 (344.34)	Sh	0.0179	3.586		15.66	35.66	55.66	75.66	105.66	155.66	255.66	355.66	455.66	555.66	655.66	755.66	855.66
	v	315.69	1191.3		3.680	3.796	3.908	4.019	4.181	4.445	4.954	5.454	5.947	6.435	6.920	7.403	7.885
	h				1201.6	1213.7	1224.5	1235.0	1250.8	1276.3	1326.5	1376.4	1426.6	1477.1	1528.7	1580.5	1633.4
130 (347.31)	Sh	0.0180	3.455		12.69	32.69	52.69	72.69	102.69	152.69	252.69	352.69	452.69	552.69	652.69	752.69	852.69
	v	318.81	1192.0		3.528	3.641	3.750	3.857	4.013	4.268	4.760	5.242	5.716	6.186	6.653	7.117	7.581
	h				1200.4	1212.7	1223.6	1234.3	1250.3	1275.8	1326.1	1376.1	1426.4	1476.9	1528.6	1580.4	1633.3
135 (350.21)	Sh	0.0180	3.333		9.79	29.79	49.79	69.79	99.79	149.79	249.79	349.79	449.79	549.79	649.79	749.79	849.79
	v	321.86	1192.7		3.388	3.497	3.603	3.707	3.859	4.105	4.580	5.045	5.502	5.955	6.405	6.853	7.303
	h				1199.2	1211.7	1222.7	1233.6	1249.7	1275.4	1325.8	1375.9	1426.2	1476.8	1528.5	1580.3	1633.2
140 (353.03)	Sh	0.0180	3.220		6.97	26.97	46.97	66.97	96.97	146.97	246.97	346.97	446.97	546.97	646.97	746.97	846.97
	v	324.83	1193.3		3.258	3.364	3.467	3.567	3.715	3.954	4.413	4.862	5.303	5.741	6.175	6.607	7.037
	h				1198.0	1210.6	1221.8	1232.9	1249.1	1275.0	1325.5	1375.7	1426.0	1476.6	1528.4	1580.2	1633.2
145 (355.76)	Sh	0.0181	3.114		4.24	24.24	44.24	64.24	94.24	144.24	244.24	344.24	444.24	544.24	644.24	744.24	844.24
	v	327.71	1193.9		3.136	3.240	3.340	3.438	3.581	3.812	4.257	4.692	5.119	5.541	5.961	6.378	6.794
	h				1196.7	1209.5	1220.9	1232.2	1248.5	1274.5	1325.1	1375.4	1425.8	1476.5	1528.3	1580.1	1633.1
150 (358.43)	Sh	0.0181	3.016			21.57	41.57	61.57	91.57	141.57	241.57	341.57	441.57	541.57	641.57	741.57	841.57
	v	330.53	1194.4			3.124	3.221	3.317	3.456	3.681	4.112	4.533	4.946	5.355	5.761	6.164	6.567
	h					1208.4	1220.0	1231.4	1248.0	1274.1	1324.9	1375.1	1425.6	1476.3	1528.1	1580.0	1633.0
155 (361.02)	Sh	0.0181	2.921			18.98	38.98	58.98	88.98	138.98	238.98	338.98	438.98	538.98	638.98	738.98	838.98
	v	333.27	1195.0			3.015	3.110	3.203	3.340	3.558	3.976	4.384	4.785	5.181	5.574	5.964	6.354
	h					1207.2	1219.1	1230.7	1247.5	1273.6	1324.5	1374.9	1425.4	1476.2	1528.0	1579.9	1632.9
160 (363.55)	Sh	0.0182	2.834			16.45	36.45	56.45	86.45	136.45	236.45	336.45	436.45	536.45	636.45	736.45	836.45
	v	335.95	1195.5			2.913	3.006	3.097	3.230	3.443	3.849	4.245	4.633	5.018	5.398	5.777	6.155
	h					1206.0	1218.3	1230.0	1246.9	1273.2	1324.1	1374.7	1425.2	1476.0	1527.9	1579.8	1632.8

Abs. Press. lb/in.² (Sat. Temp.)		Sat. Water	Sat. Steam	400°	420°	440°	460°	480°	500°	550°	600°	700°	800°	900°	1000°	1100°	1200°
165 (366.01)	Sh	0.0182	2.752	33.99	53.99	73.99	93.99	113.99	133.99	183.99	233.99	333.99	433.99	533.99	633.99	733.99	833.99
	v	338.55	1195.9	2.909	2.997	3.084	3.170	3.251	3.334	3.533	3.729	4.114	4.491	4.864	5.234	5.601	5.967
	h			1217.4	1229.3	1241.1	1251.8	1262.4	1272.8	1298.5	1323.8	1374.5	1425.0	1475.9	1527.8	1579.7	1632.7

Sh = superheat, °F.   *v* = specific volume, ft³/lb.   *h* = enthalpy, Btu/lb.

P, psia (Tsat, °F)		Sat. liq.	Sat. vap.	400	420	440	460	480	500	550	600	700	800	900	1000	1100	1200
170 (368.42)	Sh			31.58	51.58	71.58	91.58	111.58	131.58	181.58	231.58	331.58	431.58	531.58	631.58	731.58	831.58
	v	0.0182	2.674	2.816	2.903	2.988	3.071	3.151	3.232	3.426	3.617	3.991	4.357	4.720	5.079	5.436	5.791
	h	341.11	1196.3	1216.5	1228.4	1240.5	1251.3	1261.3	1272.3	1298.2	1323.5	1374.2	1424.9	1475.7	1527.6	1579.6	1632.7
175 (370.77)	Sh			29.23	49.23	69.23	89.23	109.23	129.23	179.23	229.23	329.23	429.23	529.23	629.23	729.23	829.23
	v	0.0182	2.601	2.730	2.814	2.897	2.979	3.057	3.136	3.325	3.510	3.875	4.231	4.584	4.932	5.279	5.625
	h	343.61	1196.7	1215.6	1227.6	1239.9	1250.8	1261.1	1271.9	1297.5	1323.2	1374.0	1424.7	1475.6	1527.5	1579.5	1632.6
180 (373.08)	Sh			26.92	46.92	66.92	86.92	106.92	126.92	176.92	226.92	326.92	426.92	526.92	626.92	726.92	826.92
	v	0.0183	2.532	2.648	2.731	2.812	2.892	2.968	3.045	3.229	3.410	3.765	4.112	4.455	4.794	5.132	5.468
	h	346.07	1197.2	1214.6	1226.8	1239.2	1250.2	1260.8	1271.5	1297.1	1322.8	1373.7	1424.5	1475.5	1527.4	1579.4	1632.5
185 (375.34)	Sh			24.66	44.66	64.66	84.66	104.66	124.66	174.66	224.66	324.66	424.66	524.66	624.66	724.66	824.66
	v	0.0183	2.466	2.570	2.651	2.731	2.809	2.884	2.958	3.139	3.315	3.661	3.999	4.333	4.664	4.992	5.319
	h	348.47	1197.6	1213.7	1226.0	1238.4	1249.6	1260.3	1271.0	1297.4	1322.4	1373.6	1424.3	1475.3	1527.3	1579.3	1632.4
190 (377.55)	Sh			22.45	42.45	62.45	82.45	102.45	122.45	172.45	222.45	322.45	422.45	522.45	622.45	722.45	822.45
	v	0.0183	2.404	2.496	2.576	2.654	2.731	2.804	2.877	3.053	3.225	3.563	3.893	4.218	4.540	4.860	5.179
	h	350.83	1198.0	1212.7	1225.1	1237.7	1249.0	1259.8	1270.5	1296.6	1322.1	1373.1	1424.1	1475.2	1527.1	1579.2	1632.3
195 (379.70)	Sh			20.30	40.30	60.30	80.30	100.30	120.30	170.30	220.30	320.30	420.30	520.30	620.30	720.30	820.30
	v	0.0184	2.344	2.426	2.505	2.581	2.656	2.728	2.799	2.972	3.140	3.470	3.791	4.109	4.423	4.735	5.046
	h	353.13	1198.4	1211.7	1224.2	1237.0	1248.3	1259.3	1270.0	1296.2	1321.8	1372.9	1423.9	1475.0	1527.0	1579.1	1632.2
200 (381.82)	Sh			18.18	38.18	58.18	78.18	98.18	118.18	168.18	218.18	318.18	418.18	518.18	618.18	718.18	818.18
	v	0.0184	2.288	2.360	2.437	2.512	2.585	2.656	2.726	2.895	3.059	3.381	3.697	4.005	4.311	4.616	4.919
	h	355.40	1198.7	1210.8	1223.7	1236.3	1247.9	1258.7	1269.4	1295.6	1321.4	1372.5	1423.9	1474.9	1526.6	1579.0	1632.1
205 (383.89)	Sh			16.11	36.11	56.11	76.11	96.11	116.11	166.11	216.11	316.11	416.11	516.11	616.11	716.11	816.11
	v	0.0184	2.235	2.297	2.372	2.446	2.518	2.587	2.656	2.821	2.982	3.297	3.604	3.906	4.205	4.502	4.798
	h	357.61	1199.0	1209.7	1222.5	1235.4	1247.1	1258.2	1269.0	1295.4	1321.0	1372.4	1423.5	1474.7	1526.8	1578.9	1632.1
210 (385.93)	Sh			14.07	34.07	54.07	74.07	94.07	114.07	164.07	214.07	314.07	414.07	514.07	614.07	714.07	814.07
	v	0.0184	2.183	2.237	2.311	2.384	2.454	2.522	2.589	2.751	2.909	3.216	3.516	3.812	4.104	4.395	4.683
	h	359.80	1199.4	1208.8	1221.8	1234.7	1246.5	1257.7	1268.5	1295.0	1320.7	1372.1	1423.3	1474.6	1526.8	1578.8	1632.0
215 (387.93)	Sh			12.07	32.07	52.07	72.07	92.07	112.07	162.07	212.07	312.07	412.07	512.07	612.07	712.07	812.07
	v	0.0185	2.134	2.179	2.252	2.324	2.393	2.460	2.526	2.685	2.839	3.140	3.433	3.722	4.008	4.292	4.574
	h	361.95	1199.6	1207.8	1221.0	1234.0	1245.9	1257.2	1268.0	1294.6	1320.4	1371.9	1423.1	1474.4	1526.5	1578.7	1631.9
220 (389.89)	Sh			10.11	30.11	50.11	70.11	90.11	110.11	160.11	210.11	310.11	410.11	510.11	610.11	710.11	810.11
	v	0.0185	2.086	2.124	2.196	2.267	2.335	2.400	2.465	2.621	2.772	3.067	3.354	3.637	3.916	4.193	4.469
	h	364.05	1199.9	1206.8	1220.1	1233.2	1245.2	1256.7	1267.5	1294.1	1320.0	1371.6	1422.9	1474.2	1526.4	1578.6	1631.8
225 (391.81)	Sh			8.19	28.19	48.19	68.19	88.19	108.19	158.19	208.19	308.19	408.19	508.19	608.19	708.19	808.19
	v	0.0185	2.042	2.072	2.142	2.212	2.279	2.344	2.407	2.560	2.708	2.997	3.278	3.555	3.828	4.100	4.369
	h	366.11	1200.2	1205.8	1219.2	1232.3	1244.5	1256.2	1267.1	1293.7	1319.6	1371.4	1422.7	1474.1	1526.3	1578.5	1631.7
230 (393.70)	Sh			6.30	26.30	46.30	66.30	86.30	106.30	156.30	206.30	306.30	406.30	506.30	606.30	706.30	806.30
	v	0.0186	1.9989	2.021	2.091	2.160	2.226	2.289	2.352	2.502	2.647	2.930	3.205	3.477	3.744	4.010	4.274
	h	368.16	1200.4	1204.9	1218.3	1231.6	1243.8	1255.6	1266.7	1293.3	1319.3	1371.1	1422.5	1474.0	1526.2	1578.4	1631.6

Sh = superheat, °F.   v = specific volume, ft³/lb.   h = enthalpy, Btu/lb.

TABLE C.2 (CONT.)

Temperature—Degrees Fahrenheit

Abs. Press. lb/in.² (Sat. Temp.)		Sat. Water	Sat. Steam	400°	420°	440°	460°	480°	500°	520°	550°	600°	700°	800°	900°	1000°	1100°	1200°
235 (395.56)	Sh			4.44	24.44	44.44	64.44	84.44	104.44		154.44	204.44	304.44	404.44	504.44	604.44	704.44	804.44
	v	0.0186	1.9573	1.973	2.042	2.110	2.175	2.237	2.298		2.446	2.589	2.866	3.136	3.402	3.664	3.924	4.182
	h	370.17	1200.7	1203.9	1217.5	1230.8	1243.2	1255.0	1266.2		1292.9	1319.0	1370.9	1422.3	1473.8	1526.0	1578.3	1631.6
240 (397.40)	Sh				22.60	42.60	62.60	82.60	102.60		152.60	202.60	302.60	402.60	502.60	602.60	702.60	802.60
	v	0.0186	1.9176		1.995	2.062	2.126	2.187	2.247		2.392	2.532	2.805	3.069	3.330	3.586	3.841	4.095
	h	372.16	1200.9		1216.6	1230.0	1242.5	1254.4	1265.7		1292.5	1318.6	1370.5	1422.1	1473.6	1525.9	1578.2	1631.5
245 (399.20)	Sh				20.80	40.80	60.80	80.80	100.80		150.80	200.80	300.80	400.80	500.80	600.80	700.80	800.80
	v	0.0186	1.8797		1.950	2.015	2.078	2.139	2.198		2.341	2.479	2.746	3.006	3.261	3.513	3.762	4.011
	h	374.11	1201.1		1215.6	1229.1	1241.8	1253.8	1265.2		1292.0	1318.3	1370.3	1421.9	1473.5	1525.8	1578.1	1631.4
250 (400.97)	Sh				19.03	39.03	59.03	79.03	99.03	119.03	149.03	199.03	299.03	399.03	499.03	599.03	699.03	799.03
	v	0.0187	1.8431		1.9065	1.9711	2.0334	2.0932	2.1515	2.2085	2.2920	2.4272	2.6897	2.9444	3.1949	3.4416	3.6867	3.9299
	h	376.04	1201.4		1214.6	1228.3	1241.0	1253.2	1264.7	1274.5	1291.6	1317.9	1370.0	1421.7	1473.3	1525.6	1578.0	1631.3
255 (402.71)	Sh				17.29	37.29	57.29	77.29	97.29	117.29	147.29	197.29	297.29	397.29	497.29	597.29	697.29	797.29
	v	0.0187	1.8079		1.8686	1.9286	1.9899	2.0489	2.1065	2.1626	2.2447	2.3776	2.6354	2.8855	3.1313	3.3733	3.6138	3.8524
	h	377.91	1201.6		1213.7	1227.5	1240.3	1252.6	1264.2	1274.2	1291.2	1317.5	1369.8	1421.5	1473.2	1525.5	1577.9	1631.2
260 (404.43)	Sh				15.57	35.57	55.57	75.57	95.57	115.57	145.57	195.57	295.57	395.57	495.57	595.57	695.57	795.57
	v	0.0187	1.7742		1.8246	1.8876	1.9482	2.0063	2.0631	2.1185	2.1991	2.3299	2.5833	2.8289	3.0701	3.3077	3.5437	3.7778
	h	379.78	1201.8		1212.8	1226.6	1239.5	1252.0	1263.6	1273.8	1290.8	1317.1	1369.5	1421.3	1473.0	1525.4	1577.8	1631.1
265 (406.12)	Sh				13.88	33.88	53.88	73.88	93.88	113.88	143.88	193.88	293.88	393.88	493.88	593.88	693.88	793.88
	v	0.0187	1.7416		1.7858	1.8481	1.9080	1.9654	2.0213	2.0759	2.1554	2.2840	2.5331	2.7744	3.0114	3.2446	3.4761	3.7061
	h	381.62	1202.0		1211.9	1225.7	1238.7	1251.2	1263.0	1273.4	1290.4	1316.8	1369.3	1421.1	1472.9	1525.3	1577.7	1631.1
270 (407.79)	Sh				12.21	32.21	52.21	72.21	92.21	112.21	142.21	192.21	292.21	392.21	492.21	592.21	692.21	792.21
	v	0.0188	1.7101		1.7486	1.8101	1.8692	1.9259	1.9810	2.0350	2.1131	2.2399	2.4847	2.7219	2.9548	3.1838	3.4112	3.6370
	h	383.43	1202.2		1211.0	1224.9	1238.0	1250.6	1262.5	1273.0	1290.0	1316.4	1369.0	1420.9	1472.7	1525.1	1577.6	1631.0
275 (409.44)	Sh				10.56	30.56	50.56	70.56	90.56	110.56	140.56	190.56	290.56	390.56	490.56	590.56	690.56	790.56
	v	0.0188	1.6798		1.7127	1.7735	1.8318	1.8879	1.9422	1.9956	2.0725	2.1973	2.4382	2.6714	2.9002	3.1253	3.3486	3.5704
	h	385.22	1202.3		1210.0	1224.1	1237.3	1250.0	1262.0	1272.6	1289.5	1316.1	1368.7	1420.7	1472.6	1525.0	1577.5	1630.9
280 (411.06)	Sh				8.94	28.94	48.94	68.94	88.94	108.94	138.94	188.94	288.94	388.94	488.94	588.94	688.94	788.94
	v	0.0188	1.6504		1.6780	1.7381	1.7957	1.8512	1.9048	1.9575	2.0334	2.1562	2.3932	2.6226	2.8475	3.0688	3.2883	3.5062
	h	386.99	1202.5		1209.0	1223.2	1236.5	1249.4	1261.5	1272.2	1289.1	1315.7	1368.5	1420.5	1472.4	1524.9	1577.4	1630.8
285 (412.66)	Sh				7.34	27.34	47.34	67.34	87.34	107.34	137.34	187.34	287.34	387.34	487.34	587.34	687.34	787.34
	v	0.0188	1.6232		1.6446	1.7040	1.7610	1.8157	1.8687	1.9207	1.9955	2.1165	2.3499	2.5756	2.7968	3.0143	3.2300	3.4443
	h	388.74	1202.7		1208.0	1222.3	1235.6	1248.7	1260.9	1271.8	1288.6	1315.4	1368.2	1420.3	1472.2	1524.7	1577.3	1630.7

Sh = superheat, °F.   v = specific volume, ft³/lb.   h = enthalpy, Btu/lb.

Abs. press. (Sat. temp. °F)		Sat. liq.	Sat. vap.														
290 (414.24)	Sh			5.76	25.76	45.76	65.76	85.76	105.76	135.76	185.76	285.76	385.76	485.76	585.76	685.76	785.76
	v	0.0189	1.5947	1.6122	1.6710	1.7273	1.7815	1.8338	1.8853	1.9590	2.0783	2.3080	2.5302	2.7478	2.9616	3.1738	3.3844
	h	390.47	1202.9	1207.0	1221.4	1234.8	1248.0	1260.4	1271.4	1288.2	1315.0	1367.9	1420.1	1472.1	1524.6	1577.2	1630.6
295 (415.80)	Sh			4.20	24.20	44.20	64.20	84.20	104.20	134.20	184.20	284.20	384.20	484.20	584.20	684.20	784.20
	v	0.0189	1.5684	1.5809	1.6391	1.6948	1.7484	1.8001	1.8510	1.9236	2.0413	2.2677	2.4863	2.7004	2.9108	3.1195	3.3267
	h	392.17	1203.0	1206.1	1220.5	1234.0	1247.4	1259.8	1271.0	1287.8	1314.7	1367.6	1419.9	1472.0	1524.5	1577.1	1630.5
300 (417.33)p	Sh			2.67	22.67	42.67	62.67	82.67	102.67	132.67	182.67	282.67	382.67	482.67	582.67	682.67	782.67
	v	0.0189	1.5426	1.5506	1.6082	1.6634	1.7164	1.7677	1.8172	1.8896	2.0056	2.2286	2.4447	2.6547	2.8634	3.0670	3.2707
	h	393.85	1203.2	1205.2	1219.5	1233.4	1246.6	1259.2	1270.5	1287.4	1314.4	1367.4	1419.7	1471.8	1524.4	1577.0	1630.4
310 (420.35)	Sh				19.65	39.65	59.65	79.65	99.65	129.65	179.65	279.65	379.65	479.65	579.65	679.65	779.65
	v	0.0189	1.4938		1.5495	1.6036	1.6555	1.7054	1.7546	1.8246	1.9375	2.1541	2.3631	2.5675	2.7682	2.9671	3.1645
	h	397.16	1203.5		1217.8	1231.5	1245.3	1258.0	1269.6	1286.4	1313.5	1366.9	1419.3	1471.5	1524.1	1576.8	1630.3
320 (423.29)	Sh				16.71	36.71	56.71	76.71	96.71	126.71	176.71	276.71	376.71	476.71	576.71	676.71	776.71
	v	0.0190	1.4479		1.4943	1.5473	1.5982	1.6472	1.6954	1.7637	1.8737	2.0844	2.2874	2.4857	2.6804	2.8735	3.0648
	h	400.40	1203.8		1216.0	1229.9	1244.0	1256.8	1268.6	1285.6	1312.8	1366.3	1418.9	1471.2	1523.8	1576.6	1630.1
330 (426.16)	Sh				13.84	33.84	53.84	73.84	93.84	123.84	173.84	273.84	373.84	473.84	573.84	673.84	773.84
	v	0.0190	1.4048		1.4424	1.4944	1.5445	1.5925	1.6397	1.7064	1.8138	2.0189	2.2163	2.4090	2.5981	2.7855	2.9712
	h	403.56	1204.0		1214.1	1228.2	1242.5	1255.5	1267.6	1284.7	1312.1	1365.8	1418.4	1470.8	1523.6	1576.4	1630.0
340 (428.96)	Sh				11.04	31.04	51.04	71.04	91.04	121.04	171.04	271.04	371.04	471.04	571.04	671.04	771.04
	v	0.0191	1.3640		1.3935	1.4446	1.4936	1.5409	1.5872	1.6525	1.7573	1.9572	2.1493	2.3368	2.5206	2.7027	2.8831
	h	406.65	1204.2		1212.2	1226.5	1241.0	1254.2	1266.6	1283.8	1311.4	1365.2	1418.0	1470.5	1523.3	1576.2	1629.8
350 (431.71)	Sh				8.29	28.29	48.29	68.29	88.29	118.29	168.29	268.29	368.29	468.29	568.29	668.29	768.29
	v	0.0191	1.3255		1.3472	1.3976	1.4460	1.4923	1.5377	1.6016	1.7041	1.8991	2.0863	2.2687	2.4475	2.6246	2.8000
	h	409.70	1204.4		1210.3	1224.8	1239.5	1252.9	1265.5	1282.9	1310.6	1364.7	1417.6	1470.2	1523.0	1576.0	1629.6
360 (434.39)	Sh				5.61	25.61	45.61	65.61	85.61	115.61	165.61	265.61	365.61	465.61	565.61	665.61	765.61
	v	0.0192	1.2889		1.3035	1.3532	1.4008	1.4463	1.4909	1.5536	1.6538	1.8441	2.0266	2.2044	2.3784	2.5506	2.7213
	h	412.67	1204.5		1208.5	1223.1	1238.0	1251.5	1264.5	1282.0	1309.9	1364.1	1417.2	1469.9	1522.8	1575.8	1629.4
370 (437.01)	Sh			22.99	42.99	62.99	82.99	102.99	122.99	142.99	162.99	262.99	362.99	462.99	562.99	662.99	762.99
	v	0.0192	1.2545	1.3111	1.3579	1.4028	1.4466	1.4881	1.5286	1.5675	1.6063	1.7921	1.9703	2.1435	2.3131	2.4809	2.6471
	h	415.58	1204.6	1221.4	1236.5	1250.2	1263.4	1275.2	1286.7	1298.3	1309.1	1363.6	1410.8	1469.6	1522.5	1575.8	1629.2
380 (439.59)	Sh			20.41	40.41	60.41	80.41	100.41	120.41	140.41	160.41	260.41	360.41	460.41	560.41	660.41	760.41
	v	0.0193	1.2217	1.2711	1.3173	1.3614	1.4045	1.4452	1.4850	1.5232	1.5612	1.7428	1.9168	2.0859	2.2512	2.4148	2.5768
	h	418.45	1204.7	1219.8	1235.0	1248.8	1262.2	1274.2	1286.0	1297.5	1308.4	1363.0	1416.4	1469.2	1522.2	1575.4	1629.1
390 (442.11)	Sh			17.89	37.89	57.89	77.89	97.89	117.89	137.89	157.89	257.89	357.89	457.89	557.89	657.89	757.89
	v	0.0193	1.1904	1.2332	1.2788	1.3222	1.3647	1.4046	1.4436	1.4812	1.5184	1.6961	1.8661	2.0311	2.1925	2.3521	2.5101
	h	421.27	1204.8	1218.0	1233.4	1247.4	1261.2	1273.2	1285.1	1296.7	1307.7	1362.5	1416.0	1468.9	1522.0	1575.2	1628.9
400 (444.58)	Sh			15.42	35.42	55.42	75.42	95.42	115.42	135.42	155.42	255.42	355.42	455.42	555.42	655.42	755.42
	v	0.0193	1.1609	1.1972	1.2422	1.2849	1.3269	1.3660	1.4042	1.4413	1.4777	1.6522	1.8219	1.9796	2.1367	2.2926	2.4475
	h	424.02	1204.9	1216.5	1231.6	1245.9	1259.9	1272.4	1284.3	1295.8	1307.0	1362.1	1415.5	1468.6	1521.5	1574.8	1628.8

Sh = superheat, °F. *v* = specific volume, ft³/lb. *h* = enthalpy, Btu/lb.

Physical Properties of Various Organic and Inorganic Substances

General Sources of Data for Tables on the Physical Properties, Heat Capacities, and Thermodynamic Properties in Appendices D, E, and F

1. Brown, G. G., et al., *Unit Operations*, Wiley, New York, 1956. [Heat capacity; graph for various liquids and gases (p. 587) in Appendix E.]
2. Hodgeman, Charles D., *Handbook of Chemistry and Physics*, 40th ed., Chemical Rubber Publishing Co., Cleveland, 1958.
3. Kobe, Kenneth A., and Associates, "Thermochemistry of Petrochemicals," Reprint from *Petroleum Refiner*, Gulf Publishing Co., Houston, Jan. 1949–July 1958. (Enthalpy tables D.2–D.7 and heat capacities of several gases in Table E.1, Appendix E.)
4. Lange, N. A., *Handbook of Chemistry*, 9th ed., Handbook Publishers, Sandusky, Ohio, 1956.
5. Maxwell, J. B., *Data Book on Hydrocarbons*, Van Nostrand Reinhold, New York, 1950.
6. Perry, J. H., ed., *Chemical Engineers' Handbook*, 3rd ed., McGraw-Hill, New York, 1950.
7. Rossini, Frederick D., et al., "Selected Values of Chemical Thermodynamic Properties," from *National Bureau of Standards Circular* 500, Government Printing Office, Washington, D.C., 1952.
8. Rossini, Frederick D., et al., "Selected Values of Physical and Thermodynamic Properties of Hydrocarbons and Related Compounds," American Petroleum Institute Research Project 44, Carnegie Institute of Technology, Pittsburgh, 1953.

TABLE D.1 PHYSICAL PROPERTIES OF VARIOUS ORGANIC AND INORGANIC SUBSTANCES*

Sp Gr = 20°C/4°C Unless Specified. Sp Gr for Gas Referred to Air (a) or Hydrogen (d)

Compound	Formula	Formula wt	Sp gr	Melting Temp., °K	$\Delta\hat{H}$ Fusion kcal/g mole	Normal b.p., °K	$\Delta\hat{H}$ Vap. at b.p. kcal/g mole	T_c °K	p_c atm	V_c cm³/g mole	z_c
Acetaldehyde	C_2H_4O	44.05	$0.783^{18°/4°}$	149.5		293.2		461.0			
Acetic acid	CH_3CHO	60.05	1.049	328.9	2.89	391.4	5.83	594.8	57.1	171	0.200
Acetone	C_3H_6O	58.08	0.791	178.2		329.2		508.0	47.0	213	0.238
Acetylene	C_2H_2	26.04	0.9061(A)	191.7	0.9	191.7	4.2	309.5	61.6	113	0.274
Air			1.000					132.5	37.2		
Ammonia	NH_3	17.03	$0.817^{-79°}$ 0.597(A)	195.40	1.351	239.73	5.581	405.5	111.3	72.5	0.243
Ammonium carbonate	$(NH_4)_2CO_3 \cdot H_2O$	114.11			(decomposes at 331°K)						
Ammonium chloride	NH_4Cl	53.50	$1.531^{7°}$		(decomposes at 623°K)						
Ammonium nitrate	NH_4NO_3	80.05	$1.725^{25°}$	442.8	1.3	(decomposes at 483.2°K)					
Ammonium sulfate	$(NH_4)_2SO_4$	132.14	1.769	786			(decomposes at 786°K after melting)				
Aniline	C_6H_7N	93.12	1.022	266.9		457.4		699	52.4		
Benzaldehyde	C_6H_5CHO	106.12	1.046	247.16		452.16	9.177				
Benzene	C_6H_6	78.11	0.879	278.693	2.351	353.26	7.353	562.6	48.6	260	0.274
Benzoic acid	$C_7H_6O_2$	122.12	$1.316^{28°/4°}$	395.4		523.0					
Benzyl alcohol	C_7H_8O	108.13	1.045	257.8		478.4					
Boron oxide	B_2O_3	69.64	1.85	723	5.27						
Bromine	Br_2	159.83	$3.119^{20°}$ 5.87(A)	265.8		331.78		584	102	144	0.306
1, 2-Butadiene	C_4H_6	54.09	$0.6522^{20°}$	136.7		283.3		446			

*Sources of data are listed at the beginning of Appendix D.

TABLE D.1 (CONT.)

Compound	Formula	Formula wt	Sp gr	Melting Temp., °K	ΔĤ Fusion kcal/g mole	Normal b.p., °K	ΔĤ Vap. at b.p. kcal/g mole	T_c °K	p_c atm	V_c cm³/g mole	z_c
1,3-Butadiene	C_4H_6	54.09	0.621	164.1	1.114	268.6		425	42.7	221	0.271
Butane	$n\text{-}C_4H_{10}$	58.12	0.579	134.83		272.66	5.331	425.17	37.47	255	0.274
iso-Butane	$iso\text{-}C_4H_{10}$	58.12	0.557	113.56	1.085	261.43	5.089	408.1	36.0	263	0.283
1-Butene	C_4H_8	56.10	0.60	87.81	0.9197	266.91	5.238	419.6	39.7	240	0.277
Butyl phthalate	*see* Dibutyl phthalate										
n-Butyric acid	$n\text{-}C_4H_8O_2$	88.10	0.958	267		437.1		628	52.0	290	0.293
iso-Butyric acid	$iso\text{-}C_4H_8O_2$	88.10	0.949	226		427.7		609			
Calcium arsenate	$Ca_3(AsO_4)_2$	398.06		1723							
Calcium carbide	CaC_2	64.10	2.22$^{18°}$	2573							
Calcium carbonate	$CaCO_3$	100.09	2.93	(decomposes at 1098°K)							
Calcium chloride	$CaCl_2$	110.99	2.152$^{15°}$	1055	6.78						
	$CaCl_2 \cdot H_2O$	129.01									
	$CaCl_2 \cdot 2H_2O$	147.03									
	$CaCl_2 \cdot 6H_2O$	219.09	1.68$^{17°}$	303.4	8.92	(−6H$_2$O at 473°K)					
Calcium cyanamide	$CaCN_2$	80.11	2.29								
Calcium cyanide	$Ca(CN)_2$	92.12									
Calcium hydroxide	$Ca(OH)_2$	74.10	2.24	(−H$_2$O at 853°K)							
Calcium oxide	CaO	56.08	2.62	2873	12	3123					
Calcium phosphate	$Ca_3(PO_4)_2$	310.19	3.14	1943							
Calcium silicate	$CaSiO_3$	116.17	2.915	1803	11.62						

Compound	Formula	Formula wt	Sp gr	Melting Temp., °K	$\Delta\hat{H}$ Fusion kcal/g mole	Normal b.p., °K	$\Delta\hat{H}$ Vap. at b.p. kcal/g mole	T_c °K	p_c atm	V_c cm³/g mole	z_c
Calcium sulfate (gypsum)	$CaSO_4 \cdot 2H_2O$	172.18	2.32	$(-1\frac{1}{2}H_2O$ at 301°K)							
Carbon	C	12.010	2.26	3873	11.0	4473					
Carbon dioxide	CO_2	44.01	1.53(A)	$217.0^{5.2\,atm}$	1.99	(sublimes at 195°K)		304.2	72.9	94	0.275
Carbon disulfide	CS_2	76.14	$1.26[22°/20°]$; 2.63(A)	161.1	1.05	319.41	6.40	552.0	78.0	170	0.293
Carbon monoxide	CO	28.01	0.968(A)	68.10	0.200	81.66	1.444	133.0	34.5	93	0.294
Carbon tetrachloride	CCl_4	153.84	1.595	250.3	0.60	349.9	7.17	556.4	45.0	276	0.272
Chlorine	Cl_2	70.91	2.49(A)	172.16	1.531	239.10	4.878	417.0	76.1	124	0.276
Chlorobenzene	C_6H_5Cl	112.56	1.107	228		405.26	8.73	632.4	44.6	308	0.265
Chloroform	$CHCl_3$	119.39	$1.489^{20°}$	209.5		334.2		536.0	54.0	240	0.294
Chromium	Cr	52.01	7.1			2855	72.8				
Copper	Cu	63.54	8.92	1356.2	3.11	2855	72.8				
Cumene	C_9H_{12}	120.19	0.862	177.125	1.7	425.56	8.97	636	31.0	440	0.260
Cupric sulfate	$CuSO_4$	159.61	$3.606^{15°}$	(decomposes at 873°K)							
Cyclohexane	C_6H_{12}	84.16	0.779	279.83	0.6398	353.90	7.19	553.7	40.4	308	0.274
Cyclopentane	C_5H_{10}	70.13	0.745	179.71	0.1455	322.42	6.524	511.8	44.55	260	0.27
Decane	$C_{10}H_{22}$	142.28	$0.730^{20°}$	243.3		447.0		619.0	20.8	602	0.2476
Dibutyl phthalate	$C_8H_{22}O_4$	278.34	$1.045^{21°}$			613					
Diethyl ether	$(C_2H_5)_2O$	74.12	$0.708^{25°}$	156.86	1.745	307.76	6.226	467	35.6	281	0.261
Ethane	C_2H_6	30.07	1.049(A)	89.89	0.6834	184.53	3.517	305.4	48.2	148	0.285
Ethanol	C_2H_6O	46.07	0.789	158.6	1.200	351.7	9.22	516.3	63.0	167	0.248
Ethyl acetate	$C_4H_8O_2$	88.10	0.901	189.4		350.2		523.1	37.8	286	0.252
Ethyl benzene	C_8H_{10}	106.16	0.867	178.185	2.190	409.35	8.60	619.7	37.0	360	0.260
Ethyl bromide	C_2H_5Br	108.98	1.460	154.1		311.4		504	61.5	215	0.320
Ethyl chloride	CH_3CH_2Cl	64.52	$0.903^{10°}$	134.83	1.064	285.43	5.9	460.4	52.0	199	0.274
3-Ethyl hexane	C_8H_{18}	114.22	0.7169			391.69	8.19	567.0	26.4	466	0.264

TABLE D.1 (CONT.)

Compound	Formula	Formula wt	Sp gr	Melting Temp., °K	$\Delta\hat{H}$ Fusion kcal/g mole	Normal b.p., °K	$\Delta\hat{H}$ Vap. at b.p. kcal/g mole	T_c, °K	p_c atm	V_c cm³/g mole	z_c
Ethylene	C_2H_4	28.05	0.975(A)	103.97	0.8008	169.45	3.237	283.1	50.5	124	0.270
Ethylene glycol	$C_2H_6O_2$	62.07	$1.1131^{9°}$	260	2.685	470.4	13.6				
Ferric oxide	Fe_2O_3	159.70	5.12	1833		(decomposes at 1833°K)					
Ferric sulfide	Fe_2S_3	207.90	4.3	(decomposes)							
Ferrous sulfide	FeS	87.92	4.84	1466	(decomposes)						
Formaldehyde	H_2CO	30.03	$0.815^{-20°}$	154.9		253.9	5.85				
Formic acid	CH_2O_2	46.03	1.220	281.46	3.03	373.7	5.32				
Glycerol	$C_3H_8O_3$	92.09	$1.260^{50°}$	291.36	4.373	563.2					
Helium	He	4.00	0.1368(A)	3.5	0.005	4.216	0.020	5.26	2.26	58	0.304
Heptane	C_7H_{16}	100.20	0.684	182.57	3.354	371.59	7.575	540.2	27.0	426	0.260
Hexane	C_6H_{14}	86.17	0.659	177.84	3.114	341.90	6.896	507.9	29.9	368	0.264
Hydrogen	H_2	2.016	0.06948(A)	13.96	0.028	20.39	0.216	33.3	12.8	65	0.304
Hydrogen chloride	HCl	36.47	1.268(A)	158.94	0.476	188.11	3.86	324.6	81.5	87	0.266
Hydrogen fluoride	HF	20.01	1.15	238		293		503.2			
Hydrogen sulfide	H_2S	34.08	1.1895(A)	187.63	0.568	212.82	4.463	373.6	88.9	98	0.284
Iodine	I_2	253.8	$4.93^{20°}$	386.5	3.6	457.4		826.0			
Iron	Fe	55.85	7.7	1808		3073	84.6				
Iron oxide	Fe_3O_4	231.55	5.2	1867	33.0	(decomposes at 1867° after melting)					
Lead	Pb	207.21	$11.337^{20°}$	600.6	1.22	2023	43.0				
Lead oxide	PbO	223.21	9.5	1159	2.8	1745	51				
Magnesium	Mg	24.32	1.74	923	2.2	1393	31.5				
Magnesium chloride	$MgCl_2$	95.23	$2.325^{25°}$	987	10.3	1691	32.7				
Magnesium hydroxide	$Mg(OH)_2$	58.34	2.4	(decomposes at 623°K)							

Compound	Formula	Formula wt	Sp gr	Melting Temp., °K	ΔH Fusion kcal/g mole	Normal b.p., °K	ΔH Vap. at b.p. kcal/g mole	T_c °K	p_c atm	V_c cm³/g mole	z_c
Magnesium oxide	MgO	40.32	3.65	3173	18.5	3873					
Mercury	Hg	200.61	$13.546^{20°}$								
Methane	CH_4	16.04	0.554(A)	90.68	0.225	111.67	1.955	190.7	45.8	99	0.290
Methanol	CH_3OH	32.04	0.792	175.26	0.757	337.9	8.43	513.2	78.5	118	0.222
Methyl acetate	$C_3H_6O_2$	74.08	0.933	174.3		330.3		506.7	46.3	228	0.254
Methyl amine	CH_5N	31.06	$0.699^{-11°}$	180.5		266.3^{758mm}		429.9	73.6		
Methyl chloride	CH_3Cl	50.49	1.785(A)	175.3		249		416.1	65.8	143	0.276
Methyl ethyl ketone	C_4H_8O	72.10	0.805	186.1		352.6					
Methyl cyclohexane	C_7H_{14}	98.18	0.769.	146.58	1.6134	374.10	7.58	572.2	34.32	344	0.251
Molybdenum	Mo	95.95	10.2								
Napthalene	$C_{10}H_8$	128.16	1.145	353.2		491.0					
Nickel	Ni	58.69	$8.90^{20°}$								
Nitric acid	HNO_3	63.02	1.502	231.56	2.503	359	7.241				
Nitrobenzene	$C_6H_5O_2N$	123.11	1.203	278.7		483.9					
Nitrogen	N_2	28.02	12.5(D)	63.15	0.172	77.34	1.333	126.2	33.5	90	0.291
Nitrogen dioxide	NO_2	46.01	1.448	263.86	1.753	294.46	3.520	431.0	100.0	82	0.232
Nitrogen oxide	NO	30.01	1.0367(A)	109.51	0.550	121.39	3.293	179.2	65.0	58	0.256
Nitrogen pentoxide	N_2O_5	108.02	$1.631^{8°}$	303		320					
Nitrogen tetraoxide	N_2O_4	92	$1.448^{20°}$	263.7		294.3		431.0	99.0		
Nitrogen trioxide	N_2O_3	76.02	$1.447^{2°}$	171		276.5					
Nitrous oxide	N_2O	44.02	$1.226^{-89°}$ / 1.530(A)	182.1		184.4		309.5	71.7	96.3	0.272

TABLE D.1 (CONT.)

Compound	Formula	Formula wt	Sp gr	Melting Temp., °K	ΔĤ Fusion kcal/g mole·	Normal b.p., °K	ΔĤ Vap. at b.p. kcal/g mole	T_c °K	p_c atm	V_c cm³/g mole	z_c
n-Nonane	C_9H_{20}	128.25	0.718	219.4		423.8		595	23		
n-Octane	C_8H_{18}	114.22	0.703	216.2		398.7		595.0	22.5	543	0.250
Oxalic acid	$C_2H_2O_4$	90.04	1.90	(decomposes at 459°K)							
Oxygen	O_2	32.00	1.1053(A)	54.40	0.106	90.19	1.630	154.4	49.7	74	0.290
n-Pentane	C_5H_{12}	72.15	0.630$^{18°}$	143.49	2.006	309.23	6.160	469.8	33.3	311	0.269
iso-Pentane	iso-C_5H_{12}	72.15	0.621$^{19°}$	113.1		300.9		461.0	32.9	308	0.268
1-Pentane	C_5H_{10}	70.13	0.641	107.96	1.180	303.13		474	39.9		
Phenol	C_6H_5OH	94.11	1.071$^{25°}$	315.66	2.732	454.56		692.1	60.5		
Phenyl hydrazine	$C_6H_8N_2$	108.14	1.097$^{23°}$	292.76	3.927	516.66					
Phosphoric acid	H_3PO_4	98.00	1.834$^{18°}$	315.51	2.52	($-\frac{1}{2}H_2O$ at 486°K)					
Phosphorus (red)	P_4	123.90	2.20	863	19.40	863	10.00				
Phosphorus (white)	P_4	123.90	1.82	317.4	0.60	553	11.88				
Phosphorus pentoxide	P_2O_5	141.95	2.387	(sublimes at 523°K)							
Propane	C_3H_8	44.09	1.562(A)	85.47	0.8422	231.09	4.487	369.9	42.0	200	0.277
Propene	C_3H_6	42.08	1.498(A)	87.91	0.7176	225.46	4.402	365.1	45.4	181	0.274
Propionic acid	$C_3H_6O_2$	74.08	0.993	252.2		414.4		612.5	53.0	220	0.251
n-Propyl alcohol	C_3H_8O	60.09	0.804	146		370.2		536.7	49.95		
iso-Propyl alcohol	C_3H_8O	60.09	0.785	183.5		355.4		508.8	53.0	219	0.278
n-Propyl benzene	C_9H_{12}	120.19	0.862	173.660	2.04	432.38	9.14	638.7	31.3	429	0.257
Silicon dioxide	SiO_2	60.09	2.25	1883	2.04	2503					
Sodium bisulfate	$NaHSO_4$	120.07	2.742	455							
Sodium carbonate (sal soda)	$Na_2CO_3 \cdot 10H_2O$	286.15	1.46	306.5		($-H_2O$ at 306.5°K)					

Compound	Formula	Formula wt	Sp gr	Melting Temp., °K	$\Delta \hat{H}$ Fusion kcal/g mole	Normal b.p., °K	$\Delta \hat{H}$ Vap. at b.p. kcal/g mole	T_c °K	p_c atm	V_c cm³/g mole	z_c
Sodium carbonate (soda ash)	Na_2CO_3	105.99	2.533	1127	8.0	(decomposes)					
Sodium chloride	$NaCl$	58.45	2.163	1081	6.8	1738	40.8				
Sodium cyanide	$NaCN$	49.01		835	4.0	1770	37				
Sodium hydroxide	$NaOH$	40.00	2.130	592	2.00	1663					
Sodium nitrate	$NaNO_3$	85.00	2.257	583	3.8	(decomposes at 653°K)					
Sodium nitrite	$NaNO_2$	69.00	2.168°	544		(decomposes at 593°K)					
Sodium sulfate	Na_2SO_4	142.05	2.698	1163	5.8						
Sodium sulfide	Na_2S	78.05	1.856	1223	1.6						
Sodium sulfite	Na_2SO_3	126.05	2.633^{15°}	(decomposes)							
Sodium thiosulfate	$Na_2S_2O_3$	158.11	1.667								
Sulfur (rhombic)	S_8	256.53	2.07	386	2.40	717.76	20.0				
Sulfur (monoclinic)	S_8	256.53	1.96	392	3.386	717.76	20.0				
Sulfur chloride (mono)	S_2Cl_2	135.05	1.687	193.0		411.2	8.61				
Sulfur dioxide	SO_2	64.07	2.264(A)	197.68	1.769	263.14	5.955	430.7	77.8	122	0.269
Sulfur trioxide	SO_3	80.07	2.75(A)	290.0	6.09	316.5	9.99	491.4	83.8	126	0.262
Sulfuric acid	H_2SO_4	98.08	1.834^{18°}	283.51	2.36	(decomposes at 613°K)					
Toluene	$C_6H_5CH_3$	92.13	0.866	178.169	1.582	383.78	8.00	593.9	40.3	318	0.263
Water	H_2O	18.016	1.004°	273.16	1.4363	373.16	9.7171	647.4	218.3	56	0.230
m-Xylene	C_8H_{10}	106.16	0.864	225.288	2.765	412.26	8.70	619	34.6	390	0.27
o-Xylene	C_8H_{10}	106.16	0.880	247.978	3.250	417.58	8.80	631.5	35.7	380	0.26
p-Xylene	C_8H_{10}	106.16	0.861	286.423	4.090	411.51	8.62	618	33.9	370	0.25
Zinc	Zn	65.38	7.140	692.7	1.595	1180	27.43				
Zinc sulfate	$ZnSO_4$	161.44	3.741^{15°}	(decomposes at 1013°K)							

TABLE D.2 ENTHALPIES OF PARAFFINIC HYDROCARBONS, C_1—C_6

(cal/g mole)

°K	C_1	C_2	C_3	n-C_4	i-C_4	n-C_5	n-C_6
273	0.0	0.0	0.0	0.0	0.0	0.0	0.0
291	150.8	218.1	302.2	408.6	396.4	508.0	608.3
298	210.3	305.4	423.5	572.3	556.6	711.4	851.7
300	227.1	330.6	458.7	619.7	603.0	770.2	922.1
400	1,133	1,746	2,460	3,292	3,256	4,089	4,891
500	2,175	3,460	4,944	6,559	6,531	8,131	9,709
600	3,359	5,466	7,834	10,364	10,352	12,820	15,299
700	4,681	7,730	11,094	14,624	14,632	18,070	21,540
800	6,131	10,210	14,660	19,264	19,304	23,780	28,330
900	7,697	12,890	18,500	24,230	24,320	29,900	35,580
1000	9,370	15,730	22,570	29,500	29,630	36,380	43,270
1100	11,130	18,740	26,850	35,040	35,190	43,180	51,330
1200	12,980	21,870	31,320	40,800	40,970	50,250	59,720
1300	14,910	25,130	35,930	46,780	46,960	57,570	68,390
1400	16,900	28,490	40,680	52,910	53,110	65,100	77,310
1500	18,940	31,950	45,550	59,190	59,410	72,800	86,410
1600	21,040	—	—	—	—	—	—
1800	25,350	—	—	—	—	—	—
2000	29,810	—	—	—	—	—	—
2200	34,370	—	—	—	—	—	—
2500	41,360	—	—	—	—	—	—

TABLE D.3 ENTHALPIES OF MONOOLEFINIC HYDROCARBONS, C_2—C_4

(cal/g mole)

°K	Ethylene	Propylene	1-Butene	iso-Butene	cis-2-Butene	trans-2-Butene
273	0.0	0.0	0.0	0.0	0.0	0.0
291	180.0	264	367.3	367.6	323.8	363.3
298	252	370	515	515	453	508
300	269	398	553	555	483	541
400	1,436	2,123	2,977	2,956	2,646	2,895
500	2,842	4,200	5,919	5,848	5,341	5,735
600	4,457	6,625	9,300	9,184	8,512	9,003
700	6,252	9,333	13,060	12,880	12,080	12,670
800	8,205	12,280	17,150	16,920	16,010	16,660
900	10,290	15,450	21,510	21,230	20,240	20,950
1000	12,490	18,820	26,120	25,800	24,730	25,500
1100	14,800	22,340	30,950	29,600	29,460	30,270
1200	17,200	26,010	35,960	35,580	34,380	35,230
1300	19,680	29,800	41,130	40,730	39,480	40,350
1400	22,220	33,700	46,440	45,020	44,730	45,620
1500	24,830	37,700	51,880	51,430	50,120	51,020

TABLE D.4 ENTHALPIES OF ACETYLENES AND DIOLEFINS

(cal/g mole)

°K	Acetylene	Methyl-Acetylene	Dimethyl Acetylene	Propadiene	Butadiene 1-3	Isoprene
273	0.0	0.0	0.0	0.0	0.0	0.0
291	183.8	252.2	323.5	244.1	325.2	430.0
298	256.9	352.9	453.2	342.0	456.9	603.3
300	277.9	382.0	490.5	370.3	495.0	653
400	1,410	1,974	2,552	1,965	2,664	3,507
500	2,659	3,826	5,002	3,823	5,307	6,960
600	3,994	5,904	7,804	5,914	8,327	10,890
700	5,391	8,209	10,919	8,176	11,652	15,210
800	6,856	10,646	14,306	10,639	15,221	17,230
900	8,383	13,190	17,930	13,280	19,000	24,800
1000	9,951	15,900	21,760	16,020	22,960	29,970
1100	11,565	18,730	25,770	18,870	27,090	35,360
1200	13,210	21,680	29,940	21,790	31,360	40,920
1300	14,900	24,690	34,250	24,820	35,750	46,660
1400	16,640	27,740	38,680	27,970	40,240	52,550
1500	18,390	30,900	43,210	31,140	44,830	58,550

TABLE D.5 ENTHALPIES OF SOME OXYGENATED HYDROCARBONS

(cal/g mole)

°C	Formalde-hyde	Acetalde-hyde	Methanol	Ethanol	Ethylene Oxide	Ketene
0	0.0	0.0	0.0	0.0	0.0	0.0
18	150.0	228.4	188.1	306.7	197.5	200.0
25	209.0	319.2	263.0	429.0	277.0	279.6
100	864.9	1,369	1,122	1,849	1,234	1,195
200	1,824.6	3,004	2,434	4,020	2,796	2,577
300	2,891.7	4,884	3,924	6,642	4,656	4,115
400	4,062.8	6,984	5,580	9,228	6,772	5,780
500	5,329.0	9,275	7,390	12,200	9,085	7,556
600	6,681.6	11,730	9,336	15,370	11,610	9,424
700	8,108.8	14,330	11,400	18,710	14,220	11,370
800	9,600.0	—	—	22,210	—	13,390
900	11,148	—	—	25,840	—	15,470
1000	12,743	—	—	29,600	—	17,600
1100	14,381	—	—	33,450	—	19,708
1200	16,055	—	—	37,400	—	22,000

Table D.6 Enthalpies of Nitrogen and Some of Its Oxides

(cal/g mole)

°K	N_2	NO	N_2O	NO_2	N_2O_4
273	0.0	0.0	0.0	0.0	0.0
291	125.3	128.5	163.0	157.3	331
298	174.0	178.4	227.4	219.2	463
300	187.9	191.5	2,309	235.5	498
400	883.2	904.7	3,284	1,163	2,520
500	1,588	1,628	4,345	2,168	4,760
600	2,301	2,365	5,478	3,242	7,200
700	3,024	3,120	6,674	4,375	—
800	3,766	3,894	7,924	5,555	—
900	4,532	4,684	9,226	6,772	—
1000	5,299	5,490	10,578	8,019	—
1100	6,088	6,308	11,978	9,290	—
1200	6,888	7,137	13,425	10,580	—
1300	7,700	7,976	14,920	11,886	—
1400	8,518	8,823	16,460	13,207	—
1500	9,356	9,677	18,046	14,538	—
1750	11,458	11,832	—	—	—
2000	13,600	14,014	—	—	—
2250	15,770	16,218	—	—	—
2500	17,940	18,434	—	—	—

Table D.7 Enthalpies of Sulfur Compounds

(cal/g mole)

°K	S_2	SO_2	SO_3	H_2S	CS_2	COS
223	0.0	0.0	0.0	0.0	0.0	0.0
291	138.6	168.8	215	145.2	193.0	175.7
298	192.4	235.3	300	202.0	269.1	244.8
300	207.9	254.4	320	217.4	290.9	264.2
400	1,003	1,251	1,640	1,045	1,433	1,311
500	1,829	2,329	3,115	1,907	2,655	2,443
600	2,675	3,469	4,740	2,809	3,933	3,635
700	3,535	4,661	6,490	3,754	5,252	4,878
800	4,404	5,891	8,305	4,742	6,604	6,159
900	5,279	7,150	10,200	5,771	7,980	7,572
1000	6,159	8,431	12,150	6,838	9,374	8,910
1100	7,042	9,729	14,150	7,939	10,780	10,260
1200	7,929	11,040	16,190	9,071	12,200	11,650
1300	8,819	12,360	18,260	10,230	13,630	13,050
1400	9,711	13,700	—	11,410	15,070	14,450
1500	10,605	15,040	—	12,620	16,520	15,860
1600	11,500	16,380	—	13,840	17,970	17,280
1700	12,400	17,730	—	15,080	19,430	18,720
1800	13,290	19,090	—	16,330	20,880	20,150
1900	14,190	20,440	—	—	—	—
2000	15,090	21,810	—	—	—	—
2500	19,600	28,650	—	—	—	—
3000	24,130	35,530	—	—	—	—

Heat-Capacity Information

TABLE E.1 HEAT CAPACITY EQUATIONS FOR ORGANIC AND INORGANIC COMPOUNDS*

Units:
cal/(g mole)(°K or °C)
Btu/(lb mole)(°R or °F)

Forms:
1. $C_p^\circ = a + b(T) + c(T)^2 + d(T)^3$
2. $C_p^\circ = a + b(T) + c(T)^{-2}$
3. $C_p^\circ = a + b(T) + c(T)^{-1,2}$

Compound	Formula	Mol. wt	State	Form	T	a	$b \cdot 10^2$	$c \cdot 10^5$	$d \cdot 10^9$	Temp. Range
Acetone	CH_3COCH_3	58.08	g	1	°C	17.20	4.805	-3.056	8.307	0–1200
Acetylene	C_2H_2	26.04	g	1	°C	10.14	1.4468	-1.203	4.349	0–1200
			g	1	°F	9.89	0.8273	-0.3783	0.7457	32–2200
Air			g	1	°C	6.917	0.09911	0.07627	-0.4696	0–1500
			g	1	°K	6.713	0.04697	0.1147	-0.4696	273–1800
			g	1	°F	6.900	0.02884	0.02429	-0.08052	32–2700
			g	1	°R	6.713	0.02609	0.03540	-0.08052	492–3200
Ammonia	NH_3	17.03	g	1	°C	8.4017	0.70601	0.10567	-1.5981	0–1200
			g	1	°F	8.2765	0.39006	0.035245	-0.2740	32–2200
Ammonium sulfate	$(NH_4)_2SO_4$	132.15	c	1	°K	51.6				275–328
Benzene	C_6H_6	78.11	l	1	°K	14.95	5.58	-6.022	18.54	279–350
			g	1	°C	17.700	7.875			0–1200
Boron oxide	B_2O_3	69.64	l	1	°F	16.332	4.493	-1.888	3.179	32–2200
iso-Butane	C_4H_{10}	58.12	g	1	°C	21.382	7.202	-4.519	11.92	0–1200
n-Butane	C_4H_{10}	58.12	g	1	°C	22.060	6.663	-3.697	8.360	0–1200
iso-Butene	C_4H_8	56.10	g	1	°C	19.810	6.128	-4.127	12.07	0–1200
Calcium carbide	CaC_2	64.10	c	2	°K	16.40	0.284	-2.07×10^{10}		298–720
Calcium carbonate	$CaCO_3$	100.09	c	2	°K	19.68	1.189	-3.076×10^{10}		273–1033
Calcium hydroxide	$Ca(OH)_2$	74.10	c	1	°K	21.4				276–373
Calcium oxide	CaO	56.08	c	2	°K	10.00	0.484	-1.08×10^{10}		273–1173
Carbon	C	12.01	c†	2	°K	2.673	0.2617	-1.169×10^{10}		273–1373
Carbon dioxide	CO_2	44.01	g	1	°K	6.393	1.0100	-0.3405		273–1373
					°F	8.448	0.5757	-0.2159	0.3059	273–3700
Carbon monoxide	CO	28.01	g	1	°C	6.890	0.1436	-0.02387		0–3500
					°F	6.865	0.08024	-0.007367		0–3500

Compound	Formula	Mol. wt	State	Form	T	a	$b \cdot 10^2$	$c \cdot 10^5$	$d \cdot 10^9$	Temp. Range
Carbon tetrachloride	CCl_4	153.84	l	1	°K	22.32	3.103			273–343
Chlorine	Cl_2	70.91	g	1	°C	8.031	0.3267	−0.3840		0–1200
Copper	Cu	63.54	c	1	°K	5.44	0.1462		1.547	273–1357
Cumene (isopropyl benzene)	$C_6H_5CH(CH_3)_2$	120.19	l	1	°F	50.48				50
					°F	64.91				300
Cyclohexane	C_6H_{12}	84.16	g	1	°C	33.280	12.850	−9.510	28.80	0–1200
			l	1	°F	37.05				50
					°F	41.26				150
Cyclopentane	C_5H_{10}	70.13	g	1	°C	22.500	11.860	−7.625	19.27	0–1200
			l	1	°F	30.84				50
					°F	32.95				100
			g	1	°C	17.540	9.388	−6.103	16.41	0–1200
Ethane	C_2H_6	30.07	g	1	°C	11.800	3.326	−1.390	1.740	0–1200
Ethyl alcohol	C_2H_6O	46.07	l	1	°C	24.65				0
					°C	37.96				100
			g	1	°C	14.66	3.758	−2.091	4.740	0–1200
Ethylene	C_2H_4	28.05	g	1	°C	9.740	2.741	−1.647	4.220	0–1200
Ferric oxide	Fe_2O_3	159.70	c	2	°K	24.72	1.604	-4.234×10^{10}		273–1097
Formaldehyde	H_2CO	30.03	g	1	°C	8.192	1.020	0.0000	−2.078	0–1200
Helium	He	4.00	g	1	°K	4.97		0.0000		All
n-Hexane	C_6H_{14}	86.17	l	1	°C	51.702	9.763	−5.716		20–100
			g	1	°C	32.850				
Hydrogen	H_2	2.016	g	1	°C	6.702	0.0996	−0.007804	13.78	0–1200
			g	1	°K	6.424	0.1039	−0.0078		0–3500
Hydrogen bromide	HBr	80.92	g	1	°C	6.954	−0.00542	0.2363	−1.161	0–1200
Hydrogen chloride	HCl	36.47	g	1	°C	6.962	−0.03206	0.2322	−1.036	0–1200

*Sources of data are listed at the beginning of Appendix D.
†Graphite.

TABLE E.1 (CONT.)

Compound	Formula	Mol. wt	State	Form	T	a	$b \cdot 10^2$	$c \cdot 10^5$	$d \cdot 10^9$	Temp. Range
Hydrogen cyanide	HCN	27.03	g	1	°C	8.43	0.6950	−0.2611		0–1200
Hydrogen sulfide	H_2S	34.08	g	1	°C	8.010	0.3697	0.07200	−0.7867	0–1500
Magnesium chloride	$MgCl_2$	95.23	c	1	°K	17.3	0.377			273–991
Magnesium oxide	MgO	40.32	c	2	°K	10.86	0.1197	-2.087×10^{10}		273–2073
Methane	CH_4	16.04	g	1	°C	8.200	1.307	0.08750	−2.630	0–1200
			g		°K	4.750	1.200	0.3030	−2.630	273–1500
			g		°K	3.204	1.841	0.448		273–1500
Methyl alcohol	CH_3OH	32.04	1	1	°C	18.13				0
						19.74				40
Methyl cyclohexane	C_7H_{14}	98.18	g	1	°C	10.26	1.984	−0.448	−1.92	0–700
			1	1	°F	45.17				50
					°F	53.03				200
Methyl cyclopentane	C_6H_{12}	84.16	g	1	°C	29.000	13.510	−9.016	24.09	0–1200
			1	1	°F	38.31				50
					°F	42.52				150
Nitric acid	HNO_3	63.02	g	1	°C	23.620	10.960	−7.275	20.03	0–1200
			1	1	°C	26.28				25
Nitric oxide	NO	30.01	g	1	°C	7.050	0.1957	−0.06990	0.08729	0–3500
Nitrogen	N_2	28.02	g	1	°C	6.919	0.1365	−0.02271		0–3500
					°K	6.529	0.1488	−0.02271		273–3700
					°F	6.895	0.07624	−0.007009		
Nitrogen dioxide	NO_2	46.01	g	1	°C	8.62	0.948	−0.688	1.88	0–1200
Nitrogen tetraoxide	N_2O_4	92.02	g	1	°C	18.1	2.98	−2.71		0–300
Nitrous oxide	N_2O	44.02	g	1	°C	9.000	0.9921	−0.6438	2.526	0–1200
Oxygen	O_2	32.00	g	1	°C	7.129	0.1407	−0.01791		0–3500
					°K	6.732	0.1505	−0.01791		273–3700
					°F	7.104	0.07851	−0.005528		273–3700

Compound	Formula	Mol. wt	State	Form	T	a	$b \cdot 10^2$	$c \cdot 10^5$	$d \cdot 10^9$	Temp. Range
n-Pentane	C_5H_{12}	72.15		1	°F	38.21				50
						39.66				75
Propane	C_3H_8	44.09	g	1	°C	27.450	8.148	−4.538	10.10	0–1200
Propene	C_3H_6	42.08	g	1	°C	16.260	5.398	−3.134	7.580	0–1200
			g	1	°C	14.240	4.233	−2.430	5.880	0–1200
Sodium carbonate	Na_2CO_3	105.99	c	1	°K	28.9				288–371
Sodium carbonate	$Na_2CO_3 \cdot 10H_2O$	286.15	c	1	°K	128.0				298
Sulfur	S	32.07	c‡	1	°K	3.63	0.640			273–368
			c§	1	°K	4.38	0.440			368–392
Sulfuric acid	H_2SO_4	98.08	l	1	°C	33.25	3.727			10–45
Sulfur dioxide	SO_2	64.07	g	1	°C	9.299	0.9330	−0.7418	2.057	0–1500
Sulfur trioxide	SO_3	80.07	g	1	°C	11.591	2.196	−2.041	7.744	0–1000
Toluene	$C_6H_5 \cdot CH_3$	92.13	l	1	°C	35.56				0
						43.30				100
			g	1	°C	22.509	9.292	−6.658	19.20	0–1200
			g	1	°F	20.869	5.293	−2.086		
Water	H_2O	18.016	g	1	°C	7.880	0.3200	−0.04833	3.929	32–2200
					°K	6.970	0.3464	−0.0483		0–3500

‡Rhombic.
§Monoclinic.

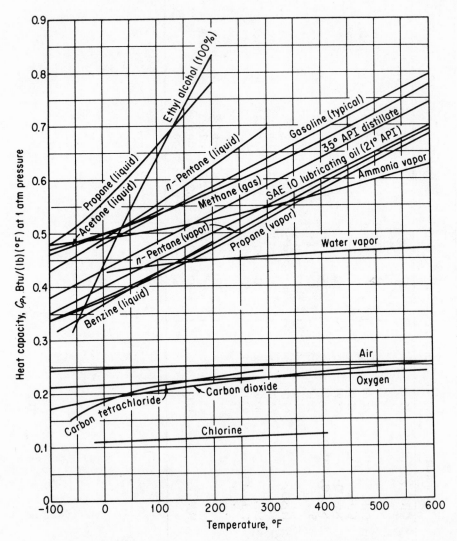

Fig. E.1. Heat capacities of liquids and gases at one atmosphere.
SOURCE: Reference 1 in Appendix D.

Heats of Formation
and Combustion

TABLE F.1 HEATS OF FORMATION AND HEATS OF COMBUSTION OF COMPOUNDS AT 25°C*

Standard States of Products for $\Delta \hat{H}_c^\circ$ Are $CO_2(g)$, $H_2O(l)$, $N_2(g)$, $SO_2(g)$, and $HCl(aq)$

Compound	Formula	Mol. wt	State	$-\Delta \hat{H}_f^\circ$ kcal/g mole	$-\Delta \hat{H}_c^\circ$ kcal/g mole
Acetic acid	CH_3COOH	60.05	l	97.800	208.34
			g		219.82
Acetaldehyde	CH_3CHO	40.052	g	39.76	284.980
Acetone	C_3H_6O	58.08	aq, 200	98.000	
			g	51.790	435.320
Acetylene	C_2H_2	26.04	g	−54.194	310.615
Ammonia	NH_3	17.032	l	16.06	
			g	11.040	91.440
Ammonium carbonate	$(NH_4)_2CO_3$	96.09	aq	225.11	
Ammonium chloride	NH_4Cl	53.50	c	75.38	
Ammonium hydroxide	NH_4OH	35.05	aq	87.59	
Ammonium nitrate	NH_4NO_3	80.05	c	87.27	
			aq	81.11	
Ammonium sulfate	$(NH_4)_2SO_4$	132.15	c	281.86	
			aq	280.38	
Benzaldehyde	C_6H_5CHO	106.12	l	21.23	
			g	9.57	
Benzene	C_6H_6	78.11	l	−11.63	780.98
			g	−19.820	789.08
Boron oxide	B_2O_3	69.64	c	302.0	
			l	297.6	
Bromine	Br_2	159.832	l	0	
			g	−7.340	
n-Butane	C_4H_{10}	58.12	l	35.29	682.51
			g	29.812	687.982
Isobutane	C_4H_{10}	58.12	l	37.87	680.93
			g	32.15	685.65

Compound	Formula	Mol. wt	State	$-\Delta \hat{H}_f^\circ$ kcal/g mole	$-\Delta \hat{H}_c^\circ$ kcal/g mole
1-Butene	C_4H_8	56.104	g	-0.280	649.757
Calcium arsenate	$Ca_3(AsO_4)_2$	398.06	c	796.0	
Calcium carbide	CaC_2	64.10	c	15.0	
Calcium carbonate	$CaCO_3$	100.09	c	288.45	
Calcium chloride	$CaCl_2$	110.99	c	190.0	
Calcium cyanamide	$CaCN_2$	80.11	c	84.0	
Calcium hydroxide	$Ca(OH)_2$	74.10	c	235.80	
Calcium oxide	CaO	56.08	c	151.9	
Calcium phosphate	$Ca_3(PO_4)_2$	310.19	c	988.9	
Calcium silicate	$CaSiO_3$	116.17	c	378.6	
Calcium sulfate	$CaSO_4$	136.15	c	342.42	
Calcium sulfate (gypsum)	$CaSO_4 \cdot 2H_2O$	172.18	aq	346.67	
Carbon	C	12.01	c	483.06	
			c Graphite (β)	0	94.052
Carbon dioxide	CO_2	44.01	g	94.052	
			l	98.69	
Carbon disulfide	CS_2	76.14	l	-21.0	256.97
			g	-27.55	263.52
Carbon monoxide	CO	28.01	g	26.416	67.636
Carbon tetrachloride	CCl_4	153.838	l	33.34	84.17
Chloroethane	C_2H_5Cl	64.52	g	25.500	92.01
Cumene (isopropylbenzene)	$C_6H_5CH(CH_3)_2$	120.19	g	25.1	339.66
			l	9.848	1246.52
Cupric sulfate	$CuSO_4$	159.61	g	-0.940	1257.31
			c	184.00	
			aq	201.51	
Cyclohexane	C_6H_{12}	84.16	l	37.34	936.88
			g	29.43	944.79
Cyclopentane	C_5H_{10}	70.130	l	25.30	786.54
			g	18.46	793.39

*Sources of data are given at the beginning of Appendix D, References 3, 6, and 7.

TABLE F.1 (CONT.)

Compound	Formula	Mol. wt	State	$-\Delta \hat{H}_f^\circ$ kcal/g mole	$-\Delta \hat{H}_c^\circ$ kcal/g mole
Ethane	C_2H_6	30.07	g	20.236	372.82
Ethyl alcohol	C_2H_5OH	46.068	l	66.356	326.700
			g	56.24	336.820
Ethyl benzene	$C_6H_5 \cdot C_2H_5$	106.16	l	2.977	1091.03
			g	−7.120	1101.13
Ethyl chloride	C_2H_5Cl	64.52	g	25.1	
Ethylene	C_2H_4	28.052	g	−12.496	337.234
Ethylene chloride	C_2H_3Cl	62.50	g	−7.500	303.90
3-Ethyl hexane	C_8H_{18}	114.22	l	59.88	1307.39
			g	50.40	1316.87
Ferric oxide	Fe_2O_3	159.70	c	196.5	
Ferric sulfide	FeS_2	*see* Iron sulfide			
Ferrosoferric oxide	Fe_3O_4	231.55	c	267.0	
Ferrous oxide	FeO	71.85	c	63.7	
Ferrous sulfide	FeS	87.92	c	22.72	
Formaldehyde	H_2CO	30.026	g	27.700	134.670
n-Heptane	C_7H_{16}	100.20	l	53.63	1151.27
			g	44.89	1160.01
n-Hexane	C_6H_{14}	86.17	l	47.52	995.01
			g	39.96	1002.570
Hydrogen	H_2	2.016	g	0	68.317
Hydrogen bromide	HBr	80.924	g	8.660	
Hydrogen chloride	HCl	36.465	g	22.063	
Hydrogen cyanide	HCN	27.026	g	−31.200	
Hydrogen sulfide	H_2S	34.082	g	−4.815	134.462
Iron sulfide	FeS_2	119.98	c	42.52	
Iron oxide	Fe_2O_3	159.68	c	196.500	
Lead oxide	PbO	223.21	c	52.40	
Magnesium chloride	$MgCl_2$	95.23	c	153.40	
Magnesium hydroxide	$Mg(OH)_2$	58.34	c	221.00	

Compound	Formula	Mol. wt	State	$-\Delta \hat{H}_f^\circ$ kcal/g mole	$-\Delta \hat{H}_c^\circ$ kcal/g mole
Magnesium oxide	MgO	40.32	c	143.84	
Methane	CH₄	16.041	g	17.889	212.80
Methyl alcohol	CH₃OH	32.042	l	57.036	173.650
Methyl chloride	CH₃Cl	50.49	g	48.100	182.590
Methyl cyclohexane	C₇H₁₄	98.182	g	19.580	183.23†
			l	45.45	1091.13
Methyl cyclopentane	C₆H₁₂	84.156	g	36.99	1099.59
			l	33.07	941.14
Nitric acid	HNO₃	63.02	g	25.50	948.72
			l	41.404	
			aq	49.372	
Nitric oxide	NO	30.01	g	−21.600	
Nitrogen dioxide	NO₂	46.01	g	−8.091	
Nitrous oxide	N₂O	44.02	g	−19.49	
n-Pentane	C₅H₁₂	72.15	l	41.36	838.80
			g	35.00	845.160
Phosphoric acid	H₃PO₄	98.00	c	306.2	
			aq (1H₂O)	305.6	
Phosphorus	P₄	123.90	c	0	
Phosphorus pentoxide	P₂O₅	141.95	c	360.0	
Propane	C₃H₈	44.09	l	28.643	526.78
			g	24.820	530.60
Propene	C₃H₆	42.078	g	−4.879	491.987
n-Propyl alcohol	C₃H₈O	60.09	g	61.0	494.40
n-Propylbenzene	C₆H₅·CH₂·C₂H₅	120.19	l	9.178	1247.19
			g	−1.870	1258.24
Silicon dioxide	SiO₂	60.09	c	203.4	
Sodium bicarbonate	NaHCO₃	84.01	c	226.0	
Sodium bisulfate	NaHSO₄	120.07	c	269.2	
Sodium carbonate	Na₂CO₃	105.99	c	270.3	
Sodium chloride	NaCl	58.45	c	98.232	

†Standard state HCl(g).

TABLE F.1 (CONT.).

Compound	Formula	Mol. wt	State	$-\Delta\hat{H}_f^\circ$ kcal/g mole	$-\Delta\hat{H}_c^\circ$ kcal/g mole
Sodium cyanide	NaCN	49.01	c	21.46	
Sodium nitrate	$NaNO_3$	85.00	c	111.54	
Sodium nitrite	$NaNO_2$	69.00	c	85.9	
Sodium sulfate	Na_2SO_4	142.05	c	330.90	
Sodium sulfide	Na_2S	78.05	c	89.2	
Sodium sulfite	Na_2SO_3	126.05	c	260.6	
Sodium thiosulfate	$Na_2S_2O_3$	158.11	c	267.0	
Sulfur	S	32.07	c (rhombic)	0	
			c (monoclinic)	−0.071	
Sulfur chloride	S_2Cl_2	135.05	l	14.4	
Sulfur dioxide	SO_2	64.066	g	70.960	
Sulfur trioxide	SO_3	80.066	g	94.450	
Sulfuric acid	H_2SO_4	98.08	l	193.91	
			aq	216.90	
Toluene	$C_6H_5 \cdot CH_3$	92.13	l	−2.867	934.50
			g	−11.950	943.58
Water	H_2O	18.016	l	68.3174	
			g	57.7979	
m-Xylene	$C_6H_4(CH_3)_2$	106.16	l	6.075	1087.92
			g	−4.120	1098.12
o-Xylene	$C_6H_4(CH_3)_2$	106.16	l	5.841	1088.16
			g	−4.540	1098.54
p-Xylene	$C_6H_4(CH_3)_2$	106.16	l	5.838	1088.16
			g	−4.290	1098.29
Zinc sulfate	$ZnSO_4$	161.45	c	233.88	
			aq	253.33	

Vapor Pressures

TABLE G.1 VAPOR PRESSURES OF VARIOUS SUBSTANCES

Antoine equation:

$$\log_{10} p^* = B - \frac{A}{C + t}$$

where p^* = vapor pressure in mm Hg
t = temperature in °C
A, B, C = constants

Name	Formula	Range, °C	A	B	C
Acetic acid	$C_2H_4O_2$	0 to 36	1651.2	7.80307	225
		36 to 170	1416.7	7.18807	211
Acetone	C_3H_6O	—	1161.0	7.02447	224
Ammonia	NH_3	−83 to +60	1002.711	7.55466	247.885
Benzene	C_6H_6	—	1211.033	6.90565	220.790
Carbon disulfide	CS_2	−10 to +160	1122.50	6.85145	236.46
Ethyl acetate	$C_4H_8O_2$	−20 to +150	1238.71	7.09808	217.0
Ethyl alcohol	C_2H_6O	—	1554.3	8.04494	222.65
Ethyl bromide	C_2H_5Br	−50 to +130	1083.8	6.89285	231.7
n-Heptane	C_7H_{16}	—	1268.115	6.90240	216.900
Methyl alcohol	CH_4O	−20 to +140	1473.11	7.87863	230.0
Sulfur dioxide	SO_2	—	1022.80	7.32776	240
Toluene	C_7H_8	—	1344.800	6.95464	219.482

SOURCE: N. A. Lange et al., *Lange's Handbook of Chemistry*, 9th ed., Handbook Publishers, Inc., Sandusky, Ohio, 1956.

Heats of Solution and Dilution

TABLE H.1 INTEGRAL HEATS OF SOLUTION AND DILUTION AT 25°C

Formula	Description		State	$-\Delta \hat{H}_f^\circ$ kcal/mole	$-\Delta \hat{H}_{soln}^\circ$ kcal/mole	$-\Delta \hat{H}_{dil}^\circ$ kcal/mole
HCl			g	22.063		
	in	1H₂O	aq	28.331	6.268	6.268
		2	aq	33.731	11.668	5.400
		3	aq	35.651	13.588	1.920
		4	aq	36.691	14.628	1.040
		5	aq	37.371	15.308	0.680
		10	aq	38.671	16.608	1.300
		20	aq	39.218	17.155	0.547
		30	aq	39.413	17.350	0.195
		40	aq	39.516	17.448	0.098
		50	aq	39.577	17.509	0.061
		100	aq	39.713	17.650	0.141
		200	aq	39.798	17.735	0.085
		300	aq	39.837	17.774	0.039
		400	aq	39.859	17.796	0.022
		500	aq	39.874	17.811	0.015
		700	aq	39.895	17.832	0.021
		1,000	aq	39.913	17.850	0.018
		2,000	aq	39.946	17.883	0.033
		3,000	aq	39.960	17.897	0.014
		4,000	aq	39.968	17.905	0.108
		5,000	aq	39.972	17.909	0.004
		7,000	aq	39.982	17.919	0.010
		10,000	aq	39.987	17.924	0.005
		20,000	aq	39.998	17.935	0.011
		50,000	aq	40.007	17.944	0.009
		100,000	aq	40.012	17.949	0.005
		∞	aq	40.023	17.960	0.011
NaOH			crystalline, II	101.99		
	in	3H₂O	aq	108.894	6.90	6.90
		4	aq	110.219	8.23	1.33
		5	aq	111.015	9.02	0.79
		10	aq	112.148	10.16	1.14
		20	aq	112.235	10.24	0.08
		30	aq	112.203	10.21	−0.03
		40	aq	112.175	10.18	−0.03
		50	aq	112.154	10.16	−0.02
		100	aq	112.108	10.12	−0.04
		200	aq	112.1	10.10	−0.02
		300	aq	112.105	10.11	0.01
		500	aq	112.117	10.13	0.02
		1,000	aq	112.139	10.15	0.02
		2,000	aq	112.162	10.17	0.02

TABLE H.1 (CONT.)

Formula	Description		State	$-\Delta \hat{H}_f^\circ$ kcal/mole	$-\Delta \hat{H}_{soln}^\circ$ kcal/mole	$-\Delta \hat{H}_{dil}^\circ$ kcal/mole
		5,000	aq	112.186	10.20	0.03
		10,000	aq	112.201	10.21	0.01
		50,000	aq	112.220	10.23	0.02
		∞	aq	112.236	10.25	0.02
H_2SO_4			liq	193.91		
	in	$0.5H_2O$	aq	197.67	3.76	3.76
		1.0	aq	200.62	6.71	2.95
		1.5	aq	202.73	8.82	2.11
		2	aq	203.93	10.02	1.20
		3	aq	205.62	11.71	1.69
		4	aq	206.83	12.92	1.21
		5	aq	207.78	13.87	0.95
		10	aq	209.93	16.02	2.15
		25	aq	211.19	17.28	1.26
		50	aq	211.44	17.53	0.25
		100	aq	211.59	17.68	0.15
		500	aq	212.25	18.34	0.66
		1,000	aq	212.69	18.78	0.44
		5,000	aq	214.09	20.18	1.40
		10,000	aq	214.72	20.81	0.63
		100,000	aq	216.29	22.38	1.57
		500,000	aq	216.69	22.78	0.40
		∞	aq	216.90	22.99	0.21

SOURCE: F. D. Rossini et al., "Selected Values of Chem. Thermo. Properties," *Natl. Bur. Std. Circ.* 500, Government Printing Office, Washington, D.C., 1952.

Enthalpy-Concentration Data

TABLE I.1 ENTHALPY-CONCENTRATION DATA FOR THE SINGLE-PHASE
LIQUID REGION AND ALSO THE SATURATED VAPOR OF THE ACETIC
ACID-WATER SYSTEM AT ONE ATMOSPHERE

| Liquid or Vapor | | Enthalpy—Btu/lb Liquid Solution | | | | | | Enthalpy |
Mole Fraction Water	Weight Fraction Water	20°C 68°F	40°C 104°F	60°C 140°F	80°C 176°F	100°C 212°F	Satu- rated Liquid	Saturated Vapor Btu/lb
0.00	0.00	93.54	111.4	129.9	149.1	169.0	187.5	361 8
0.05	0.01555	93.96	112.2	130.9	150.4	170.6	186.9	
0.10	0.03225	93.82	112.3	131.5	151.3	172.0	186.5	374.6
0.20	0.0698	92.61	111.9	131.7	152.2	173.8	185.2	395.3
0.30	0.1140	90.60	110.7	131.3	152.6	175.0	183.8	423.7
0.40	0.1667	87.84	108.9	130.6	152.9	176.1	182.9	461.4
0.50	0.231	83.96	106.3	129.1	152.7	177.1	182.4	510.5
0.55	0.268	81.48	104.5	128.1	152.5	177.5	182.3	
0.60	0.3105	78.53	102.5	126.9	152.1	178.0	182.0	573.4
0.65	0.358	75.36	100.2	125.5	151.6	178.3	181.9	
0.70	0.412	71.72	97.71	123.9	151.1	178.8	181.7	656.0
0.75	0.474	67.59	94.73	122.2	149.9	179.3	182.1	
0.80	0.545	62.88	91.43	120.2	149.7	179.9	181.6	767.3
0.85	0.630	57.44	87.56	117.8	148.9	180.5	181.6	
0.90	0.730	51.03	83.02	115.1	147.8	180.8	181.7	921.6
0.95	0.851	43.74	77.65	111.7	146.4	181.2	181.5	
1.00	1.00	36.06	71.91	107.7	143.9	180.1	180.1	1150.4

Reference states: Liquid water at 32°F and 1 atm; solid acid at 32°F and 1 atm.
Data calculated from miscellaneous literature sources and smoothed.

TABLE I.2 VAPOR-LIQUID EQUILIBRIUM DATA FOR THE ACETIC
ACID-WATER SYSTEM; PRESSURE = 1 ATMOSPHERE*

x Mole Fraction Water in the Liquid	y Mole Fraction Water in the Vapor
0.020	0.035
0.040	0.069
0.060	0.103
0.080	0.135
0.100	0.165
0.200	0.303
0.300	0.425
0.400	0.531
0.500	0.627
0.600	0.715
0.700	0.796
0.800	0.865
0.900	0.929
0.940	0.957
0.980	0.985

*From data of L. W. Cornell and R. E. Montonna,
Ind. Eng. Chem., v. 25, pp. 1331–1335 (1933).

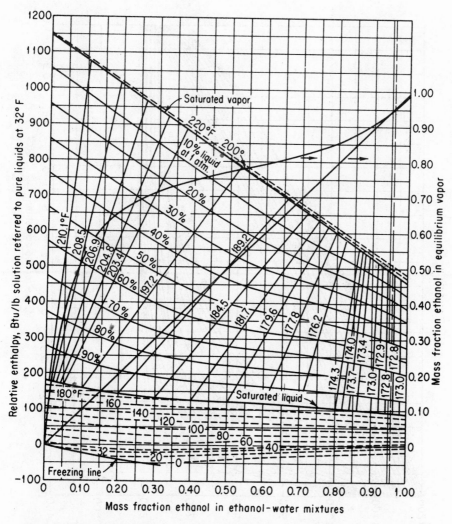

Fig. I.1. Enthalpy-composition diagram for the ethanol-water system, showing liquid and vapor phases in equilibrium at 1 atm.

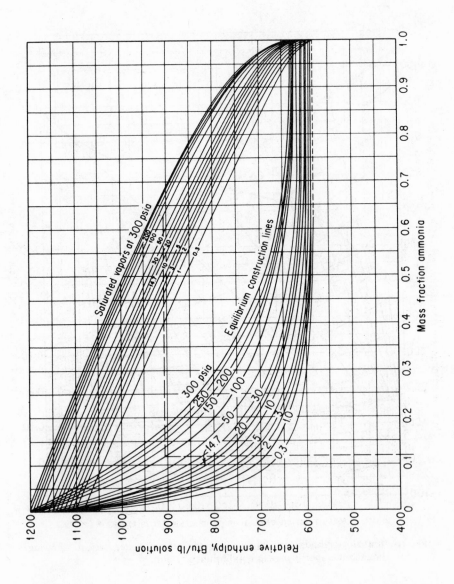

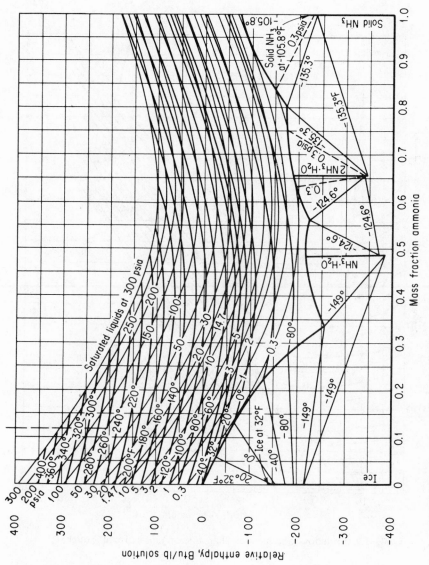

Reference states: water at 32°F and liquid ammonia at −40°F. To determine equilibrium compositions, erect a vertical line from any liquid composition at its saturation or boiling point, and locate its intersection with the appropriate equilibrium construction line. A horizontal line from this intersection will intersect the appropriate saturated vapor line at the desired equilibrium vapor composition.

Fig. I.2. Enthalpy-concentration chart for NH₃-H₂O.

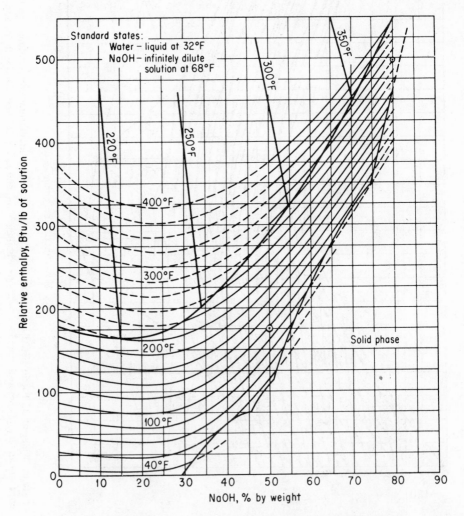

Fig. I.3. Enthalpy-concentration chart for sodium hydroxide-water.

512

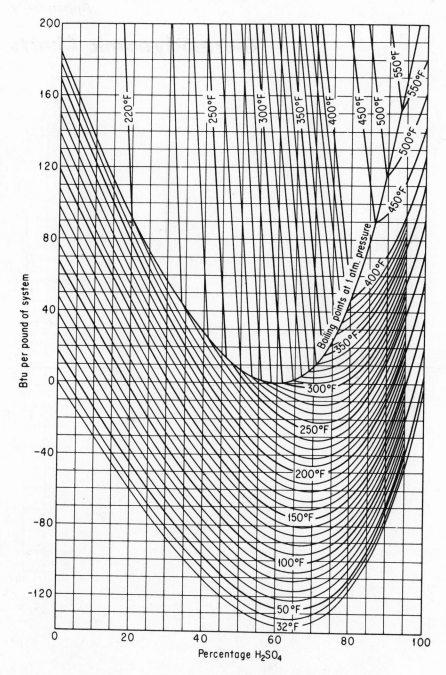

Fig. I.4. Enthalpy-concentration of sulfuric acid-water system relative to pure components. (Water and H_2SO_4 at 32°F and own vapor pressure.) Data from int. crit. tables © 1943 O. A. Hougen and K. M. Watson.

Thermodynamic Charts

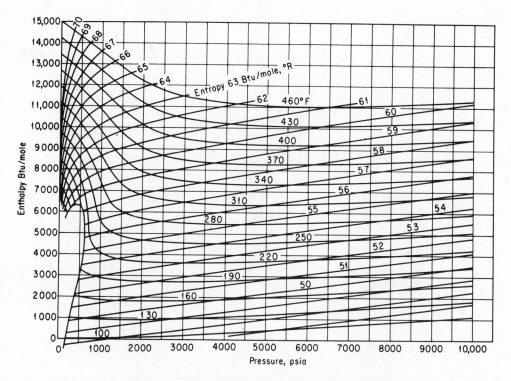

Fig. J.1. Enthalpy-pressure chart for propane.

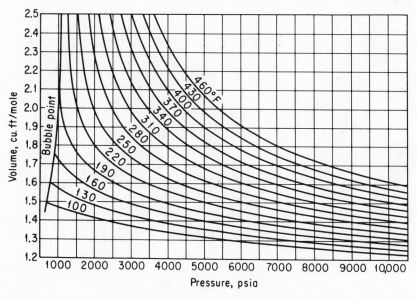

Fig. J.2. Volume-pressure chart for propane.

515

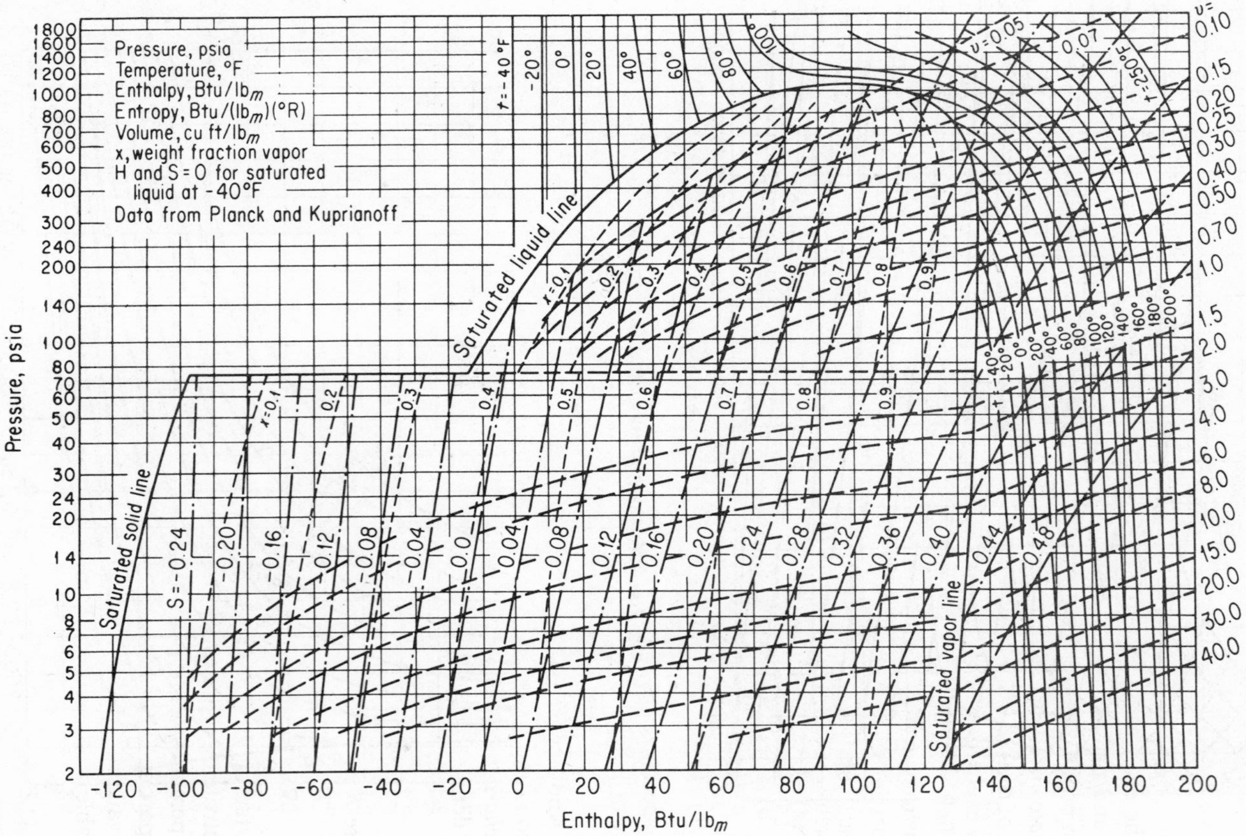

Fig. J.3. Pressure-enthalpy chart for carbon dioxide.

Enthalpy, Btu/lb$_m$

Pressure, psia

Pressure, psia
Temperature, °F
Enthalpy, Btu/lb$_m$
Entropy, Btu/(lb$_m$)(°R)
Volume, cu ft/lb$_m$
x, weight fraction vapor
H and S = 0 for saturated
   liquid at −40°F
Data from Planck and Kuprianoff

Saturated liquid line
Saturated vapor line
Saturated solid line

516

Physical Properties
of Petroleum Fractions

In the early 1930s, tests were developed which would characterize petroleum oils and petroleum fractions, so that various physical characteristics of petroleum products could be related to these tests. Details of the tests can be found in *Petroleum Products and Lubricants*, an annual publication of the Committee D-2 of the American Society for Testing Materials.[1] These tests are not scientifically exact, and hence the procedure used in the tests must be followed faithfully if reliable results are to be obtained. However, the tests have been adopted because they are quite easy to perform in the ordinary laboratory and because the properties of petroleum fractions can be predicted from the results. The tests of particular consequence to us are

(a) API gravity, which has been mentioned in Chap. 1.
(b) The Reid vapor-pressure test (D323), which is used for volatile products such as gasoline.
(c) The ASTM or Engler distillation for various petroleum fractions.

The specifications for fuels, oils, etc., are set out in terms of these tests plus many other properties, such as the flashpoint, the percent sulfur, the viscosity, etc.

Another important technique developed in the early 1930s by Watson, Nelson, and associates[2,3,4] relates petroleum properties to a factor known as the *characterization factor* (sometimes called the *UOP characterization factor*). It is defined as

$$K = \frac{(T_B)^{1/3}}{S}$$

where K = the UOP characterization factor
T = the cubic average boiling point, °R
S = the specific gravity at 60°F

This factor has been related to many of the other simple tests and properties of petroleum fractions, such as viscosity, molecular weight, critical temperature, and percentage of hydrogen, so that it is quite easy to estimate the factor for any particular sample. Furthermore, tables of the UOP characterization factor are available for a wide variety of common types of petroleum fractions as shown in Table K.1 for typical liquids.

---

[1] Report of Committee D-2, ASTM, Philadelphia, annually.
[2] R. L. Smith and K. M. Watson, *Ind. Eng. Chem.*, v. 29, p. 1408 (1937).
[3] K. M. Watson and E. F. Nelson, *Ind. Eng. Chem.*, v. 25, p. 880 (1933).
[4] K. M. Watson, E. F. Nelson, and G. B. Murphy, *Ind. Eng. Chem.*, v. 27, p. 1460 (1935).

TABLE K.1 TYPICAL UOP CHARACTERIZATION FACTORS

Type of Stock	K	Type of Stock	K
Pennsylvania crude	12.2–12.5	Propane	14.7
Mid-Continent crude	11.8–12.0	Hexane	12.8
Gulf-Coast crude	11.0–11.8	Octane	12.7
East-Texas crude	11.9	Natural gasoline	12.7–12.8
California crude	10.8–11.9	Light gas oil	10.5
Benzene	9.5	Kerosene	10.5–11.5

Van Winkle[5] discusses the relationships among the volumetric average boiling point, the molal average boiling point, the cubic average boiling point, the weight average boiling point, and the mean average boiling point, and illustrates how the *K* and other properties of petroleum fractions can be evaluated from experimental data. In Table K.2 are shown the source, boiling point basis, and any special limitations of the various charts in this appendix.

TABLE K.2 INFORMATION CONCERNING CHARTS IN APPENDIX K

1. *Specific heats of hydrocarbon liquids*
    Source: J. B. Maxwell, *Data Book on Hydrocarbons*, Van Nostrand Reinhold, New York, 1950, p. 93 (original from M. W. Kellogg Co.).
    Description: A chart of C_p (0.4 to 0.8) vs. *t* (0 to 1000°F) for petroleum fractions from 0 to 120°API.
    Boiling point basis: Volumetric average boiling point, which is equal to graphical integration of the differential ASTM distillation curve (Van Winkle's "exact method").
    Limitations: This chart is not valid at temperatures within 50°F of the pseudocritical temperatures.

2. *Vapor pressure of hydrocarbons*
    Source: Maxwell, *op. cit.*, p. 42.
    Description: Vapor pressure (0.002–100 atm) vs. temperature (50–1200°F) for hydrocarbons with normal b.p. of 100 to 1200°F (C_4H_{10} and C_5H_{12} lines shown).
    Boiling point basis: Normal boiling points (pure hydrocarbons).
    Limitations: These charts apply well to all hydrocarbon series except the lowest-boiling members of each series.

3. *Heat of combustion of fuel oils and petroleum fractions*
    Source: Maxwell, *op. cit.*, p. 180.
    Description: Heats of combustion above 60°F (17,000–20,500 Btu/lb) vs. gravity (0–60°API) with correction for sulfur and inerts included (as shown on chart).

4. *Properties of petroleum fractions*
    Source: O. A. Hougen and K. M. Watson, *Chemical Process Principles Charts*, Wiley, New York, 1946, Chart 3.
    Description: °API (−10 to 90°API) vs. b.p. (100–1000°F) with molecular weight, critical temperature, and *K* factors as parameters.

[5]M. Van Winkle, *Petroleum Refiner* (June 1955), vol. 34, pp. 136–138.

Boiling point basis: Use cubic average b.p. when using the K values; use mean average b.p. when using the molecular weights.

5. *Heats of vaporization of hydrocarbons and petroleum fractions at* 1.0 *atm pressure*
   Source: Hougen and Watson, *op. cit.,* Chart 68.
   Description: Heats of vaporization (60–180 Btu/lb) vs. mean average b.p. (100–1000°F) with molecular weight and API gravity as parameters.
   Boiling point basis: Mean average b.p.

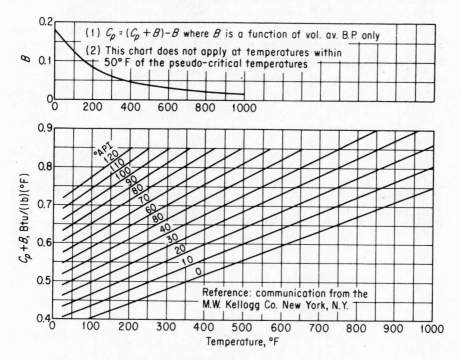

(1) $C_p = (C_p + B) - B$ where B is a function of vol. av. B.P. only

(2) This chart does not apply at temperatures within 50° F of the pseudo-critical temperatures

Reference: communication from the M.W. Kellogg Co. New York, N.Y.

Fig. K.1. Specific heats of hydrocarbon liquids.

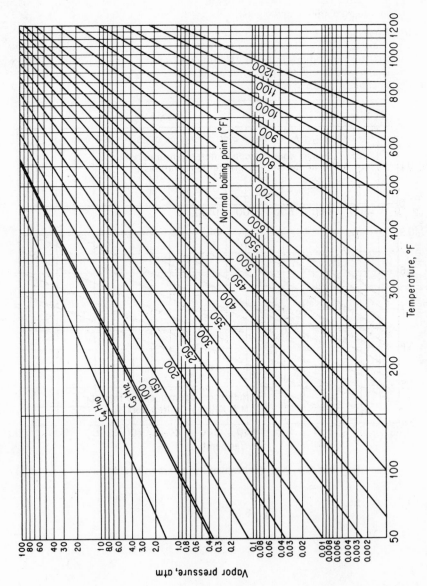

Fig. K.2. Vapor pressure of hydrocarbons.

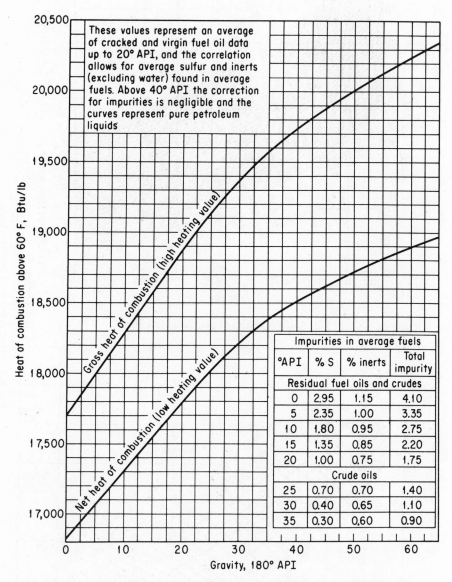

These values represent an average of cracked and virgin fuel oil data up to 20° API, and the correlation allows for average sulfur and inerts (excluding water) found in average fuels. Above 40° API the correction for impurities is negligible and the curves represent pure petroleum liquids

Gross heat of combustion (high heating value)

Net heat of combustion (low heating value)

°API	% S	% inerts	Total impurity
Impurities in average fuels			
Residual fuel oils and crudes			
0	2.95	1.15	4.10
5	2.35	1.00	3.35
10	1.80	0.95	2.75
15	1.35	0.85	2.20
20	1.00	0.75	1.75
Crude oils			
25	0.70	0.70	1.40
30	0.40	0.65	1.10
35	0.30	0.60	0.90

Heat of combustion above 60° F, Btu/lb

Gravity, 180° API

Fig. K.3. Heat of combustion of fuel oils and petroleum fractions.

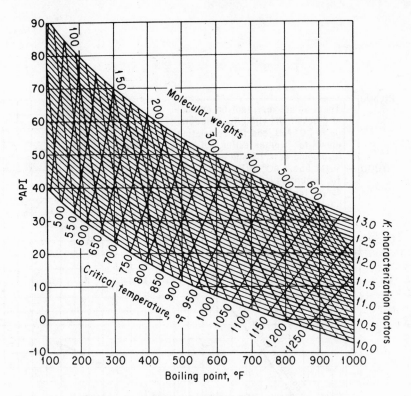

Fig. K.4. Properties of petroleum fractions.

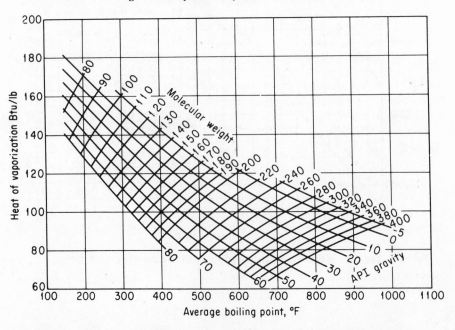

Fig. K.5. Heats of vaporization of hydrocarbons and petroleum fractions at 1.0 atm pressure.

Solution of Sets of Equations

L.1 Independent linear equations

If you write several *linear* material balances, say m in number, they will take the form

$$a_{11}x_1 + a_{12}x_2 + \cdots + a_{1n}x_n = b_1$$
$$a_{21}x_1 + a_{22}x_2 + \cdots + a_{2n}x_n = b_2 \tag{L.1}$$
$$\cdots$$
$$a_{m1}x_1 + a_{m2}x_2 + \cdots + a_{mn}x_n = b_m$$

or in compact matrix notation

$$\mathbf{ax} = \mathbf{b} \tag{L.1a}$$

where $x_1, x_2, \ldots, x_n$ represent the unknown variables, and the a_{ij} and the b_i represent the constants and known variables. As an example of Eq. (L.1), we can write the three-component mass balances corresponding to Fig. 2.5(c):

$$0.50(100) = 0.80(P) + 0.005(W)$$
$$0.40(100) = 0.05(P) + 0.925(W)$$
$$0.10(100) = 0.15(P) + 0.025(W)$$

With m equations in n unknown variables, three cases can be distinguished:

(a) There is no set of x's which satisfies Eq. (L.1).
(b) There is a unique set of x's which satisfies Eq. (L.1).
(c) There is an infinite number of sets of x's which satisfy Eq. (L.1).

Figure L.1 represents each of the three cases geometrically in two dimensions. Case 1 is usually termed inconsistent, whereas cases 2 and 3 are consistent; but to the engineer who is interested in the solution of practical problems, case 3 is as unsatisfying as case 1. Hence case 2 will be termed *determinate*, and case 3 will be termed *indeterminate*.

To ensure that a system of equations represented by (L.1) has a unique solution, it is necessary to first show that (L.1) is consistent, i.e., that the coefficient matrix $\mathbf{a}$ and the augmented matrix $[\mathbf{a}, \mathbf{b}]$ must have the same rank r. Then, if $n = r$, the system (L.1) is determinate, while, if $r < n$, as may be the case, then the number $(n - r)$ variables must be specified in some manner or determined by optimization procedures. If the equations are independent, $m = r$.

As an illustration of these ideas, consider the case of the three equations corresponding to Fig. 2.5(c), where we have more equations than unknowns ($m > h$). The matrix $[\mathbf{a}, \mathbf{b}]$ is

$$[\mathbf{a}, \mathbf{b}] = \begin{bmatrix} 0.80 & 0.005 & 50 \\ 0.50 & 0.925 & 40 \\ 0.15 & 0.025 & 10 \end{bmatrix}$$

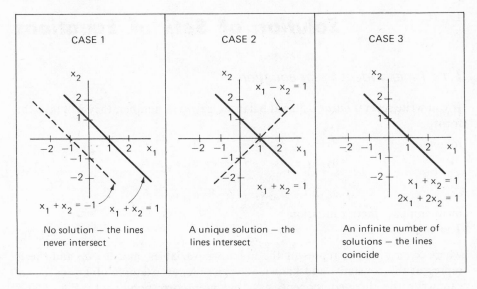

Fig. L.1. Types of solutions of linear equations.

Note how the rank of **a** (the rank of a matrix is given by the size of the largest nonzero determinant that can be formed from the matrix), the matrix composed of the first two columns, can at the most be 2, whereas the rank of [**a, b**] is 3. To obtain a consistent set of equations, one of the three material balances must be eliminated, leaving two equations in two unknowns, P and W, that have a unique solution ($m = r = 2$ and $n = 2$). It would probably be best to pick the two equations in which the coefficients were known with the greatest precision.

As another example, consider a set of 10 equations involving 16 unknowns ($m < n$) so that ($n - m) = 6$. Consequently, six more variables must be specified in some manner before the system of equations becomes determinate.

Next, suppose that you are interested in solving n linear independent equations in n unknown variables:

$$\left.\begin{array}{l} a_{11}x_1 + a_{12}x_2 + \cdots + a_{1n}x_n = b_1 \\ a_{21}x_1 + a_{22}x_2 + \cdots + a_{2n}x_n = b_2 \\ \qquad \cdots \\ a_{n1}x_1 + a_{n2}x_2 + \cdots + a_{nn}x_n = b_n \end{array}\right\} \; \mathbf{ax = b} \qquad \text{(L.1b)}$$

In general there are two ways to solve (L.1b) for $x_1, \ldots, x_n$: elimination techniques and iterative techniques. Both are easily executed by computer programs. We shall illustrate the Gauss-Jordan elimination method. The other techniques can be found in texts on matrices, linear algebra, and numerical analysis.

The essence of the Gauss-Jordan method is to transform Eq. (L.1b) into

Eq. (L.2) by elementary operations on Eq. (L.1b);

$$x_1 + 0 + \cdots + 0 = b_1'$$
$$0 + x_2 + \cdots + 0 = b_2'$$
$$\cdots$$
$$0 + 0 + \cdots + x_n = b_n'$$

(L.2)

Equation (L.2) has a solution for $x_1, \ldots, x_n$ that can be obtained by inspection.

To illustrate the elementary operations that are required, consider the following set of three equations in three unknowns:

$$1. \quad 4x_1 + 2x_2 + x_3 = 15$$
$$2. \quad 20x_1 + 5x_2 - 7x_3 = 0$$
$$3. \quad 8x_1 - 3x_2 + 5x_3 = 24$$

The augmented matrix is

$$\begin{bmatrix} 4 & 2 & 1 & 15 \\ 20 & 5 & -7 & 0 \\ 8 & -3 & 5 & 24 \end{bmatrix}$$

Take the a_{11} element as a pivot. To make it 1 and the other elements in the first column zero, carry out the following elementary operations shown in order for each row:

(a) Subtract (20/4) (Eq. (1)) from Eq. (2),
(b) Subtract (8/4) (Eq. (1)) from Eq. (3), and
(c) Multiply Eq. (1) by 1/4

to get

New eq. no.

$$\begin{bmatrix} 1 & 1/2 & 1/4 & 15/4 \\ 0 & -5 & -12 & -75 \\ 0 & -7 & 3 & -6 \end{bmatrix}$$

(1a)
(2a)
(3a)

Carry out the following elementary operations to make the pivot element a_{22} equal to 1 and the other elements in the second column equal to zero:

(d) Subtract [(1/2)/−5](Eq. (2a)) from Eq. (1a),
(e) Subtract (−7/−5) (Eq. (2a)) from Eq. (3a),
(f) Multiply Eq. (2a) by (1/−5)

to obtain

New eq. no.

$$\begin{bmatrix} 1 & 0 & -19/20 & -15/4 \\ 0 & 1 & 12/5 & 15 \\ 0 & 0 & 99/5 & 99 \end{bmatrix}$$

(1b)
(2b)
(3b)

Another series of elementary operations (left for you to propose) leads to a 1 for the element a_{33} and zeros for the other two elements in the third column:

$$\begin{bmatrix} 1 & 0 & 0 & | & 1 \\ 0 & 1 & 0 & | & 3 \\ 0 & 0 & 1 & | & 5 \end{bmatrix}$$

The solution to the original set of equations is

$$x_1 = 1$$
$$x_2 = 3$$
$$x_3 = 5$$

as can be observed from the augmentation column.

To obtain good accuracy and avoid numerical errors, the choice of the pivot should be made by scanning all the eligible coefficients and chosing the one with the greatest magnitude for the next pivot. For example, you might choose a_{21} for the first pivot, and then find that a_{31} would be the next pivot and finally a_{22} the last pivot to give

$$\begin{bmatrix} 0 & 1 & 0 & | & 3 \\ 1 & 0 & 0 & | & 1 \\ 0 & 0 & 1 & | & 5 \end{bmatrix}$$

L.2 Nonlinear independent equations

The precise criteria used to ascertain if a linear system of equations is determinate cannot be neatly extended to nonlinear systems of equations. Furthermore, the solution of sets of nonlinear equations requires the use of computer codes that may fail to solve your problem for one or more of a variety of reasons. The problem to be solved can be written as

$$\left. \begin{array}{l} f_1(x_1, \ldots, x_n) = 0 \\ f_2(x_1, \ldots, x_n) = 0 \\ \qquad \cdots \\ f_n(x_1, \ldots, x_n) = 0 \end{array} \right\} f(\mathbf{x}) = 0 \qquad (L.3)$$

Generalized linear methods first linearize Equations (L.3) at $\mathbf{x}^{(k)}$ and then iteratively solve the linear system for $\mathbf{x}^{(k+1)}$:

$$\mathbf{a}(\mathbf{x}^{(k)})(\mathbf{x}^{(k+1)} - \mathbf{x}^{(k)}) + f(\mathbf{x}^{(k)}) = 0$$

Typical methods are

Newton-Gauss-Seidel
Newton-successive over relaxation
Secant-successive over relaxation
Secant-alternating directions

A second route of approach is to minimize the sum of the squares of the individual equations by an unconstrained minimization technique. If you choose either method, it is wise to use a prepared computer code that has been verified by many test problems so that you can have some reasonable confidence in the results provided by the computer code.

Answers to Selected Problems

1.3   0.271 m³/sec

1.4   (a) 0.518 lb; (b) 21.5 ft³.

1.9   (b) 0.626 lb_f/ft^2

1.15 $0.455 /lb_m$

1.17 $1.4 \times 10^3 (ft)(lb_f)$

1.18   (b) 3.26×10^{12} erg; (d) 437(hp)(sec)

1.20 $lb_m/(ft^3)(°F)$

1.24   cm²/(sec)(cm of height)

1.25   (1) 1.248×10^6; (4) 2.38×10^7.

1.33   (e) 21.0; (f) 15,950; (g) 15,700

1.35   92.7%

1.41   1.009 to 1.031

1.44   (c) 366 g H_2SO_4/l

1.48   28.14 lb/lb mole

1.54   (a) 17.5°API; (b) 7.94 lb/gal; (c) 59.3 lb/ft³

1.57   (a) 131°F; (b) 328.2°K; (c) 591.0°R

1.63   (h) 1832°F, 2292°R, 1273°K, 1000°C

1.68 1.07×10^4 $lb_f/in.^2$

1.71   (a) 20; (b) 6.39×10^4 lb_f/ft^2; (c) 3.58 N/m²

1.76   2.04

1.79   (b) 20.65 in.

1.82   (e) 2.04 g NaCl; (f) 3.44 g NaCl

1.86   (a) 0.273 lb C(s)/lb CO_2(g)

1.90 $ 0.23/lb HCN

1.98   (a) 2.08 lb 98% H_2SO_4

1.99   (d) 474 lb

1.103   (a) 30%; (b) 77%; (c) 0.33

2.1   R = 2.8, E = 17.4

2.6   (a) 21% O_2, 79% N_2; (b) 14% CO_2, 7% O_2, 79% N_2

2.13   (a) 87.6%; (b) 25.0%

2.17   17.8% CO_2, 1.2% O_2, 8.10% N_2

2.21   (a) 3920 lb H_2O; (c) 4356 lb H_2O

2.25   296.360 g H_2O

2.30   47.8 kg

2.35 $ 51.90/ton

2.42   26.5 lb nitrocellulose/1000 lb 8% solution

2.49   44.5%

2.53 $ 1,850

2.57   3 lb R/lb F

2.60 (a) 111 lb $Ca(Ac)_2$/hr; (b) 760 lb HAc/hr
2.63 (c) 4.65%
2.66 62.5% NaCl, 37.5% H_2O

3.2 $-148°C$
3.7 488 mm Hg
3.13 4.80 kg
3.16 0.214
3.19 29,600 ft³/hr at SC
3.23 (c) 1.315 (atm)(ft³)/(lb mole)(°K)
3.24 (a) 2790 ft³ at 60°F and 30 in. Hg
3.28 net H_2/C ratio is 0.080
3.31 248 gal/hr
3.37 182 ft³/min
3.41 (a) 183 lb/lb mole; (b) Cl_2O_7
3.46 (a) 0.0059 lb/ft³; (b) 0.069
3.51 15.6%
3.56 1520 mm Hg
3.63 434°K
3.66 van der Waals: 0.862 ft³ @ 735 psia, 392°F; Compressibility: 0.877 ft³
     @ 735 psia, 392°F
3.70 Compressibility: (c) 28.7 ft³
3.76 (b) 137.0 cc/g mole; (e) 142.8 cc/g mole
3.81 From a Cox chart: 50 psia
3.86 545°K
3.89 O_2: 296 ft³ at 745 mm Hg, 25°C
3.94 (a) 6.7% CO_2, 12.4% CO, 18.1% O_2, 62.8% N_2; (b) 35.8 ft³ air at 90°F
     and 785 mm Hg/lb $H_2C_2O_4$
3.97 0.792 lb H_2O/hr
3.102 99.8 lb
3.106 (a) 103.6°F; (b) 2.68 atm
3.110 (a) 50.5%; (c) 0.192 psia
3.115 (a) 98°F; (b) 2.16×10^4 ft³ at 760 mm Hg, 100°F
3.122 (a) 750 lb H_2O; (b) 14,090 lb BDA
3.124 (a) 968 ft³ dry exit gas; (b) 11.5%

4.4 (a) intensive; (d) intensive
4.6 (a) gas; (b) solid; (c) liquid
4.10 (a) kinetic energy $= 0$; (b) potential energy $= 1.5 \times 10^5$ $(lb_f)(ft)/lb_m$
4.14 (a) 118 Btu/lb
4.17 (a) yes; (d) yes
4.21 (a) 1783 Btu; (b) 2225 Btu (ideal gas)
4.24 7.519 cal/(kg mole)(°C)
4.27 dextrose: 62.4 cal/(g mole)(°C) vs. 27.9 experimental

4.32 1.93×10^5 kJ

4.35   (a) $\Delta H = 4680$ cal/g mole; (b) $\Delta U = 3490$ cal/g mole

4.40   (b) 9110 Btu/lb mole vs. 8770 Btu/lb mole exptl.

4.44   (a) 3576 Btu; (b) 3576 Btu; (i) 0.00216 ft^3; (m) 0.34

4.47   (a) $\Delta H_{vap} = 597.4$ Btu/lb, 1275.7 psia

4.50   (d) $U_{t_2} - U_{t_1} = Q - W$

4.55 $\Delta T \cong -1.8°$F

4.61   0.218 ft^3 liquid

4.64   (a) 1852.6 Btu

4.66 $\Delta V \cong 3$ ft^3/lb

4.71   16.5 hp

4.76   (a) -5.2×10^7 (ft)(lb$_f$)/hr

4.79   324 hp

4.81   Rev. flow: $W = 0$; Rev. non-flow: $W = 1313$ Btu/lb

4.82   (c) 17.889 kcal

4.83   (b) -24.425 k cal/g mole CaO(s); (i) -26.76 kcal/g mole NaNO$_3$(s); (m) 47.72 kcal/g mole CaCN$_2$(s); (s) -68.74 kcal/g mole CS$_2$(l)

4.89   213.450 cal/g mole at 0°C

4.93   16,763 Btu/lb mole

4.95   7241 cal (evolved)

4.101 276%

4.103 2800°F

4.106 (a) 9.5% C$_6$H$_5$CH$_3$, 19.1% O$_2$, 71.4% N$_2$; (b) 8.2% C$_6$H$_5$CH$_3$, 0.33% CO$_2$, 1.4% H$_2$O, 1.2% C$_6$H$_5$CHO, 17.3% O$_2$, 71.5% N$_2$; (c) 322 gal H$_2$O/hr

5.1   20.2 lb CaCO$_3$/lb CH$_4$

5.3   (b) 2800 lb/hr

5.5   The Mg-Fe$_2$O$_3$ mixture

5.9 $3C + 8$

5.14 (e) 180.7°F; (f) 2.3 Btu/(lb)(°F)

5.18 (1) 347 lb; (2) 8.85 lb

5.22 (a) 67% (b) $Q = -580$ Btu/lb A

5.26 (a) $\mathcal{H} = 0.078$; (b) $p = 0.87$ atm; (c) dew point $= 99°$F; (d) $t_{WB} = 100°$F

5.30 (b) 75.9 lb H$_2$O/hr

5.32 $T_1 = 166°$F; $T_2 = 85.2°$F

5.37 (a) 179°F; (b) 384 lb/hr

5.41 (a) 2.57×10^8 Btu

6.1   230.5 min

6.6   34,700 lb

6.8   0.58%

6.11 40.6 min

6.15 20.2 hr

6.22 (b) $T_f = 129°$F

NOMENCLATURE*

a, b, c = constants in heat-capacity equation

$\quad a$ = constant in general

$\quad a$ = constant in van der Waals' equation, Eq. (3.11)

$\quad a$ = acceleration

$\quad \mathbf{a}$ = coefficient matrix

$\quad A$ = constant in Eq. (3.40)

$\quad A$ = area

$^\circ\text{API}$ = specific gravity of oil defined in Eq. (1.9)

$\quad \alpha S$ = absolute saturation

$\quad b$ = constant in van der Waals' equation, Eq. (3.11)

$\quad B$ = constant in Eq. (3.40)

$\quad \tilde{B}$ = rate of energy transfer accompanying $\tilde{w}$

$\quad c$ = crystalline

$\quad C$ = constant in Eq. (1.1)

$\quad C$ = number of chemical components in the phase rule

$\quad C_p$ = heat capacity at constant pressure

$\quad C_{p_m}$ = mean heat capacity

$\quad C_v$ = heat capacity at constant volume

$\quad C_S$ = humid heat defined by Eq. (5.18)

$\quad D$ = distillate product

$\quad E$ = total energy in system = $U + K + P$

$\quad E_v$ = irreversible conversion of mechanical energy to internal energy

$\quad F$ = force

$\quad F$ = number of degrees of freedom in the phase rule

$\quad F$ = feed stream

$\quad (g)$ = gas

$\quad g$ = acceleration due to gravity

$\quad g_c$ = conversion factor of $\dfrac{32.3(\text{ft})(\text{lb}_m)}{(\text{sec}^2)(\text{lb}_f)}$

$\quad h$ = distance above reference plane

$\quad h_c$ = heat-transfer coefficient in Eq. (5.24)

$\quad \hat{H}$ = enthalpy per unit mass or mole

$\quad H$ = enthalpy, with appropriate subscripts, relative to a reference enthalpy

$\quad \Delta H$ = enthalpy change, with appropriate subscripts

$\quad \Delta \hat{H}$ = enthalpy change per unit mass or mole

$\quad \Delta H_c^\circ$ = standard heat of combustion

$\quad \Delta H_f^\circ$ = standard heat of formation

ΔH_{rxn} = heat of reaction

ΔH_{soln} = heat of solution

$\quad \Delta H_0$ = constant defined in Eq. (4.50)

*Units are discussed in the text.

$\mathcal{H} =$ humidity, lb water vapor/lb dry air

$k_g' =$ mass transfer coefficient in Eq. (5.24)

$K =$ kinetic energy

(l) $=$ liquid

$l =$ distance

$m =$ mass of material

$m =$ number of equations

$\tilde{m} =$ rate of mass transport through defined surfaces

mol. wt $=$ molecular weight

$M =$ molecular weight

$n =$ number of moles

$n =$ number of unknown

$N_t =$ number of variables to be specified

$p =$ pressure

$p =$ partial pressure (with a suitable subscript for the p)

$p^* =$ vapor pressure

$\mathcal{P} =$ number of phases in the phase rule

$p_c =$ critical pressure

$p_c' =$ pseudocritical pressure

$p_r =$ reduced pressure $= p/p_c$

$p_r' =$ pseudoreduced pressure

$p_t =$ total pressure in a system

$P =$ product

$P =$ potential energy

$Q =$ heat transferred

$Q_p =$ heat evolved in a constant pressure process

$Q_v =$ heat evolved in a constant volume process

$\tilde{Q} =$ rate of heat transferred (per unit time)

$r =$ rank of a matrix in Chap. 2

$\tilde{r}_A =$ rate of generation or consumption of component A (by chemical reaction)

$R =$ universal gas constant

$R =$ recycle stream

$\mathcal{R}\mathcal{H} =$ relative humidity

$\mathcal{R}\mathcal{S} =$ relative saturation

(s) $=$ solid

sp gr $=$ specific gravity

$S =$ entropy

$S =$ cross-sectional area perpendicular to material flow

$t =$ time

$T_{DB} =$ dry-bulb temperature

$T_{WB} =$ wet-bulb temperature

$T =$ absolute temperature or temperature in general

$T_b =$ normal boiling point in °K

T_c = critical temperature (absolute)
T'_c = pseudocritical temperature
T_f = melting point in °K
T_r = reduced temperature = T/T_c
T'_r = pseudoreduced temperature
U = internal energy
v = velocity
V = system volume or fluid volume in general
$\hat{V}$ = specific volume
$\hat{V}$ = humid volume defined in Eq. (5.20)
V_c = critical volume
$\hat{V}_{c_i}$ = ideal critical volume = RT_c/p_c
$\hat{V}'_{c_i}$ = pseudocritical ideal volume
V_g = molar volume of gas
V_l = molar volume of liquid
V_r = reduced volume = V/V_c
V_{r_i} = V'_r = ideal reduced volume = $\hat{V}/\hat{V}_{c_i}$
V'_r = pseudoreduced ideal volume = $\hat{V}/\hat{V}'_{c_i}$
$\tilde{w}_A, \tilde{w}$ = rate of mass flow of component A and total mass flow, respectively, through system boundary other than a defined surface
W = work done by the system
W = waste stream
$\tilde{W}$ = rate of work done by system (per unit time)
x = unknown variable
x = mass or mole fraction in general
x = mass or mole fraction in the liquid phase for two-phase systems
y = mass or mole fraction in the vapor phase for two-phase systems
z = compressibility factor
z_c = critical compressibility factor
z'_c = pseudocritical compressibility factor
z_m = mean compressibility factor

Greek letters:
α, β, γ = constants in heat capacity equation
γ = constant in Eq. (3.18)
Δ = difference between exit and entering stream; also used for final minus initial times or small time increments
λ = molal heat of vaporization
λ_b = molal heat of vaporization at the normal boiling point
λ_f = molal latent heat of fusion at the melting point
ρ = density
ρ_A, ρ = mass of component A, or total mass, respectively, per unit volume
ρ_L = liquid density
ρ_V = vapor density

Subscripts:

A, B = components in a mixture

c = critical

i = any component

i = ideal state

t = at constant temperature

t = total

r = reduced state

$1, 2$ = system boundaries

Superscript:

= per unit mass or per mole

Index

535

J

Joule, 257

K

Kay's method, 185
Kelvin temperature, 24, 26
Kinetic energy, 9, 259
Kistyakowsky equation, 283
Kopp's rule, 270

L

Latent heat of fusion, 282,483-89
Latent heat of vaporization, 282, 285
Limiting reactant, 46
Liquid properties, 198

M

Manometer, 29
Mass, units of, 6-8
Material balance, 71ff.
  algebraic techniques, 95
  and combined energy balances, 363ff.
  component balances, 80, 82
  computer codes, 405
  independent, 84
  overall balance, 83, 84
  principles of, 73ff.
  recycle, 114
  solution of equations, Appendix L
  steady state, 73, 74
  tie components, 102
  total, 82
  unsteady state, 445ff.
  with chemical reaction, 79, 80
  without chemical reaction, 79, 80
Mean heat capacity, 273
Mechanical energy balance for flow
      process, 305
Mechanical work, 291
Melting point, 483-89
Mixing point, 96

Mixture of gases:
  air-water (*see* Humidity)
  p-V-T relations for real gases, 183
  p-V-T relations of ideal gases, 158
Mixtures, 23, 158
Moist air properties, 395ff.
Molality, 22
Molal saturation, 208
Molarity, 22
Mole:
  definition, 13
  gram, 13
  pound, 13
Molecular weight, 3, 14
Molecular weight average, 14, 20
Mole fraction, 18
Mole per cent, 18

N

Nelson and Obert charts, 173
Newton's corrections, 174
Non-flow process, 82, 257
Normal boiling point, 192, 583-89

O

Once-through balance, 116
Open system, 257
Orsat analysis, 86
Othmer plot, 285
Overall balance, 116, 365
Oxygen, excess and required, 87

P

Partial pressure, 158
Partial volume, 159
Parts per million, 22
Path function, 262
Perfect gas law, 148
Petroleum properties, 39, 40, 518-20
Phase, 218
Phase behavior:
  mixtures, 222ff.